Group Representation for Quantum Theory

Masahito Hayashi

Group Representation for Quantum Theory

Springer

Masahito Hayashi
Graduate School of Mathematics
Nagoya University
Nagoya
Japan

ISBN 978-3-319-44904-3 ISBN 978-3-319-44906-7 (eBook)
DOI 10.1007/978-3-319-44906-7

Library of Congress Control Number: 2016950886

Printed on acid-free paper

This Springer imprint is published by Springer Nature
The registered company is Springer International Publishing AG
The registered company address is: Gewerbestrasse 11, 6330 Cham, Switzerland

Preface

This book is the English edition of the Japanese book *Group Representations for Quantum Theory*, which was originally published by Kyoritsu shuppan, Tokyo, Japan in January 2014. The original Japanese book covers several topics in representation theory that is related to quantum theory. As is well known, group representation theory is a very strong tool for quantum theory, in particular, angular momentum, hydrogen-type Hamiltonian, spin–orbit interaction, quark model, quantum optics, and quantum information processing including quantum error correction. Therefore, many departments of physics have lecture courses for mathematics for physics, in particular, they have a graduate lecture course for representation theory for physics. This book conducts lecture courses on mathematics for physicists. When the contents of this book are too much for the lecture course, this book can be used by skipping several detailed parts.

To describe a big picture of application of representation theory to quantum theory, the book needs to contain the following six topics, permutation group, SU(2) and SU(d), Heisenberg representation, squeezing operation, Discrete Heisenberg representation, and the relation with Fourier transform from a unified viewpoint by including projective representation. Unfortunately, although there are so many good mathematical books for a part of six topics, no book contains all of these topics because they are too segmentalized. Further, some of them are written in an abstract way in mathematical style and, often, the materials are too segmented. At least, the notation is not familiar to people working on quantum theory. Others are good elementary books, but do not deal with topics related to quantum theory. In particular, such elementary books do not cover projective representation, which is more important in quantum theory. On the other hand, there are several books for physicists. However, these books are too simple and lack the detailed discussion. Hence, they are not useful for advanced study even in physics.

To resolve this issue, the author published the Japanese version of this book. It starts with the basic mathematics for quantum theory. Then, it proceeds to the foundation of group representation theory by including finite group. During this discussion, this book deals with mathematical formulation of boson and fermion, which are fundamental concepts in quantum theory. After this preparation, it discusses representation theory of Lie group and Lie algebra. Based on these mathematics, this book addresses bosonic system and its discretization, which are more related to quantum optics and quantum information, respectively.

However, the original Japanese version has less application to quantum physics. The author has newly added Chap. 5, the later half of Chap. 1, crucial contents of Sect. 4.4, and the discussion for Wigner functions in Chap. 7 as follows. Originally, Chap. 1 discusses only the minimum preparation of mathematics of quantum theory. For smooth connection from basic mathematics to representation theory, the author has added a section for Hamiltonian, which explains the role of Hamiltonian and the relation to representation theory. Also, the author has added a section for symmetry, which describes the details of roles of representation theory in quantum theory. Additionally, the author has added a section for the unbounded case, which is applied only to the infinite dimensional case and can be skipped when the reader is not interested in the difficulty of the infinite-dimensional case. Originally, Sect. 4. 4 briefly explained boson and fermion based on the representation theory for Lie group and Lie algebra. In this English version, the author has added several helpful examples including spin, which clarify the need for representation theory for understanding boson and fermion. Unfortunately, many existing books explain boson and fermion without group representation. The author believes that this section is helpful for the study of boson and fermion on finite-dimensional system.

Chapter 5 is devoted to applications of the representation theory to physical systems. This additional chapter starts with the spectral decomposition of the Hamiltonian on 3-dimensional physical system with a rotational invariant potential by using the group $SO(3,\mathbb{R})$. As a generalization of a part of this topic, we proceed to the spectral decomposition of the Laplacian on the general dimensional sphere, which is essentially the same as the Hamiltonian. Hydrogen-type atom is a special case of 3-dimensional physical system with a rotational invariant potential. The Hamiltonian of this case has a hidden symmetry of $SO(4,\mathbb{R})$. Indeed, this Hamiltonian has a large degeneracy, which cannot be explained by the visible symmetry of $SO(3,\mathbb{R})$. The hidden symmetry of $SO(4,\mathbb{R})$ explains such a large degeneracy. Then, it proceeds to the spin-orbit interaction, in which, the irreducible decomposition of group representation on the composite system (Clebsch Gordan coefficient) plays an important role. This kind of system frequently appears in spin magnetics. Finally, this chapter addresses quark model, which is the key concept of fundamental particle theory. The discussion on the quark model is composed of the analysis on finite-dimensional system, which can be regarded as a good exercise of fermion and group representation theory of a finite-dimensional system. To discuss quark model precisely, we need to discuss the notion of flavor, spin, and color, simultaneously. Although we need many preparations to discuss the whole topic thoroughly and consistently, this book covers all the essential preparation. We also note that several books on the quark model discuss the subject inconsistently due to a lack of basic preparation at the beginning of the books. In particular, this book explicitly writes down the wave functions of all of Baryon and Meson with $SU(3)$ symmetry while they are not given in many basic books for particle physics. This description will be helpful for students of particle physics. In the final section of Chap. 5, we discuss uncertainty relation for wake packets on various spaces, i.e., not only on the set of real numbers $\mathbb{R}$ but also on the one-dimensional and three-dimensional spheres S^1 and S^3.

Additionally, the author has newly added recent progresses for design theory as discretization of Lie group and homogeneous space in Sect. 4.5 and recent progresses for mutually unbiased bases (MUB) and symmetric informationally complete (SIC) vectors in Sect. 8.4. Another unique feature of this book is the clarification on the relation between group representation and Fourier analysis in Sects. 2.8 and 3.8. Fourier analysis is another important mathematical structure of quantum physics. In this revision, based on this relation, in Chap. 7, the author has added the description for Winger function, which is a key concept for the duality between the position and the momentum in foundation of quantum theory. Then, we discuss the uncertainty relation for Winger function. Since the uncertainty relations on the one-dimensional and three-dimensional spheres S^1 and S^3 and on Winger function were obtained recently, these topics have not been discussed in other books. So, the reader can understand the duality in the phase space more deeply.

Further, for better understanding, the author has added many figures, tables, and exercises to help the reader to understand the materials better so that this book contains 54 figures, 23 tables, and 111 exercises with solutions. This book is organized as follows. First of all, when the section or the example is too advanced, the symbol * is indicated in the title. Since such parts will be used only in the later parts with the symbol *, the reader can still understand the contents well even if the reader omit them in the first reading. The author recommends a beginner of representation theory to omit these parts in the first time. Since the symbols of representations are too complicated, this book summarizes the symbols as several tables (Tables 1-8). The reader can refer the tables to recall the symbols.

We now describe the whole structure of this English edition. As the author explained before, this book treats projective representation as well as conventional representation. Chapter 1 starts the mathematical preparation for quantum theory, and explains the back ground of group representation theory. Chapter 2 introduces group representation theory with fundamental concepts. Chapter 3 deals with general theories that do not depend on the types of Lie groups and Lie algebras. Then, it introduces the Fourier transform for Lie groups. Chapter 4 treats representations of special Lie groups and special algebras, $\mathrm{SU}(2)$, $\mathrm{SU}(1,1)$, and $\mathrm{SU}(d)$. Based on them, we discuss boson and fermion as indistinguishable particles in Sect. 4.4. We also discuss the coherent states, which are not discussed in the conventional textbooks. Using these preparations, Chap. 5 proceeds the above-mentioned applications. As advanced topics, Chap. 6 deals with representations of Lie group and Lie algebra with general form based on a root system including representations of non-compact Lie group. Since these topics are more advanced, the proofs of many theorems are omitted. The reader may need to spend a long time to understand some of these proofs. This chapter summarizes such advanced results very reasonably so that a reader can grasp the contents intuitively based on analogies with simple cases, which is another advantage of this book.

Chapter 7 deals with Heisenberg representation, which gives the bosonic system and plays an important role in quantum optics. Using the bosonic system, this chapter explains second quantization, which is a key concept of quantum field theory. In the end of Chap. 7, we discuss multi-mode squeezing, which requires knowledge for representations of Lie group and Lie algebra based on a root system, which are explained in Chaps. 3, 4, and 6. Chapter 8 deals with discrete Heisenberg representation as a discrete version of Heisenberg representations addressed in Chap. 7. This representation is useful in quantum information, especially, quantum error correcting codes, and designs of quantum circuits. Since any quantum information process is constructed based on a combination of quantum circuits, designs of quantum circuits are crucial for quantum information. Chapter 8 is organized so that the contents can be understood with the contents of Chaps. 1 and 2.

Finally, the author emphasizes the difference from existing books for representation theory as follows. There are so many books for representation theory. At least, there exist several good books containing a part of the contents of this book, in particular, the major parts of Chaps. 3, 4, and 6. However, no book contains the whole contents of this book. Further, many existing mathematical books do not adopt the notation familiar to physicist. Mathematical books containing the detail of projective representation are often too advanced. Typically, such books for representation theory do not explain the relation between the representation theory and the foundation of quantum theory, e.g., boson, fermion, second quantization, Winger function, and quantum circuits. Indeed, since representation theory requires too complicated notations, students have trouble to interpolate notations across several books by themselves. To resolve this problem, they need a single book that incorporates representation theory, which brings them a big picture of quantum theory. Therefore, the author believes that this book is helpful for students for representation theory for quantum theory.

The author is grateful when the readers would be interested in representation theory and quantum theory via this book. Finally, the author expresses the acknowledgments to all persons who cooperate to this English version. Especially, the author would like to thank Prof. Hideyuki Ishi of Nagoya University, Prof. Kwek Leong Chuan of Nanyang Technological University and National University of Singapore, Prof. Serge Richard of Nagoya University, Dr. Huangjun Zhu of University of Cologne, Prof. Soichi Okada of Nagoya University, Prof. Hiroyuki Kanno of Nagoya University, Prof. Toru Uzawa of Nagoya University, and Café-David, which is mathematics salon in Graduate School of Mathematics, Nagoya University for providing helpful comments for this English version. The author would also like to thank Dr. Claus E. Ascheron of Springer Science+Business Media for his encouragement and patience during the preparation of the manuscript.

Nagoya, Japan Masahito Hayashi

Preface to the Japanese Version

Group representational symmetry is one of most fundamental concepts in quantum theory, and has been applied to various areas in physics, e.g., particle physics, nuclear physics, condensed matter physics, and statistical physics. It also plays an important role in quantum information, which enables us fruitful information processing by using quantum phenomena. Especially, since various types of representations of various groups appear in various quantum systems, group representation theory reveals so many aspects of quantum theory. Unfortunately, besides various topics underlie group representation theory, such relations are not sufficiently recognized. In fact, such a recognition often leads us not only to deeper understanding of the topic but also to generalization of the topic. However, since the methods of operator algebra and partial differential equation have been emphasized in the area of mathematical foundation of quantum theory, that of group representation has not taken a sufficiently prominent position in this area. Hence, few books cover various methods in representation theory in a unified viewpoint. On the other hand, representation theory has been significantly developed in various directions as a part of mathematics. However, such fruitful developments are not accessible for students and researchers of quantum theory due to the problems explained later.

As useful knowledge of representation theory for quantum theory is divided into various subtopics in representation theory, they are so segmentalized that no mathematical book provides a simple collection of such knowledges from a unified viewpoint. If we deal with representation theory from mathematics, we usually specify an individual topic of representation theory. As its own characteristic features, representation theory inevitably employs very complicated symbols. When we address a topic across distinct books, we need to spend much effort to adjust the different notations among these books. Especially, representations of real Lie groups play an important role in quantum theory, and they are classified via representations of real Lie algebras. However, since many mathematical introductory books are written based on representations of complex Lie algebras, only a few mathematical introductory books emphasize real Lie algebras. In fact, although representations of real Lie algebras are obtained from representations of complex

Lie algebras via a suitable conversion, few introductory books describe this conversion carefully. In particular, many mathematicians avoid to handle representations of real Lie algebras with the above conversion because it requires more complicated notations while it is not mathematically interesting. Since there are so many elegant mathematical introductory books for individual subtopics of representation theory in the above way, we need a book to connect these individual books while few mathematical introductory books carefully describe such connection parts.

On the other hand, many books in physics care such a point; however, they omit their descriptions so much that the reader cannot understand the contents precisely. Further, projective representations play an central role in quantum theory, but many mathematical books do not cover them sufficiently. Especially, major aspects of projective representation can be treated by trivial extension, but projective representation has several blind points that requires special treatment different from conventional representation. Indeed, such subtle points are linked to variety of quantum phenomena. As another problem, many mathematical books are written in a too generalized form, and do not explain examples related to quantum theory. Also, their descriptions are far from the description of quantum theory. There are so many factors that inhibit researchers of quantum theory from accessing useful results of representation theory in this way.

The contents of this book are composed of mathematical knowledge for representation theory that are well-known for many mathematicians, and reorganized by using notations of quantum theory so that they can be easily applied to various topics in quantum system from the viewpoint of quantum theory. Since this book deals with various topics of representation theory essential for quantum system, the whole structure of representation theory will be clarified from quantum theory. This book emphasizes the similarities among several related topics so that their analogies enable the readers to easily understand some complicated topics based on related simpler topics. Hence, the reader will grasp the key points of these advanced topics of representation theory easily. That is, this book will work as a basic infrastructure to understand quantum theory from representation theory. Since the author studies quantum information mainly, the contents are related to applications to quantum information. However, as quantum information is related to foundation of quantum theory, they will be useful for understanding quantum theory beyond quantum information. Therefore, once the readers complete to read this book, they will be able to understand quantum theory much more deeply based on the representation theory. Also, they can proceed to read another book "A Group Theoretic Approach to Quantum Information."

This book is organized as follows. First of all, when the section or the example is too advanced, the symbol * is indicated in the title. Since such parts will be used only in the latter parts with the symbol *, the reader can understand the contents except for such parts even if the reader omits them. The author recommends a beginner of representation theory to omit these parts in the first time. Since the symbols of representations are too complicated, this book summarizes the symbols as several tables (Tables 1–8). The reader can refer the tables to remember the symbols.

Chapter 1 introduces basic concepts of quantum theory, measurement, state, composite system, and entanglement. It also prepares mathematical notations for quantum systems. Although these notations are specified to quantum systems, they are helpful for group representation. Hence, this book consistently deals with representation theory based on these notations. Although the second quantization is an important topic in quantum field theory, we explain it in Chapter 6 (new Chapter 7) because it needs the Bosonic system.

Chapter 2 discusses representations for group in a general framework including projective representations, which are important in quantum theory. Since projective representation is closely related to extension of group, this chapter focuses on this relation. To discuss representation theory including projective representations, we need to handle the factor system, i.e., the set of phase factors, which requires complicated notations. Since the discussions with projective representations are complicated and do not seem essential for mathematics, many mathematically standard textbooks omit them. However, since such discussions are essential for quantum theory, this chapter handles projective representations by using factor systems and we keep this style during the whole book, which is a distinct point from other related books. This chapter proceeds to the details of representations of finite groups so that it introduces the Fourier transform for finite groups, which connects analysis and algebra. As a typical example, we analyze representations of a permutation group by using Young diagrams.

Chapters 3, 4, and 5 address representation theory for Lie groups and Lie algebras. Especially, Chapter 3 deals with general theories that do not depend on the types of Lie groups and Lie algebras. Chapter 3 treats projective representations of Lie groups and Lie algebras by combining the contents of Chapter 2 although few traditional introductory textbooks touch them. Then, it introduces the Fourier transform for Lie groups including the case of projective representations. It also prepares several concepts for Chapter 6 (new Chapter 7). Also, Chapter 3 introduces complex Lie groups and complex Lie algebras.

Chapter 4 treats representations of special Lie groups and special algebras. Since representations of a Lie algebra can be classified with maximum weight, those of a Lie group can be easily treated through those of the corresponding Lie algebra. Also, Lie algebras provide several concepts important for quantum theory. Hence, Chapter 4 is organized so that it constructs representation of a Lie group via that of the corresponding Lie algebra. Chapter 4 starts with representations of Lie algebras $\mathfrak{su}(2)$ and $\mathfrak{su}(1,1)$. Because the Lie algebra $\mathfrak{su}(2)$ is compact and the Lie algebra $\mathfrak{su}(1,1)$ is not compact, they require different treatment caused in this difference. As they have unexpected common features, we handle both in a unified way. Then, we proceed to representation of the Lie algebra $\mathfrak{su}(r)$ by using Young diagrams. Especially, the representation of the Lie algebra $\mathfrak{su}(r)$ on the tensor product space is closely related to that of the permutation group on the same tensor product space. The relation is called Schur duality. We also consider how a finite subgroup can take a role of the Lie group when its representation is given. Such a problem is called design, and is discussed in this chapter.

Chapter 5 (new Chapter 6) deals with representations of a Lie group via those of the corresponding Lie algebra. This chapter discusses a noncompact Lie group as well as a compact Lie group. To discuss both, we focus on the relation between a (real) Lie algebra and a complex Lie algebra. Since representations of a noncompact group are very complicated, this chapter treats only a part of its representations that are related to quantum theory. Since such a special representation has analogies with representations of a compact Lie group, they can be more easily understood than the general case. As this chapter is composed of very advanced topics and such sections are labeled with *, The author recommends the reader to omit this chapter in the first time. Indeed, this book is organized so that the reader can understand the large part even though such advanced parts are omitted.

Chapter 6 (new Chapter 7) deals with Heisenberg representation, which gives the Bosonic system. Since Heisenberg representation is projective representation, the general theory for projective representation given in Chapter 3 plays an essential role. It also treats the representation of $\mathfrak{su}(1,1)$ that describes the squeezing operations. We discuss the multi-mode Bosonic system as well as the one-mode Bosonic system. Such a quantum system is called the continuous system, and attracts attention because it can be easily implemented in a particular sense. Further, using the Bosonic system, we explain the second quantization and the physical meaning of the tensor product state. Only sections with symbol * require knowledge given in the latter part of Chapter 5.

Chapter 7 (new Chapter 8) deals with discrete Heisenberg representation as a discrete version of the representations addressed in Chapter 6 (new Chapter 7). Although discrete Heisenberg representation is usually constructed from finite fields, there is another type of discrete Heisenberg representation based on the commutative algebra $\mathbb{Z}_d$. So, this book covers both types of discrete Heisenberg representations. Further, we address a discrete version of the representation corresponding to the squeezing operation given in Chapter 6 (new Chapter 7). This representation is closely related to quantum circuits and quantum error correction. While Chapter 7 (new Chapter 8) is related to Chapter 6 (new Chapter 7), this chapter can be read independently of Chapter 6 (new Chapter 7). Since it addresses representations of finite groups, it can be read after Chapter 2. The final section discusses mutually unbiased bases (MUB) and symmetric informationally complete (SIC) vectors, which have several applications.

In the above-mentioned way, this book summarizes knowledges of representation theory that is useful for quantum theory. Then, based on these, this book also explains the foundation of quantum theory, e.g., second quantization and quantum circuits. The author is grateful when the readers would be interested in representation theory and quantum theory via this book.

Finally, the author expresses the acknowledgments to all persons who cooperate to this book. Especially, the author would like to thank Prof. Hideyuki Ishi of Nagoya University, Dr. Wataru Kumagai of Kanagawa university, Prof. Akito Hora of Hokkaido University, and Dr. Tsuyoshi Miezaki of Yamagata University for providing helpful comments. The author would also like to thank Mr. Hideo Kotobuki and Ms. Kei Akagi of Kyoritsu Shuppan for their supports.

Nagoya, Japan
December 2013

Masahito Hayashi

Contents

About the Author

Masahito Hayashi was born in Japan in 1971. He received his B.S. degree from the Faculty of Sciences in Kyoto University, Japan, in 1994 and his M.S. and Ph.D. degrees in Mathematics from Kyoto University, Japan, in 1996 and 1999, respectively.

He worked in Kyoto University as a Research Fellow of the Japan Society of the Promotion of Science (JSPS) from 1998 to 2000, worked in the Laboratory for Mathematical Neuroscience, Brain Science Institute, RIKEN from 2000 to 2003, worked in ERATO Quantum Computation and Information Project, Japan Science and Technology Agency (JST) as the Research Head from 2000 to 2006. He also worked in the Superrobust Computation Project Information Science and Technology Strategic Core (21st Century COE by MEXT) Graduate School of Information Science and Technology, The University of Tokyo as Adjunct Associate Professor from 2004 to 2007. He worked in the Graduate School of Information Sciences, Tohoku University as Associate Professor from 2007 to 2012. In 2012, he joined the Graduate School of Mathematics, Nagoya University as Professor. He also worked in the Centre for Quantum Technologies, National University of Singapore as Visiting Research Associate Professor from 2009 till now. In 2011, he received the Information Theory Society Paper Award (2011) for Information-Spectrum Approach to Second-Order Coding Rate in Channel Coding. In 2016, he received the Japan Academy Medal from the Japan Academy and the JSPS Prize from Japan Society for the Promotion of Science.

He is a member of the Editorial Board of the International Journal of Quantum Information and International Journal on Advances in Security. His research interests include classical and quantum information theory, information-theoretic security, and classical and quantum statistical inference.

Symbols

Table 1 Symbols for groups (see Tables 3.1, 3.2, and 3.3 for other groups)

Symbol	Name (Definition)	Page
$H \times K$	Direct product group of groups H and K	28
$H \rtimes K$	Semi direct product of groups H and K	28
$H(K:H)$	Group of extension of group H by group K	34
$\mathrm{Ker}\, f$	Kernel of homomorphism f	27
$[G, G]$	Commutator subgroup of G	27
$\mathrm{H}(2r, \mathbb{Z}_d)$, $\mathrm{H}(2r, \mathbb{F}_q)$	Discrete Heisenberg group of degree r	35, 36
$C_G(M)$	Centralizer of subgroup M of group G	26
$C(G)$	Center of G ($C_G(G)$)	26
$N_G(M)$	Normalizer of M	26
$\mathrm{U}(1)$	Unit circle	23
U_k	Subgroup $\{e^{i2j/k} \mid j \in \mathbb{Z}\}$ of $\mathrm{U}(1)$	23
S_n	Permutation group of degree n	23
$\mathrm{GL}(\mathcal{H})$	General linear group on $\mathcal{H}$	37
$\mathrm{U}(\mathcal{H})$	Unitary group on $\mathcal{H}$	37
$\mathrm{SU}(\mathcal{H})$	Special unitary group on $\mathcal{H}(\mathrm{U}(\mathcal{H}) \cap SL(\mathcal{H}))$	37
G_0	Connected component of G including identity element e	70
$\pi_1(G)$	Fundamental group	95
$\tilde{G}$	Extension of commutative group C by G (universal covering group when C is $_{\pi 1}(G)$)	46, 47, 98
$\breve{G}$	Double covering group of G	97
$\overline{G}_\omega$	Extension of group G by $\mathrm{U}(1)$ defined by $\in C(G)$	49
$\mathrm{Sp}(2r, \mathbb{F}_q)$, $\mathrm{Sp}(2r, \mathbb{Z}_d)$	Discrete symplectic group	273

(continued)

Table 1 (continued)

Symbol	Name (Definition)	Page
G/H	Quotient space, quotient group (residue group) when H is normal subgroup	25
$\mathbb{Z}_k$	$\mathbb{Z}/k\mathbb{Z}$	26
$\mathbb{F}_p$	Finite field of order prime p ($\mathbb{Z}/p\mathbb{Z}$)	36
$\mathbb{F}_q$	Algebraic extension over $\mathbb{F}_p$ of degree n ($q = p^n$)	36
$\mathbb{F}_q^{(k)}$	k-dimensional subspace spanned by $e_1, \ldots, e_k$ over $\mathbb{F}_q$	277
$\mathbb{X}$	$\mathbb{F}_q$ or $\mathbb{Z}_d$	268

Table 2 Symbols for Lie algebras (see Tables 3.4, 3.5, and 3.6 for other Lie algebras)

Symbol	Name (Definition)	Page
$\mathfrak{g}_1 \oplus \mathfrak{g}_2$	Direct sum of $\mathfrak{g}_1$ and $\mathfrak{g}_2$	77
$\mathfrak{g}_1 \rtimes \mathfrak{g}_2$	Semi direct product of $\mathfrak{g}_1$ and $\mathfrak{g}_2$	90
$\mathfrak{g}_\mathbb{C}$	Complexification of $\mathfrak{g}$	77
$\mathfrak{c}(\mathfrak{g})$	Center of $\mathfrak{g}$	78
$\mathfrak{h}$	Cartan subalgebra	205
$\mathfrak{k}$	Maximal compact subalgebra	223
$\mathfrak{gl}(\mathcal{H})$	Set of linear maps from $\mathcal{H}$ to $\mathcal{H}$	37
$\mathfrak{u}(\mathcal{H})$	Set of skew-Hermitian matrices on $\mathcal{H}$	85
$\mathfrak{o}(V)$	Set of real alternative matrices on V	86

Table 3 Sets

Symbol	Name (Definition)	Page
$\hat{G}$	Set of irreducible unitary representations of G	41
$\hat{G}_\mathfrak{f}$	Set of finite-dimensional representations in $\hat{G}$	41
$\hat{G}([\{e^{i(g,g')}\}_{g,g'}])$	Set of irreducible projective unitary representations of factor system $[\{e^{i(g,g')}\}_{g,g'}] \in H(G : \mathrm{U}(1))$	48
$\hat{G}_\mathfrak{f}([\{e^{i(g,g')}\}_{g,g'}])$	Set of finite-dimensional projective representations in $\hat{G}([\{e^{i(g,g')}\}_{g,g'}])$	48
$\hat{\mathfrak{g}}$	Set of irreducible skew-Hermitian presentations of Lie algebra $\mathfrak{g}$	99
$\hat{\mathfrak{g}}_\mathfrak{f}$	Set of finite-dimensional representation in $\hat{\mathfrak{g}}$	99
$\mathcal{Y}^r$	Set of Young diagram of depth not greater than r	65
$\mathcal{Y}_n$	Set of Young diagram of size n	65
$\mathcal{Y}_n^r$	$\mathcal{Y}^r \cap \mathcal{Y}_n$	65
$Y_{U,r}(\boldsymbol{n})$	Set of Semistandard Young tableau corresponding to $\boldsymbol{n}$	65
$Y_S(\boldsymbol{n})$	Set of Standard Young tableau corresponding to $\boldsymbol{n}$	65
Φ	Root system	130, 201
$\Phi_\mathfrak{g}$	Root system of Lie algebra $\mathfrak{g}$	209
Δ	Fundamental system	130, 202
D	Unit circle	117

Table 4 Symbols for Representation

Symbol	Name (Definition)	Group (Lie algebra)	Representation space	Page
$\mathsf{f}_1 \oplus \mathsf{f}_2$	Direct sum representation	$G(\mathfrak{g})$	$\mathcal{H}_1 \oplus \mathcal{H}_2$	39, 86
$\mathsf{f}_1 \otimes \mathsf{f}_2$	Tensor product representation	$G(\mathfrak{g})$	$\mathcal{H}_1 \otimes \mathcal{H}_2$	39, 86
$\mathsf{f}_1 \bar{\otimes} \mathsf{f}_2$		$G_1 \times G_2$	$\mathcal{H}_1 \otimes \mathcal{H}_2$	44
ad	Adjoint representation	$G(\mathfrak{g})$	$\mathfrak{g}$	87
$\mathsf{f}_H \rtimes \mathsf{f}_K$	Semi direct product representation	$H \rtimes K$		50
$\mathsf{f}_H \tilde{\rtimes} \mathsf{f}_T$		$H \rtimes K$		50
$\mathsf{f}_H \rtimes \mathsf{f}_K \otimes \mathsf{f}'_K$		$H \rtimes K$		51
$\mathsf{f}_H \tilde{\rtimes} \mathsf{f}_T \otimes \mathsf{f}'_T$		$H \rtimes K$		51
$\bar{\mathsf{f}}$	Complex conjugate representation of f			53
$\tilde{\mathsf{f}}$		$\tilde{G}$		46, 98
ι		$\mathrm{SU}(2)$ ($\mathfrak{su}(2)$)	$\mathbb{R}^3$	116
W	Heisenberg representation	$\mathbb{R}^2$, $H(2,\mathbb{R})$	$L^2(\mathbb{R})$	233
U		$\mathrm{U}(r)$	$L^2(\mathbb{R})$	238, 242
W^r	r-mode Heisenberg representation	$\mathbb{R}^{2r}$, $H(2r,\mathbb{R})$	$L^2(\mathbb{R}^r)$	241
S		$\mathrm{SU}(1,1)$ ($\mathfrak{su}(1,1)$)	$L^2(\mathbb{R})$	250
S^2		$\mathrm{SU}(1,1)$ ($\mathfrak{su}(1,1)$)	$L^2(\mathbb{R}^2)$	254
S^r		$Sp(2r,\mathbb{R})$	$L^2(\mathbb{R}^r)$	259
$\tilde{\mathsf{W}}_{\mathbb{Z}}$	Pre-discrete-Heisenberg representation	$\mathbb{Z}_d^2$	$\mathbb{C}^d$	264
$\mathsf{W}_{\mathbb{Z}}$	Discrete Heisenberg representation	$\mathbb{Z}_d^2$	$\mathbb{C}^d$	264
$\mathsf{W}_{\mathbb{Z},\mathrm{H}}$		$\mathrm{H}(2,\mathbb{Z}_d)$	$\mathbb{C}^d$	265
$\tilde{\mathsf{W}}_{\mathbb{F}}$	Pre-discrete-Heisenberg representation	$\mathbb{F}_q^2$	$\mathbb{C}^q$	267
$\mathsf{W}_{\mathbb{F}}$	Discrete Heisenberg representation	$\mathbb{F}_q^2$	$\mathbb{C}^q$	267
$\mathsf{W}_{\mathbb{X}}^r$	Discrete Heisenberg representation	$\mathbb{F}_q^{2r}$, $\mathbb{Z}_d^{2r}$	$\mathbb{C}^{d^r}$, $\mathbb{C}^{q^r}$	268
$\tilde{\mathsf{W}}_{\mathbb{X}}^r$	Pre-discrete-Heisenberg representation	$\mathbb{F}_q^{2r}$, $\mathbb{Z}_d^{2r}$	$\mathbb{C}^{d^r}$, $\mathbb{C}^{q^r}$	268
$\mathsf{W}_{\mathbb{Z},\mathrm{H}}$		$\mathrm{H}(2r,\mathbb{Z}_d)$	$\mathbb{C}^{d^r}$	270
$\mathsf{W}_{\mathbb{F},\mathrm{H}}$		$\mathrm{H}(2r,\mathbb{F}_q)$	$\mathbb{C}^{q^r}$	270
$\mathsf{S}_{\mathbb{X}}^r(g)$	Metaplectic representation	$Sp(2r,\mathbb{X})$	$\mathbb{C}^{d^r}$, $\mathbb{C}^{q^r}$	278
$\mathsf{V}_{\mathbb{X}}^r$		$\mathbb{X}^{2r} \rtimes Sp(2r,\mathbb{X})$	$\mathbb{C}^{d^r}$, $\mathbb{C}^{q^r}$	282

Table 5 Vector

Symbol	Name (Definition)	Page
$\|X\rangle\rangle_{A,B}$	Vector on $\mathcal{H}_A \otimes \mathcal{H}_B$	9
$\|\lambda;j\rangle$	CONS of $\mathcal{U}_\lambda(G)$ weight vector for SU(2)	41, 121
$\|\lambda;\lambda\rangle$	Maximum weight vector	212, 225, 228
$\|\lambda;j;j'\rangle$	$\|\lambda;j\rangle \otimes \|j'\rangle$	41
$\|\lambda:\zeta\rangle$	Coherent vector	127
$\|\lambda:\zeta\rangle$	Coherent vector	222, 227, 228
$\left\|-\frac{1}{2}\mathfrak{v}:\zeta\right\rangle$	Coherent vector	261
$\|n\rangle$	Number vector	233
$\|\hat{e}_l\rangle$	Dual computational base	58
$\|\hat{e}_{\mathbb{Z}}(l)\rangle, \|\hat{e}_{\mathbb{F}}(l)\rangle$	Dual computational base	265, 268

Table 6 Symbols for vector space

Symbol	Name (Definition)	Page
$\mathbb{C}[G]$	Group algebra	59
$L^2(G)$	Set of square integrable functions of G	61, 104, 107
$L^2(\hat{G})$		61, 104, 107
$L^2(G, inv)$	Set of square integrable functions of conjugated class	63
$L^2(G, inv, \mathbb{R})$	Set of real-valued functions in $L^2(G, inv)$	63
$L^2(\hat{G}[\mathcal{E}])$		64
$\mathcal{U}_\lambda(G)$, $\mathcal{U}_\lambda$	Irreducible representation space of $\lambda \in \hat{G}$	41
$V \otimes \mathbb{C}$	Complexification of V with complex inner product	52
$V \bar{\otimes} \mathbb{C}$	Complexification of V with Hermitian inner product	52

Table 7 Symbols for matrices and operators

Symbol	Page
J	54
J_d	54
$E_{0,1}$	114
$K_{+,1}$	114
$K_{-,1}$	114
M_γ	114, 275
P_ζ	114, 275
Q_ζ	275
F_1^x	114
F_1^y	114
F_1^z	114
F_{-1}^x	116

(continued)

Table 7 (continued)

Symbol	Page
F^{y}_{-1}	116
F^{z}_{-1}	116
$E_{0,-1}$	117
$K_{+,-1}$	117
$K_{-,-1}$	117
W	118
$\Delta_{\rho}[\mathrm{Q}:\mathrm{P}]$	186
$\mathrm{X}_{\mathbb{Z}}$	278
$\mathrm{Z}_{\mathbb{Z}}$	278
$\mathrm{X}_{\mathbb{F}}(s)$	267
$\mathrm{Z}_{\mathbb{F}}(t)$	267
DFT	58
$F^{x}_{j,l}$	130
$F^{y}_{j,l}$	130
$F^{z}_{j,l}$	130
E_{j}	130
$K_{j,l;+}$	130
$K_{j,l;-}$	130
$F^{z,u}_{j}$	131
a	19, 186, 234
Q	19, 232
P	19, 232
N	234
$\mathrm{U}_{k,j}$	243

Table 8 Other symbols

Symbol	Name (Definition)	Page
$\overline{X}$	Complex conjugate matrix of X	9
X^{T}	Transposed matrix of X	9
$X^{\dagger}$	Transposed complex conjugate matrix of X	9
$\lvert G\rvert$	Order (number of elements) of G	23
$[g]$	Residue class with representative g	25
$\lvert G:H\rvert$	Index of H with respect to G	25
a^{G}	Conjugate class of a in G	26
$H\cdot K$	Product set of sets H and K	28
$\{T(k),\phi(k,k')\}$	Factor system	31
d_{λ}	Dimension or formal dimension of $U_{\lambda}(G)$	41, 102
$C^{\lambda''}_{\lambda,\lambda'}(G)$	Multiplicity	56

(continued)

Table 8 (continued)

Symbol	Name (Definition)	Page
χ_{f}	Character of representation f	41
χ_λ	Irreducible character	41
$\mathrm{supp}(\chi_{\mathsf{f}})$	Support of character χ_{f}	43
$\{\mathrm{e}^{i(g,g')}\}_{g,g'}$	Factor system	46
$H(\mathsf{f})$	Element of $H(G:\mathrm{U}(1))$ corresponding to projective representation f	46
$\lambda^* \in \hat{G}$	Label of complex conjugate representation of f_λ	53
$\mathcal{F}$	Fourier transform	63, 108
$\mathcal{F}^{-1}$	Inverse Fourier transform	63, 108
$(X,Y)_{\mathfrak{g}}$	Killing form	90
C_{f}	Casimir operator	94
μ_G	Normalized invariant measure over group G	102
μ_Θ	Normalized invariant measure over homogeneous space Θ	103
$\mu_n(r)$	Plancherel measure	67
$\mu_{\hat{G}[\mathcal{E}]}$	Plancherel measure	110
κ		118
$\mathcal{F}_{\mathrm{W}}$		255
$\omega_{\mathbb{Z}}$		264
$\tau_{\mathbb{Z}}$		264
$\omega_{\mathbb{F}}$		267
$\tau_{\mathbb{F}}$		267
$[m_1,\ldots,m_{r-1}]$	Dynkin index	130, 202
$(m_1,\ldots,m_{r-1})$	Young index	132
in		132
δ		137, 212

Chapter 1
Mathematical Foundation for Quantum System

Abstract This book introduces group representation theory in terms of quantum theory. For this purpose, this chapter introduces basic concepts of quantum theory, measurement, state, composite system, many-body system, and entanglement. It also prepares mathematical notations for quantum systems. Although these notations are specified to quantum systems, they are helpful for group representation. Hence, this book consistently deals with representation theory based on these notations.

1.1 System, State, and Measurement

In the framework of quantum theory for a microscopic system, the object of interest is called the **Quantum system** or the **System**, and is mathematically described as a complex vector space $\mathcal{H}$ with a Hermitian inner product, which is finite-dimensional or infinite-dimensional. Such a complex vector space $\mathcal{H}$ is called a **Hilbert space** even though it is finite-dimensional.[1] Since $\mathcal{H}$ is a vector space with a Hermitian inner product, we can choose a **completely orthonormal system** (**CONS**) $\{e_i\}$. Each normalized base e_i represents a state in the quantum system that is distinguished from each other. An arbitrary state of the system is given as a normalized vector $x \in \mathcal{H}$. Once a CONS has been fixed as a standard basis, the vector x describing an arbitrary state is written as a linear combination $\sum_i x_i e_i$. In quantum theory, there are two methods to describe the element $x \in \mathcal{H}$. One is the description by a ket vector $|x\rangle$, and the other is that by a bra vector $\langle x|$. Although these descriptions are defined to satisfy the linearity with respect to real coefficients, the multiplication of a complex coefficient $a \in \mathbb{C}$ is defined as

$$|ax\rangle = a|x\rangle, \ \langle ax| = \bar{a}\langle x|. \tag{1.1}$$

In particular, for a base e_i of the standard basis, $|e_i\rangle$ and $\langle e_i|$ are simplified to $|i\rangle$ and $\langle i|$, respectively. On the other hand, we define the Hermitian inner product $\langle x|y\rangle$ for $x, y \in \mathcal{H}$ to satisfy the condition $\langle ax|by\rangle = \bar{a}b\langle x|y\rangle$ for $a, b \in \mathbb{C}$. Then, the inner

[1] It is required to satisfy the **completeness** under the given inner product in the infinite-dimensional case. see Sect. 1.6.

M. Hayashi, *Group Representation for Quantum Theory*,
DOI 10.1007/978-3-319-44906-7_1

product $\langle x|y\rangle$ between x and y can be regarded as the multiplication $\langle x|\cdot|y\rangle$ of the bra vector and the ket vector.

In addition to a state, a physical quantity is also an important concept. When the system obeys the classical mechanics, the value of the physical quantity is also determined according to the state of the system, hence, the physical quantity is given as a function whose input is the state of the system. However, the physical quantity does not necessarily take a fixed value, and is defined as a Hermitian linear map on the Hilbert space.[2] In the following, for the simplicity, we employ the terminologies in the finite-dimensional case.

The most important physical quantity is the **Hamiltonian**, which is written as H. This is because it is the physical quantity describing the time evolution. Since H is a Hermitian matrix, e^{itH} is a unitary, where the real number t describes the time. Then, the map $|x\rangle \mapsto e^{itH}|x\rangle$ describes the evolution of the state of the system, i.e., the state change during time period t.

As mentioned above, a physical quantity A does not take a fixed value. What is its exact meaning? To answer this question, we need to consider the measurement of the physical quantity A. This is because the value of the physical quantity is determined via the measurement. In quantum theory, any physical quantity A is written as a Hermitian matrix or a self-adjoint operator on $\mathcal{H}$, and the Hermitian matrix or the self-adjoint operator is also written as A with this context. When a Hermitian matrix A has no multiplicity for eigenvalues, it is called **non-degenerate**. Otherwise, it is called **degenerate**. To consider the measurement, we focus on a non-degenerate Hermitian matrix A. Then, the eigenvalue decomposition of A is given as

$$A = \sum_i a_i |\phi_i\rangle\langle\phi_i|, \tag{1.2}$$

where a_i is an eigenvalue of A and ϕ_i is an eigenvector corresponding to the eigenvalue a_i.

Once the physical quantity A is measured, the possible outcomes are limited to the eigenvalues a_i of the Hermitian matrix A. The realized outcome is determined according to the state of the system. The outcome is determined only probabilistically except for several special cases. When the state is a normalized vector $x \in \mathcal{H}$, the probability to obtain the outcome a_i is given to be

$$\langle x|\phi_i\rangle\langle\phi_i|x\rangle = |\langle x|\phi_i\rangle|^2. \tag{1.3}$$

The set of the above values gives the probability distribution because the relation $\sum_i |\langle x|\phi_i\rangle|^2 = 1$ can be shown by the fact that $\{\phi_i\}_i$ is a CONS. In particular, when the state is known to be a base of the CONS $\{\phi_i\}_i$, we can identify which base is the state $|x\rangle$ via the above measurement. In general, once a CONS $\{\phi_i\}_i$ is given, it is

[2] It is defined as a self-adjoint operator in the infinite-dimensional case. The precise discussion is given in Sect. 1.6.

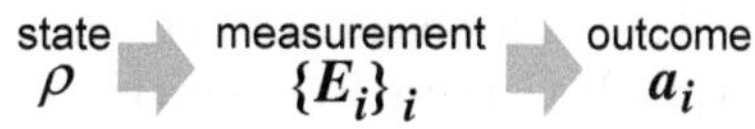

Fig. 1.1 Measurement

possible to realize the measurement whose outcome corresponding to the base ϕ_i is obtained with the probability (1.3) in the above way.

However, a Hermitian matrix might be a degenerate, i.e., might have multiplicity for eigenvalues, in general. The above fact can be generalized to the degenerate case by considering the spectral decomposition

$$A = \sum_i a_i E_i \tag{1.4}$$

instead of the eigenvalue decomposition (1.2). Then, the eigenvalues of A are $\{a_i\}$, and the eigenspace corresponding to a_i is the range of the projection E_i. When the physical quantity A is measured, the probability to obtain the outcome a_i is given to be $\langle x|E_i|x\rangle = \operatorname{Tr} E_i|x\rangle\langle x|$. Hence, in general, given an orthogonal decomposition $\oplus_i \mathcal{H}_i$ of the Hilbert space $\mathcal{H}$, a measurement can be defined by the set $\{E_i\}_i$ of the projections E_i on the Hilbert space $\mathcal{H}_i$. Such a measurement is called a **projection-valued measurement** (**PVM**). As Fig. 1.1, the outcome is given as a_i.

Here, we should remark that the above probability is determined by the one-dimensional projection $|x\rangle\langle x|$ rather than the vector $|x\rangle$. Hence, even if the vector $|x\rangle$ is changed to $|e^{i\theta}x\rangle$ by multiplying the phase factor $e^{i\theta}$, the corresponding one-dimensional projection $|x\rangle\langle x|$ does not change.

In general, it is difficult to prepare the state of the system to be a specific state $|x\rangle$ exactly due to the difficulty of control of the quantum system. The mixture of two states $|x\rangle$ and $|y\rangle$ with a ratio $(1-p) : p$ is rather natural than the exact preparation of the state $|x\rangle$. In this case, when the above measurement is applied, the probability to obtain the outcome a_i is given to be

$$(1-p)\operatorname{Tr}(E_i|x\rangle\langle x|) + p\operatorname{Tr}(E_i|y\rangle\langle y|) = \operatorname{Tr} E_i((1-p)|x\rangle\langle x| + p|y\rangle\langle y|). \tag{1.5}$$

How can we describe such a mixture of states? Since we cannot recognize the system by means other than measurements in the quantum system, it is better to identify a mixture of state with another mixture when they cannot be distinguished via any measurement. Thus, it is reasonable to define a mixed state by using the right hand side (RHS) of (1.5). That is, we describe the mixed state as a Hermitian matrix $(1-p)|x\rangle\langle x| + p|y\rangle\langle y|$. In general, the state corresponding to the mixture of several states $|\phi_i\rangle$ with the probability p_i is expressed as

$$\rho := \sum_i p_i |\phi_i\rangle\langle\phi_i|. \tag{1.6}$$

The above ρ is a Hermitian matrix, in which, the trace is 1 and the eigenvalues are not less than zero. Thanks to the above discussion, we can correctly describe the probability to obtain each outcome of each physical quantity once we know the above matrix ρ. Conversely, when a Hermitian matrix ρ satisfies the two conditions, (1) the trace is 1 and (2) the eigenvalues are not less than zero, it can be described by the RHS of (1.6) via the eigenvalue decomposition of ρ. Here, the set of the states $\{|\phi_i\rangle\}$ in the RHS of (1.6) does not necessarily form a CONS. In the following, a Hermitian matrix ρ is called a **density matrix** when it satisfies the above two conditions. In contrast, a state described by a normalized vector in $\mathcal{H}$ is called a **wave function**. When a density matrix is written as a wave function, it is called a **pure state**. Otherwise, it is called a **mixed state**.

Here, let us consider the measurement of the physical quantity A. Then, when the density matrix of ρ, the **expectation** of the outcome is given as

$$\sum_i a_i \operatorname{Tr} E_i \rho = \operatorname{Tr} A\rho. \tag{1.7}$$

Hence, the density matrix ρ can be regarded as a non-commutative extension of the probability distribution. On the other hand, the physical quantity A can be regarded as a non-commutative extension of a random variable. In probability theory, the expectation is the sum of the product of the possible values of the random variable and their probability. Since, in quantum theory, the expectation is the trace of the multiplication of a density matrix and a physical quantity, the above correspondence can be thought to be a natural non-commutative extension. Also, the **variance** is given as

$$\begin{aligned} \Delta_\rho^2 A :=& \sum_i a_i^2 \operatorname{Tr} |\phi_i\rangle\langle\phi_i|\rho - (\operatorname{Tr} A\rho)^2 \\ =& \operatorname{Tr} A^2\rho - (\operatorname{Tr} A\rho)^2 = \operatorname{Tr}(A - (\operatorname{Tr} A\rho)I)^2\rho. \end{aligned} \tag{1.8}$$

When two physical quantities A and B are commutative with each other, a simultaneous decomposition of A and B, i.e., the simultaneous measurement of A and B, is given as follows. Let a_i, $\mathcal{H}_i$, and E_i be an eigenvalue of A, the eigenspace of A associated to a_i, and the projection to $\mathcal{H}_i$, respectively. Since $B(A - a_i) = (A - a_i)B$, any vector $|x\rangle$ in $\mathcal{H}_i$ satisfies $(A - a_i)B|x\rangle = 0$. So, $B|x\rangle$ belongs to $\mathcal{H}_i$, i.e., $E_i B|x\rangle = B|x\rangle$. For any element $|y\rangle$, we have $E_i B E_i|y\rangle = BE_i|y\rangle$, i.e., $E_i B E_i = BE_i$. Since $\sum_i E_i = I$, we have

$$B = \sum_i E_i B E_i. \tag{1.9}$$

Thus, we can make the spectral decomposition $\{E_{j,i}\}_j$ of $E_i B E_i$ on $\mathcal{H}_i$ as $E_i B E_i = \sum_j b_{j,i} E_{j,i}$. Since $\sum_{j,i} E_{j,i} = I$, the decomposition $\{E_{j,i}\}_{j,i}$ forms a PVM. As (1.9) guarantees that $\sum_{j,i} b_{j,i} E_{j,i} = B$, the decomposition $\{E_{j,i}\}_{j,i}$ gives the simultaneous measurement of A and B.

In this case, the covariance of two outcomes is given as

$$\begin{aligned}&\Delta_\rho A \circ B\\ :=&\frac{1}{2}\operatorname{Tr}[(B-(\operatorname{Tr} B\rho))(A-(\operatorname{Tr} A\rho))+(A-(\operatorname{Tr} A\rho))(B-(\operatorname{Tr} B\rho))]\rho. \quad (1.10)\end{aligned}$$

Now, we consider the case when A and B are non-commutative. Although $\Delta_\rho A \circ B$ can be defined and can be regarded as a kind of correlation between the two physical quantities A and B, it cannot be thought to be a covariance because the simultaneous measurement of A and B cannot be defined.

In the above formulation, a measurement is given from a CONS. This formulation can be generalized as follows. Given a set $\{|\phi_i\rangle\}_i$ of non-normalized vectors in $\mathcal{H}$, we assume that

$$\sum_i |\phi_i\rangle\langle\phi_i| = I, \quad (1.11)$$

where I is the unit matrix. Then, the set $\{|\phi_i\rangle\langle\phi_i|\}_i$ gives a decomposition of the unit matrix I, which gives a measurement as well [58]. Such a decomposition is called a positive operator-valued measure (POVM). When the above measurement is applied to the system whose state is given as the density matrix ρ, the measurement outcome i is obtained with the probability $\langle\phi_i|\rho|\phi_i\rangle$. Such a measurement is possible within the framework of PVMs by extending the system [46, 54, 58]. Such a type of measurement can be generalized to the continuous case. That is, when normalized vectors $\{|\phi_\theta\rangle\}_{\theta\in\Theta}$ are parameterized by the set Θ and the measure on the parameter space Θ satisfies the condition

$$\int_\Theta |\phi_\theta\rangle\langle\phi_\theta|\mu(d\theta) = I, \quad (1.12)$$

the decomposition of the unit matrix is called a POVM and gives a measurement in the above-mentioned sense.

In summary, the key points of quantum theory are summarized as the following basic concepts, a physical quantity (especially the Hamiltonian), its spectral decomposition, a wave function as an element of $\mathcal{H}$, and a density matrix.

Exercise 1.1 Define the POVM $M = \{M_i\}_{i=1}^3$ and the density matrix ρ as $M_1 = \begin{pmatrix} \frac{1}{2} & \frac{i}{4} \\ -\frac{i}{4} & \frac{1}{8} \end{pmatrix}$, $M_2 = \begin{pmatrix} \frac{1}{2} & -\frac{i}{4} \\ \frac{i}{4} & \frac{1}{8} \end{pmatrix}$, $M_3 = \begin{pmatrix} 0 & 0 \\ 0 & \frac{3}{4} \end{pmatrix}$, and $\rho = \frac{1}{2}\begin{pmatrix} 1 & 1 \\ 1 & 1 \end{pmatrix}$. Calculate the distribution P_M^ρ.

Exercise 1.2 Give the spectral decomposition of $A = \begin{pmatrix} 0 & i \\ -i & 0 \end{pmatrix}$ and $B = \begin{pmatrix} 0 & 1 \\ 1 & 0 \end{pmatrix}$.

Exercise 1.3 Calculate the expectation when the measurement is the spectral decomposition of A given in Exercise 1.2 and the state is $\rho = \frac{1}{2}\begin{pmatrix} 1 & 0 \\ 0 & 1 \end{pmatrix}$

Exercise 1.4 Calculate the variance $\Delta_\rho A$ in Exercise 1.3.

Exercise 1.5 Calculate the value $\Delta_\rho A \circ B$ when A, B and ρ are given in Exercises 1.2 and 1.3, respectively.

1.2 Composite System

1.2.1 Tensor Product System

When we have two quantum systems $\mathcal{H}_A$ and $\mathcal{H}_B$ and we treat them as one quantum system, we need a description for the whole system as Fig. 1.2. For example, when the system $\mathcal{H}_A$ describes the internal freedom of a particle and the system $\mathcal{H}_B$ does its position, we need the quantum system that describing the whole freedom of the particle, whose typical example is given in Sect. 5.4. Such a quantum system is called the composite system of $\mathcal{H}_A$ and $\mathcal{H}_B$. The composite system is given by the **tensor product space** $\mathcal{H}_A \otimes \mathcal{H}_B$. When $\mathcal{H}_A$ and $\mathcal{H}_B$ have their CONSs $\{|v_i\rangle\}_{i=1}^k$ and $\{|u_j\rangle\}_{j=1}^l$, respectively, the tensor product space $\mathcal{H}_A \otimes \mathcal{H}_B$ is given as the linear space whose CONS is $\{|v_i, u_j\rangle\}_{1\le i\le k, 1\le j\le l}$. One might consider that the composite system is given as the product space $\mathcal{H}_A \oplus \mathcal{H}_B$. However, the composite system is as the tensor product space $\mathcal{H}_A \otimes \mathcal{H}_B$. To understand the reason, we consider the typical classical case. Remember that the product space $\mathcal{H}_A \oplus \mathcal{H}_B$ has the CONS $\{|v_i\rangle\}_{i=1}^k \cup \{|u_j\rangle\}_{j=1}^l$.

In the classical case, the state is given as a probability distribution over the set of events. That is, to identify the system, we need to identify the set of events. The basis of the quantum system corresponds to the set of events. In the classical case, an event of the composite system is given as a pair of events of the respective systems. That is, the set of events of the composite system is the product set of the sets of events of the respective systems. Considering the relation between the basis of the quantum system and the events of the classical system, we find that the basis of the composite system is $\{|v_i, u_j\rangle\}_{1\le i\le k, 1\le j\le l}$ and is not $\{|v_i\rangle\}_{i=1}^k \cup \{|u_j\rangle\}_{j=1}^l$. Thus, the composite system is the tensor product system $\mathcal{H}_A \otimes \mathcal{H}_B$. When we need to identify the basis of the composite system with a single number, we number it as $|e_{i+k(j-1)}\rangle = |v_i, u_j\rangle$.

Next, let us consider the case when the respective states of the systems $\mathcal{H}_A$ and $\mathcal{H}_B$ are independently prepared to be the density matrices ρ and σ. When $\rho = \sum_{i,i'=1}^k a_{i,i'}|v_i\rangle\langle v_{i'}|$ and $\sigma = \sum_{j,j'=1}^l b_{j,j'}|u_j\rangle\langle u_{j'}|$, the state of the composite system is given as the **tensor product state** $\rho \otimes \sigma := \sum_{i,i'=1}^k \sum_{j,j'=1}^l a_{i,i'} b_{j,j'} |v_i, u_j\rangle$

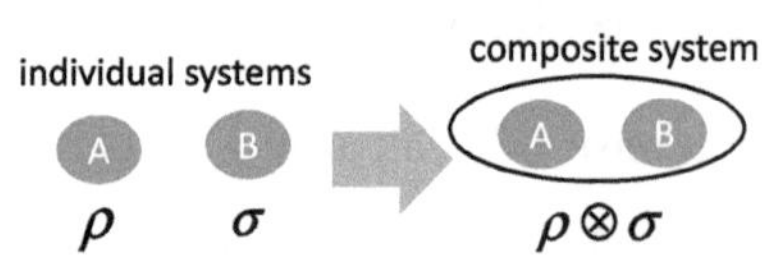

Fig. 1.2 Composite system of two quantum systems

$\langle v_{i'}, u_{j'}|$. In the following, for simplicity, we consider the case when these density matrices are diagonal, i.e., $\rho = \sum_{i=1}^{k} a_i |v_i\rangle\langle v_i|$ and $\sigma = \sum_{j=1}^{l} b_j |u_j\rangle\langle u_j|$. Since $\rho \otimes \sigma = \sum_{i=1}^{k}\sum_{j=1}^{l} a_i b_j |v_i, u_j\rangle\langle v_i, u_j|$, the event (v_i, u_j) occurs with the probability $a_i b_j$. Hence, the above independent situation does not contradict the independence in the classical case. Therefore, when the respective states are independently prepared in the quantum system, the state of the composite system is given as the tensor product state.

Here, we should remark that the tensor product $|v\rangle \otimes |u\rangle = \sum_{i=1}^{k}\sum_{j=1}^{l} c_i f_j |v_i, u_j\rangle$ of two wave functions $|v\rangle = \sum_{i=1}^{k} c_i |v_i\rangle$ and $|u\rangle = \sum_{j=1}^{l} f_j |u_j\rangle$ is independent of the choice of the bases of the respective systems. To check this fact, it is sufficient to show that the tensor product based on other CONSs $\{|\tilde{v}_i\rangle\}_{i=1}^{k}$ and $\{|\tilde{u}_j\rangle\}_{j=1}^{l}$ of the respective systems $\mathcal{H}_A$ and $\mathcal{H}_B$ equals the tensor product $|v\rangle \otimes |u\rangle = \sum_{i=1}^{k}\sum_{j=1}^{l} c_i f_j |v_i, u_j\rangle$ based on the original bases. When the wave functions $|v\rangle$ and $|u\rangle$ are given as $|v\rangle = \sum_{s=1}^{k} \tilde{c}_s |\tilde{v}_s\rangle$ and $|u\rangle = \sum_{t=1}^{l} \tilde{f}_t |\tilde{u}_t\rangle$, the tensor product based on the basis $\{|\tilde{v}_s\rangle\}_{s=1}^{k}$ and $\{|\tilde{u}_t\rangle\}_{t=1}^{l}$ is $\sum_{s,t} \tilde{c}_s \tilde{f}_t |\tilde{v}_s, \tilde{u}_t\rangle$. Now, we employ the unitary matrices U and V defined as $|\tilde{v}_s\rangle = \sum_i V_{s,i}|v_i\rangle$ and $|\tilde{u}_t\rangle = \sum_t U_{t,j}|u_j\rangle$. Then, we have $c_s = \sum_s \tilde{c}_s V_{s,i}$ and $f_t = \sum_t \tilde{f}_t U_{t,j}$, which implies that

$$\sum_{s,t} \tilde{c}_s \tilde{f}_t |\tilde{v}_s, \tilde{u}_t\rangle = \sum_{s,t,i,j} \tilde{c}_s \tilde{f}_t V_{s,i} U_{t,j} |v_i, u_j\rangle = \sum_{i,j} c_i f_j |v_i, u_j\rangle.$$

Thus, given a matrix X on $\mathcal{H}_A$ and a matrix Y on $\mathcal{H}_B$, their tensor product is defined as a linear map on the composite system $\mathcal{H}_A \otimes \mathcal{H}_B$ as follows.

$$X \otimes Y(|v\rangle \otimes |u\rangle) := (X|v\rangle) \otimes (Y|u\rangle). \tag{1.13}$$

Since the tensor product of $|v\rangle$ and $|u\rangle$ is defined independently of the choice of the bases, the definition of the linear map $X \otimes Y$ does not depend on the choice of the bases. Hence, the tensor product state $\rho \otimes \sigma$ of $\rho = \sum_{i,i'=1}^{k} a_{i,i'} |v_i\rangle\langle v_{i'}|$ and $\sigma = \sum_{j,j'=1}^{l} b_{j,j'} |u_j\rangle\langle u_{j'}|$ is defined as $\sum_{i,i'=1}^{k}\sum_{j,j'=1}^{l} a_{i,i'} b_{j,j'} |v_i, u_j\rangle\langle v_{i'}, u_{j'}|$, which does not depend on the choice of the bases. States of the composite system $\mathcal{H}_A \otimes \mathcal{H}_B$ cannot be restricted to tensor product states $\rho \otimes \sigma$ or their convex combinations $\sum_j p_j \rho_j \otimes \sigma_j$, which are called **separable states**, where p_j is a probability and ρ_j and σ_j are density matrices on $\mathcal{H}_A$ and $\mathcal{H}_B$, respectively.

1.2.2 Entangled State

We need to notice that all states on the composite system are not necessarily separable, i.e., do not necessarily have the form $\sum_j p_j \rho_j \otimes \sigma_j$. For example, the pure state corresponding to the vector $\sum_j \sqrt{p_j} |v_j, u_j\rangle$ cannot be described as a convex combination of tensor product states unless p_j is a deterministic distribution. For

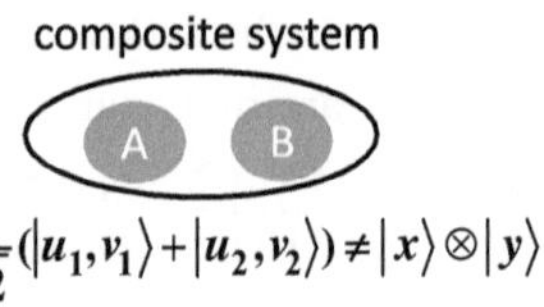

Fig. 1.3 Entangled state between two quantum systems

example, the state $\frac{1}{\sqrt{2}}(|u_1, v_1\rangle + |u_2, v_2\rangle)$ cannot be written with any convex combination (Fig. 1.3). It is not a separable state. Such a state is called an **entangled state**, which has been studied as a resource of quantum information processing. In particular, such a property is called **entanglement**.

When the state of the composite system $\mathcal{H}_A \otimes \mathcal{H}_B$ is given as the density matrix

$$\rho = \sum_{i,i'=1}^{k} \sum_{j,j'=1}^{l} c_{i,j,i',j'} |v_i, u_j\rangle\langle v_{i'}, u_{j'}|,$$

there exists a density matrix $\mathrm{Tr}_{\mathcal{H}_B} \rho$ such that

$$\mathrm{Tr}(\mathrm{Tr}_{\mathcal{H}_B} \rho)X = \mathrm{Tr}\, \rho(X \otimes I_{\mathcal{H}_B})$$

for any matrix X on the system $\mathcal{H}_A$. Thus, it is natural to consider that the state of the smaller system $\mathcal{H}_A$ is given as the density matrix $\mathrm{Tr}_{\mathcal{H}_B} \rho$. The state $\mathrm{Tr}_{\mathcal{H}_B} \rho$ is called the **partial trace** of ρ, and is calculated as

$$\mathrm{Tr}_{\mathcal{H}_B} \rho = \sum_{i,i'=1}^{k} \sum_{j=1}^{l} c_{i,j,i',j} |v_i\rangle\langle v_{i'}|.$$

When the density matrix ρ is diagonal under the basis $\{|v_i, u_j\rangle\}_{i,j}$, the partial trace coincides with the marginal distribution of the distribution composed of the diagonal elements. The partial trace can be defined when ρ is not necessarily a density matrix. When it is a density matrix, the partial trace is also called the **reduced density matrix** of ρ. In the following, when there is no possibility for confusion, $\mathrm{Tr}_{\mathcal{H}_B}$ is simplified to Tr_B.

Now, we consider the linear space of Hermitian matrices on $\mathcal{H}_A$ as the normed linear space $\mathcal{B}(\mathcal{H}_A)$ with respect to the matrix norm. The partial trace $\mathrm{Tr}_{\mathcal{H}_B}$ can be regarded as a linear map from $\mathcal{B}(\mathcal{H}_A \otimes \mathcal{H}_B)$ to $\mathcal{B}(\mathcal{H}_A)$. The norm of the linear map is $\min\{\dim \mathcal{H}_B, \dim \mathcal{H}_A\}$ because the relation

$$\frac{\| \mathrm{Tr}_{\mathcal{H}_B} X\|}{\|X\|} \le \min\{\dim \mathcal{H}_B, \dim \mathcal{H}_A\} \tag{1.14}$$

holds for a Hermitian matrix X on $\mathcal{H}_A \otimes \mathcal{H}_B$.

In this book, if there is no possibility for confusion, for a matrix X on $\mathcal{H}_A$ and a matrix Y on $\mathcal{H}_B$, the matrices $X \otimes I_{\mathcal{H}_B}$ and $I_{\mathcal{H}_A} \otimes Y$ on $\mathcal{H}_A \otimes \mathcal{H}_B$ are simplified to X and Y. Hence, $X \otimes Y$ is written as XY. Once standard bases of the systems $\mathcal{H}_A$ and $\mathcal{H}_B$ are fixed, we use the following simplified notations. Since two quantum systems are addressed, the ket vectors of their standard bases are distinguished as $|j\rangle_A, |j\rangle_B$. Hence, the basis of the composite system $\mathcal{H}_A \otimes \mathcal{H}_B$ is $|k\rangle_A \otimes |j\rangle_B$, which is simplified to $|k, j\rangle_{A,B}$.

In the following, given a matrix $X = (x_{k,j})$, we denote the **complex conjugate matrix** (The entries are given as the complex conjugate of the original entries.) under the standard basis by $\overline{X}$, and the **transposed matrix** under the standard basis by X^T. Then, we denote the **complex conjugate transposed matrix** by $X^\dagger$. Although the matrices $\overline{X}$ and X^T depend on the choice of the standard basis, the matrix $X^\dagger$ depends only on the Hermitian matrix and does not depend on the choice of the standard basis. Using the matrix $X = (x_{k,j})$, we denote the vector $\sum_{k,j} x_{k,j}|k, j\rangle_{A,B}$ on the composite system $\mathcal{H}_A \otimes \mathcal{H}_B$ by $|X\rangle\rangle_{A,B}$. Then, we have

$$Y \otimes Z|X\rangle\rangle_{A,B} = |YXZ^T\rangle\rangle_{A,B}. \tag{1.15}$$

The inner product of two vectors $|X\rangle\rangle_{A,B}$ and $|Y\rangle\rangle_{A,B}$ is given as

$$_{A,B}\langle\langle X|Y\rangle\rangle_{A,B} = \operatorname{Tr} X^\dagger Y. \tag{1.16}$$

Hence, the vector $|X\rangle\rangle_{A,B}$ is a wave function, i.e., its norm is 1 if and only if

$$\operatorname{Tr} X^\dagger X = 1. \tag{1.17}$$

We also have the following formula for the partial trace;

$$\operatorname{Tr}_B |X\rangle\rangle_{A,B\ A,B}\langle\langle X| = XX^\dagger, \quad \operatorname{Tr}_A |X\rangle\rangle_{A,B\ A,B}\langle\langle X| = X^T\overline{X}. \tag{1.18}$$

Further, the states $\left|\frac{1}{\sqrt{2}}\begin{pmatrix} 1 & 0 \\ 0 & 1 \end{pmatrix}\right\rangle\Big\rangle_{A,B}$, $\left|\frac{1}{\sqrt{2}}\begin{pmatrix} 1 & 0 \\ 0 & -1 \end{pmatrix}\right\rangle\Big\rangle_{A,B}$, $\left|\frac{1}{\sqrt{2}}\begin{pmatrix} 0 & 1 \\ 1 & 0 \end{pmatrix}\right\rangle\Big\rangle_{A,B}$, and $\left|\frac{1}{\sqrt{2}}\begin{pmatrix} 0 & 1 \\ -1 & 0 \end{pmatrix}\right\rangle\Big\rangle_{A,B}$ are called Bell states. Since they form a CONS on $\mathbb{C}^2 \otimes \mathbb{C}^2$ (see Exercise 1.7), they give a PVM, i.e., a measurement on the composite system $\mathbb{C}^2 \otimes \mathbb{C}^2$. The measurement is called Bell measurement. To realize this measurement, the measurement device needs to handle the quantum correlation between two systems (Fig. 1.4).

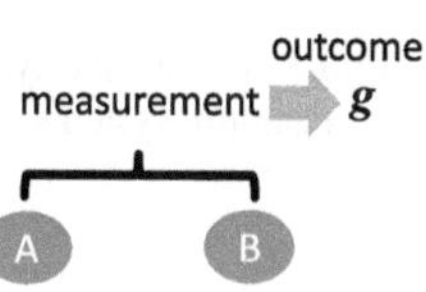

Fig. 1.4 Bell measurement

Exercise 1.6 Assume that X is a 2×2 matrix. Show that the state $|X\rangle\rangle_{A,B}\ {}_{A,B}\langle\langle X|$ is entangled if and only if $\det X \neq 0$.

Exercise 1.7 Show that $\left|\frac{1}{\sqrt{2}}\begin{pmatrix} 1 & 0 \\ 0 & 1 \end{pmatrix}\right\rangle\rangle_{A,B}$, $\left|\frac{1}{\sqrt{2}}\begin{pmatrix} 1 & 0 \\ 0 & -1 \end{pmatrix}\right\rangle\rangle_{A,B}$, $\left|\frac{1}{\sqrt{2}}\begin{pmatrix} 0 & 1 \\ 1 & 0 \end{pmatrix}\right\rangle\rangle_{A,B}$, and $\left|\frac{1}{\sqrt{2}}\begin{pmatrix} 0 & 1 \\ -1 & 0 \end{pmatrix}\right\rangle\rangle_{A,B}$ forms a CONS on $\mathbb{C}^2 \otimes \mathbb{C}^2$.

1.3 Many-Body System

Consider the case when there are n particles and their quantum systems are given as $\mathcal{H}_i$ $(i = 1, \ldots, n)$ (Fig. 1.5). When they can be distinguished from each other, the composite system of these n particles is given as $(((\mathcal{H}_1 \otimes \mathcal{H}_2) \cdots) \otimes \mathcal{H}_n)$. Since the tensor product space does not depend on the order of the tensor products, the above tensor product space is the same as $(\mathcal{H}_1(\cdots(\mathcal{H}_{n-1} \otimes \mathcal{H}_n)))$. Hence, we denote it by $\mathcal{H}_1 \otimes \mathcal{H}_2 \otimes \cdots \otimes \mathcal{H}_n$. In particular, when each system $\mathcal{H}_i$ is isometric to $\mathcal{H}$, the tensor product space is simplified to $\mathcal{H}^{\otimes n}$.

However, when these particles are the same particles and cannot be distinguished from each other, we cannot use the above notation. Such a difficult case will be addressed as bosonic system or fermionic system in the latter chapter (Sects. 4.4.2 and 7.4). Even if these n particles are the same particles, they might be distinguished from each other. For example, when the system $\mathcal{H}_i$ describes the internal freedom of the particle and the position of each particle can be distinguished with probability 1, the composite system is given as $\mathcal{H}_1 \otimes \mathcal{H}_2 \otimes \cdots \otimes \mathcal{H}_n$.

When the state of each system $\mathcal{H}_i$ is independently prepared to be the density matrix ρ_i, the state of the composite system is given as $(((\rho_1 \otimes \rho_2) \cdots) \otimes \rho_n)$. Since the tensor product of the matrices does not depend on the order of the tensor product, the density matrix given by the tensor product is written as $\rho_1 \otimes \rho_2 \cdots \otimes \rho_n$. In particular, when $\rho_i = \rho$, i.e., the state ρ is independently prepared in n quantum systems, the density matrix of the composite system is simplified to $\rho^{\otimes n}$, and is called an n-fold **tensor product state** of ρ. This notation can be applied to the case when ρ and ρ_i are not restricted to density matrices.

Given a matrix A on the tensor product system $\mathcal{H}_1 \otimes \mathcal{H}_2 \otimes \cdots \otimes \mathcal{H}_n$, we denote the partial trace of A with respect to the system $\mathcal{H}_i$ by $\mathrm{Tr}_{\mathcal{H}_i} A$. Conversely, we denote the partial trace with respect to the all systems except for the system $\mathcal{H}_i$ by $\mathrm{Tr}_{\check{\mathcal{H}}_i} A$. In particular, when A is a density matrix ρ, $\mathrm{Tr}_{\check{\mathcal{H}}_i} \rho$ is simplified to $\rho_{\mathcal{H}_i}$ or ρ_i. On the other hand, given a matrix A on the system $\mathcal{H}_i$, the matrix $I^{\otimes i-1} \otimes A \otimes I^{\otimes n-i}$ on

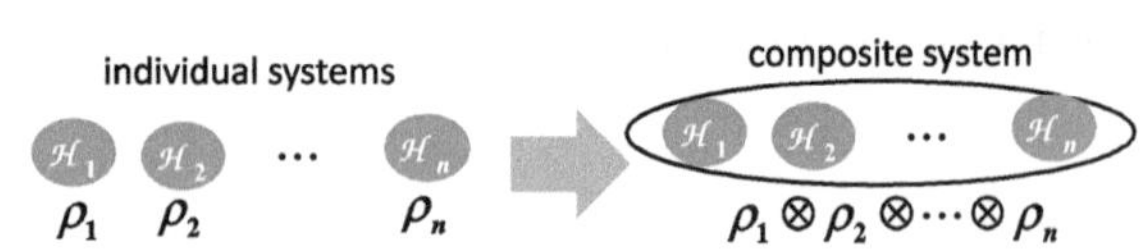

Fig. 1.5 Composite system of n quantum systems

the composite system $\mathcal{H}_1 \otimes \mathcal{H}_2 \otimes \cdots \otimes \mathcal{H}_n$ is simplified to A_i. We also abbreviate the matrix $\sum_{i=1}^n A_i$ to $A^{(n)}$.

As mentioned above, to consider the quantum system, we need to discuss a physical quantity (especially Hamiltonian), its spectral decomposition, a wave function, and a density matrix. However, even when the respective systems $\mathcal{H}_i$ are 2-dimensional, the composite system has the dimension 2^n and requires very complicated treatment. So, we have serious troubles to handle physical quantities and density matrices. However, when they have invariance with respect to physical transformation, the number of parameters to describe them can be reduced so that their mathematical treatment is simplified. The purpose of this book is to provide the systematic method to treat physical quantities and density matrices by reducing the freedom under the suitable physical invariance.

Exercise 1.8 Give the spectral decomposition of $\rho^{\otimes n}$ when $\rho = \frac{1}{3}E_0 + \frac{2}{3}E_1$ on $\mathbb{C}^2$.

1.4 Hamiltonian

1.4.1 Dynamics and Hamiltonian

In the isolated quantum system, time evolution of the state of the system is given by the time span t and the **Hamiltonian** H as

$$\rho \mapsto U\rho U^\dagger, \tag{1.19}$$

where the unitary matrix U is defined as $U := e^{itH}$ (Fig. 1.6). However, when we cannot estimate the time t, we need to take the average with respect to T as

$$\frac{1}{T}\int_0^T e^{itH}\rho e^{-itH} dt. \tag{1.20}$$

In particular, taking the limit $T \to \infty$, we have the state $\sum_j E_j \rho E_j$ when the Hamiltonian H has the spectral decomposition $\sum_j h_j E_j$ (see Exercise 1.9). Then, only eigenvectors are stable. Hence, it is an important topic to calculate eigenvectors of the Hamiltonian H. That is, we need to the following equation;

$$H|x\rangle = E|x\rangle \tag{1.21}$$

with eigenvalue E, which is called **Schrödinger equation**.

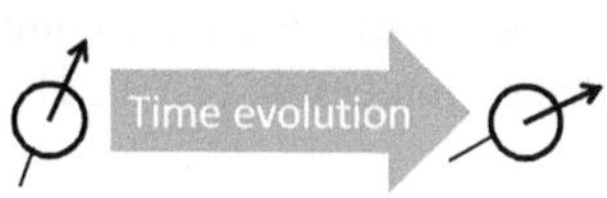

Fig. 1.6 Time evaluation

Hamiltonian has another role in quantum theory. When the system is correlated to a huge system called an environment system or a heat bath during a long time, the state of the system approaches to the state $e^{-\beta H} / \operatorname{Tr} e^{-\beta H}$ called the thermal equilibrium state or thermal state, where the parameter β is called the inverse temperature. On the other hand, under the limit that the inverse temperature β goes to infinity, the state converges to an eigenvector of H associated with the minimum eigenvalue. Especially, when the eigenspace is the one-dimensional, the state is called the ground state. Such a state can be realized by keeping the system at a lower temperature. Therefore, it is a central issue to derive the ground state for a given Hamiltonian. In general, the number of particles in the given system is not necessarily one. When the particle is the type of fermion, more than one particles cannot share the same state. That is, when one particle takes the ground state, another particle takes a state orthogonal to the ground state. Hence, to clarify all of states of these particles, we need to calculate several eigenvectors of the Hamiltonian H associated to lower eigenvalues.

However, in general, it is often difficult to describe the exact form of the Hamiltonian. In some cases, the main term H_0 of the Hamiltonian is given and the fluctuation ΔH is given as a small constant ϵ times of another Hermitian matrix H_1. In particular, when H_0 and H_1 are commutative with each other, we have common eigenvectors as a basis. Then, we can take a eigenvector system of H independently of the coefficient ϵ.

Exercise 1.9 Show that $\lim_{T\to\infty} \frac{1}{T}\int_0^T e^{itH}\rho e^{-itH} dt = \sum_j E_j \rho E_j$ when $H = \sum_j h_j E_j$.

1.4.2 Simultaneous Diagonalization

Now, we consider the diagonalization of a given Hamiltonian H on a given composite system $\mathcal{H}_1 \otimes \mathcal{H}_2$. Usually, this task needs our complicated operation for the composite system $\mathcal{H}_1 \otimes \mathcal{H}_2$. However, when a non-degenerate Hermitian matrix A on $\mathcal{H}_2$ is commutative with H as a Hermitian matrix on $\mathcal{H}_1 \otimes \mathcal{H}_2$, this task can be decomposed to several operations on the subsystems $\mathcal{H}_1$ and $\mathcal{H}_2$ as follows. First, we diagonalize A as $\sum_i a_i |x_i\rangle\langle x_i|$. When $|x_i\rangle\langle x_i|$ is identified with $|x_i\rangle\langle x_i| \otimes I_{\mathcal{H}_2}$, this diagonalization can be regarded as the diagonalization of a matrix on the composite system $\mathcal{H}_1 \otimes \mathcal{H}_2$. Then, we have a Hermitian matrix $H_i := \operatorname{Tr}_{\mathcal{H}_2} |x_i\rangle\langle x_i|H|x_i\rangle\langle x_i|$ on $\mathcal{H}_1$. Now, we diagonalize H_i as $H_i = \sum_j h_{j,i} E_{j,i}$. Since

$$|x_i\rangle\langle x_i|H|x_i\rangle\langle x_i| = \sum_j h_{j,i}(E_{j,i} \otimes |x_i\rangle\langle x_i|), \tag{1.22}$$

we obtain the spectral decomposition of H as $H = \sum_{j,i} h_{j,i} E_{j,i} \otimes |x_i\rangle\langle x_i|$. from the discussion before (1.10). This method is employed in Sect. 5.1.

Notice that this method does not work when the Hermitian matrix A is degenerate because (1.22) does not hold in general. To resolve this problem, we consider a set of Hermitian matrices $\{A_k\}_k$ on $\mathcal{H}_2$ such that the simultaneous measurement of the Hermitian matrices $\{A_k\}_k$ is composed of rank-one projections, i.e., has the form $\{|x_i\rangle\langle x_i|\}_i$. In this case, we can show that $|x_i\rangle\langle x_i|H|x_i\rangle\langle x_i| = H|x_i\rangle\langle x_i|$, which implies that

$$H = \sum_i |x_i\rangle\langle x_i|H|x_i\rangle\langle x_i|. \tag{1.23}$$

Then, we can define the Hermitian matrix H_i on $\mathcal{H}_1$ in the same way. Using the spectral decomposition of H_i, we obtain that of H in the above way.

Exercise 1.10 Give the simultaneous diagonalization of $A \otimes I$ and $I \otimes B$, where A and B are given in Exercise 1.2.

1.4.3 Relation to Representation

Due to the above discussion, it is important to derive the spectral decompositions of the Hamiltonian H and the related Hermitian matrices. When the system $\mathcal{H}$ is too large and/or is too complicated, it is very difficult to find their spectral decompositions. That is, it is almost impossible to derive them without reducing the problem to an easier problem.

When a set of Hermitian matrices $\{A_j\}_j$ satisfies the following condition, the condition almost determines the structure of these Hermitian matrices. There exists a set of real numbers $c_{k,j,l}$ such that

$$i[A_j, A_k] = \sum_l c_{k,j,l} A_l, x \tag{1.24}$$

where $[X, Y] := XY - YX$. Indeed, the condition vanishes a larger part of the freedom of these Hermitian matrices. Then, the difficulty of the spectral decomposition is transformed to an algebraic problem. When we use terminologies roughly, this commutation condition (1.24) is called a representation of Lie algebra, and brings very powerful methods. This method is related to the fact that the set of skew-Hermitian matrices is closed with respect to the commutation relation.

Next, we focus on the set of unitary matrices. The set of unitary matrices is closed with respect to the matrix product. Now, we assume that a subset $\mathcal{S}$ of the set of unitary matrices are closed with respect to the matrix product. When the Hamiltonian H is invariant for $\mathcal{S}$, i.e., $UHU^\dagger = H$ for any element $U \in \mathcal{S}$, the algebraic structure of $\mathcal{S}$ for the matrix product determines a larger part of the structure of H. At least, the structure vanishes a larger part of freedom of the Hilbert space $\mathcal{H}$. Hence, the problem of the analysis of the Hamiltonian can be reduced to much more simple problem. When we use terminologies roughly, the condition for the subset $\mathcal{S}$ is called a representation of a group. A representation of a group is a powerful tool as well as a representation of Lie algebra. To discuss a representation of a group, we need more rigorous formulation. So, in the next chapter, we start our study of a representation of a group with the formal definition of a group. Since a representation of Lie algebra requires much more preparation, we can start its discussion in the middle in Chap. 3.

1.5 Relation to Symmetry

As above-mentioned, symmetry plays an important role in the derivation of Schrödinger equation. In fact, the three dimensional space has natural symmetry with respect to the three-dimensional rotation and displacement. The symmetry of three-dimensional rotation is crucial to discuss quantum system. These symmetries are related to the freedom of the physical space and appear even in classical mechanics.

However, quantum mechanics has a different type of symmetry as Fig. 1.7. Since quantum mechanical object is composed of microscopic particles, we cannot distinguish them when they are the same kind of particles, like, electrons. This problem appears when we have more than one particles while this problem does not occur for the single particle case. Indeed, in quantum theory, it is the most important thing to identify the Hilbert space $\mathcal{H}$ to describe the quantum system because we cannot discuss anything without identifying the Hilbert space $\mathcal{H}$. Hence, we need a proper Hilbert space $\mathcal{H}$ to describe the possible physical freedom reflecting the indistinguishability. Since the indistinguishability is related to a special kind of symmetry, this description is closely related to the symmetry, which never appear in the classical mechanics. The symmetry is rooted in group representation theory and requires many preparations. Indeed, at least, we have two kinds of indistinguishable particles, boson and fermion. We need different type of symmetry dependently of the type of indistinguishable particles. Hence, we discuss both types of indistinguishable particles in Sect. 4.4.2 after many discussions for group representation theory. Then, we apply this description of the Hilbert space $\mathcal{H}$ to several important real physical systems in Chaps. 5 and 7.

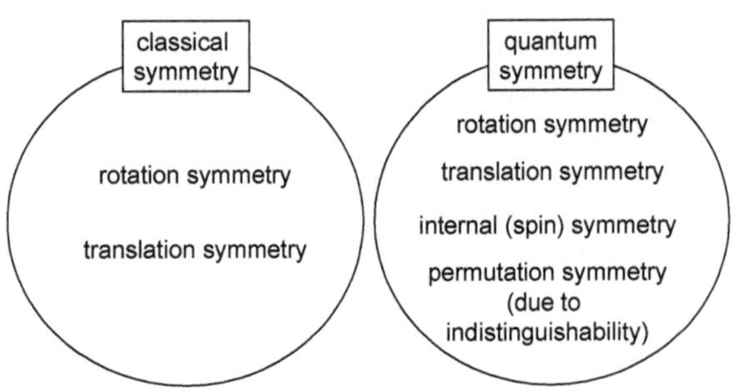

Fig. 1.7 Classical and quantum symmetries

As another relation between quantum system and symmetry, we have discrete symmetry of available unitaries. When we artificially control the quantum system, we need to consider the set of possible unitaries, which usually forms a finite set because it is quite hard to prepare the devices corresponding to infinitely many unitaries. In this case, it is important to discuss what subset of unitaries generates the given subset of unitaries. To discuss this issue, we need algebraic structure of the set based on the matrix product. In this context, representation theory plays a crucial role for the design of quantum operation. To protect quantum state from the noises, we often the quantum noise based on this algebraic structure. Hence, this algebraic structure brings us an essential utility for quantum technologies. Since this kind of symmetry has analogy with boson, we discuss it in Chap. 8 after Chap. 7 that discusses the detail of boson.

1.6 Remark for Unbounded Case*

Finally, we remark several mathematical issues for precise description. If a reader is not interested in such mathematical issues, he/she can skip this part. When the system $\mathcal{H}$ is infinite-dimensional, we need more careful discussions. Firstly, the system $\mathcal{H}$ needs to satisfy the **completeness** under the given inner product, i.e., the system $\mathcal{H}$ needs to contain the limit of Cauchy sequences in $\mathcal{H}$. When it does not satisfy the condition, we need to attach the **completion** for a given vector space $\tilde{\mathcal{H}}$ with Hermitian inner product, which is explained below. First, consider the set of Cauchy sequences $\{x_n\}$ in $\tilde{\mathcal{H}}$. Second, we introduce the equivalence of the set as follows. Given two Cauchy sequences $\{x_n\}$ and $\{y_n\}$, we define equivalence $\{x_n\} \sim \{y_n\}$ when the sequence $\{\|x_n - y_n\|\}$ goes to zero. Finally, we take the quotient space composed of sequences with respect to this equivalence. So, the resultant vector space satisfies the completeness.

Even though the system $\mathcal{H}$ satisfies the completeness, we need to care about the operator norm $\|H\| := \sup_{x\in\mathcal{H}} \frac{\|Hx\|}{\|x\|}$ for an operator H on $\mathcal{H}$. When the operator norm $\|H\|$ takes finite value, the operator is called **bounded**. Otherwise, it is called **unbounded** and we need another mathematical delicate problem. In the infinite-dimensional case, there is a possibility that an operator H on $\mathcal{H}$ is unbounded. In such a case, the operator H is called unbounded and requires a more delicate treatment. When we omit this careful treatment, the discussion contains several inconsistencies.[3]

When the operator norm of an operator H is infinite, we need to restrict the domain of H to a subspace $\mathcal{D}(H)$ of $\mathcal{H}$ so that $\|Hx\| < \infty$ for $x \in \mathcal{D}(H)$. Then, we define the graph norm $\|x\|_H := \sqrt{\|x\|^2 + \|Hx\|^2}$. When the domain $\mathcal{D}(H)$ is closed in the sense of the graph norm, H is called a **closed operator**. Now, we define the **adjoint operator** $H^\dagger$ of H on the domain

$$D(H^\dagger) := \{x \in \mathcal{H} | \exists y \in \mathcal{H} \text{ such that } \langle y|z\rangle = \langle x|Hz\rangle \text{ for } \forall z \in D(H)\}$$

[3]Even if a reader ignores such inconsistencies, so serious problem will not be caused when the Hilbert space $L^2(\mathbb{R}^d)$ is discussed. So, it is possible to skip this section.

as

$$\langle H^\dagger x|z\rangle = \langle x|Hz\rangle \tag{1.25}$$

for $z \in D(H)$. When $D(H) \subset D(H^\dagger)$ and $Hx = H^\dagger x$ for $x \in D(H)$, H is called **symmetric**. When $D(H) = D(H^\dagger)$ and $Hx = H^\dagger x$ for $x \in D(H)$, H is called **self-adjoint**.

In general, it is quite difficult to directly define a self-adjoint operator H on the proper domain $D(H)$ because the domain cannot be described explicitly in many cases. Hence, we often define a self-adjoint operator indirectly as follows. Firstly, we usually define an operator $\tilde{H}$ on a dense subspace $D(\tilde{H})$ of $\mathcal{H}$ so that $\tilde{H}$ is symmetric. Then, we take the completion of the subspace $D(\tilde{H})$ in the sense of the graph norm of $\tilde{H}$. The completion (the resultant Hilbert space of the completion) is different from $\mathcal{H}$ because we consider the graph norm of $\tilde{H}$ not the norm of the original Hilbert space $\mathcal{H}$. Then, we denote the completion by $D(H)$ and the operator $\tilde{H}$ on $D(H)$ by H. Since the operator H is automatically a closed operator, it is called the **closed extension** of $\tilde{H}$. When $\tilde{H}$ is symmetric, the closed extension H is also symmetric. However, in general, the closed extension H is not necessarily self-adjoint even though $\tilde{H}$ is symmetric. Indeed, when an operator $\tilde{H}$ has a closed extension H that is self-adjoint, it is called **essentially self-adjoint**.

For a self-adjoint operator H, we have the following lemma.

Lemma 1.1 *For a self-adjoint operator H, there exists a PVM E on $\mathbb{R}$ such that*

$$\langle y|Hx\rangle = \int_{\mathbb{R}} s\langle y|E(ds)x\rangle. \tag{1.26}$$

Conversely, we have the following lemma.

Lemma 1.2 *Given a PVM E taking the outcome in $\mathbb{R}$, we define the operator X_E as*

$$X_E x := \int_{\mathbb{R}} sE(ds)x \tag{1.27}$$

on

$$D(X_E) := \left\{x \in \mathcal{H}\middle| \int_{\mathbb{R}} s^2\langle x|E(ds)x\rangle < \infty\right\}. \tag{1.28}$$

Then, X_E is a self-adjoint operator.

Due to this lemma, for any self-adjoint operator X and any real-valued function f, we can define the self-adjoint operator $f(X)$ as $\int_{\mathbb{R}} f(s)E_X(ds)$, where E_X is the PVM defined for the self-adjoint operator X in Lemma 1.1.

Proof Assume that, for an element $x \in \mathcal{H}$, there exists an element $y \in \mathcal{H}$ such that

$$\langle x|X_E z\rangle = \langle y|z\rangle \tag{1.29}$$

for any element $z \in D(X_E)$. Hence, it is sufficient to show that $\int_{\mathbb{R}} s^2 \langle x|E(ds)x\rangle < \infty$. Firstly, we define the bounded subset $S_R := \{s \in \mathbb{R} | |s| < R\}$. Then, we have

$$\langle y|z\rangle = \langle x|X_E z\rangle = \lim_{R\to\infty} \left\langle x \middle| \int_{S_R} sE(ds)z \right\rangle = \lim_{R\to\infty} \left\langle \int_{S_R} sE(ds)x \middle| z \right\rangle \tag{1.30}$$

for any element $z \in D(X_E)$. Since $D(X_E)$ is a dense subset of $\mathcal{H}$, $\int_{S_R} sE(ds)x$ converges to y as $R \to \infty$. Thus, $\int_{\mathbb{R}} s^2 \langle x|E(ds)x\rangle = \langle \int_{S_R} sE(ds)x | \int_{S_R} sE(ds)x\rangle < \infty$, which is the desired statement. ■

When a Hamiltonian is given as a self-adjoint operator H, the unitary matrix U describing the time evolution is given as $U := \int_{\mathbb{R}} e^{its} E(ds)$.

Next, we consider the case when the Hilbert space $\mathcal{H}$ is given as the set of square summable sequences with an index $n = 0, 1, 2, \ldots, \infty$. An infinite-dimensional matrix $\{h_{i,j}\}_{i,j\geq 0}$ is called a band-diagonal matrix with width l when $h_{i,j} = 0$ for $|i - j| > l$.

Lemma 1.3 *For two band-diagonal matrices* $\{h_{i,j}\}_{i,j\geq 0}$ *and* $\{g_{i,j}\}_{i,j\geq 0}$ *with width* l*, the sum* $\{h_{i,j} + g_{i,j}\}_{i,j\geq 0}$ *is a band-diagonal matrix with width* l*, and the product* $\{f_{i,j}\}_{i,j\geq 0}$ *is a band-diagonal matrix with width* $2l$*, where* $f_{i,j} := \sum_{j'=i-l}^{i+l} h_{i,j} g_{j',j}$.

The above lemma shows that the set of band-diagonal matrices are closed with respect to the sum and the matrix multiplication. A band-diagonal matrix $\{h_{i,j}\}_{i,j\geq 0}$ is called **Hermitian** when $h_{i,j} = \overline{h_{j,i}}$.

Lemma 1.4 ([68, Theorem 2.3]). *For a band-diagonal matrix* $\{h_{i,j}\}_{i,j\geq 0}$*, we define an operator* H *on*

$$D(H) := \left\{ \{a_k\}_{k=0}^{\infty} \in \mathcal{H} \middle| \sum_{n=0}^{\infty} \Big| \sum_{k=-l}^{l} h_{nn+k} a_{n+k} \Big|^2 < \infty \right\} \tag{1.31}$$

as

$$(Ha)_n := \sum_{k=-l}^{l} h_{nn+k} a_{n+k} \tag{1.32}$$

for $a \in D(H)$*. When* $\{h_{i,j}\}_{i,j\geq 0}$ *is Hermitian, the operator* H *is self-adjoint.*

This lemma shows that any band-diagonal matrix can be regarded as a self-adjoint operator.

Proof Since $a \in D(H)$ satisfies $H^\dagger a = Ha$, it is sufficient to show that for an element $x \in \mathcal{H} \setminus D(H)$, no element $y \in \mathcal{H}$ satisfies that

$$\langle x|Hz\rangle = \langle y|z\rangle \tag{1.33}$$

for $z \in D(H)$. When $y \in \mathcal{H}$ satisfies the above condition, the conditions with $z = \{\delta_{m,n}\}_{n=0}^{\infty}$ implies that $y_m = \sum_{k=-l}^{l} h_{m,m+k}x_{m+k}$. Since $\sum_{n=0}^{\infty} |y_n|^2 < \infty, x \in D(H)$. This, we obtain the desired statement. ■

Lemma 1.5 *Let $\{h_{i,j}\}_{i,j\geq 0}$ be a Hermitian band-diagonal matrix. When a dense subspace $D(\tilde{H})$ of $\mathcal{H}$ contains the sequence $\{\delta_{m,n}\}_{n=0}^{\infty}$ for any non-negative integer m and is included in $D(H)$, the operator $\tilde{H}$ defined by the Hermitian band-diagonal matrix $\{h_{i,j}\}_{i,j\geq 0}$ on $D(\tilde{H})$ is symmetric. Also, the closed extension of $\tilde{H}$ is H.*

Proof The first part is trivial because it is included in $D(H)$. Since the domain of the closed extension of $\tilde{H}$ is included in $D(H)$, it is enough to show that any element of $D(H)$ is contained in the domain of the closed extension of $\tilde{H}$. Any finite sequence $\{a_n\}_{n=0}^{m}$ is contained in the domain of the closed extension of $\tilde{H}$. When a sequence $\{a_n\}_{n=0}^{\infty}$ satisfies $\sum_{n=0}^{\infty} |\sum_{k=-l}^{l} h_{n,n+k}a_{n+k}|^2 < \infty$ and $\sum_{n=0}^{\infty} |a_n|^2 < \infty$, the sequence $\{a_n\}_{n=0}^{\infty}$ belongs to the domain of the closed extension of $\tilde{H}$, which implies the desired statement. ■

We often consider the case when the Hilbert space $\mathcal{H}$ is given as $L^2(\mathbb{R})$, which is defined as follows. Firstly, we consider the set $\mathscr{S}(\mathbb{R})$ of rapidly decreasing functions as

$$\mathscr{S}(\mathbb{R}) := \left\{\phi \in C^{\infty}(\mathbb{R}) \middle| \sup_{x\in\mathbb{R}} |x^{\alpha} \frac{d^{\beta}}{dx^{\beta}} f(x)|^2 < \infty, \forall \alpha, \beta \in \mathbb{Z}_+ \right\}, \tag{1.34}$$

where $C^{\infty}(\mathbb{R})$ is the set of smooth functions. Then, we define the space $L^2(\mathbb{R})$ as the completion of the L^2 norm defined as $\|\phi\|^2 := \int_{\mathbb{R}} |\phi(x)|^2 dx$. The multiplication operator $\phi(x) \mapsto x\phi(x)$ and the differential operator $\phi(x) \mapsto \frac{d}{dx}\phi(x)$ are defined in the following way. So, we can define the symmetric operators $\tilde{Q}$ and $\tilde{P}$ on $\mathscr{S}(\mathbb{R})$ by

$$(\tilde{\mathsf{Q}}\phi)(x) := x\phi(x), \quad (\tilde{\mathsf{P}}\phi)(x) := -i\frac{d}{dx}\phi(x). \tag{1.35}$$

The set $\mathscr{S}(\mathbb{R})$ contains the n-th Hermitian function $\psi_n(x) := \frac{1}{\sqrt{n!2^n\sqrt{\pi}}} e^{\frac{x^2}{2}} \frac{d^n}{dx^n} e^{-x^2}$, which is identified with $|n\rangle$. We define the operators $\tilde{\mathsf{a}}$ and $\tilde{\mathsf{a}}^{\dagger}$ as $\tilde{\mathsf{a}} := \frac{1}{\sqrt{2}}(\tilde{\mathsf{Q}} + i\tilde{\mathsf{P}})$ and $\tilde{\mathsf{a}}^{\dagger} := \frac{1}{\sqrt{2}}(\tilde{\mathsf{Q}} - i\tilde{\mathsf{P}})$, respectively. Since $\tilde{\mathsf{a}}|n\rangle = \sqrt{n}|n-1\rangle$ and $\tilde{\mathsf{a}}^{\dagger}|n\rangle = \sqrt{n+1}|n+1\rangle$, the operators $\tilde{\mathsf{a}}$ and $\tilde{\mathsf{a}}^{\dagger}$ have band-diagonal forms. So, the operators $\tilde{Q}$ and $\tilde{P}$ have Hermitian band-diagonal forms. Due to Lemmas 1.4 and 1.5, the closed extensions Q and P of $\tilde{Q}$ and $\tilde{P}$ are self-adjoint. Now, we consider a two-variable ordered polynomial f. That is, when the degree is 2, $f(a, b)$ is written as linear sum of a, b, a^2, ab, ba, and b^2. Then, an operator $f(\tilde{Q}, \tilde{P})$ is defined on $\mathscr{S}(\mathbb{R})$ as a linear sum of multiplications of $\tilde{P}$ and $\tilde{Q}$, and it has a band-diagonal form. In this case, the ordered polynomial f equals the opposite ordered polynomial of f if and only if $f(a, b)$ is written as linear sum of $a, b, a^2, ab + ba$, and b^2. When this condition is additionally assumed, the operator $f(\tilde{Q}, \tilde{P})$ is symmetric and has a Hermitian band-diagonal form. Due to Lemma 1.5, the closed extension $f(Q, P)$ of $f(\tilde{Q}, \tilde{P})$ is self-adjoint. So, the operators a and $\mathsf{a}^\dagger$ are defined as the closed operators in this way.

Further, we can define the set $\mathscr{S}(\mathbb{R}^d)$ of d-dimensional space rapidly decreasing functions in the same way. Taking the completion, we can define the d-dimensional L^2 space $L^2(\mathbb{R}^d)$. Then, we can define self-adjoint operators Q_j, P_j and $f(\mathsf{Q}_1, \ldots, \mathsf{Q}_d, \mathsf{P}_1, \ldots, \mathsf{P}_d)$ in the above way when f is $2d$-variable ordered polynomial and it equals the opposite ordered polynomial of f.

Now, we consider the case with the Coulomb potential. Firstly, we define the self-adjoint operator $\sum_{j=1}^3 \mathsf{Q}_j^2$ on a dense subspace of $L^2(\mathbb{R}^3)$. Then, according to the method explained after Lemma 1.2, we define the self-adjoint operator$-(\sum_{j=1}^3 \mathsf{Q}_j^2)^{-\frac{1}{2}}$, which is called the Coulomb potential. This operator is also constructed by a closed extension of an operator defined in the domain $\mathscr{S}(\mathbb{R}^3)$.

Lemma 1.6 *Let f be a smooth function densely defined on $\mathbb{R}^d$. Then, we can define the self-adjoint operator $f(\mathsf{Q}_1, \ldots, \mathsf{Q}_d)$ according to the method explained after Lemma 1.2. We also define the operator $f(\tilde{\mathsf{Q}}_1, \ldots, \tilde{\mathsf{Q}}_d)$ on $\mathscr{S}(\mathbb{R}^d)$. Then, the closed extension of $f(\tilde{\mathsf{Q}}_1, \ldots, \tilde{\mathsf{Q}}_d)$ is $f(\mathsf{Q}_1, \ldots, \mathsf{Q}_d)$.*

Proof Since the domain of $f(\mathsf{Q}_1, \ldots, \mathsf{Q}_d)$ includes $\mathscr{S}(\mathbb{R}^d)$, the domain of $f(\mathsf{Q}_1, \ldots, \mathsf{Q}_d)$ includes the domain of the closed extension of $f(\tilde{\mathsf{Q}}_1, \ldots, \tilde{\mathsf{Q}}_d)$. So, it is sufficient to show that any element g of the domain of $f(\mathsf{Q}_1, \ldots, \mathsf{Q}_d)$ is expressed as the limit of sequence in $\mathscr{S}(\mathbb{R}^d)$ in the sense of the graph norm.

Given $R > 0$, we choose a function $\tilde{g}_R \in \mathscr{S}(\mathbb{R}^d)$ such that $\tilde{g}_R(x) = 1$ for $|f(x)| < R$ and $\tilde{g}_R(x) = 0$ for $|f(x)| > 2R$. We also choose a sequence of functions $g_n \in \mathscr{S}(\mathbb{R}^d)$ such that $\|g - g_n\| \to 0$ as $n \to \infty$. Now, we choose an integer n_R such that $\|\tilde{g}_R(g - g_{n_R})\| < \frac{1}{R^2}$.

Then, we have $\tilde{g}_m g_{n_m} \in \mathscr{S}(\mathbb{R}^d)$. Now, we show that $\tilde{g}_m g_{n_m}$ converges to g in the sense of graph norm as follows. We have

$$\|\tilde{g}_m g_{n_m} - g\| \le \|\tilde{g}_m g_{n_m} - \tilde{g}_m g\| + \|\tilde{g}_m g - g\| \to 0. \tag{1.36}$$

and

$$\begin{aligned}
&\|f(\mathsf{Q}_1, \ldots, \mathsf{Q}_d)(\tilde{g}_m g_{n_m} - g)\| \\
\le &\|f(\mathsf{Q}_1, \ldots, \mathsf{Q}_d)(\tilde{g}_m g_{n_m} - \tilde{g}_m g)\| + \|f(\mathsf{Q}_1, \ldots, \mathsf{Q}_d)(\tilde{g}_m g - g)\| \\
\le &2m\|(\tilde{g}_m g_{n_m} - \tilde{g}_m g)\| + \sqrt{\int_{|s|>n_m} |g(s)|^2 ds} \\
\le &2m\frac{1}{m^2} + \sqrt{\int_{|s|>m} |g(s)|^2 ds} \to 0.
\end{aligned} \tag{1.37}$$

■

Although the self-adjoint operator $f(\mathsf{Q}_1, \ldots, \mathsf{Q}_d, \mathsf{P}_1, \ldots, \mathsf{P}_d)$ is defined only on a dense subspace of the Hilbert space $\mathcal{H}$, to avoid the complexity of description, we simply say that the self-adjoint operator $f(\mathsf{Q}_1, \ldots, \mathsf{Q}_d, \mathsf{P}_1, \ldots, \mathsf{P}_d)$ is defined on the Hilbert space $\mathcal{H}$ in this book when the self-adjoint operator is defined in the above way.

Exercise 1.11 Show that the functions $xf(x)$ and $\frac{d}{dx}(x)$ for a function f in $\mathscr{S}(\mathbb{R}^d)$ also belong to $\mathscr{S}(\mathbb{R}^d)$.

Exercise 1.12 Consider two functions f and g in $\mathscr{S}(\mathbb{R}^d)$. Show that the product $f(x)g(x)$ also belongs to $\mathscr{S}(\mathbb{R}^d)$.

Chapter 2
Group Representation Theory

Abstract This chapter discusses representations for group in a general framework including projective representations, which are important in quantum theory. Since a projective representation is closely related to extension of group, this chapter focuses on this relation. To discuss representation theory including projective representations, we need to handle the factor system, i.e., the set of phase factors, which requires complicated notations. The discussions with projective representations are complicated and are often omitted in introductory book for mathematics. However, they are more essential for quantum theory. So, this chapter handles projective representations by using factor systems and we keep this style during the whole book. In the latter part of this chapter, we proceed to the details of representations of finite groups so that it introduces the Fourier transform for finite groups, which connects analysis and algebra. As a typical example, we analyze representations of a permutation group by using Young diagrams.

2.1 Group and Homogeneous Space

2.1.1 Group

It is natural that physical rule does not depend on the coordinate or the origin. The change of the origin is generated by the replacement, and that of the coordinate is by the rotation or the reflection. Hence, in physics, we often consider various transformations, e.g., rotation, reflection, and replacement. Important physical quantities and density matrices often satisfy the invariance or other proper properties with respect to these transformations. To describe such useful properties, we need to employ the mathematical theory for such a transformation.

In general, when a transformation g_1 and another transformation g_2 are given, we can apply the transformation g_2 after the transformation g_1, which can be regarded as another transformation. That is, there exists a transformation combining the transformation g_1 and the transformation g_2. Further, when the transformation g_1 is invertible, we have the inverse transformation of g_1. Hence, we impose the following mathematical properties to the set G of invertible transformations. For any two elements

M. Hayashi, *Group Representation for Quantum Theory*,
DOI 10.1007/978-3-319-44906-7_2

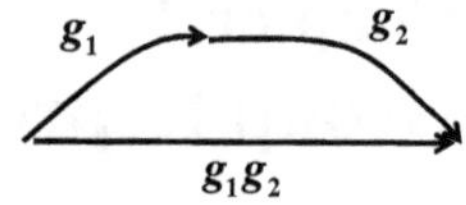

Fig. 2.1 Product

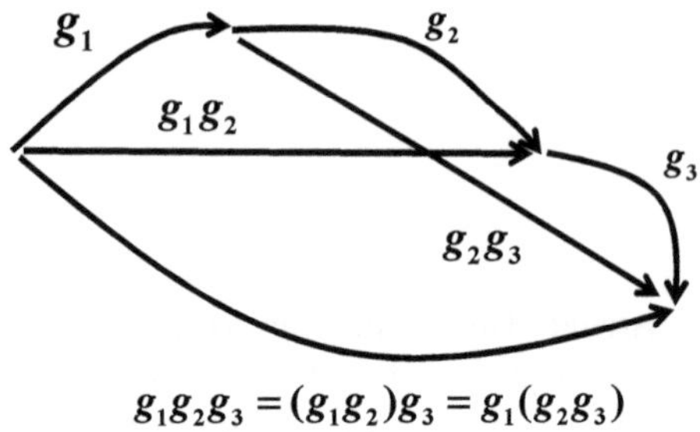

Fig. 2.2 Associative law

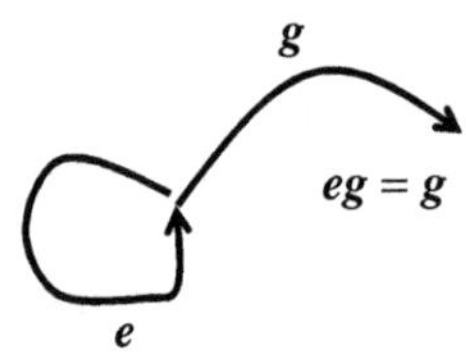

Fig. 2.3 Identity element 1

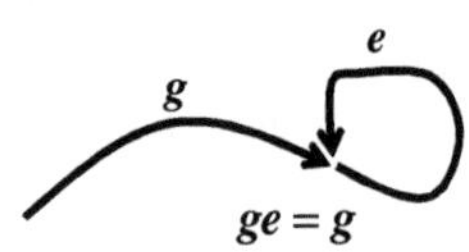

Fig. 2.4 Identity element 2

$g_1, g_2 \in G$, we have the product $g_1 g_2 \in G$, as Fig. 2.1, that satisfies the following conditions.

Associative law: Any three elements $g_1, g_2, g_3 \in G$ satisfy the condition $(g_1 g_2)g_3 = g_1(g_2 g_3)$ as Fig. 2.2.

Identity element: There uniquely exists an element e of G satisfying the following condition. Any element $g \in G$ satisfies $eg = ge = g$ as Figs. 2.3 and 2.4. (The element e is called the identity element, and corresponds to the situation that we apply no transformation.)

Inverse element: For any element $g \in G$, there uniquely exists an element $g^{-1} \in G$ such that $g^{-1}g = gg^{-1} = e$ as Fig. 2.5. (Such an element is called the inverse element of g.)

For example, we can easily check that the set of rotations and the set of replacements satisfy these conditions. In mathematics, when the set G satisfies the above conditions, it is called a **Group**, which has been studied very extensively. In the following, to address a group precisely, we prepare several mathematical concepts required in the latter discussions. When a subset H of G forms a group with respect

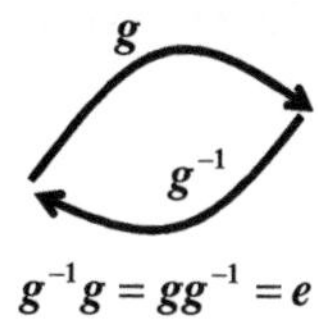

Fig. 2.5 Inverse element

to the same product of G it is called a **subgroup** of G. When the number of elements of the group G is finite, it is called a **finite group** and the number of elements is called the **order** of the group G and is denoted by $|G|$. As a typical example of a finite group, we consider the **permutation group** S_n of degree n, which is defined as the set of permutations among n letters. The order of S_n is $n!$. When an element of S_n is written as the exchange between two letters a and b, it is written as (a, b) and is called a **transposition**. It is known that any permutation is given as products of transpositions. When an element of S_n is written as the product of an even number of transposition, it is called an **even permutation**. The set of even permutations among n letters is called the **alternating group** of degree n and is written as A_n. We say that a subset G' of the group G **generates** G when all of elements of G can be written as products of elements of G' and their inverses. In this case, the subset G' is called a **generator** of G. A group G is called **finitely generated** when there exists a subset G' with finite elements of G that generates G.

For an element $g \in G$, the minimum natural integer n satisfying $g^n = e$ is called the **order** of g. The order of an element $g \in G$ is a divisor of the order of G.

The group G is called a **commutative group** when it satisfies the following commutative law.

Commutative law: Any two elements $g_1, g_2 \in G$ satisfy the condition $g_1 g_2 = g_2 g_1$.

For example, when the Hamiltonian H is given as $|1\rangle\langle 1|$, the time evolution is given by the unitary $e^{itH} = |0\rangle\langle 0| + e^{it}|1\rangle\langle 1|$. By considering the diagonal element of the basis $|1\rangle$, the time evolution is identified with the unit circle $\mathrm{U}(1) := \{z \in \mathbb{C} | \, |z| = 1\} = \{e^{it} | \, t \in \mathbb{R}\}$. Since $e^{it_1} e^{it_2} = e^{i(t_1+t_2)}$, the unit circle is a commutative group with respect to the multiplication. The subset $\mathrm{U}_k := \{e^{i2\pi j/k} | j \in \mathbb{Z}\}$ of $\mathrm{U}(1)$ is also a subgroup. Since any element of U_k can be written as the n-th power of $e^{i2\pi/k}$ with a natural number n less than k, the subgroup U_k is called a **cyclic group** and $e^{i2\pi/k}$ is a **generator** of U_k. As another commutative group, we have **Klein four-group**, which is composed of four elements e, i, j, k with products $ij = ji = k$, $jk = kj = i$, $ki = ik = j$, $i^2 = j^2 = k^2 = e$. This group is denoted by $\mathbb{Z}_2 \times \mathbb{Z}_2$.

Exercise 2.1 List up all of elements of the permutation group S_3. Calculate the orders of all elements.

Exercise 2.2 Show that the subset $\{e, (1,2)(3,4), (1,3)(2,4), (1,4)(2,3)\}$ of S_4 is a subgroup.

Fig. 2.6 Action of group

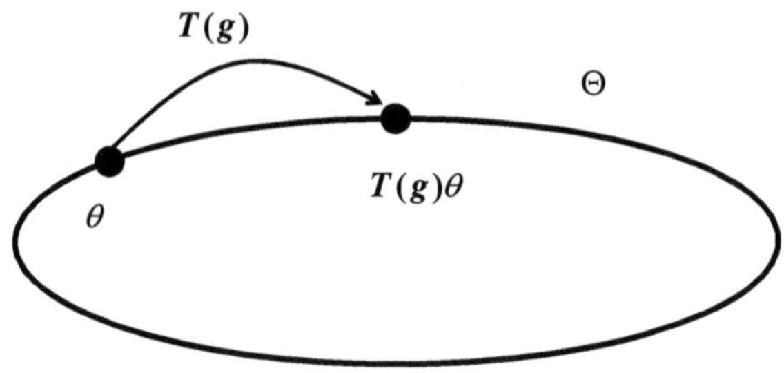

2.1.2 Homogeneous Space

Now, we assume that the target of interest is written by the set Θ and that the transformation on the set Θ is given as a group G. That is, for a given element $g \in G$, we have a transformation $T(g)$ on the set Θ as Fig. 2.6, which satisfies the following condition.

Action of group: The relation $T(g_1) \circ T(g_2) = T(g_1 g_2)$ holds and the transformation $T(e)$ is the identity map.

Here, we can show that $T(g)$ is a bijection. In the following, T is called an **action** of the group G on the set Θ. Such a situation often occurs among natural phenomena. A subset Θ' of Θ is called an invariant subset with respect to the action T on the set Θ when the following condition holds.

Invariance: The relation $T(g)\theta' \in \Theta'$ holds for any elements $g \in G$ and $\theta' \in \Theta'$.

Therefore, when Θ has an invariant non-trivial proper subset, there exist two elements θ and θ' of Θ such that θ and θ' cannot be mapped to each other via the action $T(g)$ with any element $g \in G$. Conversely, when Θ has no invariant non-trivial proper subset, the action T is called **transitive** and the image of an element $\theta \in \Theta$ coincides with the whole set Θ, i.e., the following transitive law holds.

Transitive law: For any two elements $\theta, \theta' \in \Theta$, there exists an element $g \in G$ such that $T(g)\theta = \theta'$.

When there exists a transitive action T for the set Θ, the set Θ is called a **homogeneous space**. In general, even though the action T is transitive, the set Θ is not identical to the set G.

In the following, we describe the structure of the homogeneous space Θ based on the action of G via constructing a homogeneous space essentially identical to Θ. For this purpose, we need to define the equivalence as homogeneous spaces. Two actions T_1, T_2 on two sets Θ_1, Θ_2 are called **equivalent** with each other when there exists a map T_3 from Θ_1 to Θ_2 that satisfies the condition (2.1) as Fig. 2.7.

$$T_3 \circ T_1(g) = T_2(g) \circ T_3. \tag{2.1}$$

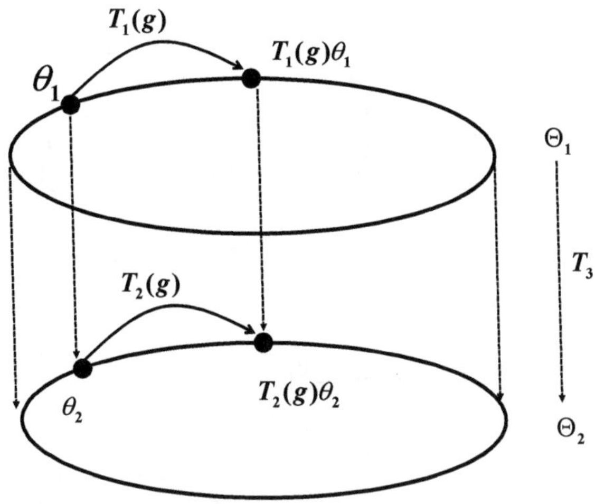

Fig. 2.7 Equivalence between two actions

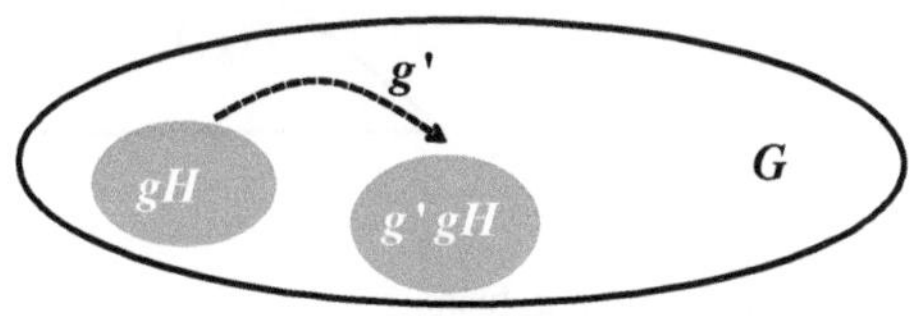

Fig. 2.8 Residue class and quotient space

In particular, when these actions are transitive as well as equivalent, we write $\Theta_1 \cong \Theta_2$.

From the condition of action of group, we can show that the subset $\{g \in G|T(g)\theta_0 = \theta_0\}$ of G is a subgroup for a given $\theta_0 \in \Theta$ (Exercise 2.6). Since the subgroup does not change the element θ_0, it is called a **stabilizer** of θ_0. In fact, the homogeneous space Θ can be reconstructed from the group G and the stabilizer H. For an element $g \in G$, the subset $[g] := gH$ is called a **residue class** with respect to the subgroup H. (In general, for a subset $B \subset G$, the subsets $gB := \{gh|h \in B\}$ and $Bg := \{hg|h \in B\}$ are defined.) Then, the set of all of residue classes is written as $G/H = \{gH|g \in G\}$ and is called the **quotient space**. When the number of elements of G/H is finite, the number is written as $|G : H|$ and is called the **index** of H with respect to G. The map $g \mapsto gH$ is called the **projection** with respect to the residue G/H. Now, for an element $g' \in G$, we can naturally define the action $gH \mapsto (g'g)H$ on the quotient space G/H as Fig. 2.8. When the subgroup H is the stabilizer of the element θ_0 with respect to the action T on the set Θ, the homogeneous space Θ is equivalent to the quotient space G/H as homogeneous spaces. Although the stabilizer H depends on the choice of the element θ_0, the quotient space G/H is equivalent to the quotient space based on another choice of the element θ_0.

Next, for an element $g \in G$, we define the action $g' \mapsto gg'g^{-1}$ on the group G. When a subgroup H is invariant with respect to this action, the subgroup H is called an **invariant subgroup** or a **normal subgroup**. This condition is equivalent to the condition that $gH = Hg$ for any element $g \in G$ as Fig. 2.9. For example, for a

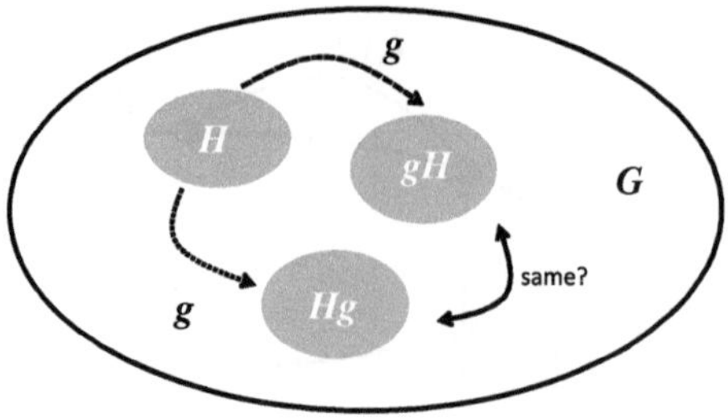

Fig. 2.9 Residue class and quotient space

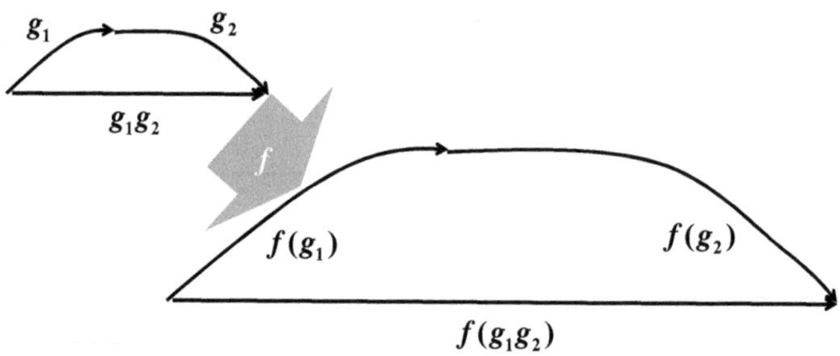

Fig. 2.10 Homomorphism

subgroup $M \subset G$, the subset $C_G(M) := \{h \in G | gh = hg, \ \forall g \in M\}$ is a subgroup of G and is called a **centralizer** of M. In particular, $C_G(G)$ is a commutative normal subgroup of G. So, we simplify it to $C(G)$ and call it the **center** of G. Since the center $C(G)$ is a commutative group, any subgroup of $C(G)$ is also a normal subgroup of G. For a subset $M \subset G$, the subset $N_G(M) := \{g \in G | gM = Mg\}$ is a subgroup of G, and is called the **normalizer** of M. Especially, when M is a subgroup of G, $N_G(M)$ is a normal subgroup of M. For an element $a \in G$, the set $a^G := \{gag^{-1} | g \in G\}$ is called the **conjugacy class** of G containing a. For example, the conjugacy class containing $a \in C(G)$ is $\{a\}$.

For a normal subgroup H, the quotient space G/H forms a group with respect to the multiplication $gH \cdot g'H := (gg')H$ and is called the **residue group** or the **quotient group** for H. The residue group for the center $C(G)$ is called the **index group**. A map f from a group G_1 to a group G_2 is called a **homomorphism** when the condition $f(g_1)f(g_2) = f(g_1g_2)$ holds for any two elements $g_1, g_2 \in G_1$ as Fig. 2.10. In addition, when f is a bijection, f is called an **isomorphism**. When there is an isomorphism from a group G_1 to a group G_2, the two groups G_1 and G_2 are called **isomorphic** to each other and we write $G_1 \cong G_2$. For example, the set $\mathbb{Z}$ of integers is a normal subgroup of the set $\mathbb{R}$ of real numbers with respect to addition, and the quotient group $\mathbb{R}/\mathbb{Z}$ is isomorphic to $\mathrm{U}(1)$. Similarly, the quotient group $\mathbb{Z}_k := \mathbb{Z}/k\mathbb{Z}$ is isomorphic to the cyclic group U_k. Hence, the group $\mathbb{Z}_k$ is also called the **cyclic group**. In particular, when $G_1 = G_2$, an isomorphic map f from

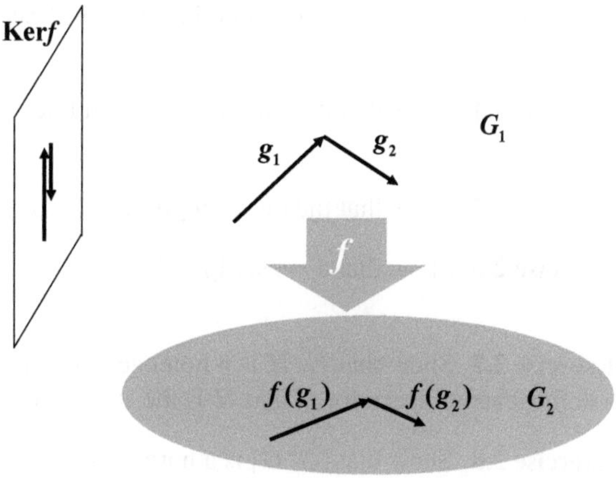

Fig. 2.11 Kernel of homomorphism f

G_1 to G_2 is called an **automorphism** of G_1. An action T of a group G_1 on a group G_2 is called an **automorphic action** of the group G_1 on the group G_2 when $T(g)$ is an automorphism of G_2 for any $g \in G_1$. For a homomorphism f from a group G_1 to a group G_2, the **kernel** of f is defined as $f^{-1}(e)$ and is written as Ker f (Fig. 2.11). It can be shown to be a normal subgroup of G_1. Then, we have $G_1/\operatorname{Ker} f \cong G_2$, which is called the **homomorphism theorem**.

When a group G do not have any subgroup except for $\{e\}$ and G, the group G is called **simple**. When a group G has a normal subgroup, we can analyze the structure of the group G via the analysis on the structure of the normal subgroup. That is, the analysis of the group can be reduced into that of simple groups. For example, for a group G, the subgroup $[G, G]$ of G generated by the set $\{g^{-1}h^{-1}gh | g, h \in G\}$ is a normal subgroup of G and is called the **commutator subgroup** of G.

Lemma 2.1 *The following conditions are equivalent for a subgroup H of a group G.*

(1) *The subgroup H is a normal subgroup and G/H is a commutative group.*
(2) $[G, G] \subset H$.

Proof **(1)⇒(2)**: Assume that the subgroup H satisfies the condition **(1)**. Since $[a][b] = [b][a]$ holds for $a, b \in G$, $[a]^{-1}[b]^{-1}[a][b]$ is the identity element. This fact implies that $a^{-1}b^{-1}ab \in H$. Hence, we obtain the condition **(2)**.
(2)⇒(1): Assume that the subgroup H satisfies the condition **(2)**. Since the relation $d^{-1}cd = cc^{-1}d^{-1}cd \in c[G, G] \subset H$ holds for $c \in H$ and $d \in G$, the subgroup H is a normal subgroup. Further, since any elements $a, b \in G$ satisfies $ab = baa^{-1}b^{-1}ab$ and $a^{-1}b^{-1}ab \in [G, G] \subset H$, we obtain $[a][b] = [b][a]$. ∎

Example 2.1 The subgroup $\mathbb{Z}_2 \times \mathbb{Z}_2$ given in Exercise 2.2 is a normal subgroup of S_4 (see Exercises 2.3 and 2.4). Then, we have the quotient group $S_4/(\mathbb{Z}_2 \times \mathbb{Z}_2)$, which is isometric to S_3 (see Exercise 2.5).

Exercise 2.3 Show that the subgroup given in Exercise 2.2 is isometric to Klein four-group $\mathbb{Z}_2 \times \mathbb{Z}_2$.

Exercise 2.4 Show that the subgroup given in Exercise 2.2 is a normal subgroup of S_4.

Exercise 2.5 Show that the quotient group $S_4/(\mathbb{Z}_2 \times \mathbb{Z}_2)$ is isometric to S_3.

Exercise 2.6 Show that a subset $\{g \in G|T(g)\theta_0 = \theta_0\}$ of a group G is a subgroup of G.

Exercise 2.7 Show that G/H is a homogeneous space equivalent to Θ when Θ is a homogeneous space for G and H is the stabilizer of an element $\theta_0 \in \Theta$.

Exercise 2.8 Show that $[G, G]$ is a normal subgroup of a group G.

2.2 Extension of Group

2.2.1 *General Case*

When two different sets of transformations are given, we often consider the pair of two transformations. When these two transformations act independently, we can easily handle these transformations. However, these two transformations effect each other in general. In such a case, it is very difficult to describe the set of the pair of two transformations. Group extension is a concept to handle such a complicated situation.

Given two different groups H and K, the product set $H \cdot K$ is a group under the product $(h, k)(h', k') := (hh', kk')$, which is called the **direct product (group)** of H and K, and is denoted by $H \times K$. (In this book, the direct product as a set is denoted by $H \cdot K$ and the direct product as a group is denoted by $H \times K$.) The direct product corresponds to the case when the transformations given by the groups H and K act independently. For example, Klein four-group is isometric to the product group $\mathbb{Z}_2 \times \mathbb{Z}_2$. So, it is written as $\mathbb{Z}_2 \times \mathbb{Z}_2$.

Next, we consider the case when the transformation by the group K effects the action of the group H. To handle such a situation, we focus on an automorphic action T of the group K on the group H. Then, the following product is defined for the direct product set $H \cdot K$ by reflecting the structure of the automorphic action T.

$$(h, k)(h', k') := (hT(k)(h'), kk'). \tag{2.2}$$

This product is different from the product in the direct product group $H \times K$. The group based on this product is called a **semi direct product** of H and K and is denoted by $H \rtimes K$. When the elements $h \in H$ $k \in K$ are identified with (h, e) and (e, k), respectively, we have

Fig. 2.12 Regular hexahedron group

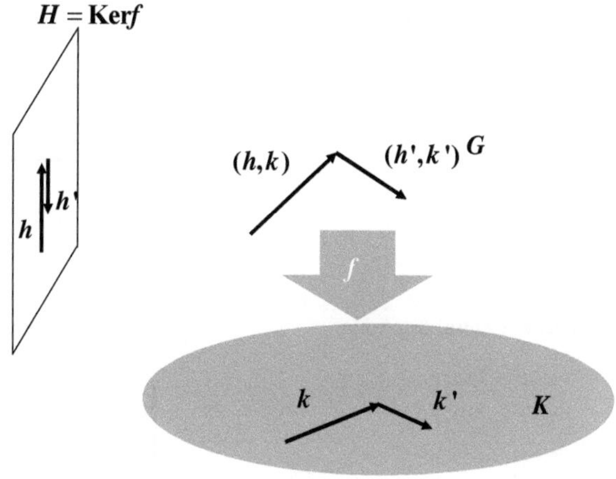

$$hk = (h, k), \quad khk^{-1} = T(k)(h). \tag{2.3}$$

Further, when H and K are identified with the subsets $\{(h, e)\}_{h \in H}$ and $\{(e, k)\}_{k \in K}$, respectively, H is a normal subgroup of $H \rtimes K$ and K is a subgroup $H \rtimes K$.

Example 2.2 Consider the case when $H = \mathbb{Z}_2 \times \mathbb{Z}_2 = \{e, i, j, k\}$ and $K = S_3$. Define the action T of S_3 on $\mathbb{Z}_2 \times \mathbb{Z}_2$ as the permutation among three elements $\{i, j, k\}$. Then, the semi direct product $(\mathbb{Z}_2 \times \mathbb{Z}_2) \rtimes S_3$ is isometric to S_4 because the map $h \mapsto khk^{-1}$ for $h \in \mathbb{Z}_2 \times \mathbb{Z}_2$ and $k \in S_3$ behaves as the permutation on the three elements $\{(1, 2)(3, 4), (1, 3)(2, 4), (1, 4)(2, 3)\}$ with the correspondence $(1, 2)(3, 4) \mapsto 3$, $(1, 3)(2, 4) \mapsto 2$, $(1, 4)(2, 3) \mapsto 1$ (see Exercise 2.9). On the other hand, the group of rotations preserving the regular hexahedron is called the regular hexahedron group. The group $(\mathbb{Z}_2 \times \mathbb{Z}_2) \rtimes S_3$ is isometric to the regular hexahedron group as follows (Fig. 2.12). The elements of i, j, k correspond to the π rotations of x, y, z axises. The elements of S_3 correspond to the rotations permuting x, y, z axises. Using this correspondence, we have the above isometric relation, because the rotation except for the π rotations of x, y, z axises permutes these three rotations.

The direct product and the semi direct product are methods to give a group G that contains H as a normal subgroup and satisfies the condition $G/H \cong K$. Even if a group structure for the direct product set $H \cdot K$ is restricted to satisfy these conditions, it is not limited to the direct product and the semi direct product. In general, such a group G is called an **extension** of the group H by the group K. Further, an extension of a group is not necessarily given as the direct product or the semi direct product. In fact, studying the extensions of a group, we can understand what transformation is possible for a given set G of transformations caused by the groups H and K under the condition $G/H \cong K$. When the quotient G/H is given by a homomorphism f, this situation is illustrated as Fig. 2.13. Especially, when the group H is the center $C(G)$ of an extension G of the group H as Fig. 2.14, the

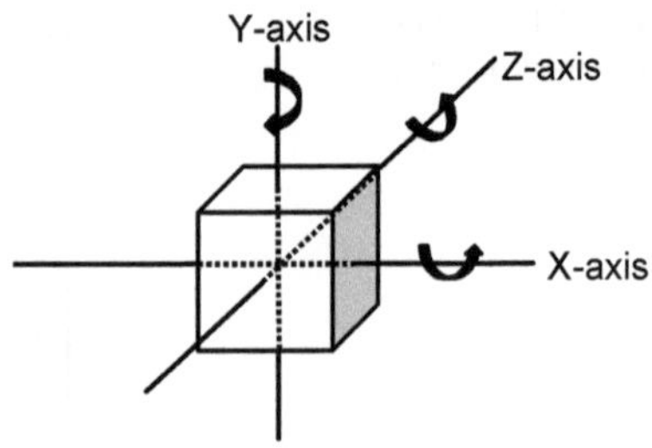

Fig. 2.13 Relation among G, H, and K

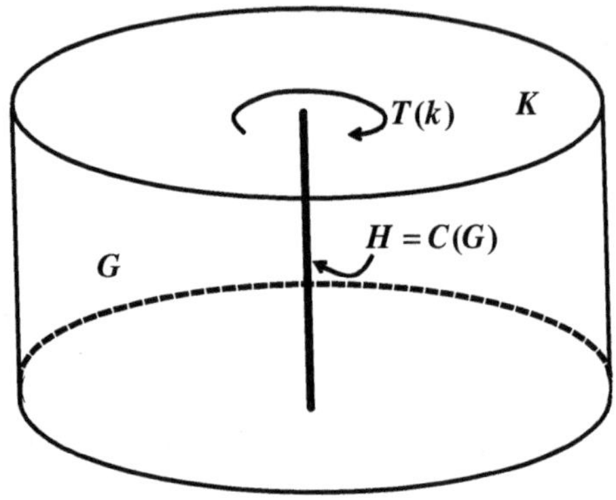

Fig. 2.14 Central extension

extension G plays an important role and is called a **central extension** of the group H by the group K.

Let G be an extension of the group H. When we choose a representative $g_k \in G$ for an element $k \in K$, the residue class is given as $g_k H$ and the group G is written as $g_{k_1} H \cup g_{k_2} H \cup \cdots \cup g_{k_n} H$. Now, for an element $k \in K$, we define the automorphism $T(k)$ of the group H as

$$T(k)(h) := g_k h g_k^{-1}, \quad \forall k \in K, \forall h \in H. \tag{2.4}$$

Then, for two elements $k, k' \in K$, we define the element $\phi(k, k') \in H$ by

$$g_k g_{k'} = \phi(k, k') g_{kk'}. \tag{2.5}$$

Then, for two elements $h, h' \in H$, we have

$$(hg_k)(h'g_{k'}) = hg_k h' g_k^{-1} g_k g_{k'} = hT(k)(h')\phi(k, k') g_{kk'}. \tag{2.6}$$

Thus, since G is a group, the combination of (2.4) and (2.5) yields the relations

$$T(k)(T(k')(h)) = \phi(k, k')T(kk')(h)\phi(k, k')^{-1} \tag{2.7}$$
$$\phi(k, k')\phi(kk', k'') = T(k)(\phi(k', k''))\phi(k, k'k''). \tag{2.8}$$

When the above two conditions are satisfied, $\{T(k), \phi(k, k')\}$ is called a **factor system** of K. The factor system depends on the choice of the representative g_k. Due to the above discussion, the factor system is determined by the extension of the group. Conversely, we have the following lemma [9, Theorem 8.1].

Lemma 2.2 *When a factor system $\{T(k), \phi(k, k')\}$ satisfies the conditions (2.7) and (2.8), the element $\hat{e}_H := \phi(e, k)^{-1} \in H$ does not depend on $k \in K$. Then, we can define the following product for the set $H \cdot K$ by replacing hg_k by (h, k);*

$$(h, k)(h', k') := (hT(k)(h')\phi(k, k'), kk'). \tag{2.9}$$

The product satisfies the condition for the group, in which, the identify element is $e_{H,K}:=(\hat{e}_H, e)$ and the inverse element of (h, k) is $(\hat{e}_H\phi(k^{-1}, k)^{-1}T(k^{-1})(h)^{-1}, k^{-1})$. This group contains the subset $\{(h, e)\}_{h\in H}$ as a normal subgroup. Hence, the subset $\{(h_k, k)\}_{k\in K}$ satisfies the conditions (2.4) and (2.5) for g_k when the element $h_k \in H$ is chosen as an arbitrary element of H.

Hence, the factor system $\{T(k), \phi(k, k')\}$ satisfies the conditions (2.7) and (2.8) if and only if there exists an element $g_k \in G$ satisfying (2.4) and (2.5) for an element $k \in K$. In particular, when the identity element e is chosen as the representative g_e, $T(e)$ is the identity map and the relation $\phi(k, e) = \phi(e, k) = e$ holds. So, the identity element is given as (e, e). This condition is satisfied when h_e is chosen to be e in Lemma 2.2. Such a factor system is called a **normalized factor system**[1]

Further, when $\phi(k, k') = e$, the product (2.9) is the product of the semi direct product group. So, the extension of the group is a generalization of the semi direct product group.

Proof of Lemma 2.2 The condition (2.8) with $k' = e$ implies that

$$\phi(k, e) = T(k)(\phi(e, k'')),$$

which yields that $T(k)^{-1}(\phi(k, e)) = \phi(e, k'')$. That is,

$$T(k)^{-1}(\phi(k, e)^{-1}) = (T(k)^{-1}(\phi(k, e)))^{-1} = \phi(e, k'')^{-1}. \tag{2.10}$$

Since k and k'' are chosen independently, both sides are independent of k and k''. Hence, we have

$$\begin{aligned}&(h, k)(\hat{e}_H, e) = (h, k)(T(k)^{-1}(\phi(k, e)^{-1}), e)\\ =&(hT(k)(T(k)^{-1}(\phi(k, e)^{-1}))\phi(k, e), ke)\\ =&(h\phi(k, e)^{-1}\phi(k, e), k) = (h, k).\end{aligned} \tag{2.11}$$

The condition (2.8) with $k = e$ implies that

[1]The book [99] calls a normalized factor system a factor system.

$$T(e)(T(k')(h)) = \phi(e,k')T(k')(h)\phi(e,k')^{-1}.$$

Since any element h' can be written as $T(k')(h)$, we have

$$\phi(e,k')^{-1}T(e)(h')\phi(e,k') = h'. \tag{2.12}$$

Hence,

$$(\hat{e}_H, e)(h,k) = (\phi(e,k)^{-1}, e)(h,k) = (\phi(e,k)^{-1}T(e)(h)\phi(e,k), e) = (h,k). \tag{2.13}$$

Thus, $(\hat{e}_H, e)$ satisfies the condition for the identity element.

Next, we show that the inverse element of (h,k) is $(\hat{e}_H\phi(k^{-1},k)^{-1}T(k^{-1})(h)^{-1}, k^{-1})$.

$$\begin{aligned}&(\hat{e}_H\phi(k^{-1},k)^{-1}T(k^{-1})(h)^{-1}, k^{-1})(h,k)\\ =&(\hat{e}_H\phi(k^{-1},k)^{-1}T(k^{-1})(h)^{-1}T(k^{-1})(h)\phi(k^{-1},k), k^{-1}k) = (\hat{e}_H, e).\end{aligned} \tag{2.14}$$

The condition (2.8) with $k' = k^{-1}$ and $k'' = k$ implies that

$$T(k)(\phi(k^{-1},k)) = \phi(k,k^{-1})\phi(e,k)\phi(k,e)^{-1}. \tag{2.15}$$

Hence, we have

$$\begin{aligned}&T(k)(\hat{e}_H\phi(k^{-1},k)^{-1}T(k^{-1})(h)^{-1})\\ =&T(k)(\hat{e}_H)T(k)(\phi(k^{-1},k))^{-1}T(k)(T(k^{-1})(h^{-1}))\\ \stackrel{(a)}{=}&\phi(k,e)^{-1}(\phi(k,k^{-1})\phi(e,k)\phi(k,e)^{-1})^{-1}\phi(k,k^{-1})T(e)(h^{-1})\phi(k,k^{-1})^{-1}\\ =&\phi(e,k)^{-1}T(e)(h^{-1})\phi(k,k^{-1})^{-1}\\ =&\phi(e,k)^{-1}T(e)(h^{-1})\phi(e,k)\phi(e,k)^{-1}\phi(k,k^{-1})^{-1}\\ \stackrel{(b)}{=}&h^{-1}\hat{e}_H\phi(k,k^{-1})^{-1},\end{aligned} \tag{2.16}$$

where (a) and (b) follow from (2.15) and (2.12), respectively. Thus,

$$\hat{e}_H\phi(k^{-1},k)^{-1}T(k^{-1})(h)^{-1} = T(k)^{-1}(h^{-1}\hat{e}_H\phi(k,k^{-1})^{-1}). \tag{2.17}$$

Therefore,

$$\begin{aligned}&(h,k)(\hat{e}_H\phi(k^{-1},k)^{-1}T(k^{-1})(h)^{-1}, k^{-1})\\ =&(h,k)(T(k)^{-1}(h^{-1}\hat{e}_H\phi(k,k^{-1})^{-1}), k^{-1})\\ =&(hh^{-1}\hat{e}_H\phi(k,k^{-1})^{-1}\phi(k,k^{-1}), kk^{-1}) = (\hat{e}_H, e).\end{aligned} \tag{2.18}$$

Combining (2.14) and (2.18), we find that the inverse element of (h, k) is $(\hat{e}_H \phi(k^{-1}, k)^{-1} T(k^{-1})(h)^{-1}, k^{-1})$. For the associative rule, see Exercise 2.10. ∎

Exercise 2.9 Show that the map $h \mapsto khk^{-1}$ given in Example 2.2 gives the permutation on the three elements $\{(1, 2)(3, 4), (1, 3)(2, 4), (1, 4)(2, 3)\}$ with the correspondence $(1, 2)(3, 4) \mapsto 3$, $(1, 3)(2, 4) \mapsto 2$, $(1, 4)(2, 3) \mapsto 1$.

Exercise 2.10 Show the associative rule for the product given in Lemma 2.2.

2.2.2 Central Extension of a Commutative Group

Since a central extension of a commutative group plays an important role in the latter section, we discuss the structure of the set of central extensions. When the group H is a commutative group, the extension of the group H by the group K based on the factor system $\{T(k), \phi(k, k')\}$ is a central extension if and only if $T(k)$ is the identity map for any element $k \in K$. In the following, $T(k)$ is fixed to be the identity map because we discuss only the central extension. Hence, when the group H is a commutative group, a factor system describing a central extension can be obtained from the set $\{\phi(k, k')\} \subset H$ satisfying the condition:

$$\phi(k, k'k'')\phi(k', k'') = \phi(k, k')\phi(kk', k'') \tag{2.19}$$

because (2.8) is simplified to (2.19). This condition is equivalent to the existence of the element $g_k \in G$ satisfying (2.5) for an element $k \in K$. Then, the product can be simplified as $(h, k)(h', k') = (hh'\phi(k, k'), kk')$. This book addresses a central extension mainly for the case when H is a commutative group because such a case is closely related to a projective representation. So, the subset $\{\phi(k, k')\} \subset H$ satisfying (2.19) is called a **factor system**. In particular, when $\phi(k, e) = \phi(e, k) = e$, it is a normalized factor system, however, we do not assume this condition for a factor system.

Now, we discuss the set $Z(K : H)$ of factor systems $\{\phi(k, k')\}_{k,k'}$. This set has the structure of group with the product $\{\phi'(k, k')\}_{k,k'}\{\phi(k, k')\}_{k,k'} :=\{\phi'(k, k')\phi(k, k')\}_{k,k'}$. When $\{e\}$ expresses the factor system satisfying the condition $\phi(k, k') = e$ for any two elements $k, k' \in K$, $\{e\}_{k,k'}$ is the identity element of $Z(K : H)$. The central extension corresponding to $\{e\}_{k,k'}$ is the direct product group of $H \times K$. Next, we focus on the subgroup $B(K : H)$ of $Z(K : H)$ defined by the set of factor systems $\{h_{k'} h_k h_{kk'}^{-1}\}_{k,k'}$ by using $\{h_k\}_k$. Two factor systems $\{\phi(k, k')\}$ and $\{\phi'(k, k')\}$ are called equivalent to each other when there exists a subset $\{h_k\}_{k \in K} \subset H$ such that $\phi'(k, k') = h_k h_{k'} h_{kk'}^{-1} \phi(k, k')$. This equivalence condition holds if and only if the two factor systems $\{\phi(k, k')\}$ and $\{\phi'(k, k')\}$ belong to the same element of the quotient group $H(K : H) := Z(K : H)/B(K : H)$. So, the central extensions corresponding to $\{\phi(k, k')\}$ and $\{\phi'(k, k')\}$ are isometric to each other if and only if they are equivalent to each other. Thus, there is a one-to-one correspondence between an element of the quotient group $H(K:H)$ and a central extension of the commutative group H by the group K. In particular, when the factor system is given by an element of B, the constructed central extension is isomorphic to the direct product group $H \times K$. Hence, $H(K:H)$ is called the **group of extension** of the group H by the group K.[2] Using this relation $H(K:H) = Z(K:H)/B(K:H)$, we can characterize the order of an element of the group of extension $H(K:H)$ as follows [9, Theorem 8.5].

Theorem 2.1 *The order of an element of the group of extension $H(K:H)$ is a divisor of the order of H when H is a finite group. Also, the order is a divisor of the order of K when K is a finite group. That is, the order is a common divisor of the orders of H and K when H and K are finite groups. In particular, when the orders of H and K are coprime, a central extension of H by K is limited to the direct product $H \times K$.*

Proof Assume that H is a finite group. Since $\phi(k,k')^{|H|} = e$ for $k, k' \in K$, $\{h_{k'}h_k h_{kk'}^{-1}\}_{k,k'}^n$ is the identity element of Z. Hence, the order of an element of $H(K:H)$ is a divisor of the order of H.

Assume that K is a finite group. Using $b_k := \prod_{k'' \in K} \phi(k,k'')$, we have

$$\phi(k,k')^{|K|} b_{kk'} = \phi(k,k')^{|K|} \prod_{k'' \in K} \phi(kk',k'') = \prod_{k'' \in K} (\phi(k,k')\phi(kk',k''))$$
$$\stackrel{(a)}{=} \prod_{k'' \in K} (\phi(k',k'')\phi(k,k'k'')) = (\prod_{k'' \in K} \phi(k',k''))(\prod_{k'' \in K} \phi(k,k'k'')) = b_{k'} b_k,$$

where (a) follows from (2.19). So, we have $\phi(k,k')^{|K|} = b_{k'} b_k b_{kk'}^{-1} \in B$. Hence, the order of an element of the group of extension $H(K:H)$ is a divisor of the order of K. ∎

Exercise 2.11 Show the relation (2.3) with respect to the semi direct product $H \rtimes K$.

Exercise 2.12 Show that H is a normal subgroup of the semi direct product $H \rtimes K$.

Exercise 2.13 Show that the factor system $\{T(k), \phi(k,k')\}$ gives a central extension if and only if $T(k)$ is the identify map for an element $k \in K$ when H is a commutative group.

2.2.3 Examples for Central Extensions

In the following, we list examples for central extensions.

---

[2] Even when the automorphism $T(k)$ is not necessarily the identity, when $T(k)T(k') = T(kk')$, we can define the group of extension by replacing the condition (2.19) by the condition (2.8). In this case, the extension corresponding to $\{e\}_{k,k'}$ is the direct semi product group of $H \rtimes K$, not the direct product group of $H \times K$. Also, we can show Theorem 2.1 in the same way.

Example 2.3 (*Cyclic group* $\mathbb{Z}_{p^n}$) We focus on the cyclic group $\mathbb{Z}_{p^n}$ for a prime number p. For $m < n$, the subgroup $\{kp^m | kp^m \in \mathbb{Z}_p^n\}$ of $\mathbb{Z}_{p^n}$ is a normal subgroup isomorphic to $\mathbb{Z}_{p^{n-m}}$. Then, $\mathbb{Z}_{p^n}/\mathbb{Z}_{p^{n-m}}$ is isomorphic to $\mathbb{Z}_{p^m}$. Hence, $\mathbb{Z}_{p^n}$ is a central extension of $\mathbb{Z}_{p^{n-m}}$ by $\mathbb{Z}_{p^m}$. For two elements $k_1, k_2 \in \mathbb{Z}_{p^m}$, the factor system is given as

$$\phi(k_1, k_2) = \begin{cases} 0 \text{ when } k_1 + k_2 \geq p^m \\ 1 \text{ when } k_1 + k_2 < p^m. \end{cases} \tag{2.20}$$

Thus, for the group $\mathbb{Z}_{p^n}$, we can construct a sequence of normal subgroups $\{0\} \subset \mathbb{Z}_{p^1} \subset \mathbb{Z}_{p^2} \subset \cdots \subset \mathbb{Z}_{p^n}$.

Example 2.4 (*Discrete Heisenberg group on* $\mathbb{Z}_d$ *with degree* 1) Assume that d is an odd number. The set $\mathbb{Z}_d^2 \cdot \mathrm{U}_d$ forms a group with the product $(s, t, \omega_d^n) \cdot (s', t', \omega_d^{n'}) := (s + s', t + t', \omega_d^{n+n'+st'-ts'})$, where ω_d is the d-th root of 1. This group is called the discrete Heisenberg group on $\mathbb{Z}_d$ with degree 1 and is written as $\mathrm{H}(2, \mathbb{Z}_d)$. The subset $\{(0, 0, \omega_d^n)\}$ is a normal subgroup isomorphic to U_d, hence it is also written as U_d. Also the quotient group $\mathrm{H}(2, \mathbb{Z}_d)/\mathrm{U}_d$ is isomorphic to $\mathbb{Z}_d^2$. When we choose the representative $(s, t, 1) \in \mathrm{H}(2, \mathbb{Z}_d)$ of the element of $\mathrm{H}(2, \mathbb{Z}_d)/\mathrm{U}_d$ corresponding to an element $(s, t) \in \mathbb{Z}_d^2$, we have $(s, t, 1) \cdot (s', t', 1) = (s + s', t + t', \omega_d^{st'-ts'})$. That is, $\mathrm{H}(2, \mathbb{Z}_d)$ is a central extension of U_d by $\mathbb{Z}_d^2$, and the factor system is given as $\phi((s, t), (s', t')) = \omega_d^{st'-ts'}$. The center of $\mathrm{H}(2, \mathbb{Z}_d)$ is U_d.

Assume that d is an even number. The set $\mathbb{Z}_d^2 \cdot \mathrm{U}_{2d}$ forms a group with the product $(s, t, \omega_{2d}^n) \cdot (s', t', \omega_{2d}^{n'}) := (s + s', t + t', \omega_{2d}^{n+n'+st'-ts'})$. The group is called the discrete Heisenberg group on $\mathbb{Z}_d$ with degree 1 and is written as $\mathrm{H}(2, \mathbb{Z}_d)$. Similar to the odd case, the normal subgroup $\{(0, 0, \omega_{2d}^n)\}$ is denoted by U_{2d}, and we have $\mathbb{Z}_d^2 \cong \mathrm{H}(2, \mathbb{Z}_d)/\mathrm{U}_{2d}$. Then, $\mathrm{H}(2, \mathbb{Z}_d)$ is a central extension of U_{2d} by $\mathbb{Z}_d^2$, and the factor system can be calculated similarly. The center of $\mathrm{H}(2r, \mathbb{Z}_d)$ is U_{2d}.

Example 2.5 (*Discrete Heisenberg group on* $\mathbb{Z}_d$ *with multi-degree*) Assume that d is an odd number. The subset $\mathbb{Z}_d^{2r} \cdot \mathrm{U}_d$ forms a group with the product $(\boldsymbol{s}, \boldsymbol{t}, \omega_d^n) \cdot (\boldsymbol{s}', \boldsymbol{t}', \omega_d^{n'}) := (\boldsymbol{s} + \boldsymbol{s}', \boldsymbol{t} + \boldsymbol{t}', \omega_d^{n+n'+(\boldsymbol{s},\boldsymbol{t}')_{\mathbb{Z}}-(\boldsymbol{t},\boldsymbol{s}')_{\mathbb{Z}}})$, where $(\boldsymbol{s}, \boldsymbol{t})_{\mathbb{Z}} := \sum_{j=1}^r s_j t_j$. The group is called the discrete Heisenberg group on $\mathbb{Z}_d$ with degree r and is written as $\mathrm{H}(2r, \mathbb{Z}_d)$. The normal subgroup $\{(0, 0, \omega_d^n)\}$ is written as U_d. The quotient group $\mathrm{H}(2r, \mathbb{Z}_d)/\mathrm{U}_d$ is isomorphic to $\mathbb{Z}_d^{2r}$. When we choose the representative $(\boldsymbol{s}, \boldsymbol{t}, 1) \in \mathrm{H}(2r, \mathbb{Z}_d)$ of the element of $\mathrm{H}(2r, \mathbb{Z}_d)/\mathrm{U}_d$ corresponding to an element $(\boldsymbol{s}, \boldsymbol{t}) \in \mathbb{Z}_d^{2r}$, we have $(\boldsymbol{s}, \boldsymbol{t}, 1) \cdot (\boldsymbol{s}', \boldsymbol{t}', 1) = (\boldsymbol{s} + \boldsymbol{s}', \boldsymbol{t} + \boldsymbol{t}', \omega_d^{(\boldsymbol{s},\boldsymbol{t}')_{\mathbb{Z}}-(\boldsymbol{t},\boldsymbol{s}')_{\mathbb{Z}}})$. That is, $\mathrm{H}(2r, \mathbb{Z}_d)$ is a central extension of U_d by $\mathbb{Z}_d^{2r}$. The factor system is given as $\phi((\boldsymbol{s}, \boldsymbol{t}), (\boldsymbol{s}', \boldsymbol{t}')) = \omega_d^{(\boldsymbol{s},\boldsymbol{t}')_{\mathbb{Z}}-(\boldsymbol{t},\boldsymbol{s}')_{\mathbb{Z}}}$. The center of $\mathrm{H}(2r, \mathbb{Z}_d)$ is U_d.

Assume that d is an even number. The set $\mathbb{Z}_d^{2r} \cdot \mathrm{U}_{2d}$ forms a group with the product $(\boldsymbol{s}, \boldsymbol{t}, \omega_{2d}^n) \cdot (\boldsymbol{s}', \boldsymbol{t}', \omega_{2d}^{n'}) := (\boldsymbol{s} + \boldsymbol{s}', \boldsymbol{t} + \boldsymbol{t}', \omega_{2d}^{n+n'+(\boldsymbol{s},\boldsymbol{t}')_{\mathbb{Z}}-(\boldsymbol{t},\boldsymbol{s}')_{\mathbb{Z}}})$. The group is called the discrete Heisenberg group on $\mathbb{Z}_d$ with degree r and is written as $\mathrm{H}(2r, \mathbb{Z}_d)$. Similar to the odd case, the normal subgroup $\{(0, 0, \omega_{2d}^n)\}$ is denoted by U_{2d}, and we have $\mathbb{Z}_d^{2r} \cong \mathrm{H}(2r, \mathbb{Z}_d)/\mathrm{U}_{2d}$. Then, $\mathrm{H}(2r, \mathbb{Z}_d)$ is a central extension of U_{2d} by $\mathbb{Z}_d^{2r}$. The factor system can be similarly calculated. The center of $\mathrm{H}(2r, \mathbb{Z}_d)$ is U_{2d}.

Example 2.6 (*Discrete Heisenberg group on $\mathbb{F}_q$ with degree* 1) Let q be a power p^n of a prime p, and $\mathbb{F}_q$ be the n-th algebraic extension of the finite filed $\mathbb{F}_p := \mathbb{Z}/p\mathbb{Z}$. Assume that p is an odd prime. The set $\mathbb{F}_q^2 \cdot \mathrm{U}_p$ forms a group with the product $(s, t, \omega_p^n) \cdot (s', t', \omega_p^{n'}) := (s + s', t + t', \omega_p^{n+n'+tr(st'-ts')})$, where trz is defined as follows. Consider the linear space $\mathbb{F}_p^m$ isomorphic to $\mathbb{F}_q$. Let M_z be the matrix on the finite filed $\mathbb{F}_p$ corresponding to the linear map $x \mapsto zx$ defined by the multiplication under the finite filed $\mathbb{F}_q$. Then, we denote $\operatorname{Tr} M_z$ by trz. The group is called the discrete Heisenberg group on $\mathbb{F}_q$ with degree 1, and is written as $\mathrm{H}(2, \mathbb{F}_q)$. The subset $\{(0, 0, \omega_p^n)\}$ is a normal subgroup isomorphic to U_p, and is written as U_p. The quotient group $\mathrm{H}(2, \mathbb{F}_q)/\mathrm{U}_p$ is isomorphic to $\mathbb{F}_q^2$. When we choose the representative $(s, t, 1) \in \mathrm{H}(2, \mathbb{F}_q)$ of the element of $\mathrm{H}(2, \mathbb{F}_q)/\mathrm{U}_p$ corresponding to an element $(s, t) \in \mathbb{F}_q^2$, we have $(s, t, 1) \cdot (s', t', 1) = (s + s', t + t', \omega_p^{tr(st'-ts')})$. That is, $\mathrm{H}(2, \mathbb{F}_q)$ is a central extension of U_p by $\mathbb{F}_q^2$. The factor system is $\phi((s, t), (s', t')) = \omega_p^{tr(st'-ts')}$. The center of $\mathrm{H}(2, \mathbb{F}_q)$ is U_p.

Assume that $p = 2$. The set $\mathbb{F}_q^2 \cdot \mathrm{U}_4$ forms a group with the inner product $(s, t, \omega_4^n) \cdot (s', t', \omega_4^{n'}) := (s + s', t + t', \omega_4^{n+n'+tr(st'-ts')})$. The group is called the discrete Heisenberg group on $\mathbb{F}_q$ with degree 1, and is written as $\mathrm{H}(2, \mathbb{F}_q)$. Similar to the odd prime power case, the normal subgroup $\{(0, 0, \omega_4^n)\}$ is denoted by U_4, and we have $\mathbb{F}_q^2 \cong \mathrm{H}(2, \mathbb{F}_q)/\mathrm{U}_4$. Then, $\mathrm{H}(2, \mathbb{F}_q)$ is a central extension of U_4 by $\mathbb{F}_q^2$. The factor system can be calculated in the same way. Also, $\mathrm{H}(2, \mathbb{F}_q)$ is the center of U_4.

Example 2.7 (*Discrete Heisenberg group on $\mathbb{F}_q$ with multi-degree*) Assume that p is an odd prime. The set $\mathbb{F}_q^{2r} \cdot \mathrm{U}_p$ forms a group with the product $(\boldsymbol{s}, \boldsymbol{t}, \omega_p^n) \cdot (\boldsymbol{s}', \boldsymbol{t}', \omega_p^{n'}) := (\boldsymbol{s} + \boldsymbol{s}', \boldsymbol{t} + \boldsymbol{t}', \omega_p^{n+n'+(\boldsymbol{s},\boldsymbol{t}')_{\mathbb{F}}-(\boldsymbol{t},\boldsymbol{s}')_{\mathbb{F}}})$, where $(\boldsymbol{s}, \boldsymbol{t})_{\mathbb{F}} := \sum_{j=1}^r tr(s_j t_j)$. The group is called the discrete Heisenberg group on $\mathbb{F}_q$ with degree r, and is written as $\mathrm{H}(2r, \mathbb{F}_q)$. Then, the normal subgroup $\{(0, 0, \omega_p^n)\}$ is denoted by U_p, and the quotient group $\mathrm{H}(2r, \mathbb{F}_p)/\mathrm{U}_p$ is isomorphic to $\mathbb{F}_p^{2r}$. When we choose the representative $(\boldsymbol{s}, \boldsymbol{t}, 1) \in \mathrm{H}(2r, \mathbb{F}_q)$ of the element of $\mathrm{H}(2r, \mathbb{F}_q)/\mathrm{U}_p$ corresponding to an element $(\boldsymbol{s}, \boldsymbol{t}) \in \mathbb{F}_p^{2r}$, we have $(\boldsymbol{s}, \boldsymbol{t}, 1) \cdot (\boldsymbol{s}', \boldsymbol{t}', 1) = (\boldsymbol{s} + \boldsymbol{s}', \boldsymbol{t} + \boldsymbol{t}', \omega_p^{(\boldsymbol{s},\boldsymbol{t}')_{\mathbb{F}}-(\boldsymbol{t},\boldsymbol{s}')_{\mathbb{F}}})$, which implies that $\phi((\boldsymbol{s}, \boldsymbol{t}), (\boldsymbol{s}', \boldsymbol{t}')) = \omega_p^{(\boldsymbol{s},\boldsymbol{t}')_{\mathbb{F}}-(\boldsymbol{t},\boldsymbol{s}')_{\mathbb{F}}}$. That is, $\mathrm{H}(2r, \mathbb{F}_q)$ is the central extension of U_p by $\mathbb{F}_q^{2r}$. The factor system is $\phi((s, t), (s', t')) = \omega_p^{tr(st'-ts')}$. The center of $\mathrm{H}(2r, \mathbb{F}_q)$ is U_p.

Assume that the prime p is 2. The set $\mathbb{F}_q^{2r} \cdot \mathrm{U}_4$ forms a group with the product $(\boldsymbol{s}, \boldsymbol{t}, \omega_4^n) \cdot (\boldsymbol{s}', \boldsymbol{t}', \omega_4^{n'}) := (\boldsymbol{s} + \boldsymbol{s}', \boldsymbol{t} + \boldsymbol{t}', \omega_4^{n+n'+(\boldsymbol{s},\boldsymbol{t}')_{\mathbb{F}}-(\boldsymbol{t},\boldsymbol{s}')_{\mathbb{F}}})$. The group is called the discrete Heisenberg group on $\mathbb{F}_q$ with degree r, and is written as $\mathrm{H}(2r, \mathbb{F}_q)$. Similar to the odd prime power case, the normal subgroup $\{(0, 0, \omega_4^n)\}$ is denoted by U_4, and we have $\mathbb{F}_q^{2r} \cong \mathrm{H}(2r, \mathbb{F}_q)/\mathrm{U}_4$. Then, $\mathrm{H}(2r, \mathbb{F}_q)$ is a central extension of U_4 by $\mathbb{F}_q^{2r}$. The factor system can be similarly calculated. The center of $\mathrm{H}(2r, \mathbb{F}_q)$ is U_4.

2.3 Representation and Projective Representation

2.3.1 *Definitions of Representation and Projective Representation*

Next, we consider what transform can be realized of a given quantum system $\mathcal{H}$ based on a given group G. For this purpose, we need several additional notations. Let $\mathfrak{gl}(\mathcal{H})$ be the set of linear maps on a finite-dimensional Hilbert space $\mathcal{H}$, and $\mathrm{GL}(\mathcal{H})$ be the set of invertible elements of $\mathfrak{gl}(\mathcal{H})$. The set $\mathrm{GL}(\mathcal{H})$ has a group structure when the product is defined as the composition of maps. So, it is called the **general linear group** on $\mathcal{H}$. In particular, the subset of elements of $\mathrm{GL}(\mathcal{H})$ with determinant 1 is called the **special linear group** on $\mathcal{H}$ and is written as $\mathrm{SL}(\mathcal{H})$. An element of $\mathrm{GL}(\mathcal{H})$ is called a **unitary map** when it preserves the inner product. The set of unitary maps is written as $\mathrm{U}(\mathcal{H})$, and its subgroup $\mathrm{U}(\mathcal{H}) \cap \mathrm{SL}(\mathcal{H})$ is as $\mathrm{SU}(\mathcal{H})$. The set $\mathrm{U}(\mathcal{H})$ ($\mathrm{SU}(\mathcal{H})$) has a group structure when the product is given as the composition of maps. So, it is called the (special) unitary group on $\mathcal{H}$. A homomorphism f from a group G to a group $\mathrm{GL}(\mathcal{H})$ is called a **representation** of G on $\mathcal{H}$, and $\mathcal{H}$ is called its **representation space**. In particular, a representation f on $\mathcal{H}$ is called a **unitary representation** when the image of f is included in $\mathrm{U}(\mathcal{H})$. A unitary representation plays an important role in quantum theory as follows. A unitary representation f of group G gives a transformation $\rho \mapsto \mathsf{f}(g)\rho\mathsf{f}(g)^\dagger$ for density matrices by an element $g \in G$ as Fig. 2.15. That is, a unitary representation f determines an action of a group G on the set of density matrices. However, a physical transformation by a group G appearing in quantum theory cannot be necessarily limited to a unitary representation f.

To cover all of physical transformations by a group G, we need to consider a larger class of actions of a group G. A map f from a group G to the group $\mathrm{GL}(\mathcal{H})$ is called a **projective representation** when it satisfies the following condition. For any elements $g, g' \in G$, there exists a real number $\theta(g, g') \in \mathbb{R}$ such that

$$\mathsf{f}(g)\mathsf{f}(g') = e^{i\theta(g,g')}\mathsf{f}(gg'). \tag{2.21}$$

In particular, it is called a **projective unitary representation** when the image of f is included in $\mathrm{U}(\mathcal{H})$. One might consider that the phase factor $e^{i\theta(g,g')}$ breaks the physical symmetry. However, the phase factor does not appear in the action on the set of the density matrices. Hence, even when f is not a representation but a projective unitary

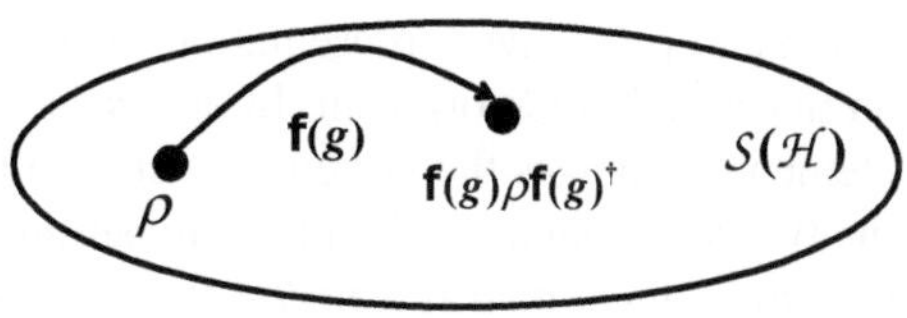

Fig. 2.15 Transformation for density matrix

representation, we have the relation $\mathsf{f}(g)\mathsf{f}(g')\rho(\mathsf{f}(g)\mathsf{f}(g'))^\dagger = \mathsf{f}(gg')\rho\mathsf{f}(gg')^\dagger$. That is, the phase factor $e^{i\theta(g,g')}$ does not appear in the action with respect to the density matrix ρ. On the other hand, when a representation or a projective representation is not unitary, it does not transform a density matrix to another density matrix. Thus, in quantum system, actions of a group are written by projective unitary representations, which contain unitary representations.

Since possible unitary representation and possible projective unitary representations are limited to a small class due to the property of the group of interest, the transformation with respect to the given quantum system derives various restrictions for the states of the quantum system. Sometimes, these limitations simplify our analysis. This book explains this kind of restrictions for the states based on the properties of the group. Although only a unitary representation and a projective unitary representation give physical transformations of quantum systems, a non-unitary representation or a non-unitary projective representation often plays an important role for the analysis of a unitary representation or a projective unitary representation. Due to the above reason, this book covers a representation and a projective representation that are not necessarily unitary. The role of representations is not limited to analysis of physical transformations. A density matrix or a physical quantity is often determined by the properties of a representation. For example, given a (projective) unitary representation f of G, an important physical quantity is often given as the Hermitian matrix $-i \log \mathsf{f}(g)$ with use of an element $g \in G$ or the Hermitian matrix $-i \lim_{t\to 0} \frac{\mathsf{f}(g_t)-I}{t}$ with use of a 1-dimensional subgroup $\{g_t\}_{t\in\mathbb{R}}$ of G. As explained in Sect. 4.4.1, given a non-unitary (projective) representation f of $\mathrm{GL}(n,\mathbb{C})$, a density matrix ρ is often given as $\mathsf{f}(g)$ with use of $g \in G$. In quantum theory, we are usually interested in the matrix forms of a physical quantity and a density matrix. In the above case, their matrix forms are often given via the properties of the group of interest. In this way, a non-unitary (projective) representation f often plays an important role in quantum theory.

On the other hand, since unitary operators can be defined even on an infinite-dimensional Hilbert space $\mathcal{H}$, we can define the set $\mathrm{U}(\mathcal{H})$ even for an infinite-dimensional Hilbert space $\mathcal{H}$. Hence, we can define a unitary representation and a projective unitary representation for an infinite-dimensional Hilbert space $\mathcal{H}$ in the same way as the finite-dimensional case. However, a non-unitary representation is not useful in the infinite-dimensional case, even though it can be defined in the infinite-dimensional case. Hence, we discuss a non-unitary representation and a non-unitary projective representation only in the finite-dimensional case. That is, when $\mathcal{H}$ is finite-dimensional, a representation and a projective representation are not limited to unitary. However, when $\mathcal{H}$ is infinite-dimensional, they are limited to unitary.

In the following, we summarize mathematically important concepts of representation and projective representation. A representation f is called **faithful** when it is an injective homomorphism. In this case, any element of G except for the identity element e changes the system. Given two representations f_1 and f_2 on $\mathcal{H}_1$ and $\mathcal{H}_2$, the two representations f_1 and f_2 are **isomorphic** to each other when there exists an invertible linear map A from $\mathcal{H}_1$ to $\mathcal{H}_2$ such that

$$A\mathsf{f}_1(g) = \mathsf{f}_2(g)A, \quad \forall g \in G. \tag{2.22}$$

Then, A is called an **isomorophism** (with respect to representation of the group G). As mentioned in Sect. 2.8, any representation of a finite group G is isomorphic to a unitary representation of G. Similarly, two unitary representations are called **(unitarily) isomorphic** to each other when there exists a unitary operator A satisfying (2.22). The concept of "isomorphic" can be extended to the case of projective representations. Two projective representations f_1 and f_2 are called **isomorphic** to each other when there exists a linear map A satisfying (2.22). Two projective unitary representations f_1 and f_2 are called **(unitarily) isomorphic** to each other when there exists a unitary operator A satisfying (2.22).

Given two representations f_1 and f_2 of group G on $\mathcal{H}_1$ and $\mathcal{H}_2$, the representation $\mathsf{f}_1 \oplus \mathsf{f}_2$ defined below on the direct sum space $\mathcal{H}_1 \oplus \mathcal{H}_2$ is called the **direct sum representation** of f_1 and f_2, and the representation $\mathsf{f}_1 \otimes \mathsf{f}_2$ defined below on the tensor product space $\mathcal{H}_1 \otimes \mathcal{H}_2$ is called the **tensor product representation** of f_1 and f_2.

$$\mathsf{f}_1 \oplus \mathsf{f}_2(g) := \mathsf{f}_1(g) \oplus \mathsf{f}_2(g), \quad \mathsf{f}_1 \otimes \mathsf{f}_2(g) := \mathsf{f}_1(g) \otimes \mathsf{f}_2(g). \tag{2.23}$$

When f_1 and f_2 are projective representations, the tensor product projective representation $\mathsf{f}_1 \otimes \mathsf{f}_2$ can be defined in the same way. However, as explained in the end of Sect. 2.4, the direct sum projective representation $\mathsf{f}_1 \oplus \mathsf{f}_2$ can be defined only under the restricted case.

When a representation f on the direct sum space $\mathcal{H}_1 \oplus \mathcal{H}_2$ is given as a direct sum representation $\mathsf{f}_1 \oplus \mathsf{f}_2$, we can easily discuss its property. In general, to discuss the matrix form of the representation $\mathsf{f}(g)$, we need to calculate the matrix entries. However, when the representation f is a direct sum representation, there does not exist a non-zero matrix entry that expresses the transformation from the space $\mathcal{H}_1$ to the space $\mathcal{H}_2$. Similarly, there does not exist a non-zero matrix entry of the opposite direction.

For example, when the dimensions of the spaces $\mathcal{H}_1$ and $\mathcal{H}_2$ are d_1 and d_2, respectively, we need to consider $(d_1 + d_2)^2$ entries. However, since $2d_1d_2$ entries are 0 as explained in (2.24), it is enough to consider the remaining $d_1^2 + d_2^2$ entries.

$$\mathsf{f}(g) = \begin{pmatrix} \mathsf{f}_1(g) & 0 \\ 0 & \mathsf{f}_2(g) \end{pmatrix}. \tag{2.24}$$

Further, since the remaining $d_1^2 + d_2^2$ entries can be divided into d_1^2 entries and d_2^2 entries, the treatment is much easier than the case when the representation f is not a direct sum representation. The difficulty of analysis of a representation on a higher dimensional space depends on whether it is a direct sum representation.

A (projective) representation f of a group G on $\mathcal{H}$ is called **reducible** when there exists a subspace $\mathcal{K}$ of $\mathcal{H}$ satisfying the following invariance condition except for $\{0\}$ nor $\mathcal{H}$.

Invariance: $\mathsf{f}(g)u \in \mathcal{K}, \quad \forall u \in \mathcal{K}, \forall g \in G.$

In this case, a subspace $\mathcal{K}$ is called an **invariant subspace** of $\mathcal{H}$. We can define $\mathsf{f}(g)|_{\mathcal{K}}$ as a restriction of $\mathsf{f}(g)$ to the invariant subspace $\mathcal{K}$. The map $\mathsf{f}|_{\mathcal{K}} : g \mapsto \mathsf{f}(g)|_{\mathcal{K}}$ is a representation of G on $\mathcal{K}$, and is called the **subrepresentation** of f on $\mathcal{K}$. Conversely, the representation f is called **irreducible** when there does not exist a non-trivial subspace $\mathcal{K}$ satisfying the above invariance condition. The representation f is called **decomposable** when there exist two non-trivial subspaces $\mathcal{K}_1$ and $\mathcal{K}_2$ satisfying the above invariance condition and the relation $\mathcal{H} = \mathcal{K}_1 \oplus \mathcal{K}_2$, i.e., the representation f is given as the direct sum representation $\mathsf{f}_1 \oplus \mathsf{f}_2$ of the subrepresentations f_1 and f_2 of f on the respective subspaces. In particular, in the case of (projective) unitary representation, the above given invariant subspaces $\mathcal{K}_1$ and $\mathcal{K}_2$ are orthogonal. Conversely, the representation f is called **indecomposable** when it is not decomposable.

Although any irreducible (projective) representation (unitary representation) f is indecomposable, we have the following lemma as a converse statement.

Lemma 2.3 *An indecomposable (projective) unitary representation* f *is irreducible. That is, a reducible (projective) unitary representation* f *is decomposable, and is given as the direct sum representation of the subrepresentations on the invariant subspace.*

Proof We show its contraposition. Assume that a (projective) unitary representation f of a group G on $\mathcal{H}$ is reducible. Then, there exists a non-trivial subspace $\mathcal{K}$ of $\mathcal{H}$ that is invariant for the action of G. As shown later, the orthogonal complement $\mathcal{K}^{\perp}$ of $\mathcal{K}$ is also an invariant subspace for the action of G. Hence, we can conclude that a (projective) unitary representation f is decomposable.

Since any two vectors $v_1 \in \mathcal{K}$ and $v_2 \in \mathcal{K}^{\perp}$ satisfy $\|\mathsf{f}(g)(v_1 + v_2)\|^2 = \|v_1 + v_2\|^2$, we have $\langle \mathsf{f}(g)v_1, \mathsf{f}(g)v_2 \rangle = 0$. Since $\mathsf{f}(g)\mathsf{f}(g^{-1}) = e^{i\theta(g,g^{-1})} I$ ($\mathsf{f}(g)\mathsf{f}(g^{-1}) = I$ in the case of projective representation), the image of $\mathcal{K}$ with respect to the linear map $\mathsf{f}(g)$ is $\mathcal{K}$. Hence, $\mathsf{f}(g)v_2$ is contained the orthogonal complement $\mathcal{K}^{\perp}$ of $\mathcal{K}$. Thus, $\mathcal{K}^{\perp}$ is an invariant subspace for the action of G. ∎

However, there exists a non-unitary representation that is indecomposable and is not irreducible. For example, the following representation of the group $\mathbb{R}$ with addition on the space $\mathbb{C}^2$ is indecomposable and is not irreducible.

$$\mathsf{f}(t) := \begin{pmatrix} 1 & t \\ 0 & 1 \end{pmatrix}, \quad \forall t \in \mathbb{R}. \tag{2.25}$$

A (projective) representation f on $\mathcal{H}$ is called **completely reducible** when it is given as a direct sum representation of (projective) irreducible representations. Hence, when $\mathcal{H}$ is finite-dimensional, Lemma 2.3 guarantees that a (projective) unitary representation f is completely reducible. When the representation is completely reducible, all of matrix entries of the representation can be calculated from all of the matrix entries of irreducible subrepresentations of this representation. Hence, to discuss (projective) unitary representation on the finite-dimensional space, it is enough to discuss irreducible subrepresentations.

Let $\hat{G}$ be the set of indexes describing the irreducible unitary representations of the group G. In this notation, we identify representations isomorphic to each other. For $\lambda \in \hat{G}$, we denote the corresponding representation space by $\mathcal{U}_\lambda(G)$, and the corresponding representation by f_λ. Especially, we denote the set of finite-dimensional representations by $\hat{G}_{\mathsf{f}}$. The dimension of the representation space $\mathcal{U}_\lambda(G)$ is written as d_λ, and its CONS is written as $\{|\lambda; j\rangle\}_{j=1}^{d_\lambda}$. If there is no possibility for confusion, $\mathcal{U}_\lambda(G)$ is simplified to $\mathcal{U}_\lambda$.

Hence, when a unitary representation f is completely reducible, the representation space $\mathcal{H}$ has the following isomorphic representation:

$$\mathcal{H} \cong \bigoplus_{\lambda \in \hat{G}} \mathcal{U}_\lambda(G) \otimes \mathbb{C}^{n_\lambda}, \tag{2.26}$$

where n_λ is the multiplicity of the irreducible representation space $\mathcal{U}_\lambda(G)$ in the space $\mathcal{H}$. Then, f is written as $\oplus_{\lambda \in \hat{G}} n_\lambda \mathsf{f}_\lambda$, and the RHS of (2.26) is called the **irreducible decomposition** of the representation f. Under the above irreducible decomposition, $\{|j\rangle\}_{j=1}^{n_\lambda}$ is the CONS of the space $\mathbb{C}^{n_\lambda}$ corresponding to the multiplicity of $\mathcal{U}_\lambda(G)$. Then, we abbreviate $|\lambda; j\rangle \otimes |j'\rangle$ to $|\lambda; j; j'\rangle$.

For a finite-dimensional representation f, the function $\chi_{\mathsf{f}}(g) := \mathrm{Tr}\, \mathsf{f}(g)$ is called a **character**. When g and g' are contained in the same conjugate class, their characters are the same function. That is, the relation

$$\chi_{\mathsf{f}}(g') = \chi_{\mathsf{f}}(g^{-1} g' g), \quad \forall g, g' \in G \tag{2.27}$$

holds for any representation f. Especially, the character of an irreducible unitary representation f_λ is called an **irreducible character**, and is written as χ_λ.

An irreducible decomposition gives the direct sum decomposition $\oplus_i \mathcal{H}_i$ of the representation space. By denoting the projection to the subspace $\mathcal{H}_i$ by E_i, $\{E_i\}$ forms a projection-valued measurement. Assume that a density matrix ρ is given as the matrix $\mathsf{f}(g)$ by using an element $g \in G$ and a completely reducible representation f of G. When the projection-valued measurement $\{E_i\}$ is applied to the system with the state ρ, the probability to obtain the outcome i is $\mathrm{Tr}\, \mathsf{f}(g) E_i$. The probability $\mathrm{Tr}\, \mathsf{f}(g) E_i$ is nothing less than the character $\chi_{\mathsf{f}_i}(g)$ of the irreducible representation f_i on $\mathcal{H}_i$. A character has an important physical meaning in this way.

2.3.2 *Schur's Lemma*

We have Schur's lemma for an irreducible representation as follows.

Lemma 2.4 *Assume that* f *is a (projective) unitary representation of a group* G *on finite-dimensional space* $\mathcal{H}$. *Then, the following two conditions are equivalent.*

(1) *The (projective) unitary representation* f *is irreducible.*

(2) *When a matrix A on $\mathcal{H}$ satisfies*

$$\mathsf{f}(g)A = A\mathsf{f}(g) \tag{2.28}$$

for any element $g \in G$, the relation $A = cI$ holds with a constant $c \in \mathbb{C}$.

Hence, Schur's lemma (Lemma 2.4) guarantees that the matrix $\mathsf{f}(g)$ is a constant for any element g of the center $C(G)$.

An important physical quantity A often satisfies the condition (2.28). In this case, we say that A is invariant with respect to the representation f. When the representation f is irreducible, Lemma 2.4 guarantees that the physical quantity A is determined to be a fixed value.

Proof **(1)⇒(2)**: Since A is a matrix with complex entries, it has an eigenvalue c. Hence, the matrix $A - cI$ has a non-trivial Kernel. The Kernel is an invariant subspace for the representation f. The irreducibility implies that the Kernel is $\mathcal{H}$.
(2)⇒(1): We show the contraposition of **(2)⇒(1)**. Assume that **(1)** does not hold. Then, f is reducible. So, Lemma 2.3 guarantees that f is decomposable. That is, there exist two orthogonal invariant subspaces $\mathcal{K}_1$ and $\mathcal{K}_2$ of $\mathcal{H}$. Hence, $\mathcal{H}$ has a direct sum decomposition $\mathcal{H} = \mathcal{K}_1 \oplus \mathcal{K}_2$. Thus, when we define the matrix A to be the pair of the constant 1 on $\mathcal{K}_1$ and the constant 0 on $\mathcal{K}_2$, the relation (2.28) holds, but the relation $A = cI$ does not hold. So, the condition **(2)** does not hold. ■

Lemma 2.4 can be extended to an irreducible (projective) unitary representation on an infinite-dimensional space $\mathcal{H}$. The proof of the part **(2)⇒(1)** is still valid even in the infinite-dimensional case. The other part **(1)⇒(2)** can be shown as follows.

Proof Firstly, we show the case when A is a self-adjoint operator. We have $\mathsf{f}(g)A\mathsf{f}(g)^\dagger = A$ for any element $g \in G$. When the spectral decomposition of A is given as $\int E(d\lambda)$, the integral $\int \mathsf{f}(g)E(d\lambda)\mathsf{f}(g)^\dagger$ is a spectral decomposition of the operator $\mathsf{f}(g)A\mathsf{f}(g)^\dagger$ because $\mathsf{f}(g)$ is a unitary operator. Hence, the relation $E([a, b]) = \mathsf{f}(g)E([a, b])\mathsf{f}(g)^\dagger$ holds for any closed interval $[a, b] \subset \mathbb{R}$. Thus, the range of the operator $E([a, b])$ is an invariant subspace. The irreducibility guarantees that the above range coincides with $\mathcal{H}$ or $\{0\}$. Hence, we find that A is a constant operator on $\mathcal{H}$. When A is a unitary or a constant times of a unitary, we can show that A is a constant operator on $\mathcal{H}$ by considering the spectral decomposition with the same way.

Even when A is not necessarily self-adjoint, $A^\dagger A$ is self-adjoint. So, the relation $\mathsf{f}(g^{-1})A^\dagger\mathsf{f}(g^{-1})^\dagger = A^\dagger$ holds for any element $g \in G$. Hence, we can show the relation $\mathsf{f}(g)AA^\dagger\mathsf{f}(g)^\dagger = AA^\dagger$ for any element $g \in G$ by replacing g^{-1} by g. Since $AA^\dagger$ is self-adjoint, $AA^\dagger$ is a constant operator cI on $\mathcal{H}$. If A has a non-trivial kernel, since the kernel is an invariant subspace, the irreducibility of f guarantees that the kernel coincides with the whole space $\mathcal{H}$. Thus, we consider the case when the kernel of A is $\{0\}$. In this case, the image of A is $\mathcal{H}$ because $AA^\dagger = cI$. So, A is a constant times of a unitary operator. Therefore, the analysis of this case can be reduced to the case of unitary operators. Similarly, we can show the same fact when the kernel of $A^\dagger$ is $\{0\}$. ■

We define the support $\mathrm{supp}(\chi_{\mathsf{f}}) := \{g \in G | \chi_{\mathsf{f}}(g) \neq 0\}$ of the character χ_{f}. Then, we obtain the following lemma as an application of Schur's lemma.

Lemma 2.5 *Assume that the index group $G/C(G)$ of a group G is a commutative group. Further, we assume that χ_{f} is an irreducible character and the relation $\mathsf{f}(g) \neq \mathsf{f}(e)$ holds for $g \neq e \in C(G)$. Then, we have* $\mathrm{supp}(\chi_{\mathsf{f}}) = C(G)$.

Proof Since Lemma 2.1 guarantees that the commutator group of G is contained in $C(G)$, we have $g'gg'^{-1}g^{-1} \in C(G)$ for $g \in G \setminus C(G)$ and $g' \in G$. Since g is not contained in $C(G)$, there exists an element $g' \in G$ such that $z := g'gg'^{-1}g^{-1} \neq e$. Schur's lemma guarantees that $\mathsf{f}(z)$ is a constant ω. The assumption implies that $\omega \neq 1$. Hence, $\chi_{\mathsf{f}}(g) = \chi_{\mathsf{f}}(g'gg'^{-1}) = \chi_{\mathsf{f}}(zg) = \omega\chi_{\mathsf{f}}(g)$, which implies that $\chi_{\mathsf{f}}(g) = 0$. ■

Lemma 2.5 can be generalized as follows.

Lemma 2.6 *Assume that the index group $G/C(G)$ of a group G is a commutative group. Also, we assume that χ_{f} is an irreducible character of normal subgroup N of G. Additionally, we suppose that an element $g \in C(G) \cap N$ satisfies $\mathsf{f}(g) \neq I$ if and only if it is the identity element. Then, we have* $\mathrm{supp}(\chi_{\mathsf{f}}) = C(N)$.

Proof Since Lemma 2.1 guarantees that the commutator group of G is contained in $C(G)$, the relation $g'gg'^{-1}g^{-1} \in C(G) \cap N$ holds for $g \in N \setminus C(N)$ and $g' \in G$. Since g is not contained in $C(G)$, there exists an element $g' \in G$ such that $z := g'gg'^{-1}g^{-1} \neq e$. Because $z \in N$, Schur's lemma guarantees that $\mathsf{f}(z)$ is a constant $\omega \neq 1$. So, similar to Lemma 2.5, we obtain the desired argument. ■

Schur's lemma can be generalized to the completely reducible case as follows.

Lemma 2.7 *Assume that a (projective) representation f of a group G on the space $\mathcal{H}$ is irreducibly decomposed as (2.26). Then, the following two conditions for a matrix A on $\mathcal{H}$ are equivalent to each other.*

(1) *The matrix A can be written in the following way by using a matrix A_λ on $\mathbb{C}^{n_\lambda}$:*

$$A = \bigoplus_{\lambda} I_\lambda \otimes A_\lambda, \tag{2.29}$$

where I_λ is the identity matrix on $\mathcal{U}_\lambda$.

(2) *The relation*

$$\mathsf{f}(g)A = A\mathsf{f}(g) \tag{2.30}$$

holds for any element $g \in G$.

When there is no pair of isomorphic subspaces in the irreducible decomposition (2.26), the meaning of this lemma is clear. In this case, when A satisfies the commutative condition **(2)**, A can be written to be the form $\sum_\lambda a_\lambda E_\lambda$ by using the projection

E_λ to the irreducible space $\mathcal{U}_\lambda(G)$. That is, A has a fixed value on each irreducible space. However, when there is a pair of isomorphic subspaces in the irreducible decomposition (2.26), we cannot derive the same conclusion.

Proof Since the part **(1)⇒(2)** is trivial, we show only the part **(2)⇒(1)** as follows. We firstly show **(1)** for the case when $\mathcal{H}$ consists of only one kind of irreducible representation, i.e., $\mathcal{H} = \mathcal{U}_\lambda \otimes \mathbb{C}^{n_\lambda}$. When the vector $u \otimes v \in \mathcal{U}_\lambda \otimes \mathbb{C}^{n_\lambda}$ satisfies $A(u \otimes v) = \sum_i u_i \otimes v_i$, u_i is a constant times of u because of the condition **(2)** and Lemma 2.4. Hence, $A(u \otimes v)$ can be written as $u \otimes v'$ by using $v' \in \mathbb{C}^{n_\lambda}$. That is, A satisfies **(1)**.

Next, we show the general case. In this case, we have $\mathcal{H} = \oplus_{\lambda \in \hat{G}} \mathcal{U}_\lambda \otimes \mathbb{C}^{n_\lambda}$, and denote the projection to $\mathcal{U}_\lambda \otimes \mathbb{C}^{n_\lambda}$ by P_λ. The relation $P_\lambda A P_{\lambda'} \mathsf{f}(g) = \mathsf{f}(g) P_\lambda A P_{\lambda'}$ holds for $\lambda \neq \lambda'$ and $g \in G$. However, since $\mathcal{U}_\lambda$ is not isomorphic to $\mathcal{U}_{\lambda'}$, we have $P_\lambda A P_{\lambda'} = 0$. Hence, the matrix A can be written as $\oplus_{\lambda \in \hat{G}} A_\lambda$ by using the matrix A'_λ on $\mathcal{U}_\lambda \otimes \mathbb{C}^{n_\lambda}$. Finally, applying the above discussion to each matrix A_λ, we obtain the desired argument. ■

2.3.3 *Representations of Direct Product Group*

Consider two given representations f_1 and f_2 of two groups G_1 and G_2 on $\mathcal{H}_1$ and $\mathcal{H}_2$. When the group G_1 does not cause any change on the space $\mathcal{H}_2$ and the group G_2 does not cause any change on the space $\mathcal{H}_1$, it is natural to consider the action on the tensor product space $\mathcal{H}_1 \otimes \mathcal{H}_2$ (the composite system) as the following representation of the direct product group $G_1 \times G_2$. That is, we define the (unitary) representation $\mathsf{f}_1 \bar{\otimes} \mathsf{f}_2$ of the direct product group $G_1 \times G_2$ on the tensor product space $\mathcal{H}_1 \otimes \mathcal{H}_2$ as follows.

$$(\mathsf{f}_1 \bar{\otimes} \mathsf{f}_2)(g_1, g_2) := \mathsf{f}_1(g_1) \otimes \mathsf{f}_2(g_2), \quad \forall (g_1, g_2) \in G_1 \times G_2. \tag{2.31}$$

Then, we have the following formula for characters

$$\chi_{\mathsf{f}_1 \bar{\otimes} \mathsf{f}_2}(g_1, g_2) = \chi_{\mathsf{f}_1}(g_1) \chi_{\mathsf{f}_2}(g_2). \tag{2.32}$$

When f_1 is reducible, $\mathsf{f}_1 \bar{\otimes} \mathsf{f}_2$ is also reducible. That is, when $\mathsf{f}_1 \bar{\otimes} \mathsf{f}_2$ is irreducible, f_1 is irreducible. As the converse statement, we have the following lemma.

Lemma 2.8 *When the representations f_1 and f_2 are irreducible, the (unitary) representation $\mathsf{f}_1 \bar{\otimes} \mathsf{f}_2$ is also irreducible. Conversely, any irreducible (unitary) representation of the direct product group $G_1 \times G_2$ can be written as $\mathsf{f}_1 \bar{\otimes} \mathsf{f}_2$ by using irreducible (unitary) representations f_1 and f_2 of the groups G_1 and G_2. The same fact holds for projective (unitary) representations.*

Now, given a representation f of G on $\mathcal{H}$, we consider the following representation on the set $\mathfrak{gl}(\mathcal{H})$ of matrices on $\mathcal{H}$ of $G \times G$. For any element $(g_1, g_2) \in G \times G$, we define a map $\hat{\mathsf{D}}(\mathsf{f})(g_1, g_2)$ on $X \in \mathfrak{gl}(\mathcal{H})$ as

$$\hat{\mathsf{D}}(\mathsf{f})(g_1, g_2)(X) := \mathsf{f}(g_1)X\mathsf{f}(g_2)^T, \tag{2.33}$$

which satisfies the condition for a representation of $G \times G$. Thanks to Lemma 2.8 and its above discussion, the representation $\hat{\mathsf{D}}(\mathsf{f})$ is irreducible if and only if f is irreducible. In the irreducible case, we can symbolically express this relation as follows.

$$\mathfrak{gl}(\mathcal{U}_\lambda) \cong \mathcal{U}_\lambda \otimes \mathcal{U}_\lambda. \tag{2.34}$$

Proof Any irreducible subspace of the group subgroup $\{e\} \times G_2$ can be written as $\{v_1' \otimes v_2 | v_2 \in \mathcal{H}_2\}$ by choosing a suitable element $v_1' \in \mathcal{H}_1$. Hence, an irreducible subspace of $\mathcal{H}_1 \otimes \mathcal{H}_2$ includes the subset $\{v_1' \otimes v_2 | v_2 \in \mathcal{H}_2\}$ unless the irreducible subspace is $\{0\}$. Similarly, an irreducible subspace of the subgroup $G_1 \times \{e\}$ is written as $\{v_1 \otimes v_2' | v_1 \in \mathcal{H}_1\}$ by choosing a suitable element $v_2' \in \mathcal{H}_2$. Hence, when an irreducible subspace of $\mathcal{H}_1 \otimes \mathcal{H}_2$ contains $v_1' \otimes v_2'$, it includes the subset $\{v_1 \otimes v_2' | v_1 \in \mathcal{H}_1\}$. Since an irreducible subspace of $\mathcal{H}_1 \otimes \mathcal{H}_2$ includes the subset $\{v_1' \otimes v_2 | v_2 \in \mathcal{H}_2\}$ unless it is $\{0\}$, it is coincides with the whole space $\mathcal{H}_1 \otimes \mathcal{H}_2$.

Let f be an irreducible (unitary) representation of the direct product group $G_1 \times G_2$ on the space $\mathcal{H}$. We define the (unitary) representation f_1 of the group G_1 on the space $\mathcal{H}$ by $g_1 \mapsto \mathsf{f}(g_1, e)$. Let $\mathcal{H}_1$ be an irreducible subspace of f_1 that is different from $\{0\}$. We denote the representation on $\mathcal{H}_1$ by f_{λ_1}. Then, the irreducibility guarantees that the subset $\{\mathsf{f}(e, g_2)\mathcal{H}_1\}$ generates the space $\mathcal{H}$. Hence, we obtain the irreducible decomposition $\mathcal{H} \cong \mathcal{U}_\lambda(G_1) \otimes \mathbb{C}^{n_\lambda}$. Since $\mathsf{f}(e, g_2) : \mathcal{H}_1 \to \mathsf{f}(e, g_2)\mathcal{H}_1$ is an isomorophic map for (unitary) representation of G_1, the map $\mathsf{f}(e, g_2)$ maps $v_1 \otimes \mathbb{C}^{n_\lambda}$ to $v_1 \otimes \mathbb{C}^{n_\lambda}$ for any element $v_1 \in \mathcal{U}_\lambda(G_1)$. Define the (unitary) subrepresentation f_2 of G_2 on $v_1 \otimes \mathbb{C}^{n_\lambda}$ as $g_2 \mapsto \mathsf{f}(e, g_2)$, we have $\mathsf{f} \cong \mathsf{f}_\lambda \bar{\otimes} \mathsf{f}_2$. Since f_2 is not an irreducible (unitary) representation of G_2, f also is not an irreducible (unitary) representation of $G_1 \times G_2$. By contradiction, we find that f_2 is also irreducible. Since f is $\mathsf{f}_1 \bar{\otimes} \mathsf{f}_2$, we obtain the desired argument. ∎

Exercise 2.14 Show that any irreducible representation of G has a one-dimensional representation space when G is a commutative group.

2.4 Projective Representation and Extension of Group

2.4.1 Factor System of Projective Representation

As explained in the previous section, any transformation on a quantum system can be written by using a projective unitary representation. However, since the phase factor appears in a projective representation, its treatment is more complicated than that of

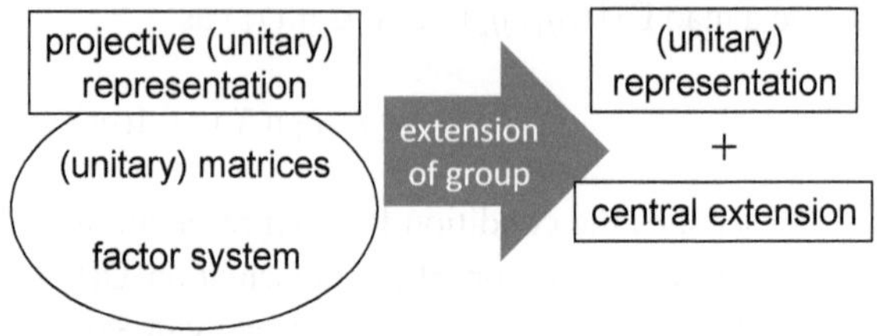

Fig. 2.16 Relation between central extension and projective representation

a representation. In fact, a projective representation of a group G can be reduced to a representation of a central extension of the group G as Fig. 2.16. The purpose of this section is to investigate this reduction.

A projective (unitary) representation f of group G can be reduced to a (unitary) representation of a central extension of the commutative group $\mathrm{U}(1)$ by the group G as follows. Remember that a central extension of G is a larger group than G. As given in (2.21), we define the complex number $e^{i\theta(g,g')} := \mathsf{f}(g)\mathsf{f}(g')\mathsf{f}(gg')^{-1} = \mathsf{f}(gg')^{-1}\mathsf{f}(g)\mathsf{f}(g')$. In fact, we find that $\{e^{i\theta(g,g')}\}_{g,g'}$ satisfies the condition (2.19) for a factor system because $\mathsf{f}(gg'g'')^{-1}\mathsf{f}(g)\mathsf{f}(g'g'')(\mathsf{f}(g'g'')^{-1}\mathsf{f}(g')\mathsf{f}(g'')) = \mathsf{f}(gg'g'')^{-1}\mathsf{f}(gg')(\mathsf{f}(gg')^{-1}\mathsf{f}(g)\mathsf{f}(g'))\mathsf{f}(g'')$. Then, we call $\{e^{i\theta(g,g')}\}_{g,g'}$ the **factor system** of the projective (unitary) representation f. Let $\tilde{G}$ be an extension of a commutative group $\mathrm{U}(1)$ by G whose factor system is $\{e^{i\theta(g,g')}\}_{g,g'}$. Based on a projective (unitary) representation f of G, we can define the (unitary) representation of $\tilde{G}$ by $\tilde{\mathsf{f}}(e^{i\phi}, g) := e^{i\phi}\mathsf{f}(g)$. When the factor system is $\{1\}_{g,g'}$, the extension $\tilde{G}$ is the direct product group $G \times \mathrm{U}(1)$. Then, the projective (unitary) representation f is a (unitary) representation. When f and f' are projective unitary representations and $\mathsf{f}'(g) = \mathsf{f}(g)e^{i\theta(g)}$, both actions for a density matrix ρ satisfy that $\mathsf{f}(g)\rho\mathsf{f}(g)^\dagger = \mathsf{f}'(g)\rho\mathsf{f}'(g)^\dagger$. So, these actions are the same action. Due to this reason, two projective (unitary) representations f and f' are called **equivalent** to each other when there exist a matrix (unitary operator) A and a phase factor $e^{i\theta(g)}$ such that $\mathsf{f}'(g) = e^{i\theta(g)}A\mathsf{f}(g)A^{-1}$. Then, the factor system $\{e^{i\theta'(g,g')}\}_{g,g'}$ of the projective (unitary) representation f' is given by $e^{i\theta'(g,g')} = e^{i\theta(g)}e^{i\theta(g)}e^{-i\theta(gg')}e^{i\theta(g,g')}$. That is, $\{e^{i\theta'(g,g')}\}_{g,g'}$ is a factor system equivalent to $\{e^{i\theta(g,g')}\}_{g,g'}$. Then, the (unitary) representation f' of G and the (unitary) representation $\tilde{\mathsf{f}}$ of the group $\tilde{G}$ satisfy the relation $\tilde{\mathsf{f}}(\omega_g e^{i\phi}, g) := e^{i\phi}\mathsf{f}'(g)$. Hence, an element of the group of extension $H(G : \mathrm{U}(1))$ (the set of extensions of $\mathrm{U}(1)$ by G) uniquely corresponds to a equivalent class of projective (unitary) representations. The element of the group of extension $H(G : \mathrm{U}(1))$ is written as $H(\mathsf{f})$. In particular, when a projective (unitary) representation f is equivalent to a (unitary) representation, i.e., the factor system $H(\mathsf{f})$ is the identity element of the group $H(G : \mathrm{U}(1))$, the projective (unitary) representation f is called **reducible** to a (unitary) representation. Conversely, a projective (unitary) representation f is called a **proper projective** (unitary) representation when it is not reducible to a (unitary) representation. When the set $\{e^{i\theta(g,g')}\}_{g,g'}$ is included in a subgroup C of the group $\mathrm{U}(1)$, the set $\{e^{i\theta(g,g')}\}_{g,g'}$ can be regarded as a factor system corresponding to an extension of C by G. Hence, the set $\{e^{i\theta(g,g')}\}_{g,g'}$ can be regarded as an element of the group of extension $H(G : C)$. When the identity element e of G satisfies $\mathsf{f}(e) = I$, the factor system $H(\mathfrak{f})$ given by the projective representation $\mathfrak{f}$ is a normalized factor system.

In the following, we study the properties of a projective (unitary) representation $\mathfrak{f}$ of a group G via the order of the element $H(\mathfrak{f}) \in H(G : \mathrm{U}(1))$.

Lemma 2.9 *Let $\mathfrak{f}$ be a projective (unitary) representation of a group G on an k-dimensional space $\mathcal{H}$. Then, the order of $H(\mathfrak{f}) \in H(G : \mathrm{U}(1))$ is a divisor of k.*

Proof Firstly, we define another projective (unitary) representation $\mathfrak{f}'(g) := \mathfrak{f}(g)(\det \mathfrak{f}(g))^{-1/k}$ from the given projective (unitary) representation $\mathfrak{f}$. The projective (unitary) representation $\mathfrak{f}'$ satisfies the condition $\det \mathfrak{f}'(g) = 1$ for any element $g \in G$. Then, (2.21) implies that $\mathfrak{f}'(g_2)\mathfrak{f}'(g_1)\mathfrak{f}'(g_2 g_1)^{-1} = e^{i\theta'(g_2,g_1)}$. Since the determinant of the LHS is 1, we have $e^{ik\theta'(g_2,g_1)} = 1$. Therefore, the order of $H(\mathfrak{f}) \in H(G : \mathrm{U}(1))$ is a divisor of k. ∎

Combining the above lemma and Theorem 2.1, we have the following corollary.

Corollary 2.1 *Let $\mathfrak{f}$ be a projective (unitary) representation of a finite group G on $\mathcal{H}$. When the order G is coprime to the dimension k of $\mathcal{H}$, $H(\mathfrak{f}) \in H(G : \mathrm{U}(1))$ is the identity element, i.e., $\mathfrak{f}$ is a (unitary) representation.*

2.4.2 *Irreducibility and Projective Representation*

In the previous subsection, we have seen that any projective representation of a group G can be described by a representation of a central extension $\tilde{G}$ of a commutative finite group C by the group G. In the following, we investigate how an irreducible projective (unitary) representation of G can be characterized by an irreducible (unitary) representation $\tilde{\mathfrak{f}}$ of a central extension $\tilde{G}$.

In this case, Schur's lemma guarantees that $\tilde{\mathfrak{f}}(c)$ is a constant for any element $c \in C$. Letting m be the order of C, we have $\tilde{\mathfrak{f}}(c)^m = 1$. Then, for any element $g \in G = \tilde{G}/C$, we define

$$\mathfrak{f}(g) := \tilde{\mathfrak{f}}(\tilde{g}) \tag{2.35}$$

by choosing an element $\tilde{g} \in \tilde{G}$ among the equivalent class g. Then, since $\tilde{\mathfrak{f}}(c)$ is a constant, $\mathfrak{f}$ is a projective (unitary) representation of G. Given two elements $\tilde{g}, \tilde{g}' \in \tilde{G}$ among the equivalent class g, $\tilde{\mathfrak{f}}(\tilde{g}')$ is a constant times of $\tilde{\mathfrak{f}}(\tilde{g})$. Since the condition (2.28) for $\mathfrak{f}$ is equivalent to the condition (2.28) for $\tilde{\mathfrak{f}}$, $\mathfrak{f}$ is an irreducible projective (unitary) representation. From an irreducible (unitary) representation of a central extension $\tilde{G}$ by G, we can construct an irreducible projective (unitary) representation of group G. Hence, the factor system of a projective representation $\mathfrak{f}$ is an element of group of extension $H(G : C)$. Conversely, when $\mathfrak{f}$ is an irreducible projective (unitary) representation and a (unitary) representation $\tilde{\mathfrak{f}}$ of $\tilde{G}$ satisfies (2.35), the (unitary) representation $\tilde{\mathfrak{f}}$ is irreducible. In general, if $\tilde{\mathfrak{f}}$ is irreducible and $\mathfrak{f}$ is defined by (2.35), the factor system of $\mathfrak{f}$ depends on the representation $\tilde{\mathfrak{f}}$. Using this fact, we obtain the following corollary of Lemma 2.9.

Corollary 2.2 *Let $\mathfrak{f}$ be a projective (unitary) representation on $\mathcal{H}$. Assume that the condition (2.35) holds for an irreducible (unitary) representation $\tilde{\mathfrak{f}}$ of an extension $\tilde{G}$ of a commutative group C by G. When the order of the commutative group C is coprime to the dimension k of $\mathcal{H}$, $H(\mathfrak{f}) \in H(G : C)$ is the identity element, i.e., $\mathfrak{f}$ is an irreducible (unitary) representation.*

Next, let us consider the case when a (unitary) representation $\tilde{\mathfrak{f}}$ of a group $\tilde{G}$ is not necessarily irreducible. In this case, since Schur's lemma does not hold, $\tilde{\mathfrak{f}}(c)$ is not necessarily a constant. Hence, there does not necessarily exist a projective (unitary) representation $\mathfrak{f}$ satisfying (2.35). In general, the matrix entries of $\tilde{\mathfrak{f}}(c)$ depend only on the irreducible component.

Let $\hat{G}([\{e^{i\theta(g,g')}\}_{g,g'}])$ be the set of labels of irreducible projective unitary representations whose factor system is equivalent to $[\{e^{i\theta(g,g')}\}_{g,g'}] \in H(G : \mathrm{U}(1))$, and $\hat{G}_{\mathrm{f}}([\{e^{i\theta(g,g')}\}_{g,g'}])$ be the set of labels of finite-dimensional projective unitary representations. When $\tilde{\mathfrak{f}}$ is an irreducible unitary representation, since $\tilde{\mathfrak{f}}(c)$ is a constant for $c \in C(\tilde{G})$, the center $C(\tilde{G})$ does not change any density matrix. Hence, the group of transformations for density matrices is not $\tilde{G}$ but rather $G = \tilde{G}/C(\tilde{G})$. Therefore, even when $\tilde{\mathfrak{f}}$ is not irreducible and $\tilde{\mathfrak{f}}(c)$ is a constant for $c \in C(\tilde{G})$, the group $G = \tilde{G}/C(\tilde{G})$ expresses the essentially physical transforms. Then, there exists a projective unitary representation $\mathfrak{f}$ of G satisfying (2.35) and $\mathfrak{f}$ describes the action for density matrices.

When we choose irreducible projective unitary representations $\mathfrak{f}_{\lambda_1}, \ldots, \mathfrak{f}_{\lambda_l}$ whose factor systems are $\{e^{i\theta(g,g')}\}_{g,g'}$ for $\lambda_1, \ldots, \lambda_l \in \hat{G}([\{e^{i\theta(g,g')}\}_{g,g'}])$, their direct sum representation $\mathfrak{f}_{\lambda_1} \oplus \cdots \oplus \mathfrak{f}_{\lambda_l}$ is a projective unitary representation. For example, when C is the cyclic group U_2 with order 2, $H(G : \mathrm{U}_2)$ has only two elements. Now, let $\tilde{G}$ be central extension of U_2 by G and $\tilde{\mathfrak{f}}_1, \ldots, \tilde{\mathfrak{f}}_k$ be representations of $\tilde{G}$. When the projective representations $\mathfrak{f}_1, \ldots, \mathfrak{f}_k$ of G are proper projective representations, their factor systems are equivalent to each other. Hence, we can add phase factors to the projective representations so that their suitable factor system is the same. After this replacement, the direct sum representation $\mathfrak{f}_1 \oplus \cdots \oplus \mathfrak{f}_k$ is a projective representation.

2.4.3 *Extension by* U(1)

Next, we consider an extension of the group G by $\mathrm{U}(1)$, which is different from an extension of $\mathrm{U}(1)$ by the group G because the roles of G and $\mathrm{U}(1)$ are opposite to the previous discussion. Since this kind of extensions play a certain role in the latter discussion, we discuss them here. We choose an element ω of $C(G)$ and denote its order by m. Using the element ω, we define an extension of G by $\mathrm{U}(1)$ as follows. That is, for two elements $(e^{i\theta_1}, g_1), (e^{i\theta_2}, g_2) \in \mathrm{U}(1) \cdot G$, their product is defined as

$$(e^{i\theta_1}, g_1) \cdot (e^{i\theta_2}, g_2) := \begin{cases} (e^{i(\theta_1+\theta_2)}, g_1 g_2) & \text{if} \theta_1 + \theta_2 < 2\pi/m \\ (e^{i(\theta_1+\theta_2-2\pi)}, g_1 g_2 \omega) & \text{if} \theta_1 + \theta_2 \geq 2\pi/m. \end{cases}$$

Then, the center of the defined group is $C(G) \rtimes \mathrm{U}(1)$. The group defined above is written as $\overline{G}_\omega$. Let f be an irreducible representation of G on $\mathcal{H}$. Schur's lemma guarantees that there exists an integer $l(\mathsf{f})$ such that $\mathsf{f}(\omega) = e^{i2\pi l(\mathsf{f})/m} I$ for a generator ω of $C(G)$. Note that a generator of $C(G)$ is an element of $C(G)$, but it is not necessarily unique. A representation of an extension of $\overline{G}_\omega$ of G is given as $\mathsf{f}(e^{i\theta_1}, g) = \mathsf{f}(g) e^{i\theta_1(nm+l(\mathsf{f}))}$ by using an integer $n \in \mathbb{Z}$.

2.5 Semi Direct Product and Its Representation

2.5.1 *From $H \rtimes K$ to K and H*

Let us study the relation between a (projective) (unitary) representation of a semi direct product $H \rtimes K$ and (projective) (unitary) representations of the groups K and H. We discuss how to construct a (projective) (unitary) representation of the groups K and H from that of the semi direct product $H \rtimes K$. Consider a (unitary) representation f on $\mathcal{H}$ of the semi direct product $H \rtimes K$ based on an action T of a group K on a group H. Define (unitary) representations f_H and f_K of the subgroups H and K by restricting the domain of the homomorphism f to the subgroups H and K, respectively. Then, the relation (2.3) guarantees that

$$\mathsf{f}(h, k) = \mathsf{f}_H(h)\mathsf{f}_K(k) \tag{2.36}$$

$$\mathsf{f}_H(T(k)(h)) = \mathsf{f}_K(k)\mathsf{f}_H(h)\mathsf{f}_K(k^{-1}). \tag{2.37}$$

The map $\mathsf{f}_{H,k} : H \ni h \mapsto \mathsf{f}_H(khk^{-1})$ defined for any element $k \in K$ is a (unitary) representation of H.

In the same way, given a projective (unitary) representation f of the semi direct product $H \rtimes K$, we define the projective (unitary) representations f_H and f_K of the subgroups H and K. Then, due to (2.3), there exist real numbers $\theta(h, k)$ and $\tilde{\theta}(h, k)$ such that

$$e^{i\theta(h,k)}\mathsf{f}(h, k) = \mathsf{f}_H(h)\mathsf{f}_K(k) \tag{2.38}$$

$$e^{i\tilde{\theta}(h,k)}\mathsf{f}_H(T(k)(h)) = \mathsf{f}_K(k)\mathsf{f}_H(h)\mathsf{f}_K(k^{-1}). \tag{2.39}$$

Then, for any $k \in K$, the map $\mathsf{f}_{H,k} : H \ni h \mapsto \mathsf{f}_H(khk^{-1})$ is a projective (unitary) representation of H. In this way, we can construct a (projective) (unitary) representation of H and K from a (projective) (unitary) representation of $H \rtimes K$.

2.5.2 From K and H to $H \rtimes K$*

Next, we investigate the opposite direction, i.e., how to construct a (projective) (unitary) representation of $H \rtimes K$ from (projective) (unitary) representations of H and K. In particular, we discuss the case when these (projective) (unitary) representations are irreducible.

Construction 1: Let f_H be a (unitary) representation of H on $\mathcal{H}$. Assume that a (unitary) representation $\mathsf{f}_{H,k} : H \ni h \mapsto \mathsf{f}_H(khk^{-1})$ is isomorphic to f_H for any element $k \in K$. Then, there exists a (unitary) matrix $\mathsf{f}_K(k)$ such that $\mathsf{f}_{H,k}(h) = \mathsf{f}_K(k)\mathsf{f}_H(h)\mathsf{f}_K(k)^{-1}$. The map $k \mapsto \mathsf{f}_K(k)$ is a projective (unitary) representation of K, and is called the **associated representation** of K with respect to f_H. Here, we assume that the associated representation f_K of K with respect to f_H is a representation of K. Hence, defining $\mathsf{f}(h, k) := \mathsf{f}_H(h)\mathsf{f}_K(k)$, we find that f satisfies the condition (2.37) and forms a (unitary) representation of $H \rtimes K$. The representation is called the **semi direct product (unitary) representation** of f_H and f_K, and is written as $\mathsf{f}_H \rtimes \mathsf{f}_K$. In particular, when f_H is irreducible, $\mathsf{f}_H \rtimes \mathsf{f}_K$ is also irreducible.

Construction 2: Let f_H be a (unitary) representation of H on $\mathcal{H}$. Define the subset T consisting of elements $t \in K$ such that the (unitary) representation $\mathsf{f}_{H,t}$ is isomorphic to f_H. The subset T is a subgroup of K and is called the **inertia group** of the representation f_H. Here, we assume that the associated representation f_T of T with respect to f_H is a representation of T. Now, we focus on the representation space $\mathcal{H}_{[k]}$ of the representation $\mathsf{f}_{H,k}(h)$ dependently of the element $[k]$ of the homogeneous space K/T. The representation $\mathsf{f}_{H,k}(h)$ does not depend on the choice of the representative k. Choosing the representatives $k_1, \ldots, k_n, \ldots$ of all of elements of K/T, we define a linear map $U_{[k_n]}$ from $\mathcal{H}$ to $\mathcal{H}_{[k_n]}$ as $\mathsf{f}_{H,k_n}(h) = U_{[k_n]}\mathsf{f}_H(h)U_{[k_n]}^{-1}$, where the representative of the equivalent class containing the identity element e should be chosen to be e. We define the unitary matrix $\mathsf{f}_K(k') := \oplus_{[k]\in K/T} U_{[k'k]}\mathsf{f}_T(t(k', [k]))U_{[k]}^{-1}$ on the representation space $\oplus_{[k]\in K/T}\mathcal{H}_{[k]}$ for an element $k' \in K$, where $t(k', [k])$ is given as $k_2^{-1}k'k_1$ by using the above-defined representatives k_1 and k_2 of the equivalent classes $[k]$ and $[k'k]$. Then, f_K forms a (unitary) representation of the group K. Also, the map $h \mapsto \tilde{\mathsf{f}}_H(h) := \oplus_{[k]\in K/T}\mathsf{f}_{H,k}(h)$ forms a (unitary) representation of H. Hence, we can define the (unitary) representation $\tilde{\mathsf{f}}_H \rtimes \mathsf{f}_K$ of the semi direct product $H \rtimes K$. Since the the representation $\tilde{\mathsf{f}}_H \rtimes \mathsf{f}_K$ depends only on f_H and f_T, it is denoted by $\mathsf{f}_H \tilde{\rtimes} \mathsf{f}_T$. When f_H is irreducible, $\mathsf{f}_H \tilde{\rtimes} \mathsf{f}_T$ is also irreducible. When the subgroup T consists only of the identity element, the representation f_T is limited to the trivial representation, and is written as $\mathsf{f}_H \tilde{\rtimes} 1$.

Construction 3: Assume that a (unitary) representation $\mathsf{f}_{H,k} : H \ni h \mapsto \mathsf{f}_H(khk^{-1})$ is isomorphic to f_H for any element $k \in K$. Consider the case when the associated representation f_K of K with respect to a representation f_H of H on $\mathcal{H}'$ defined in **Construction 1** is not necessarily a representation. Let $\mathsf{f}'_K(k)$ be the projective (unitary) representation of K on $\mathcal{H}'$ whose factor system is the complex conjugate of the factor system of $\mathsf{f}_K(k)$. Then, $\mathsf{f}_K \otimes \mathsf{f}'_K$ forms a (unitary) representation of K on $\mathcal{H} \otimes \mathcal{H}'$. Hence, we can define the (unitary) representation $(\mathsf{f}_H \otimes I) \rtimes (\mathsf{f}_K \otimes \mathsf{f}'_K)$ of $H \rtimes K$. Since the (unitary) representation $(\mathsf{f}_H \otimes I) \rtimes (\mathsf{f}_K \otimes \mathsf{f}'_K)$ depends only on f_H,

$\mathfrak{f}_K$, and $\mathfrak{f}'_K$, it is written as $\mathfrak{f}_H \rtimes \mathfrak{f}_K \otimes \mathfrak{f}'_K$. In particular, when $\mathfrak{f}_H$ and $\mathfrak{f}'_K$ are irreducible, $\mathfrak{f}_H \rtimes \mathfrak{f}_K \otimes \mathfrak{f}'_K$ is irreducible, too.

Construction 4: We define the subgroup T of K as the same way as **Construction 2**. Consider the case when the associated representation $\mathfrak{f}_T$ of T with respect to a representation $\mathfrak{f}_H$ of H on $\mathcal{H}'$ is not necessarily a representation. Let $\mathfrak{f}'_T$ be the projective (unitary) representation of T whose factor system is the complex conjugate of the factor system the projective representation $\mathfrak{f}_T(t)$. Then, $\mathfrak{f}_T \otimes \mathfrak{f}'_T$ is a (unitary) representation of T. Hence, we can define the (unitary) representation $(\mathfrak{f}_H \otimes I)\tilde{\rtimes}(\mathfrak{f}_T \otimes \mathfrak{f}'_T)$ of $H \rtimes K$ on $\oplus_{[k]\in K/T}\mathcal{H}_{[k]} \otimes \mathcal{H}'$. Since it depends only on $\mathfrak{f}_H$, $\mathfrak{f}_T$, and $\mathfrak{f}'_T$, it is written as $\mathfrak{f}_H \tilde{\rtimes} \mathfrak{f}_T \otimes \mathfrak{f}'_T$. In particular, when $\mathfrak{f}_H$ and $\mathfrak{f}'_T$ are irreducible, $\mathfrak{f}_H \tilde{\rtimes} \mathfrak{f}_T \otimes \mathfrak{f}'_T$ is also irreducible.

We obtain the following generalization of Lemma 2.8.

Theorem 2.2 *Any irreducible (unitary) representation* $\mathfrak{f}$ *of* $H \rtimes K$ *is written as* $\mathfrak{f}_H \tilde{\rtimes} \mathfrak{f}_T \otimes \mathfrak{f}'_T$ *by using an irreducible representation* $\mathfrak{f}_H$ *of* H*, the associated representation* $\mathfrak{f}_T$ *of* T *with respect to* $\mathfrak{f}_H$*, and an irreducible projective (unitary) representation* $\mathfrak{f}'_T$ *of* T*, where* T *is the inertia group of* $\mathfrak{f}_H$ *and the factor system of projective (unitary) representation* $\mathfrak{f}'_T$ *is the complex conjugate of the factor system of* $\mathfrak{f}_T$.

Proof We regarded the representation $\mathfrak{f}$ as the representation of H, and focus on its irreducible decomposition. We choose one irreducible representation $\mathfrak{f}_H$ among irreducible representations appearing in the irreducible decomposition, and denote its representation space by $\mathcal{H}$. Let m be the multiplicity of $\mathfrak{f}_H$ in the irreducible decomposition and T be the inertia group of the representation $\mathfrak{f}_H$. Also, we denote the associated representation of T with respect to $\mathfrak{f}_H$ by $\mathfrak{f}_T$. Here, we consider the tensor product space $\mathcal{H} \otimes \mathbb{C}^m$. Due to the definition of the inertia group, for $|u\rangle \otimes |v\rangle$ and $t \in T$, there exists a vector $|u'\rangle$ such that $\mathfrak{f}(t)|u\rangle \otimes |v\rangle = |\mathfrak{f}_T(t)u\rangle \otimes |u'\rangle$. Thus, letting $\mathfrak{f}'_T(t)$ be the matrix mapping $|u\rangle$ to $|u'\rangle$, we can show that $\mathfrak{f}'_T$ is an irreducible projective (unitary) representation of T.

In the following, we show this fact. We notice that $\mathfrak{f}_T \otimes \mathfrak{f}''_T$ a (unitary) representation of T. We choose an irreducible projective (unitary) subrepresentation $\mathfrak{f}''_T$ of $\mathfrak{f}'_T$ with respect to T. Since $\mathfrak{f}_T \otimes \mathfrak{f}''_T$ is a (unitary) representation, the factor system of $\mathfrak{f}''_T$ is the complex conjugate of that of $\mathfrak{f}_T$. Recalling the action on the homogeneous space K/T given in Construction 2 (or Construction 4), we define the irreducible (unitary) representation $\mathfrak{f}_H \tilde{\rtimes} \mathfrak{f}_T \otimes \mathfrak{f}''_T$ of $H \rtimes K$, which is a (unitary) subrepresentation of $\mathfrak{f}$ with respect to T. Due to the irreducibility of $\mathfrak{f}$, $\mathfrak{f}$ coincides with $\mathfrak{f}_H \tilde{\rtimes} \mathfrak{f}_T \otimes \mathfrak{f}''_T$. Now, we denote the representation space of $\mathfrak{f}''_T$ by $\mathcal{H}''$, and we consider the irreducible decomposition of the representation space $\mathfrak{f}_H \tilde{\rtimes} \mathfrak{f}_T \otimes \mathfrak{f}''_T$ with respect to the representation $\mathfrak{f}_H$ of the subgroup H. Due to the discussions in Construction 4, any representation subspace isomorphic to $\mathfrak{f}_H$ is included in $\mathcal{H} \otimes \mathcal{H}''$. Hence, we find that $\mathcal{H}''$ coincides with $\mathbb{C}^m$, which implies that $\mathfrak{f}'_T$ equals $\mathfrak{f}''_T$. Thus, $\mathfrak{f}$ equals $\mathfrak{f}_H \tilde{\rtimes} \mathfrak{f}_T \otimes \mathfrak{f}'_T$. ∎

2.6 Real Representation and Complex Conjugate Representation

2.6.1 Real Linear Space and Its Complexification

Since quantum theory mainly discusses complex Hilbert spaces, we rarely treat real vector spaces. However, to discuss several physical-meaningful representations, we often need to handle real vector spaces. In this section, we treat representations on real vector spaces and the relation between real vector spaces and complex vector spaces. For this purpose, we need to define several kinds of complexification of a real vector space.

For a real linear space V with inner product, we define the complex-multiplication on the real vector space $V \oplus V$ as follows.

$$(x+iy)(a,b) := (xa - yb, ay + xb), \quad \forall (a,b) \in V \oplus V,\ \forall x+iy \in \mathbb{C}.$$

We let $V \otimes \mathbb{C}$ be the above-defined complex linear space with the following inner product.

$$((a',b'),(a,b)) := (a' \cdot a - b' \cdot b) + i(a' \cdot b + b' \cdot a), \quad \forall (a',b'),(a,b) \in V \oplus V.$$

Also we let $V \bar{\otimes} \mathbb{C}$ be the above-defined complex linear space with the following inner product.

$$\langle (a',b') | (a,b) \rangle := (a' \cdot a + b' \cdot b) + i(a' \cdot b - b' \cdot a), \quad \forall (a',b'),(a,b) \in V \oplus V.$$

Then, we have $\langle z'v' | zv \rangle = \bar{z}' z \langle v' | v \rangle$ and $(z'v', zv) = z'z(v', v)$ for $v, v' \in V \bar{\otimes} \mathbb{C}$ and $z, z' \in \mathbb{C}$.

2.6.2 Real Representation

Up to the previous section, we have discussed only representations whose representation space is a complex space. For a given real linear space V, we denote the set of invertible linear maps on V by $\mathrm{GL}(V)$. A homomorphism f from a group G to $\mathrm{GL}(V)$ is called a **real representation** of G on V. In this case, V is called the **representation space** of the real representation f. In general, "a representation" means a representation whose representation space is a complex vector space. When we need to distinguish it from a real representation, we call it a complex representation.

In the case of a real representation, similar to the case of complex representation, we can define invariance, irreducibility, reducibility, completely reducibility, decomposability, irreducible decomposition, character, irreducible character, and multiplicity. When a real representation f preserves the inner product, it is called **orthogonal representation**. Similarly, we can define a projective representation.

However, Schur's lemma does not hold for a real representation. That is, there exists an irreducible real representation f of a group G on a real linear space V that satisfies the following condition. There exists a real matrix A on V that is not a constant but satisfies the relation $\mathsf{f}(g)A = A\mathsf{f}(g)$ for any element $g \in G$. For an example, we define the representation f of the group $G = \mathrm{U}(1)$ on the real linear space $V = \mathbb{R}^2$ by $\mathsf{f}(e^{i\theta}) := \begin{pmatrix} \cos\theta & \sin\theta \\ -\sin\theta & \cos\theta \end{pmatrix}$. Then, the matrix $A := \mathsf{f}(e^{i\theta_0})$ satisfies the above condition.

Hence, the irreducibility with respect to a real representation is not useful. Thus, in this book, even though f is a real representation on a real linear space V, the irreducibility of f means the irreducibility of the complex representation on the complex linear space $V \bar{\otimes} \mathbb{C}$.

2.6.3 Complex Conjugate Representation

Given a representation f of a group G on a finite-dimensional complex Hilbert space $\mathcal{H}$, the representation f defined by

$$\bar{\mathsf{f}}(g) := \overline{\mathsf{f}(g)} \tag{2.40}$$

is called the **complex conjugate representation** of f. (In general, we cannot define the complex conjugate operator for an operator A on an infinite-dimensional complex Hilbert space $\mathcal{H}$. Hence, we discuss complex conjugate representations only of finite-dimensional representations.) The definition of a complex conjugate representation depends on the choice of the standard basis. However, even when we define the complex conjugate representation based on another basis, it is isomorphic to the complex conjugate representation based on the original basis. When the representation f is isomorphic to $\bar{\mathsf{f}}$, the representation f is called a **self-conjugate presentation**.[3] Then, the character is a real-valued function. Further, given a label $\lambda \in \hat{G}$, the complex conjugate representation of the irreducible representation f_λ is an irreducible representation, whose label is written as $\lambda^* \in \hat{G}$. That is, f_{λ^*} is the complex conjugate representation of f_λ.

Now, we modify the discussion in (2.33) by using complex conjugate representation. Given a projective representation f of G on $\mathcal{H}$, we consider the following representation on the set $\mathfrak{gl}(\mathcal{H})$ of matrices on $\mathcal{H}$ of $G \times G$. For any element $(g_1, g_2) \in G \times G$, we define a map $\mathsf{D}(\mathsf{f})(g_1, g_2)$ on $X \in \mathfrak{gl}(\mathcal{H})$ as

[3] Such a representation is usually called a **self-adjoint representation**. However, since 'an adjoint representation' will be used in the different meaning in this book, we express such a representation self-conjugate.

$$\mathsf{D}(\mathsf{f})(g_1, g_2)(X) := \mathsf{f}(g_1)X\mathsf{f}(g_2)^\dagger. \tag{2.41}$$

Then, D satisfies the condition for a representation of $G \times G$ even when D is proper projective representation. Thanks to Lemma 2.8 and its above discussion, the representation $\mathsf{D}(\mathsf{f})$ is irreducible if and only if f is irreducible. When f is a unitary projective representation, $\mathsf{D}(\mathsf{f})$ is also a unitary representation under the inner product $\langle X, Y\rangle := \operatorname{Tr} XY^\dagger$. In the irreducible case, we can symbolically express this relation as follows.

$$\mathfrak{gl}(\mathcal{U}_\lambda) \cong \mathcal{U}_\lambda \otimes \mathcal{U}_{\lambda^*}. \tag{2.42}$$

A real linear map A from a complex Hilbert space $\mathcal{H}_1$ to a complex Hilbert space $\mathcal{H}_2$ is called **conjugate linear** when the relation $A(c|v\rangle) = \overline{c}A(|v\rangle)$ holds for any complex number $c \in \mathbb{C}$ and any vector $|v\rangle \in \mathcal{H}_1$. Two (unitary) representations f_1 and f_2 on $\mathcal{H}_1$ and $\mathcal{H}_2$ are called **conjugate isomorphic** when there exists a conjugate linear map A such that $A\mathsf{f}_1(g) = \mathsf{f}_2(g)A$ for any element $g \in G$. Then, the conjugate linear map A is called a **conjugate isomorophism** for (unitary) representation of G. When two (unitary) representations f_1 and f_2 of G are conjugate isomorphic, f_2 is isomorphic to the conjugate representation $\overline{\mathsf{f}_1}$ of f_1. In fact, an irreducible real representation f on V can be regarded as complex representation on $V \bar{\otimes} \mathbb{C}$. The complex representation is a self-adjoint representation because its conjugate representation has the same form.

Conversely, given a complex representation f on the complex Hilbert space $\mathcal{H}$, f can be regarded as a real representation and is called a real representation when $\mathcal{H}$ is regarded as a real vector space. When there exists a real vector subspace $V \subset \mathcal{H}$ such that $\mathcal{H} = V_{\mathbb{C}}$ and the real vector subspace V is invariant for the real representation f, we can define a real representation $\mathsf{f}_{\mathbb{R}}$ on the real linear space V as a restriction of the real representation f. In this case, the complex representation f is called **underlying real form**. However, a self-adjoint complex representation is not necessarily underlying real form.

When f_1 is a projective (unitary) representation and its factor system is given as $\mathcal{E} := [\{e^{i\theta(g,g')}\}_{g,g'}] \in H(G : U(1))$, its conjugate representation $\overline{\mathsf{f}_1}$ has the factor system $\mathcal{E}^{-1} = [\{e^{-i\theta(g,g')}\}_{g,g'}] \in H(G : U(1))$. We can define the concept of "conjugate isomorophic" for the projective case. When a projective (unitary) representation has a conjugate isomorphic projective (unitary) representation, its factor system $\mathcal{E}$ satisfies $\mathcal{E}^{-1} = \mathcal{E}$, i.e., it has order 2 as an element of the group $H(G : U(1))$.

On a $2d$-dimensional real representation space V, using the two-dimensional matrix $J = \begin{pmatrix} 0 & -1 \\ 1 & 0 \end{pmatrix}$, we define the matrix

$$J_d := I_d \otimes J. \tag{2.43}$$

Then, considering the complex-valued vector $(x_1 + ix_{d+1}, \ldots, x_d + ix_{2d})$ for $x \in V$, we can regard V as a d-dimensional complex linear space $\mathcal{V}$. When a real representation f forms a complex representation on a complex linear space $\mathcal{V}$, it is called **complexifiable**. Here, we should remark that $\mathcal{V}$ is different from $V \otimes \mathbb{C}$. A real representation f is complexifiable if and only if

$$\mathsf{f}(g)J_d = J_d\mathsf{f}(g). \tag{2.44}$$

Exercise 2.15 Show that the character of a self-adjoint representation takes real numbers.

Exercise 2.16 Show that (2.44) is holds if and only if a real representation f is complexifiable.

2.7 Representation on Composite System

In Sect. 2.3, we have mentioned that, the treatment of a representation can be simplified by its irreducible decomposition because the treatment can be divided into each irreducible representation. However, in the physical context, the representation of our interest is usually given in a form with a tensor product representation.

As the simplest example, we consider the case when the first quantum system is given as the representation space $\mathcal{U}_\lambda(G)$ of the irreducible representation f_λ of a group G and the group G does not act on the second quantum system. In this case, we are interested in the behavior of a vector $|X\rangle\rangle \in \mathcal{U}_\lambda(G) \otimes \mathbb{C}^d$ under the action G. In the following, we focus on the case when the dimension of the second quantum system is larger than the dimension d_λ of $\mathcal{U}_\lambda(G)$.

Now, we remember that a vector in the tensor product space $\mathcal{U}_\lambda(G) \otimes \mathbb{C}^d$ can be written as $|X\rangle\rangle$ by using a $d_\lambda \times d$ matrix X. We employ a polar decomposition of a matrix X. That is, we choose suitable CONSs $\{|e_i\rangle\}$ and $\{|f_i\rangle\}$ of $\mathcal{U}_\lambda(G)$ and $\mathbb{C}^d$ such that $|X\rangle\rangle = \sum_{j=1}^{d_\lambda} a_j |e_j\rangle|f_j\rangle$. Let $\mathcal{K}$ be the subspace of $\mathbb{C}^d$ generated by $\{f_j\}_{j=1}^{d_\lambda}$. Then, the vector $\mathsf{f}_\lambda(g)|X\rangle\rangle$ is contained in $\mathcal{U}_\lambda(G) \otimes \mathcal{K}$ for any element $g \in G$. Hence, to discuss the orbit of $|X\rangle\rangle$ by the action of G, we do not need to discuss the whole space $\mathcal{U}_\lambda(G) \otimes \mathbb{C}^d$ and only need to treat the subspace $\mathcal{U}_\lambda(G) \otimes \mathcal{K}$. Since the subspace is isomorphic to $\mathcal{U}_\lambda(G) \otimes \mathbb{C}^{d_\lambda}$, it is sufficient to treat at most the dimension d_λ.

In general, when a representation of a group G is completely reducible, as given in (2.26), the representation space has the irreducible decomposition $\oplus_{\lambda \in \hat{G}} \mathcal{U}_\lambda(G) \otimes \mathbb{C}^{n_\lambda}$. However, to address the orbit of a specific vector by the action of G, the above discussion guarantees that it is enough to discuss the representation subspace $\oplus_{\lambda \in \hat{G}} \mathcal{U}_\lambda(G) \otimes \mathbb{C}^{d_\lambda}$. As will be explained in Sect. 2.8, restricting our analysis in the subspace enables us a deeper analysis.

Next, we consider the case when two quantum systems are given as two representation spaces $\mathcal{U}_\lambda(G)$ and $\mathcal{U}_{\lambda'}(G)$ of two irreducible representations f_λ and $\mathsf{f}_{\lambda'}$ of a group G. Then, we have the tensor product representation $\mathsf{f}_\lambda \otimes \mathsf{f}_{\lambda'}$ on the tensor product space $\mathcal{U}_\lambda(G) \otimes \mathcal{U}_{\lambda'}(G)$ corresponding to the composite system. When these two representations f_λ and $\mathsf{f}_{\lambda'}$ are finite-dimensional unitary representations, the tensor product representation is completely reducible due to Lemma 2.3. When the tensor product representation $\mathsf{f}_\lambda \otimes \mathsf{f}_{\lambda'}$ is completely reducible, we define the **multiplicity coefficient** $C^{\lambda''}_{\lambda,\lambda'}(G)$ that describes the multiplicity of $\mathcal{U}_{\lambda''}(G)$ in the space $\mathcal{U}_\lambda(G) \otimes \mathcal{U}_{\lambda'}(G)$ as follows

$$\mathcal{U}_\lambda(G) \otimes \mathcal{U}_{\lambda'}(G) \cong \bigoplus_{\lambda'' \in \hat{G}} \mathcal{U}_{\lambda''}(G) \otimes \mathbb{C}^{C^{\lambda''}_{\lambda,\lambda'}(G)}. \tag{2.45}$$

When G is a commutative group, its irreducible representation is one-dimensional (Exercise 2.14). That is, the relation $\mathsf{f}_\lambda : G \mapsto \mathrm{U}(1)$ holds for $\lambda \in \hat{G}$. Then, the map $g \mapsto \mathsf{f}_\lambda(g)\mathsf{f}_{\lambda'}(g)$ is a representation for $\lambda, \lambda' \in \hat{G}$. We denote the label of the representation by $\lambda + \lambda' \in \hat{G}$. The set $\hat{G}$ forms a commutative group with this product. In particular, the multiplicity coefficient satisfies the following formula.

$$C^{\lambda''}_{\lambda,\lambda'}(G) = \begin{cases} 1 \text{ when } \lambda'' = \lambda + \lambda' \\ 0 \text{ otherwise.} \end{cases} \tag{2.46}$$

As a special case, we consider the case when the second representation space $\mathcal{U}_{\lambda'}(G)$ is $\mathcal{U}_{\lambda^*}(G)$.

In this case, the relation $\mathsf{f}_{\lambda^*} = \bar{\mathsf{f}}_\lambda$ holds under a suitable choice of the standard basis $\{|j\rangle_A\}, \{|j\rangle_B\}$. By using the standard basis, any vector on the composite system can be written as $|X\rangle\rangle$ with a square matrix X. Then, we have the following lemma.

Lemma 2.10 *The following conditions are equivalent to each other for any vector in the composite system* $|X\rangle\rangle$.

(1) $|X\rangle\rangle$ *is a constant times of* $|I\rangle\rangle$.
(2) *The relation*

$$\mathsf{f}(g) \otimes \bar{\mathsf{f}}(g)|X\rangle\rangle = |X\rangle\rangle, \quad \forall g \in G \tag{2.47}$$

holds for any element $g \in G$.

Proof **(1)⇔(2)**: This part can be shown from the relation $\mathsf{f}(g) \otimes \bar{\mathsf{f}}(g)|I\rangle\rangle = |\mathsf{f}(g)\bar{\mathsf{f}}(g)^T\rangle\rangle = |I\rangle\rangle$.
(2)⇔(1): Assume the condition **(2)**. Since the relation $X = \mathsf{f}_\lambda(g)X\mathsf{f}_{\lambda^*}(g)^T = \mathsf{f}_\lambda(g)X\mathsf{f}_\lambda(g)^{-1}$ holds, Schur's lemma restricts the matrix X to a constant times of the identity matrix. Hence, we obtain the condition **(1)**. ■

We also have the following lemma.

Lemma 2.11 *The following conditions are equivalent to each other for a representation* f *of a group* G *on a space* $\mathcal{H}$.

(1) *The representation space* $\mathcal{H}$ *contains the irreducible representation space* $\mathcal{U}_\lambda$ *with multiplicity* 1.

(2) *There uniquely exists a wave function, (a normalized vector) $|X\rangle\rangle$ in the composite system $\mathcal{H} \otimes \mathcal{U}_{\lambda^*}$ such that*

$$\mathsf{f}(g) \otimes \mathsf{f}_{\lambda^*}(g)|X\rangle\rangle = |X\rangle\rangle, \quad \forall g \in G. \tag{2.48}$$

Proof **(2)⇔(1)**: Assume the condition **(2)**. The irreducibility of $\mathcal{U}_{\lambda^*}$ guarantees that $\mathrm{Tr}_{\mathcal{H}} |X\rangle\rangle\langle\langle X|$ is a constant times of the identity matrix. Hence, $\mathrm{Tr}_{\mathcal{U}_{\lambda^*}} |X\rangle\rangle\langle\langle X|$ is a constant times of the projection $P_{\mathcal{K}}$ to the range $\mathcal{K}$ of $\mathrm{Tr}_{\mathcal{U}_{\lambda^*}} |X\rangle\rangle\langle\langle X|$. Since $\mathrm{Tr}_{\mathcal{U}_{\lambda^*}} |X\rangle\rangle\langle\langle X|$ is invariant with respect to the representation f, the space $\mathcal{K}$ is also invariant with respect to the representation f. Hence, we have $P_{\mathcal{K}}X = \mathsf{f}|_{\mathcal{K}}(g)P_{\mathcal{K}}X(\mathsf{f}_{\lambda^*}(g))^T$, which implies that $\mathsf{f}|_{\mathcal{K}}(g) = P_{\mathcal{K}}X((\mathsf{f}_{\lambda^*}(g))^T)^{-1}(P_{\mathcal{K}}X)^{-1} = P_{\mathcal{K}}X\overline{\mathsf{f}_{\lambda^*}(g)}(P_{\mathcal{K}}X)^{-1}$. This relation shows that the representation $\mathsf{f}|_{\mathcal{K}}(g)$ is isomorphic to the representation $\mathsf{f}_{\lambda}(g)$. Thus, $\mathcal{H}$ contains the irreducible representation space $\mathcal{U}_{\lambda}$ at least with the multiplicity 1. If the multiplicity is strictly larger than 1, there exist more than two wave functions satisfying the invariance (2.48), which contradicts the condition **(2)**. Therefore, we can conclude that $\mathcal{H}$ contains the irreducible representation space $\mathcal{U}_{\lambda}$ only with the multiplicity 1, which yields the condition **(1)**.
(1)⇔(2): Assume the condition **(1)**. When X is a constant times of the identity matrix, $|X\rangle\rangle$ satisfies (2.48). So, it is sufficient to show the uniqueness of $|X\rangle\rangle$ to satisfy (2.48). Assume that $|X\rangle\rangle$ satisfies (2.48). Since $\mathrm{Tr}_{\mathcal{U}_{\lambda^*}} |X\rangle\rangle\langle\langle X|$ is invariant with respect to the representation f, the space $\mathcal{K}$ is also invariant with respect to the representation f. Due to the condition **(1)**, the space $\mathcal{K}$ is $\mathcal{U}_{\lambda}$. That is, the support of $|X\rangle\rangle$ is limited to $\mathcal{U}_{\lambda} \otimes \mathcal{U}_{\lambda^*}$. In this case, the condition **(1)** is equivalent to the condition $\mathsf{f}_{\lambda}(g)X\mathsf{f}_{\lambda}(g)^{\dagger} = X$. Schur's lemma guarantees that X is a constant times of the identity matrix. Hence, we obtain the condition **(2)**. ∎

Applying Lemma 2.11 to a wave function on the composite system consisting of three quantum systems, we obtain the following lemma.

Lemma 2.12 *The following four conditions are equivalent to each other for three irreducible unitary representations $\lambda_1, \lambda_2, \lambda_3 \in \hat{G}$ of a group G.*

(1) *The tensor product space $\mathcal{U}_{\lambda_1} \otimes \mathcal{U}_{\lambda_3}$ contains the space $\mathcal{U}_{\lambda_2}$ as an irreducible component (with multiplicity 1).*
(2) *There (uniquely) exists a wave function as $|x\rangle$ invariant for the action of G in the composite system $\mathcal{U}_{\lambda_1} \otimes \mathcal{U}_{\lambda_3} \otimes \mathcal{U}_{\lambda_2^*}$ as*

$$\mathsf{f}_{\lambda_1}(g) \otimes \mathsf{f}_{\lambda_3}(g) \otimes \mathsf{f}_{\lambda_2^*}(g)|x\rangle = |x\rangle, \quad \forall g \in G. \tag{2.49}$$

(3) *The tensor product space $\mathcal{U}_{\lambda_1^*} \otimes \mathcal{U}_{\lambda_2}$ contains the space $\mathcal{U}_{\lambda_3}$ as an irreducible component (with multiplicity 1).*
(4) *The tensor product space $\mathcal{U}_{\lambda_2^*} \otimes \mathcal{U}_{\lambda_1}$ contains the space $\mathcal{U}_{\lambda_3^*}$ as an irreducible component (with multiplicity 1).*

Proof Lemma 2.11 yields **(1)⇔(2)**, **(3)⇔(2)**, and **(4)⇔(2)**. ∎

Thus, when $\mathcal{U}_{\lambda_1} \otimes \mathcal{U}_{\lambda_3}$ contains the space $\mathcal{U}_{\lambda_2}$ as an irreducible component (with multiplicity 1), $\mathcal{U}_{\lambda_1^*} \otimes \mathcal{U}_{\lambda_2}$ and $\mathcal{U}_{\lambda_1} \otimes \mathcal{U}_{\lambda_2^*}$ contain the spaces $\mathcal{U}_{\lambda_3}$ and $\mathcal{U}_{\lambda_3^*}$, respectively. The following lemma discusses how these subspaces are contained.

Lemma 2.13 *Assume that $\mathcal{U}_{\lambda_1} \otimes \mathcal{U}_{\lambda_3}$ contains the space $\mathcal{U}_{\lambda_2}$ as an irreducible component with multiplicity* 1. *We have the following items.*

(1) *The range of the following map from $\mathcal{U}_{\lambda_3}$ to $\mathcal{U}_{\lambda_1^*} \otimes \mathcal{U}_{\lambda_2}$ is the subspace $\mathcal{U}_{\lambda_3}$ of $\mathcal{U}_{\lambda_1^*} \otimes \mathcal{U}_{\lambda_2}$, and the map gives an isomorophism for representation of G.*

$$|u\rangle \mapsto (I_{\lambda_1^*} \otimes Q_{\lambda_2})|u\rangle \otimes |I_{\lambda_1}\rangle\rangle, \quad \forall |u\rangle \in \mathcal{U}_{\lambda_3}, \tag{2.50}$$

where Q_{λ_2} is the projection to $\mathcal{U}_{\lambda_2}$.

(2) *The range of the map $|u\rangle (\in \mathcal{U}_{\lambda_3}) \mapsto \langle u|I_{\lambda_1}\rangle\rangle$ from $\mathcal{U}_{\lambda_3}$ to $\mathcal{U}_{\lambda_1} \otimes \mathcal{U}_{\lambda_2^*}$ is the subspace $\mathcal{U}_{\lambda_3^*}$ of $\mathcal{U}_{\lambda_1} \otimes \mathcal{U}_{\lambda_2^*}$, and the map gives an isomorophism for representation of G.*

Proof **(1)**: When $|u\rangle \neq 0$, we have $\mathrm{Tr}_{\mathcal{U}_{\lambda_1^*}} |u\rangle\langle u| \otimes |I_{\lambda_1}\rangle\rangle\langle\langle I_{\lambda_1}| = |u\rangle\langle u| \otimes I_{\lambda_1}$, which implies that $(I_{\lambda_1^*} \otimes Q_{\lambda_2})|u\rangle \otimes |I_{\lambda_1}\rangle\rangle \neq 0$. Hence, the linear map given in (2.50) is a homomorphism for representation of G. The irreducibility guarantees that it is an isomorophism. So, we obtain Item **(1)**.

Item **(2)** can be shown by the same way. ∎

Here, we should be careful when f_λ and $\mathsf{f}_{\lambda'}$ are projective (unitary) representations. When the factor systems of f_λ and $\mathsf{f}_{\lambda'}$ are $\mathcal{E}$ and $\mathcal{E}' \in H(G : U(1))$, respectively, the factor system of the projective (unitary) representation $\mathsf{f}_\lambda \otimes \mathsf{f}_{\lambda'}$ is $\mathcal{E} \cdot \mathcal{E}' \in H(G : U(1))$.

2.8 Fourier Transform for Finite Group

2.8.1 Discrete Fourier Transform

Firstly, we focus on the group $\mathbb{Z}_d$ as a typical example of commutative group. Let $\mathcal{H}$ be the d-dimensional Hilbert space spanned by the standard basis $\{|0\rangle, \ldots, |d-1\rangle\}$. The unitary matrix $\mathsf{DFT} := \frac{1}{\sqrt{q}} \sum_{l,j \in \mathbb{F}_q} e^{i2\pi lj/d} |l\rangle\langle j|$ on $\mathcal{H}$ is called the **discrete Fourier transform** and plays an important role in quantum information processing.

Regarding the basis $|j\rangle$ as elements of cyclic group $\mathbb{Z}_d$, we define the representation f of the cyclic group $\mathbb{Z}_d$ by $\mathsf{f}(k)|j\rangle := |j+k\rangle$, $(\forall j, k \in \mathbb{Z}_d)$. Then, we have the irreducible decomposition of $\mathcal{H}$ as $\mathcal{H} = \mathcal{H}_0 \oplus \cdots \oplus \mathcal{H}_{d-1}$, where $\mathcal{H}_l$ is the one-dimensional subspace generated by the dual base

$$|\hat{e}_l\rangle := \frac{1}{\sqrt{d}} \sum_{k \in \mathbb{Z}_d} e^{-2\pi i lk/d} |k\rangle = \mathsf{DFT}^\dagger |l\rangle, \tag{2.51}$$

and the action of f_l on $\mathcal{H}_l$ is given by $\mathsf{f}_l(k)|\hat{e}_l\rangle := e^{2\pi i lk/d}|e_l\rangle$. Since $\mathbb{Z}_d$ is a commutative group, its irreducible representation is limited a one-dimensional representation, which is one of f_0, ..., and f_{d-1}. Hence, the set $\hat{\mathbb{Z}}_d$ of labels corresponding to the irreducible representations of $\mathbb{Z}_d$ is $\{0, \dots, d-1\}$. Since $\mathsf{DFT}|v\rangle = \sum_{l=0}^{d-1}\langle \hat{e}_l|v\rangle|l\rangle$, the discrete Fourier transform can be regarded as a conversion operation from a vector $|v\rangle := \sum_{j=0}^{d-1} v_j|j\rangle$ in $\mathcal{H}$ to a function $l \mapsto \langle \hat{e}_l|v\rangle$ on $\hat{U}_d$, where the vector $|v\rangle$ is regarded a function from $\mathbb{Z}_d$ to $\mathbb{C}$. In the following, we investigate how to extend this characterization to the case of non-commutative groups. In fact, a finite group plays an important role for the description of the asymmetry in a quantum system as well as a continuous group (Lie group), which is discussed in the next chapter. This section addresses a generalization of discrete Fourier transform for finite groups from a viewpoint of representation theory.

2.8.2 Character and Orthogonality

In this subsection, we discuss a finite group G and its representation, which has several good properties. For any representation f of G on $\mathcal{H}$, there exists an inner product of $\mathcal{H}$ such that $\mathsf{f}(g)$ is a unitary matrix [98, Remark in Sect. 1.3]. Such an inner product is given by

$$\sum_{g\in G}\frac{1}{|G|}\langle \mathsf{f}(g)v|\mathsf{f}(g)v'\rangle \tag{2.52}$$

because the relation $\sum_{g\in G}\frac{1}{|G|}\langle \mathsf{f}(g)v|\mathsf{f}(g)v'\rangle = \sum_{g\in G}\frac{1}{|G|}\langle \mathsf{f}(g)\mathsf{f}(g')v|\mathsf{f}(g)\mathsf{f}(g')v'\rangle$ holds for any element $g' \in G$ and two vectors v, v'. Hence, using this inner product, without loss of generality, we can assume that a representation f of a finite group G is unitary. This fact holds even for a projective representation because the relation $\langle \mathsf{f}(g)\mathsf{f}(g')v|\mathsf{f}(g)\mathsf{f}(g')v'\rangle = \langle e^{i\theta(g,g')}\mathsf{f}(gg')v|e^{i\theta(g,g')}\mathsf{f}(gg')v'\rangle = \langle \mathsf{f}(gg')v|\mathsf{f}(gg')v'\rangle$ holds for $g, g' \in G$ and two vectors v, v'. Hence, Lemma 2.3 guarantees that any representation f of a finite group G on the Hilbert space $\mathcal{H}$ is completely reducible. Therefore, due to the discussions in Sect. 2.7, it is enough to consider the case when the representation space is the direct sum product of $\mathcal{U}_\lambda \otimes \mathbb{C}^{d_\lambda}$. This subsection and the next subsection address various methods to describe the representation space $\mathcal{U}_\lambda \otimes \mathbb{C}^{d_\lambda}$, which bring mathematically easier analyses. The character of the representation on a representation space $\oplus_\lambda \mathcal{U}_\lambda \otimes \mathbb{C}^{m_\lambda}$ is $\sum_\lambda m_\lambda \chi_\lambda$, where $m_\lambda = 0$ means that $\mathcal{H}$ contains no irreducible representation isomorphic to f_λ. We employ this relation in the following discussion.

To study the structure of representations of a finite group G, we consider the complex linear space $\mathbb{C}[G]$ whose CONS is composed of elements g of the finite group G. When we define the following product in the complex linear space $\mathbb{C}[G]$, it is called the **group algebra** of G.

$$\left(\sum_g a_g g, \sum_g b_g g\right) \in \mathbb{C}[G] \times \mathbb{C}[G] \mapsto \sum_g \sum_{g'} a_g b_{g'} g g' \in \mathbb{C}[G]. \tag{2.53}$$

This product satisfies the associative law and the **bilinearity**, which means the linearity for both inputs. The identity element e of G works the identify element with respect to the product of the group algebra $\mathbb{C}[G]$. Such a linear space with an associative product and the bilinearity is called a **linear algebra**. For example, the set $\mathfrak{gl}(\mathcal{U}_\lambda)$ of linear maps on $\mathcal{U}_\lambda$ forms a linear algebra when the product is defined by the composition of maps. As defined in (2.41), the space $\mathfrak{gl}(\mathcal{U}_\lambda)$ also has the representation $\mathsf{D}(\mathsf{f}_\lambda)$ of $G \times G$. Hence, their direct sum $L^2(\hat{G}) := \bigoplus_{\lambda \in \hat{G}} \mathfrak{gl}(\mathcal{U}_\lambda) = \bigoplus_{\lambda \in \hat{G}} \mathcal{U}_\lambda \otimes \mathcal{U}_{\lambda^*}$ also has the representation of $G \times G$ as well as that of G. Also, the space $L^2(\hat{G})$ forms a linear algebra. Here, a linear map from a linear algebra V to another linear algebra W is called a **homomorphism** for linear algebra when it preserves the structure of the product. The space $L^2(\hat{G})$ is isomorphic to $\oplus_\lambda \mathcal{U}_\lambda \otimes \mathbb{C}^{d_\lambda}$ with respect to the application of G from the left hand side. That is, the space $L^2(\hat{G})$ is isomorphic to representation space $\oplus_\lambda \mathcal{U}_\lambda \otimes \mathbb{C}^{d_\lambda}$.

Now, we define the homomorphism f from $\mathbb{C}[G]$ to $L^2(\hat{G})$ as follows.

$$\mathsf{f}\left(\sum_g a_g g\right) = \bigoplus_\lambda \sum_g a_g \mathsf{f}_\lambda(g). \tag{2.54}$$

Since the homomorphism is injective as shown later, the dimensions of $\mathbb{C}[G]$ is less than that of $L^2(\hat{G})$, which yields

$$|G| \le \sum_{\lambda=1}^{k} d_\lambda^2. \tag{2.55}$$

Because the opposite inequality can be shown from Lemma 2.14, we obtain

$$|G| = \sum_{\lambda=1}^{k} d_\lambda^2, \tag{2.56}$$

which implies that $\mathbb{C}[G]$ is isomorphic to $L^2(\hat{G})$. Hence, the character coincides with $\sum_\lambda d_\lambda \chi_\lambda$. When we focus on the representation corresponding to the action of G from the left hand side on $\mathbb{C}[G]$, the character of the representation takes the value 0 for a non-identity element g because $\{g'\}_{g' \in G}$ forms a basis of $\mathbb{C}[G]$ and the relation $gg' \neq g'$ holds for non-identity element g. Hence, we obtain

$$\sum_\lambda d_\lambda \chi_\lambda(g) = 0. \tag{2.57}$$

Proof of the injectivity of f. Assume that $\mathsf{f}(\sum_g a_g g) = 0$. The left-hand-side application of any element of G to $\sum_g a_g g$ is the application of the zero matrix. Hence, the relation $\sum_g a_g \mathsf{f}(g) = 0$ holds for any unitary representation f. Since $\mathbb{C}[G]$ can be regarded as a representation space with respect to the left-hand-side application, the application of $\sum_g a_g g$ to $\mathbb{C}[G]$ is written by the zero matrix. That is, the relation $\sum_g a_g g = (\sum_g a_g g)e = 0$ holds for the identity element e of $\mathbb{C}[G]$. Hence, we obtain the injectivity of f. ∎

Considering the correspondence between an element $\sum_g a_g g$ of $\mathbb{C}[G]$ and the function $g \mapsto a_g$ from G to $\mathbb{C}$, we find that $\mathbb{C}[G]$ is isomorphic to the set $L^2(G)$ of functions from G to $\mathbb{C}$.

We define the inner product in $L^2(G)$ as $\langle h_1, h_2 \rangle := \sum_{g \in G} \frac{1}{|G|} \overline{h_1(g)} h_2(g)$, $(\forall h_1, h_2 \in L^2(G))$. Choosing a CONS $|\lambda, j\rangle$ of $\mathcal{U}_\lambda$, we define $R_{\lambda, j', j} \in L^2(G)$ as $R_{\lambda, j', j}(g) := \langle \lambda, j' | \mathsf{f}(g) | \lambda, j \rangle$. Given a unitary representation f of G and a matrix A on its representation space, the matrix

$$\sum_{g \in G} \mathsf{f}(g)^{-1} A \mathsf{f}(g) \tag{2.58}$$

is commutative with $\mathsf{f}(g')$ for any $g' \in G$. Using this fact, we can show the **Plancherel formula** as follows.

Lemma 2.14 *The following orthogonality holds.*

$$\sum_{g \in G} \frac{1}{|G|} \overline{R_{\lambda, j, j'}(g)} R_{\tilde{\lambda}, \tilde{j}, \tilde{j}'}(g) = \frac{1}{d_\lambda} \delta_{\lambda, \tilde{\lambda}} \delta_{j, \tilde{j}} \delta_{j', \tilde{j}'}. \tag{2.59}$$

Proof We choose the inner product of the space $\oplus_\lambda \mathcal{U}_\lambda$ such that the space $\mathcal{U}_\lambda$ is orthogonal to the space $\mathcal{U}_{\tilde{\lambda}}$ for different labels λ and $\tilde{\lambda}$. Then, we denote the unitary representation of G on the space $\oplus_\lambda \mathcal{U}_\lambda$ by $\tilde{\mathsf{f}}$. Hence, we have

$$\begin{aligned}
\sum_{g \in G} \frac{1}{|G|} \overline{R_{\lambda, j, j'}(g)} R_{\tilde{\lambda}, \tilde{j}, \tilde{j}'}(g) &= \sum_{g \in G} \frac{1}{|G|} \langle \lambda; j' | \mathsf{f}_\lambda(g)^{-1} | \lambda, j \rangle \langle \tilde{\lambda}; \tilde{j} | \mathsf{f}_{\tilde{\lambda}}(g) | \tilde{\lambda}; \tilde{j}' \rangle \\
&= \langle \lambda; j' | \left(\sum_{g \in G} \frac{1}{|G|} \tilde{\mathsf{f}}(g)^{-1} | \lambda; j \rangle \langle \tilde{\lambda}; \tilde{j} | \tilde{\mathsf{f}}(g) \right) | \tilde{\lambda}; \tilde{j}' \rangle.
\end{aligned}$$

Since the matrix $\left(\sum_{g \in G} \frac{1}{|G|} \tilde{\mathsf{f}}(g)^{-1} | \lambda, j \rangle \langle \tilde{\lambda}; \tilde{j} | \tilde{\mathsf{f}}(g) \right)$ is commutative with $\tilde{\mathsf{f}}(g)$, Schur's lemma guarantees that the matrix is written as $\oplus_{\lambda'} c_{\lambda'} I_{\lambda'}$, where $I_{\lambda'}$ is the identity on the space $\mathcal{U}_{\lambda'}$. When $\lambda \neq \tilde{\lambda}$, the LHS of (2.59) is zero because the vector $|\lambda; j\rangle$ is orthogonal to the vector $|\tilde{\lambda}; \tilde{j}\rangle$.

When $\lambda = \tilde{\lambda}$, we have

$$\left(\sum_{g \in G} \frac{1}{|G|} \tilde{\mathsf{f}}(g)^{-1} |\lambda; j\rangle\langle\tilde{\lambda}; \tilde{j}| \tilde{\mathsf{f}}(g) \right) = \frac{\langle\tilde{\lambda}; \tilde{j}|\lambda; j\rangle}{d_\lambda} I_\lambda,$$

because the left hand side is a constant times of the identity matrix and the above observation yields that the traces of both sides of this equation are the same. Hence, we obtain (2.59). ■

Since $R_{\lambda,j,j'} \in L^2(G)$, this lemma shows that $\dim L^2(G) \ge \sum_\lambda d_\lambda^2$. Because $L^2(G)$ is isomorphic to $\mathbb{C}[G]$, we obtain the inequality opposite to (2.55). Hence, $R_{\lambda,j,j'}$ forms a CONS of $L^2(G)$.

By considering the trace of the matrix $(R_{\lambda,j,j'})_{j,j'}$, the orthogonality discussed in Lemma 2.14 yields the following orthogonality for characters.

$$\begin{aligned} &\sum_{g \in G} \frac{1}{|G|} \chi_{\lambda'}(g) \chi_\lambda(g^{-1}) = \sum_{g \in G} \frac{1}{|G|} \chi_{\lambda'}(g) \overline{\chi_\lambda(g)} \\ =&\sum_{g \in G} \frac{1}{|G|} \chi_{\lambda'}(g) \chi_{\lambda^*}(g) = \begin{cases} 1 \text{ if } \lambda = \lambda' \\ 0 \text{ if } \lambda \neq \lambda'. \end{cases} \end{aligned} \tag{2.60}$$

Similarly, we can characterize the projection to the irreducible subspace $\mathcal{U}_\lambda$ as follows.

$$\begin{aligned} &\sum_{g \in G} \frac{d_\lambda}{|G|} \mathsf{f}_{\lambda'}(g) \chi_\lambda(g^{-1}) = \sum_{g \in G} \frac{d_\lambda}{|G|} \mathsf{f}_{\lambda'}(g) \overline{\chi_\lambda(g)} \\ =&\sum_{g \in G} \frac{d_\lambda}{|G|} \mathsf{f}_{\lambda'}(g) \chi_{\lambda^*}(g) = \begin{cases} I_\lambda \text{ if } \lambda = \lambda' \\ 0 \ \ \text{ if } \lambda \neq \lambda'. \end{cases} \end{aligned} \tag{2.61}$$

When the representation space $\mathcal{H}$ of the unitary representation f is given by $\oplus_{\lambda \in \hat{G}} \mathcal{U}_\lambda \otimes \mathbb{C}^{m_\lambda}$, the orthogonality (2.60) yields

$$\sum_{g \in G} \frac{|\mathsf{f}(g)|^2}{|G|} = \sum_{\lambda \in \hat{G}} m_\lambda. \tag{2.62}$$

Hence, we obtain the following lemma.

Lemma 2.15 *The following conditions are equivalent to each other for a representation* f

(1) f *is an irreducible representation.*
(2) $\sum_{g \in G} \frac{|\mathsf{f}(g)|^2}{|G|} = 1$.

The orthogonality (2.60) derives the coefficient $C^{\lambda''}_{\lambda,\lambda'}$ given in (2.45) as follows.

$$C^{\lambda''}_{\lambda,\lambda'} = \sum_{g\in G} \frac{1}{|G|}\chi_\lambda(g)\chi_{\lambda'}(g)\overline{\chi_{\lambda''}(g)} = \sum_{g\in G} \frac{1}{|G|}\chi_\lambda(g)\chi_{\lambda'}(g)\chi_{\lambda''*}(g). \tag{2.63}$$

Since (2.63) implies that $C^{\lambda'*}_{\lambda,\lambda''*}$ and $C^{\lambda*}_{\lambda',\lambda''*}$ equal the RHS of (2.63), we have

$$C^{\lambda''}_{\lambda,\lambda'} = C^{\lambda'*}_{\lambda,\lambda''*} = C^{\lambda*}_{\lambda',\lambda''*}. \tag{2.64}$$

The relations (2.56) and (2.57) yield another orthogonality:

$$\begin{aligned}&\sum_\lambda \frac{d_\lambda}{|G|}\operatorname{Tr}(\mathfrak{f}_\lambda(g))^\dagger \mathfrak{f}_\lambda(g') = \sum_\lambda \frac{d_\lambda}{|G|}\operatorname{Tr}\mathfrak{f}_\lambda(g^{-1}g')\\ =&\sum_\lambda \frac{d_\lambda}{|G|}\chi_\lambda(g^{-1}g') = \begin{cases} 1 \text{ if } g = g' \\ 0 \text{ if } g \neq g'. \end{cases}\end{aligned} \tag{2.65}$$

Let $L^2(G, inv)$ be the set of functions f in $L^2(G)$ whose value $f(g)$ depends only on the conjugate class of g. That is, $L^2(G, inv)$ is the set of functions in $L^2(G)$ satisfying (2.27). Especially, let $L^2(G, inv, \mathbb{R})$ be the set of functions satisfying (2.27). Then, we have the following lemma.

Lemma 2.16 *The set $\{\chi_\lambda\}_\lambda$ of irreducible characters forms a CONS of $L^2(G, inv)$. Especially, the set $\{\chi_\lambda + \chi_{\lambda^*}\}_\lambda$ forms a CONS of $L^2(G, inv, \mathbb{R})$.*

2.8.3 Fourier Transform

Next, we define the inner product $\langle A, B\rangle_{L^2(\hat{G})} := \sum_{\lambda=1}^k \operatorname{Tr} A_\lambda^\dagger B_\lambda$ for any two elements $A = (A_1, \ldots, A_k)$ and $B = (B_1, \ldots, B_k)$ of $L^2(\hat{G}) = \bigoplus_{\lambda\in\hat{G}} \mathfrak{gl}(\mathcal{U}_\lambda)$. Then, we define the **Fourier transform** $\mathcal{F}[\phi] = (\hat{\phi}_1, \ldots, \hat{\phi}_k) \in L^2(\hat{G})$ for a function $\phi \in L^2(G)$ by $(\mathcal{F}[\phi])_\lambda := \sqrt{d_\lambda}\sum_{g\in G}\frac{1}{|G|}\mathfrak{f}_\lambda(g)^\dagger\phi(g)$. The relation (2.65) yields the relation $\|\phi\|_{L^2(G)} = \|\mathcal{F}\phi\|_{L^2(\hat{G})}$ for $\phi \in L^2(G)$. By defining the **Inverse Fourier transform** $\mathcal{F}^{-1}[A]$ for $A \in L^2(\hat{G})$ as $\mathcal{F}^{-1}[A](g) := \sum_\lambda \sqrt{d_\lambda}\operatorname{Tr}\mathfrak{f}_\lambda(g)A_\lambda$, the relation (2.65) implies that

$$\mathcal{F}^{-1}\circ\mathcal{F}[\phi] = \phi. \tag{2.66}$$

That is, the relation between $\mathcal{F}$ and $\mathcal{F}^{-1}$ is summarized as Fig. 2.17.

Hence, the behavior of an vector of any representation space of G can be characterized as an element of $L^2(G)$ via the Fourier transform. This property plays an important role in the estimation of the unknown group action.

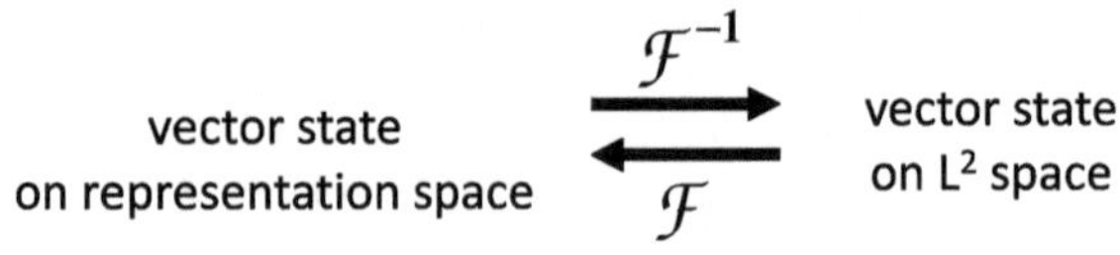

Fig. 2.17 Relation between $\mathcal{F}$ and $\mathcal{F}^{-1}$

The above discussion can be extended to the case of projective representations. Using the factor system $\mathcal{E} := [\{e^{i\theta(g,g')}\}_{g,g'}] \in H(G : U(1))$, we modify the product of the group algebra defined in (2.53) as follows.

$$\left(\sum_g a_g g, \sum_g b_g g\right) \in \mathbb{C}[G] \times \mathbb{C}[G] \mapsto \sum_g \sum_{g'} a_g b_{g'} e^{i\theta(g,g')} g g' \in \mathbb{C}[G]. \quad (2.67)$$

Using the set $\hat{G}[\mathcal{E}]$ of labels of irreducible projective representation of G, we define the linear space $L^2(\hat{G}[\mathcal{E}]) := \bigoplus_{\lambda \in \hat{G}[\mathcal{E}]} \mathfrak{gl}(\mathcal{U}_\lambda)$. Similar to (2.54), we define the homomorphism f from $\mathbb{C}[G]$ to $L^2(\hat{G}[\mathcal{E}])$ and can show the inequality (2.55). Since we can show the commutativity between $\mathsf{f}(g)$ and the matrix defined in (2.58), we can show Plancherel formula, i.e., Lemma 2.14. Hence, (2.56) holds. Similarly, Plancherel formulas yields the relations (2.60)–(2.62), (2.65), and Lemma 2.15. Therefore, we can define the Fourier transform and the Inverse Fourier transform. Since the relation (2.65) holds, we have (2.66). That is, the contents in this section can be extended to the case of projective unitary representations with the above modification.

2.9 Representation of Permutation Group and Young Diagram

2.9.1 *Young Diagram and Young Tableau*

In this section, we focus on the set of permutation among the integers $1, \ldots, n$. This set forms a group when the product is defined by the composition of two permutations. So, the set is called the **permutation group** of degree n and is written as S_n. The order of S_n is $n!$. Since the permutation group describes operations for n particles with the same kind, it is a very natural object in quantum theory.

To discuss representations of the permutation group S_n of degree n, we introduce Young diagram and Young tableau as follows. The **Young diagram** with depth r is defined to be a collection of r rows like 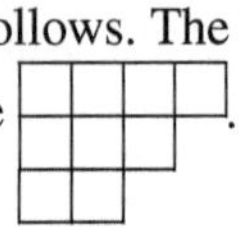. (This picture describes the case with $r = 3$.) The i-th row has n_i boxes. A Young diagram is described by a vector $\boldsymbol{n} = (n_1, \ldots, n_r)$ that consists of a monotone decreasing sequence of r non-negative integers. The integer $|\boldsymbol{n}| := \sum_{j=1}^{r} n_j$ is called its **size**. We let $\mathcal{Y}^r$ be the set of Young diagram whose depth is not greater than r and $\mathcal{Y}_n$ be the set of Young diagram whose size is not greater than n. Then, we define the set $\mathcal{Y}_n^r := \mathcal{Y}^r \cap \mathcal{Y}_n$. When $|\boldsymbol{n}|$ boxes of a Young diagram $\boldsymbol{n}$ have $|\boldsymbol{n}|$ different integers from 1 to $|\boldsymbol{n}|$ like $\begin{array}{|c|c|c|c|}\hline 2&3&6&4\\\hline 7&1&8\\\cline{1-3} 5&9\\\cline{1-2}\end{array}$, it is called a **Young tableau** and is written as T. In particular, it is called a **Standard Young tableau** when the integer is monotone decreasing from left to right and from up to down like $\begin{array}{|c|c|c|c|}\hline 9&8&6&1\\\hline 7&4&3\\\cline{1-3} 5&2\\\cline{1-2}\end{array}$. The set of Standard Young tableau corresponding to $\boldsymbol{n}$ is denoted by $Y_S(\boldsymbol{n})$.

On the other hand, when each box of a Young diagram $\boldsymbol{n}$ has an integer such that the number is monotone decreasing from left to right and it is strictly monotone decreasing from up to down like $\begin{array}{|c|c|c|c|}\hline 3&3&3&3\\\hline 2&2&2\\\cline{1-3} 1&1\\\cline{1-2}\end{array}$ and $\begin{array}{|c|c|c|c|}\hline 3&3&2&2\\\hline 2&2&1\\\cline{1-3} 1&1\\\cline{1-2}\end{array}$, it is called a **Semistandard Young tableau**. Here, the same number can be used in these boxes. Usually, a Semistandard Young tableau is not a Young tableau. Let $Y_{U,r}(\boldsymbol{n})$ be the set of Semistandard Young tableaus given by a Young diagram $\boldsymbol{n}$ whose numbers are restricted to the integers $1, \ldots, r$.

It is relatively easy to evaluate the cardinality $|\mathcal{Y}_n^r|$ of Young diagrams with depth r and size n. When we remove the condition that n_i is monotone decreasing, the number of sequences $(n_1, \ldots, n_r)$ of non-negative integers is $\binom{n+r-1}{r-1}$, which is not greater than $(n+1)^{r-1}$. Since we impose the condition that n_i is monotone decreasing, the relation

$$|\mathcal{Y}_n^r| \le \binom{n+r-1}{r-1} \le (n+1)^{r-1} \tag{2.68}$$

holds. Now, we consider a correspondence between a sequence $(n_1, \ldots, n_r)$ of non-negative integers and a sequence $(n_1, \ldots, n_r)$ of non-negative integers with the monotonically decreasing condition. Applying a suitable permutation, the former can be converted to the latter. One of the latter sequence corresponds to $r!$ former sequences at most. Hence, we have

$$|\mathcal{Y}_n^r| \ge \binom{n+r-1}{r-1} / r!. \tag{2.69}$$

Let us consider the cardinality $|\mathcal{Y}_n|$ of Young diagrams with size n by removing the condition for the depth. The cardinality $|\mathcal{Y}_n|$ is called the **Partition function**, and has been studied for a long time. When the size n is large, we have the following asymptotic formula (Hardy–Ramanujan–Rademacher formula) [43, 86], [19, p. 169], [85, p. 925].

$$|\mathcal{Y}_n| \cong \frac{1}{4n\sqrt{3}} \exp\left(\pi\sqrt{\frac{2n}{3}}\right). \tag{2.70}$$

2.9.2 *Permutation Group and Young Diagram*

A permutation $\sigma \in S_n$ is called a **horizontal permutation** of a Young tableau T when it permutates the numbers only in the horizontal direction in the given Young tableau T. Similarly, a permutation $\sigma \in S_n$ is called a **vertical permutation** of the Young tableau T when it permutates the numbers only in the vertical direction in the given Young tableau T. The set of horizontal permutations and the set of vertical permutations are subgroups of S_n, which are called the **horizontal permutation group** and the vertical permutation group and are written as $H(T)$ and $V(T)$, respectively. Given a Young tableau T with size n, we define the **Young horizontal symmetrizer** $c_T^H := \sum_{\sigma \in H(T)} \sigma$ and the **Young vertical symmetrizer** $c_T^V := \sum_{\sigma \in V(T)} \operatorname{sgn}(\sigma)\sigma$ as an element of $\mathbb{C}[S_n]$, where $\operatorname{sgn}(\sigma)$ is defined as follows. Also, define the **Young symmetrizer** $c_T := c_T^H c_T^V$. A permutation is called a **transposition** when it permuates only two integers. A transposition exchanging j and j' is written as (j, j'). Any permutation $\sigma \in S_n$ can be written as products of transpositions. We set $\operatorname{sgn}(\sigma) = 1$ when the number of the above transpositions is an even number, and set $\operatorname{sgn}(\sigma) = -1$ when the number is an odd number. We can show that this definition does not depend on the choice of the product form of transpositions. It is also known that the subspace $\mathbb{C}[S_n]c_T := \{xc_T | x \in \mathbb{C}[S_n]\}$ is an irreducible representation space for the action of the group S_n from the left hand side. The representation of S_n on the space depends only on the Young diagram $\boldsymbol{n} = (n_1, \ldots, n_r)$ and does not depend on the choice of the Young tableau T. Hence, we let $\mathcal{U}_{\boldsymbol{n}}(S_n)$ be the representation space that is isometric to this representation.

Now, we define a vector $\boldsymbol{\delta} := (r-1, r-2, \cdots, 0)$. We also define the action of the element $\sigma \in S_r$ (not S_n) for an r-dimensional vector $\boldsymbol{v} := (v_1, \cdots, v_r)$ as $\sigma(\boldsymbol{v}) := (v_{\sigma(1)}, \cdots, v_{\sigma(r)})$. Additionally, we define $\boldsymbol{n}! := n_1! \cdots n_r!$ for an r-dimensional vector $\boldsymbol{n}$ that consists of non-negative integers. The dimension of $\mathcal{U}_{\boldsymbol{n}}(S_n)$ equals the number of Standard Young tableaus corresponding to the Young diagram $\boldsymbol{n}$ and is known to be given as [37, Corollary 9.1.5]

$$d_{S_n,\boldsymbol{n}} := \dim \mathcal{U}_{\boldsymbol{n}}(S_n) = \sum_{\sigma \in S_r} \operatorname{sgn}(\sigma) \frac{n!}{[\boldsymbol{n} + \boldsymbol{\delta} - \sigma(\boldsymbol{\delta})]!} \tag{2.71}$$

$$= \frac{n!}{(\boldsymbol{n} + \boldsymbol{\delta})!} \prod_{i,j \in \{1,\ldots,r\}: i<j} (n_i - n_j + j - i), \tag{2.72}$$

where the sum in (2.71) means the sum of $\sigma \in S_r$ satisfying that $\boldsymbol{n} + \boldsymbol{\delta} - \sigma(\boldsymbol{\delta}) \geq 0$. Let $\boldsymbol{n}^T$ be the Young diagram that is the transpose of the Young diagram $\boldsymbol{n}$ by exchanging the horizontal and vertical directions. Then, when the Standard Young tableau corresponds to $\boldsymbol{n}$, the transposed Standard Young tableau does to $\boldsymbol{n}^T$. So, we have

$$\dim \mathcal{U}_{\boldsymbol{n}}(S_n) = \dim \mathcal{U}_{\boldsymbol{n}^T}(S_n). \tag{2.73}$$

Exercise 2.17 Assume that $r = 2$ and that k is an integer between n and $n/2$ so that $n - k$ is an even number. Show the following relation using (2.72);

$$d_{S_n,(\frac{n+k}{2},\frac{n-k}{2})} = \frac{k+1}{n+1}\binom{n+1}{\frac{n+k}{2}+1}. \tag{2.74}$$

2.9.3 Plancherel Measure

Now, using the formula (2.56), we define the **Plancherel measure** $\mu_n(r) := \sum_{\boldsymbol{n}\in\mathcal{Y}_n:n_1=r} \frac{d^2_{S_n,\boldsymbol{n}}}{n!}$, where we used the fact that the order of the permutation group S_n is $n!$. The relation (2.73) yields that $\mu_n(r) = \sum_{\boldsymbol{n}\in\mathcal{Y}_n:n_1=r} \frac{d^2_{S_n,\boldsymbol{n}^T}}{n!} = \sum_{\boldsymbol{n}\in\mathcal{Y}_n^r} \frac{d^2_{S_n,\boldsymbol{n}}}{n!}$. That is, $\mu_n(r)$ expresses the probability that the depth of the Young diagram is not greater than r when the the Young diagram $\boldsymbol{n}$ is generated with the probability $\frac{d^2_{S_n,\boldsymbol{n}}}{n!}$.

Let R be the random variable subject to the Plancherel measure μ_n. Then, it is known that the random variable $\frac{R}{\sqrt{n}}$ converges to 2 in probability under the limit $n \to \infty$. More precisely, the random variable $R - 2\sqrt{n}$ behaves with the order $n^{1/6}$ as mentioned below. Letting u be the solution of the Painlevé II equation: $\frac{d^2u}{dx^2}(x) = 2u(x)^3 + xu(x)$, we define the Tracy–Widom distribution function [102]: $F_{TW}(t) := \exp(-\int_t^\infty (x-t)u(x)^2 dx)$. Then, the asymptotic behavior of Plancherel measure μ_n can be characterized as [12]

$$\lim_{n\to\infty} \mu_n \left\{ r \left| \frac{r - 2\sqrt{n}}{n^{1/6}} \le t \right. \right\} = F_{TW}(t). \tag{2.75}$$

Chapter 3
Foundation of Representation Theory of Lie Group and Lie Algebra

Abstract This chapter deals with general theories that do not depend on the types of Lie groups and Lie algebras. As generalizations, it addresses projective representations of Lie groups and Lie algebras by combining the contents of Chap. 2. Then, it introduces the Fourier transform for Lie groups including the case of projective representations. It also prepares several concepts for Chap. 6. Also, this chapter introduces complex Lie groups and complex Lie algebras, which are helpful for real Lie groups and real Lie algebras.

3.1 Lie Group

3.1.1 Basic Examples

As an example of a parametrized group, we consider the set of rotations in the 2-dimensional space:

$$\mathrm{SO}(2,\mathbb{R}) := \left\{ \begin{pmatrix} \cos\theta & \sin\theta \\ -\sin\theta & \cos\theta \end{pmatrix} \middle| 0 \le \theta < 2\pi \right\}. \tag{3.1}$$

The group can be written as the set of the 2-dimensional orthogonal matrices whose determinant is 1. An element of the group can be characterized by the continuous-valued parameter θ. A group is called a **Lie group** when an element of the group is characterized by continuous-valued parameters $\theta = (\theta^1, \ldots, \theta^l)$. The group of orthogonal matrices on d-dimensional real linear space is called the orthogonal group and is written as $\mathrm{O}(d,\mathbb{R})$. Any matrix g in the group $\mathrm{O}(d,\mathbb{R})$ has determinant 1 or -1. Especially, the set of elements in $\mathrm{O}(d,\mathbb{R})$ with determinant 1 is a subgroup. It is written as $\mathrm{SO}(d,\mathbb{R})$ and is called the special linear group. As an example of an important transformation in quantum system, we can list the set of rotations on the 3-dimensional space, which is written as $\mathrm{SO}(3,\mathbb{R})$.

Since the determinant is a continuous function of a matrix, the subset of matrices with determinant 1 is completely separated from the subset of matrices with determinant -1 in $\mathrm{O}(d,\mathbb{R})$. Conversely, an element of $\mathrm{SO}(d,\mathbb{R})$ has determinant 1, and

M. Hayashi, *Group Representation for Quantum Theory*,
DOI 10.1007/978-3-319-44906-7_3

all of its elements are continuously connected to each other. Such a Lie group is called a connected Lie group. In general, a Lie group is not necessarily connected and consists of several connected components. For example, the group $\mathrm{O}(d, \mathbb{R})$ has the connected component of matrices with determinant 1 and the connected component of matrices with determinant -1. Especially, for a Lie group G, we denote the connected component containing the identity element by G_0. For example, we have $\mathrm{O}(d, \mathbb{R})_0 = \mathrm{SO}(d, \mathbb{R})$ (see Exercises 3.3). Then, the following lemma is known.

Lemma 3.1 *The connected component G_0 of a Lie group G is a Lie group. Hence, we have $gg' \in G_0$ for any two elements $g, g' \in G_0$.*

A coordinate transformation in the d-dimensional real linear space is given as an invertible real square matrix of size d. The set of such matrices forms a group, so called the general linear group $\mathrm{GL}(d, \mathbb{R})$. The set of invertible linear maps on a complex Hilbert space $\mathcal{H}$ is written as $\mathrm{GL}(\mathcal{H})$. The set of elements of determinant 1 in $\mathrm{GL}(d, \mathbb{R})$ forms a group so called the special linear group $\mathrm{SL}(d, \mathbb{R})$. The groups $\mathrm{O}(d, \mathbb{R})$ and $\mathrm{SO}(d, \mathbb{R})$ are bounded closed sets, i.e., they are bounded closed subsets of set of all $d \times d$ matrices, which is a d^2-dimensional Euclidean space. Such a group is called a **compact Lie group**. Since $\mathrm{GL}(d, \mathbb{R})$ or $\mathrm{SL}(d, \mathbb{R})$ is not bounded, either is not compact. They are called non-compact groups. In fact, any compact Lie group is isomorphic to a subgroup of $\mathrm{GL}(\mathcal{H})$ [66, Corollary 4.22]. A larger part of properties of a Lie group is determined by the local structure around the identity element even though it is not compact. It is also known as Ado's Theorem that the local structure of any Lie group is isomorphic to a subgroup of $\mathrm{GL}(\mathcal{H})$ [104, Theorem 3.17.8]. This book discusses any Lie group as a subgroup of $\mathrm{GL}(\mathcal{H})$.

Physical transformation cannot be limited to linear transformations on the space–time, e.g., a rotation. For example, a displacement of the particle on the 3-dimensional space is a physical transformation. The group of these displacements is written as the 3-dimensional real linear space $\mathbb{R}^3$. A pair of a rotation in the 3-dimensional system and a displacement in the 3-dimensional system is also a transformation in the 3-dimensional space. More generally, we focus on a pair of an element of the d-dimensional special linear group $\mathrm{SO}(d, \mathbb{R})$ and an element of $\mathbb{R}^d$. In this combination, the transformations by $\mathrm{SO}(d, \mathbb{R})$ and $\mathbb{R}^d$ are not independent of each other, but effect each other. When a pair of a displacement $a' \in \mathbb{R}^d$ and a rotation $g' \in \mathrm{SO}(d, \mathbb{R})$ is applied to a particle with position $x \in \mathbb{R}^d$, the position is transformed as follows.

$$x \mapsto (g', a')x := g'(x) + a'. \tag{3.2}$$

When we apply another pair of another displacement $a \in \mathbb{R}^d$ and another rotation $g \in \mathrm{SO}(d, \mathbb{R})$ after the above transformation, the position is transformed as follows.

$$g'x + a' \mapsto (g, a)(g'x + a') = g(g'(x) + a') + a = gg'(x) + ga' + a. \tag{3.3}$$

When the product of two elements (g, a) and (g', a') as $(g, a)(g', a') := (gg', ga' + a)$, the direct product set $\mathbb{R}^d \cdot \mathrm{SO}(d, \mathbb{R})$ forms a semi direct product $\mathbb{R}^d \rtimes \mathrm{SO}(d, \mathbb{R})$ and is called the Euclidean group. The transformation (3.2) is a natural transformation by the semi direct product $\mathbb{R}^d \rtimes \mathrm{SO}(d, \mathbb{R})$. Similarly, the semi direct product $\mathbb{R}^d \rtimes \mathrm{GL}(d, \mathbb{R})$ is called the affine group and is written as $\mathrm{Aff}(n, \mathbb{R})$.

When our transformation rotates the time and the position in the 3-dimensional space, the transformation can be regarded as a rotation in the 4-dimensional space–time. The special relativity provides such a rotation as a Lorentz transformation as follows. The set of Lorentz transformations is given as the following Lie group and is called the Lorentz group.

$$\mathrm{O}(1,3) := \left\{ g \in \mathrm{GL}(4, \mathbb{R}) | g^T I_{1,3} g = I_{1,3} \right\}, \tag{3.4}$$

$$I_{1,3} := \begin{pmatrix} 1 & 0 & 0 & 0 \\ 0 & -1 & 0 & 0 \\ 0 & 0 & -1 & 0 \\ 0 & 0 & 0 & -1 \end{pmatrix}, \tag{3.5}$$

where $I_{p,q}$ is the diagonal matrix of size $p+q$ satisfying that the initial p diagonal entries are 1 and the remaining q diagonal entries are -1. The Lorentz group is divided into four connected components, and each component contains one of the following matrices.

$$\begin{pmatrix} 1 & 0 & 0 & 0 \\ 0 & 1 & 0 & 0 \\ 0 & 0 & 1 & 0 \\ 0 & 0 & 0 & 1 \end{pmatrix}, \begin{pmatrix} -1 & 0 & 0 & 0 \\ 0 & 1 & 0 & 0 \\ 0 & 0 & 1 & 0 \\ 0 & 0 & 0 & 1 \end{pmatrix}, \begin{pmatrix} 1 & 0 & 0 & 0 \\ 0 & -1 & 0 & 0 \\ 0 & 0 & 1 & 0 \\ 0 & 0 & 0 & 1 \end{pmatrix}, \begin{pmatrix} -1 & 0 & 0 & 0 \\ 0 & -1 & 0 & 0 \\ 0 & 0 & 1 & 0 \\ 0 & 0 & 0 & 1 \end{pmatrix}. \tag{3.6}$$

The most important connected component is the component containing the identity element, which is written as $\mathrm{O}(1,3)_0 = \mathrm{SO}(1,3)_0$. Since it is complicated to handle the rotation between the time and the position in the 3-dimensional space, we firstly discuss the space–time with the 2-dimensional space and the one-dimensional time. In this case, the Lorentz group is written as

$$\mathrm{O}(1,2) := \left\{ g \in \mathrm{GL}(3, \mathbb{R}) | g^T I_{1,2} g = I_{1,2} \right\}. \tag{3.7}$$

Similarly, the most important connected component is the component containing the identity element, i.e., $\mathrm{O}(1,2)_0 = \mathrm{SO}(1,2)_0$.

Although $\mathrm{SO}(3)$ and $\mathrm{SO}(1,2)_0$ are physically important, it is complicated to handle them because they are written as square matrices of size 3. If we treat the same transformation with square matrices of size 2, our calculation becomes simpler. For this purpose, we denote the set of invertible complex square matrices of size d by $\mathrm{GL}(d, \mathbb{C})$ and the subset consists of complex square matrices with determinant 1 by $\mathrm{SL}(d, \mathbb{C})$. We also denote the set of complex unitary matrices of size d by $\mathrm{U}(d)$ and define $\mathrm{SU}(d) := \mathrm{U}(d) \cap \mathrm{SL}(d, \mathbb{C})$. We also define $\mathrm{U}(p,q) := \{g \in \mathrm{GL}(p+q, \mathbb{C}) | g I_{p,q} g^\dagger = I_{p,q}\}$, and $\mathrm{SU}(p,q) := \mathrm{U}(p,q) \cap \mathrm{SL}(d, \mathbb{C})$. In fact, $\mathrm{SO}(3)$ and $\mathrm{SO}(1,2)_0$ are deeply related to $\mathrm{SU}(2)$ and $\mathrm{SU}(1,1)$, respectively, e.g., the isomorphic relations

$$\mathrm{SO}(3) \cong \mathrm{SU}(2)/\mathbb{Z}_2, \quad \mathrm{SO}(1,2)_0 \cong \mathrm{SU}(1,1)/\mathbb{Z}_2 \tag{3.8}$$

hold, as shown in Sect. 4.1.2 and Exercise 4.1. Since the group $\mathbb{Z}_2$ is contained in SU(2) and SU(1, 1) as the normal subgroup $\{\pm I\}$, SU(2) and SU(1, 1) are central extensions of SO(3) and $\mathrm{SO}(1,2)_0$, respectively. Hence, the discussion in the initial part of Sect. 2.4.2 gives a one-to-one correspondence between projective representations of SO(3) ($\mathrm{SO}(1,2)_0$) and SU(2) (SU(1, 1)), we can discuss SU(2) and SU(1, 1) instead of SO(3) and $\mathrm{SO}(1,2)_0$, respectively. Similarly, as shown in the beginning of Sect. 5.2.2, the following relation holds between SL(2, C) and $\mathrm{SO}(1,3)_0$.

$$\mathrm{SO}(1,3)_0 \cong \mathrm{SL}(2,\mathbb{C})/\mathbb{Z}_2. \tag{3.9}$$

Exercise 3.1 Show that SO(2, $\mathbb{R}$) is connected.

Exercise 3.2 Show that given an element $g \in \mathrm{SO}(d,\mathbb{R})$ there exists another element $g' \in \mathrm{SO}(d,\mathbb{R})$ such that $g'gg'^{-1}$ is block-diagonal, with each block either a 2×2 rotation matrix or 1.

Exercise 3.3 Show that $\mathrm{SO}(d,\mathbb{R})$ is connected.

3.1.2 Symmetry in Analytical Mechanics

Next, we discuss the symmetry in analytical mechanics. When we consider a dynamical system of r real parameters, analytical mechanics discusses these r real parameters and their conjugate parameters. That is, we need to discuss $2r$ parameters as $(\boldsymbol{q},\boldsymbol{p}) := (q^1,\ldots,q^r,p^1,\ldots,p^r)$. In such a system, only the following elements g in GL($2r$, $\mathbb{R}$) are allowed as possible transformations, so called symplectic transformations.

$$gJ_rg^T = J_r, \tag{3.10}$$

where J_n is the matrix defined in (2.43) in the previous chapter. The condition can be rewritten as follows by using the symplectic inner product

$$\langle(\boldsymbol{q},\boldsymbol{p}),(\boldsymbol{q}',\boldsymbol{p}')\rangle := \boldsymbol{q}'\cdot\boldsymbol{p} - \boldsymbol{q}\cdot\boldsymbol{p}'.$$
$$\langle(\boldsymbol{q},\boldsymbol{p}),(\boldsymbol{q}',\boldsymbol{p}')\rangle = \langle g(\boldsymbol{q},\boldsymbol{p}), g(\boldsymbol{q}',\boldsymbol{p}')\rangle. \tag{3.11}$$

The set of elements g in GL($2r$, $\mathbb{R}$) satisfying the condition is called the symplectic group and is written as Sp($2r$, $\mathbb{R}$). This condition is illustrated in Fig. 3.1.

Let us consider the reason of the condition from the viewpoint of quantum theory. For simplicity, we consider the case of $r = 1$. The group composed of displacements of the position in the 2-dimensional space $\mathbb{R}^2$ can be regarded as itself $\mathbb{R}^2$. To discuss the transformation in quantum system by the group $\mathbb{R}^2$, we need to a projective unitary

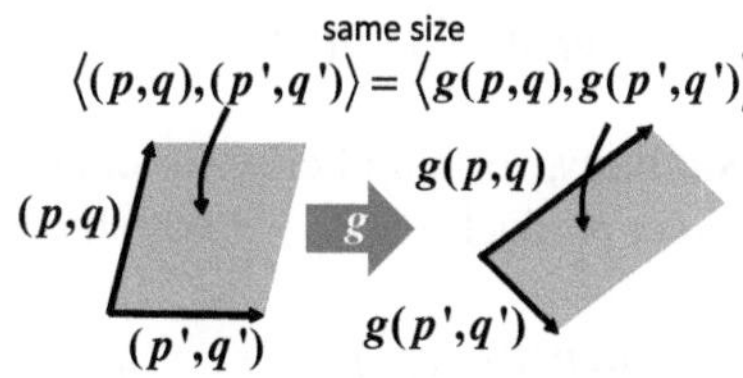

Fig. 3.1 Condition for symplectic group

representation of $\mathbb{R}^2$ and we cannot restrict it to unitary representations of $\mathbb{R}^2$. To address its projective unitary representation, we need to consider central extensions of U(1) by $\mathbb{R}^2$. As shown in Sect. 3.3.3, for any central extension of U(1) by $\mathbb{R}^2$, there exists a real number λ such that

$$(q, p, e^{i\theta}) \cdot (q', p', e^{i\theta'}) = (q + q', p + p', e^{i(\theta+\theta'+\lambda\langle(q,p),(q',p')\rangle/2)}) \qquad (3.12)$$

for $(q, p, e^{i\theta}), (q', p', e^{i\theta'}) \in \mathbb{R}^2 \cdot \mathrm{U}(1)$. That is, the factor system $e^{i\lambda\langle(q,p),(q',p')\rangle/2}$ of any projective representation of $\mathbb{R}^2$ can be characterized by the above central extension of U(1) by $\mathbb{R}^2$. To avoid the variety of central extensions dependently of the parameter λ, we treat the central extension of $\mathbb{R}$ by $\mathbb{R}^2$. That is, we consider the central extension of $\mathbb{R}$ by $\mathbb{R}^2$ as the direct product set $\mathbb{R}^2 \cdot \mathbb{R}$ with the product

$$(q, p, \theta) \cdot (q', p', \theta') := (q + q', p + p', \theta + \theta' + \langle(q, p), (q', p')\rangle/2) \qquad (3.13)$$

for $(q, p, \theta), (q', p', \theta') \in \mathbb{R}^2 \cdot \mathbb{R}$. The group is called the Heisenberg group of degree 1 and is written as $\mathrm{H}(2, \mathbb{R})$. Any projective representation of $\mathbb{R}^2$ is given from a representation of the Heisenberg group $\mathrm{H}(2, \mathbb{R})$ of degree 1. For example, the projective representation f with the factor system $e^{i\lambda\langle(q,p),(q',p')\rangle/2}$ is given by the representation $(q, p, \theta) \mapsto \mathsf{f}(q, p)e^{i\lambda\theta}$ of the Heisenberg group $\mathrm{H}(2, \mathbb{R})$.

Now, we define the action of $\mathrm{GL}(2, \mathbb{R})$ on the Heisenberg group $\mathrm{H}(2, \mathbb{R})$ as $g(q, p, \theta) := (g(q, p), \theta)$ for $g \in \mathrm{GL}(2, \mathbb{R})$. When an element $g \in \mathrm{GL}(2, \mathbb{R})$ satisfies

$$g(q, p, \theta) \cdot g(q', p', \theta') = g((q, p, \theta) \cdot (q', p', \theta')) \qquad (3.14)$$

if and only if the element g belongs to $\mathrm{Sp}(2, \mathbb{R})$. That is, to reserve the composition under the projective representation f of $\mathbb{R}^2$, i.e., to satisfy $\mathsf{f}(g(q, p))\mathsf{f}(g(q', p')) = \mathsf{f}(g(q + q', p + p'))$, we need to restrict the linear transformation on $\mathbb{R}^2$ to elements of $\mathrm{Sp}(2, \mathbb{R})$. For any extension of U(1) by $\mathbb{R}^{2r}$, there exists a square matrix λ of size $2r$ such that

$$(\boldsymbol{q}, \boldsymbol{p}, e^{i\theta}) \cdot (\boldsymbol{q}', \boldsymbol{p}', e^{i\theta'}) = (\boldsymbol{q} + \boldsymbol{q}', \boldsymbol{p} + \boldsymbol{p}', e^{i(\theta+\theta'+\langle\lambda(\boldsymbol{q},\boldsymbol{p}),\lambda(\boldsymbol{q}',\boldsymbol{p}')\rangle/2)}) \qquad (3.15)$$

for $(\boldsymbol{q}, \boldsymbol{p}, e^{i\theta}), (\boldsymbol{q}', \boldsymbol{p}', e^{i\theta'}) \in \mathbb{R}^{2r} \cdot \mathrm{U}(1)$. When the matrix λ is invertible, by applying a suitable coordinate transformation, the product is simplified as

$$(\boldsymbol{q}, \boldsymbol{p}, e^{i\theta}) \cdot (\boldsymbol{q}', \boldsymbol{p}', e^{i\theta'}) = (\boldsymbol{q} + \boldsymbol{q}', \boldsymbol{p} + \boldsymbol{p}', e^{i(\theta+\theta'+\langle(\boldsymbol{q},\boldsymbol{p}),(\boldsymbol{q}',\boldsymbol{p}')\rangle/2)}). \tag{3.16}$$

Similarly, for simplicity, we consider central extension of $\mathbb{R}$ by $\mathbb{R}^{2r}$, which is the direct product set $\mathbb{R}^{2r} \cdot \mathbb{R}$ with the product

$$(\boldsymbol{q}, \boldsymbol{p}, \theta) \cdot (\boldsymbol{q}', \boldsymbol{p}', \theta') := (\boldsymbol{q} + \boldsymbol{q}', \boldsymbol{p} + \boldsymbol{p}', \theta + \theta' + \langle(\boldsymbol{q}, \boldsymbol{p}), (\boldsymbol{q}', \boldsymbol{p}')\rangle/2) \tag{3.17}$$

for $(\boldsymbol{q}, \boldsymbol{p}, \theta), (\boldsymbol{q}', \boldsymbol{p}', \theta') \in \mathbb{R}^2 \cdot \mathbb{R}$. The group is called the Heisenberg group of degree r and is written as $\mathrm{H}(2r, \mathbb{R})$. Hence, to reserve the composition under the projective representation $\mathfrak{f}$ of $\mathbb{R}^{2r}$, i.e., to satisfy $\mathfrak{f}(g(\boldsymbol{q}, \boldsymbol{p}))\mathfrak{f}(g(\boldsymbol{q}', \boldsymbol{p}')) = \mathfrak{f}(g(\boldsymbol{q} + \boldsymbol{q}', \boldsymbol{p} + \boldsymbol{p}'))$, we need to restrict the linear transformation on $\mathbb{R}^{2r}$ to elements of $\mathrm{Sp}(2r, \mathbb{R})$. In this way, the property required by a projective representation restricts our linear transformations to symplectic transformations. Thus, to consider groups of the above linear transformation on the space $\mathbb{R}^{2r}$, it is sufficient to treat the semi direct product $\mathrm{H}(2r, \mathbb{R}) \rtimes \mathrm{Sp}(2r, \mathbb{R})$. Chapter 8 addresses unitary representations of this group.

3.1.3 Complex Lie Group

Although we have discussed Lie groups with real number parameters up to the previous subsection. Now, we consider Lie groups with complex parameter. Additionally, we assume that the product is given as a holomorphic function for the complex parameter. Such a Lie group is called a **complex Lie group**. On the other hand, when a Lie group is not necessarily a complex Lie group and it is needed to be distinguished from a complex Lie group, it is called a real Lie group. In fact, a complex Lie group has simpler structure than a real Lie group because the holomorphic condition is so strong.

The most of physical meaningful transformations are not complex Lie groups and but real Lie groups. Hence, we do not need to treat complex Lie groups from the physical viewpoint. However, real Lie groups have too complicated structure to directly treat without considering complex Lie groups. So, we consider complex Lie groups, whose typical examples are summarized in the following Table 3.1.

Many real Lie groups have their corresponding complex Lie groups as their own complexifications, respectively. However, each complex Lie group has more than one corresponding real Lie groups so that a complicated treatment is required.

3.1.4 Other Examples of Real Lie Groups

Although we have listed several real Lie groups up to now, there are so many remaining real Lie groups, which are listed below. Compact real Lie groups corresponding to complex Lie groups in Table 3.1 are given in Table 3.2. (The last 2 examples in Table 3.2 do not have their corresponding compact real Lie groups in Table 3.1.) We

Table 3.1 Complex Lie group

Name of group	Notation	Definition
Complex general linear group	$\mathrm{GL}(n, \mathbb{C})$	$\{n \times n$ complex regular matrix $\}$
Complex special linear group	$\mathrm{SL}(n, \mathbb{C})$	$\{g \in \mathrm{GL}(n, \mathbb{C}) \mid \det g = 1\}$
Complex orthogonal group	$\mathrm{O}(n, \mathbb{C})$	$\{g \in \mathrm{GL}(n, \mathbb{C}) \mid g^T g = I\}$
Complex special orthogonal group	$\mathrm{SO}(n, \mathbb{C})$	$\mathrm{O}(n, \mathbb{C}) \cap \mathrm{SL}(n, \mathbb{C})$
Complex symplectic group	$\mathrm{Sp}(2n, \mathbb{C})$	$\{g \in \mathrm{GL}(2n, \mathbb{C}) \mid g^T J_n g = J_n\}$

Table 3.2 Compact Lie group

Name of group	Notation	Definition	Dimension	Fundamental group
Unitary group	$\mathrm{U}(n)$	$\{g \in \mathrm{GL}(n, \mathbb{C}) \mid g^\dagger g = I\}$	n^2	1
Special unitary group	$\mathrm{SU}(n)$	$\mathrm{U}(n) \cap \mathrm{SL}(n, \mathbb{C})$	$n^2 - 1$	1
Real orthogonal group	$\mathrm{O}(n, \mathbb{R})$	$\mathrm{O}(n, \mathbb{C}) \cap \mathrm{GL}(n, \mathbb{R})$	$n(n-1)/2$	Non-connected
Real special linear group	$\mathrm{SO}(n, \mathbb{R})$	$\mathrm{SO}(n, \mathbb{C}) \cap \mathrm{GL}(n, \mathbb{R})$	$n(n-1)/2$	$\mathbb{Z}_2$ $(n \geq 3)$ $\mathbb{Z}$ $(n = 2)$
Hyper-unitary group	$\mathrm{Sp}(n)$	$\mathrm{U}(2n) \cap \mathrm{Sp}(2n, \mathbb{C})$	$n(2n+1)$	1
	$\mathrm{D}(n)$	diagonal matrices in $\mathrm{U}(n)$	n	$\mathbb{Z}_2^n$
	$\mathrm{Sd}(n)$	diagonal matrices in $\mathrm{SU}(n)$	$n-1$	$\mathbb{Z}_2^{n-1}$

denote the set of real regular matrices of size n by $\mathrm{GL}(n, \mathbb{R})$. The definition of the fundamental group is given in Sect. 3.6.

Non-compact real Lie groups are listed in Table 3.3. Since a complex Lie group usually has more than one corresponding non-compact real Lie groups, the table of non-compact real Lie groups (Table 3.3) is very complicated. In the lower dimensional case, several groups in Table 3.3 are the same or isomorphic to each other although all of them are not isomorphic to each other in the higher dimensional case.

$$\mathrm{SU}(1, 1) \cong \mathrm{SL}(2, \mathbb{R}) = \mathrm{Sp}(2, \mathbb{R}) \tag{3.18}$$

3.2 Lie Algebra

The previous section addressed real Lie groups and complex Lie groups. Although complex Lie groups can be more easily treated than real Lie groups, complex Lie groups still have difficulty in their own treatment. The structures of real Lie groups and complex Lie groups are almost determined by their local structures around their

Table 3.3 Non-compact real Lie group

Name of group	Notation	elements of group	Dimension
Real general linear group	$GL(n, \mathbb{R})$	$\{n \times n$ real regular matrices$\}$	n^2
Non-definite unitary group	$U(p, q)$	$\left\{ g \,\middle\vert\, \begin{array}{l} g \in GL(p+q, \mathbb{C}) \\ g^\dagger I_{p,q} g = I_{p,q} \end{array} \right\}$	$(p+q)^2$
	$U^*(2m)$	$\left\{ g \,\middle\vert\, \begin{array}{l} g \in GL(2m, \mathbb{C}) \\ \overline{g} J_m = J_m g \end{array} \right\}$	$(2m)^2$
Real special linear group	$SL(n, \mathbb{R})$	$GL(n, \mathbb{R}) \cap SL(n, \mathbb{C})$	$n^2 - 1$
Real non-definite unitary group	$SU(p, q)$	$SL(p+q, \mathbb{C}) \cap U(p, q)$	$(p+q)^2 - 1$
	$SU^*(2m)$	$SL(n, \mathbb{R}) \cap U^*(2m)$	$(2m)^2 - 1$
Non-definite orthogonal group	$O(p, q)$	$GL(p+q, \mathbb{R}) \cap U(p, q)$	$\frac{(p+q)(p+q-1)}{2}$
Special non-definite orthogonal group	$SO(p, q)$	$SL(p+q, \mathbb{R}) \cap U(p, q)$	$\frac{(p+q)(p+q-1)}{2}$
	$SO^*(2m)$	$SO(2m, \mathbb{C}) \cap U^*(2m)$	$m(2m-1)$
Real symplectic group	$Sp(2n, \mathbb{R})$	$GL(2n, \mathbb{R}) \cap Sp(2n, \mathbb{C})$	$n(2n+1)$
Non-definite symplectic group	$Sp(p, q)$	$U(2p, 2q) \cap Sp(2p + 2q, \mathbb{C})$	$n(2n+1)$
Heisenberg group	$H(2n, \mathbb{R})$		$2n+1$
Affine group	$Aff(n, \mathbb{R})$		$n^2 + n$

identity element, respectively. Further, their local structures around their identity element are much simpler than their global structures. Hence, it is better to discuss only their local structures before discussing the whole structures of the groups. For this purpose, we introduce the concept of "Lie algebra", which reflects the local structure of each Lie group. This section gives only the formal definition of Lie algebra, and the latter section explains the relation with Lie group. A real (complex) linear space V is called a real (complex) Lie algebra when there is a map $[,] : V \times V \to V$ (called a **commutator**) that satisfies the following conditions.[1]

[1] In physics, we identify elements of a real Lie algebra and a Hermitian matrix as a physical quantity. However, in our definition, the real Lie algebra of $SU(d)$ consists of skew-Hermitian matrices. That is, in physics, the products of $\sqrt{-1}$ and the elements of Lie algebra defined here are regarded as elements of a Lie algebra.

Bilinearlity: For any element $X \in V$, the maps $Y \mapsto [X, Y]$ and $Y \mapsto [Y, X]$ are linear.
Skew-symmetry (skew-symmetricity): $[X, Y] = -[Y, X]$.
Jacobi law: $[X, [Y, Z]] + [Y, [Z, X]] + [Z, [X, Y]] = 0$.

For example, the set $\mathfrak{gl}(V)$ of linear maps on a real (complex) linear space V form a real (complex) Lie algebra with the following commutator

$$[X, Y] := XY - YX. \tag{3.19}$$

Similarly, the set $\mathfrak{u}(\mathcal{H})$ of skew-Hermitian matrices on a Hilbert space $\mathcal{H}$ forms a real Lie algebra with the commutator (3.19). When a Lie algebra $\mathfrak{g}$ is given as a subset of matrices with the commutator (3.19), two elements $X, Y \in g$ satisfy $[X, Y] = 0$ if and only if they are commutative with each other as matrices. Hence, even when a Lie algebra $\mathfrak{g}$ is not given as a subset of matrices, two elements $X, Y \in \mathfrak{g}$ are called **commutative** when $[X, Y] = 0$. We need to notice that a complex Lie algebra is a real Lie algebra but a real Lie algebra is not a complex Lie algebra in general.

A real (complex) linear map f from a real (complex) Lie algebra $\mathfrak{g}_1$ to a real (complex) Lie algebra $\mathfrak{g}_2$ is called a **homomorphism for real (complex) Lie algebras** when $[f(X), f(Y)] = f([X, Y])$, $(\forall X, Y \in \mathfrak{g}_1)$. Especially, when f is injective, f is called an isomorphism. When there exists an isomorphism from a real (complex) Lie algebra $\mathfrak{g}_1$ to a real (complex) Lie algebra $\mathfrak{g}_2$, $\mathfrak{g}_1$ and $\mathfrak{g}_2$ are called isomorphic to each other as Lie algebras, whose relation is written as $\mathfrak{g}_1 \cong \mathfrak{g}_2$. In this book, we simply denote a real Lie algebra by a Lie algebra, and a homomorphism for real Lie algebras by a homomorphism for Lie algebras. In general, even though a map from a complex Lie algebra to another complex Lie algebra is a homomorphism for real Lie algebras, it is not necessarily a homomorphism for complex Lie algebras.

For two real (complex) Lie algebras $\mathfrak{g}_1$ and $\mathfrak{g}_2$, their direct sum $\mathfrak{g}_1 \oplus \mathfrak{g}_2$ forms a real (complex) Lie algebra with the commutator $[X_1 \oplus X_2, Y_1 \oplus Y_2] := [X_1, Y_1] \oplus [X_2, Y_2]$. Since a real Lie algebra $\mathfrak{g}$ forms a real linear space, we define the complexification as the linear space by replacing the real coefficients by the complex coefficients. Then, the commutator can be extended to the complexification so that it forms a complex Lie algebra, which is called the **complexification** of $\mathfrak{g}$ and is denoted by $\mathfrak{g}_{\mathbb{C}}$. In general, even though two real Lie algebras $\mathfrak{g}_1$ and $\mathfrak{g}_2$ satisfy $\mathfrak{g}_{1,\mathbb{C}} \cong \mathfrak{g}_{2,\mathbb{C}}$, the relation $\mathfrak{g}_1 \cong \mathfrak{g}_2$ does not necessarily hold, whose counterexample will be given later.

Here, we define several fundamental concepts for real (complex) Lie algebras. For this purpose, for real (complex) linear subspaces $\mathfrak{a}$ and $\mathfrak{b}$ of a real (complex) Lie algebra $\mathfrak{g}$, we introduce the notations $[\mathfrak{a}, \mathfrak{b}] := \{[A, B] | A \in \mathfrak{a}, B \in \mathfrak{b}\}$, $\mathfrak{g}^{(0)} := \mathfrak{g}$, $\mathfrak{g}^{(n)} := [\mathfrak{g}, \mathfrak{g}^{(n-1)}]$, and $D^{(0)}\,\mathfrak{g} := \mathfrak{g}$, $D^{(n)}\,\mathfrak{g} := [D^{(n-1)}\,\mathfrak{g}, D^{(n-1)}\,\mathfrak{g}]$. Then, $\mathfrak{a}$ is called a **real (complex) Lie subalgebra** when $[\mathfrak{a}, \mathfrak{a}] \subset \mathfrak{a}$. In particular, it is called an **ideal** when $[\mathfrak{g}, \mathfrak{a}] \subset \mathfrak{a}$. The quotient space $\mathfrak{g} / \mathfrak{a}$ forms a real (complex) Lie algebra and is called a quotient real (complex) Lie algebra. A real (complex) Lie algebra $\mathfrak{g}$ is called **simple** when its ideal is restricted to $\{0\}$ or itself. As mentioned later, the simplicity means the impossibility of a non-trivial decomposition. For any linear subspace $\mathfrak{k}$ of $\mathfrak{g}$, a linear subspace $\mathfrak{c}_{\mathfrak{g}}(\mathfrak{k}) := \{X \in \mathfrak{g} \,|[X, Y] = 0, \forall Y \in \mathfrak{k}\}$ forms a Lie subalgebra of $\mathfrak{g}$ and is called the **centralizer** of $\mathfrak{k}$ with respect to $\mathfrak{g}$. When $\mathfrak{k}$ is the Lie algebra $\mathfrak{g}$, $\mathfrak{c}_{\mathfrak{g}}(\mathfrak{g})$ forms an ideal of $\mathfrak{g}$, which is called the **center** of $\mathfrak{g}$ and is simplified to $\mathfrak{c}(\mathfrak{g})$.

On the other hand, when $\mathfrak{g}^{(1)} = 0$, all of elements of $\mathfrak{g}$ are commutative with each other and $\mathfrak{g}$ is called **commutative**. When there exists an integer k such that $\mathfrak{g}^{(k)} = 0$, $\mathfrak{g}$ is called **nilpotent**. Moreover, when there exists an integer k such that $D^{(k)}\,\mathfrak{g} = 0$, $\mathfrak{g}$ is called **solvable**. In the nilpotent case, $\mathfrak{g}^{(1)}$ is an ideal. Hence, we find that the concepts "nilpotent" and "simple" are opposite to each other. When a solvable ideal of $\mathfrak{g}$ is limited to $\{0\}$, $\mathfrak{g}$ is called **semi simple**. Even though $\mathfrak{g}$ is not semi simple, we have the following characterization. Let $\{\mathfrak{g}_i\}_i$ be the set of all solvable ideals of $\mathfrak{g}$. Then, $\cup_i \mathfrak{g}_i$ is also a solvable ideal of $\mathfrak{g}$, which is called the **radical** and is denoted by *rad* $\mathfrak{g}$. Then, the quotient real (complex) Lie algebra $\mathfrak{g}\,/rad\,\mathfrak{g}$ is semi simple.

Given a semi simple real (complex) Lie algebra $\mathfrak{g}$, we can choose simple real (complex) Lie subalgebras $\mathfrak{g}_1, \ldots, \mathfrak{g}_k$ such that $\mathfrak{g} = \oplus_{i=1}^{k} \mathfrak{g}_i$. Hence, the analysis of a semi simple real (complex) Lie algebra can be reduced to that of simple real (complex) Lie algebras.

Finally, given two real (complex) Lie subalgebras $\mathfrak{g}_3$ and $\mathfrak{g}_4$ of a real (complex) Lie algebra $\mathfrak{g}$, the real (complex) linear subspace $\mathfrak{g}_3 + \mathfrak{g}_4 := \{X_3 + X_4 | X_3 \in \mathfrak{g}_3, X_4 \in \mathfrak{g}_4\}$ forms a real (complex) Lie subalgebra of $\mathfrak{g}$. In particular, when $\mathfrak{g}_3 \cap \mathfrak{g}_4 = \{0\}$, $\mathfrak{g}_3 + \mathfrak{g}_4$ is isomorphic to $\mathfrak{g}_3 \oplus \mathfrak{g}_4$.

3.3 Relation Between Lie Group and Lie Algebra I

3.3.1 *Infinitesimal Transformation and Lie Algebra*

As mentioned in Sect. 3.2, an element of real (complex) Lie algebra corresponds to an infinitesimal transformation in a real (complex) Lie group around the identity element I. The following discussion concretely gives this relation. Remember that a real (complex) Lie group G is defined as a subgroup of $\mathrm{GL}(n, \mathbb{C}) = \mathrm{GL}(\mathbb{C}^n)$. We focus on a one-dimensional differentiable subset $\{g(t) \in G | t \in \mathbb{R}\}$ satisfying $g(0) = I$, which is called a **path** of G. Then, the real (complex) Lie algebra $\mathfrak{g}(G)$ corresponding to G is defined as the set of the limits $X = \lim_{t \to 0} \frac{g(t)-I}{t}$ as Fig. 3.2. As another definition of the real (complex) Lie algebra $\mathfrak{g}(G)$ corresponding to G, we often adopt the definition $\mathfrak{g}(G) := \{X \in \mathfrak{gl}(\mathbb{C}^n) | e^{tX} \in G, \forall t \in \mathbb{R}\}$, where $e^X := \exp(X) := \sum_{n=0}^{\infty} \frac{X^n}{n!}$.

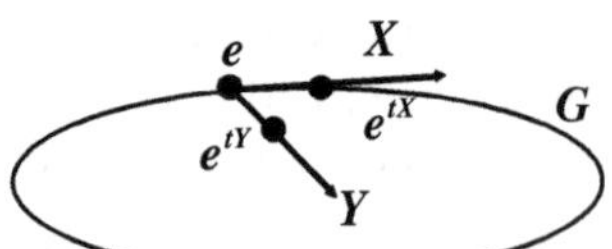

Fig. 3.2 Relation between Lie algebra and Lie group

Since any two elements $X, Y \in \mathfrak{g}(G)$ satisfy

$$e^{tY}e^{tX} = e^{t(X+Y)+\frac{t^2}{2}[Y,X]+O(t^3)} \tag{3.20}$$

as the limit $t \to 0$, we can show that

$$e^{tY}e^{tX}e^{-tY}e^{-tX} = e^{t^2[Y,X]+O(t^3)}. \tag{3.21}$$

Using this fact, we have

$$\begin{aligned}[Y, X] &= \lim_{t\to 0} \frac{e^{tY}e^{tX} - e^{tX}e^{tY}}{t^2} \\ &= \lim_{t\to 0} \frac{e^{tY}e^{tX}e^{-tY}e^{-tX} - I}{t^2} = \lim_{t\to 0} \frac{e^{tY}Xe^{-tY} - X}{t},\end{aligned} \tag{3.22}$$

which naturally derives the commutator of Lie algebra from the product of Lie group. Hence, the set of matrices X given as the above limits forms a real (complex) Lie algebra with the commutator (3.19).

Conversely, when a real (complex) Lie algebra $\mathfrak{g}$ is a real (complex) Lie subalgebra of $\mathfrak{gl}(\mathcal{H})$ with the commutator (3.19), the set

$$G(\mathfrak{g}) := \{e^{X_1} \cdots e^{X_m} | X_1, \ldots, X_m \in \mathfrak{g}\}$$

forms a real (complex) Lie group, whose product is the conventional matrix multiplication. When G is compact or nilpotent, we can simplify it as $G(\mathfrak{g}) = \{\exp(X)|X \in \mathfrak{g}\}$. This definition depends on the choice of the Hilbert space $\mathcal{H}$. That is, even though $\mathfrak{g}_1 \subset \mathfrak{gl}(\mathcal{H}_1)$ is isomorphic to $\mathfrak{g}_2 \subset \mathfrak{gl}(\mathcal{H}_2)$ as a Lie algebra, the Lie group $G(\mathfrak{g}_1)$ is not necessarily isomorphic to the Lie group $G(\mathfrak{g}_2)$. For example, as mentioned later, $\mathfrak{su}(2)$ and $\mathfrak{so}(3, \mathbb{R})$ are isomorphic to each other, SU(2) and SO(3, $\mathbb{R}$) are not isomorphic to each other.

When the Lie group is simple as a group, the corresponding Lie algebra is simple. However, the converse is not true in general. For example, although $\mathfrak{su}(2)$ is a simple real Lie algebra, SU(2) is not a simple group because it has the normal subgroup U_2. When two Lie groups G and H correspond to Lie algebras $\mathfrak{g}$ and $\mathfrak{h}$, respectively, and H is a subgroup G, then $\mathfrak{h}$ is a Lie subalgebra of $\mathfrak{g}$. Additionally, when H is a normal subgroup of G, $\mathfrak{h}$ is an ideal of $\mathfrak{g}$.

Then, the following theorem holds.

Theorem 3.1 *Given a homomorphism f from a Lie group G_1 to a Lie group G_2, the map*

$$f(X) := \lim_{t\to 0} \frac{f(e^{tX}) - I}{t}, \quad \forall X \in \mathfrak{g}(G_1).$$

is a homomorphism from the Lie algebra $\mathfrak{g}(G_1)$ to the Lie algebra $\mathfrak{g}(G_2)$.

However, we cannot necessarily make a homomorphism for Lie groups from a homomorphism for Lie algebras.

Proof The relations (3.21) and (3.22) imply that

$$
\begin{aligned}
f([X,Y]) &= \lim_{t\to 0} \frac{f(e^{t^2[X,Y]})-I}{t^2} \\
&= \lim_{t\to 0} \frac{f(e^{tY}e^{tX}e^{-tY}e^{-tX})-I}{t^2} = \lim_{t\to 0} \frac{f(e^{tY})f(e^{tX})f(e^{-tY})f(e^{-tX})-I}{t^2} \\
&= \lim_{t\to 0} \frac{e^{tf(Y)}e^{tf(X)}e^{-tf(Y)}e^{-tf(X)}-I}{t^2} = [f(X), f(Y)]
\end{aligned}
$$

for $X, Y \in \mathfrak{g}$. Hence, we obtain the desired argument. ∎

Although our aim is the characterization of real Lie groups, the process of this aim is summarized as follows.

Characterization of complex Lie algebras	$\Longrightarrow$	Characterization of real Lie algebras	$\Longrightarrow$	Characterization of real Lie groups

Exercise 3.4 Show that $\mathfrak{g}(G_1) \oplus \mathfrak{g}(G_2) = \mathfrak{g}(G_1 \times G_2)$ based on the above definition of direct sum for Lie algebras.

Exercise 3.5 Show (3.20) and (3.21).

3.3.2 Examples

Table 3.4 summarizes the complex Lie algebras corresponding to the complex Lie groups given in Table 3.1 in Sect. 3.1, where the notations of the Lie algebras are given as the Fraktur of the notations of the corresponding Lie groups.

Table 3.5 lists the real Lie algebras corresponding to the compact Lie group given in Table 3.2, where $\mathfrak{gl}(n,\mathbb{R})$ is $\mathfrak{gl}(\mathbb{R}^n)$. In the latter section, we define the compactness for a real Lie algebra when a real Lie algebra is defined as a linear space with a commutator. All of examples in Table 3.5 are compact under this definition.

Table 3.4 Complex Lie algebra

Notation of Lie algebra	Definition
$\mathfrak{gl}(n,\mathbb{C})$	$\mathfrak{gl}(\mathbb{C}^n)$
$\mathfrak{sl}(n,\mathbb{C})$	$\{X \in \mathfrak{gl}(n,\mathbb{C}) \mid \operatorname{Tr} X = 0\}$
$\mathfrak{o}(n,\mathbb{C})$	$\{X \in \mathfrak{gl}(n,\mathbb{C}) \mid X^T = -X\}$
$\mathfrak{sp}(2n,\mathbb{C})$	$\{X \in \mathfrak{gl}(2n,\mathbb{C}) \mid X^T J_n + J_n X = 0\}$

Table 3.5 Compact real Lie algebra

Notation of Lie algebra	Definition	Dimension
$\mathfrak{u}(n)$	$\{X \in \mathfrak{gl}(n, \mathbb{C}) \mid X^\dagger = -X\}$	n^2
$\mathfrak{su}(n)$	$\mathfrak{u}(n) \cap \mathfrak{sl}(n, \mathbb{C})$	$n^2 - 1$
$\mathfrak{o}(n, \mathbb{R})$	$\mathfrak{gl}(n, \mathbb{R}) \cap \mathfrak{o}(n, \mathbb{C})$	$n(n-1)/2$
$\mathfrak{so}(n, \mathbb{R})$	$= \mathfrak{o}(n, \mathbb{R})$	$n(n-1)/2$
$\mathfrak{sp}(n)$	$\mathfrak{u}(2n) \cap \mathfrak{sp}(2n, \mathbb{C})$	$n(2n+1)$
$\mathfrak{d}(n)$	Diagonal matrices in $\mathfrak{u}(n)$	n
$\mathfrak{sd}(n)$	Diagonal matrices in $\mathfrak{su}(n)$	$n-1$

Table 3.6 Non-compact real Lie algebra

Symbol of Lie algebra	Definition	Dimension
$\mathfrak{gl}(n, \mathbb{R})$	$\mathfrak{gl}(\mathbb{R}^n)$	n^2
$\mathfrak{u}(p, q)$	$\{X \in \mathfrak{gl}(p+q, \mathbb{C}) \mid X^\dagger I_{p,q} + I_{p,q} X = 0\}$	$(p+q)^2$
$\mathfrak{u}^*(2m)$	$\{X \in \mathfrak{gl}(2m, \mathbb{C}) \mid \overline{X} J_m = J_m X\}$	$(2m)^2$
$\mathfrak{sl}(n, \mathbb{R})$	$\mathfrak{sl}(n, \mathbb{C}) \cap \mathfrak{gl}(n, \mathbb{R})$	$n^2 - 1$
$\mathfrak{su}(p, q)$	$\mathfrak{sl}(p+q, \mathbb{C}) \cap \mathfrak{u}(p, q)$	$(p+q)^2 - 1$
$\mathfrak{su}^*(2m)$	$\mathfrak{sl}(n, \mathbb{C}) \cap \mathfrak{u}^*(2m)$	$(2m)^2 - 1$
$\mathfrak{o}(p, q)$	$\mathfrak{gl}(p+q, \mathbb{R}) \cap \mathfrak{u}(p, q)$	$(p+q)(p+q-1)/2$
$\mathfrak{so}(p, q)$	$\mathfrak{sl}(p+q, \mathbb{R}) \cap \mathfrak{u}(p, q)$	$(p+q)(p+q-1)/2$
$\mathfrak{so}^*(2m)$	$\mathfrak{so}(2m, \mathbb{C}) \cap \mathfrak{u}^*(2m)$	$m(2m-1)$
$\mathfrak{sp}(2n, \mathbb{R})$	$\mathfrak{gl}(2n, \mathbb{R}) \cap \mathfrak{sp}(2n, \mathbb{C})$	$n(2n+1)$
$\mathfrak{sp}(p, q)$	$\mathfrak{u}(2p, 2q) \cap \mathfrak{sp}(2p+2q, \mathbb{C})$	$n(2n+1)$
$\mathfrak{h}(2n, \mathbb{R})$		$2n+1$
$\mathfrak{aff}(n, \mathbb{R})$		$n^2 + n$

Table 3.6 lists real Lie algebras corresponding to non-compact Lie groups given in Table 3.3. Similar to the compactness, in the latter section, we define the non-compactness for a real Lie algebra when a real Lie algebra is defined as a linear space with a commutator. All of examples in Table 3.6 are non-compact under this definition.

In the above table, for $z \in \mathbb{R}, x \in \mathbb{R}^n, \vec{z} \in \mathbb{R}^{2n}, A \in \mathfrak{gl}(n, \mathbb{R})$, the elements $X_h(\vec{z}, z)$ and $X_a(A, x)$ are defined as

$$X_h(\vec{z}, z) := \begin{pmatrix} 0 & -x^T & z \\ 0 & 0_n & y \\ 0 & 0^T & 0 \end{pmatrix} \tag{3.23}$$

$$X_a(A, x) := \begin{pmatrix} A & x \\ 0 & 0 \end{pmatrix}, \tag{3.24}$$

where $\vec{z}$ is defined as $\vec{z} := (x_1, y_1, \ldots, x_d, y_d)^T$ for $x, y \in \mathbb{R}^n$. Note that the real Lie algebra $\mathfrak{o}(n, \mathbb{R})$ is isomorphic to the real Lie algebra $\mathfrak{so}(n)$.

The following relation holds between a real Lie algebra and a complex Lie algebra when $p + q = n$.

$$\begin{aligned}
\mathfrak{gl}(n, \mathbb{C}) &\cong \mathfrak{u}(n)_{\mathbb{C}} \cong \mathfrak{gl}(n, \mathbb{R})_{\mathbb{C}} \cong \mathfrak{u}(p, q)_{\mathbb{C}}, \quad \mathfrak{gl}(2m, \mathbb{C}) \cong \mathfrak{u}^*(2m)_{\mathbb{C}} \\
\mathfrak{sl}(n, \mathbb{C}) &\cong \mathfrak{su}(n)_{\mathbb{C}} \cong \mathfrak{sl}(n, \mathbb{R})_{\mathbb{C}} \cong \mathfrak{su}(p, q)_{\mathbb{C}}, \quad \mathfrak{sl}(2m, \mathbb{C}) \cong \mathfrak{su}^*(2m)_{\mathbb{C}} \\
\mathfrak{o}(n, \mathbb{C}) &= \mathfrak{so}(n, \mathbb{C}) \cong \mathfrak{o}(n, \mathbb{R})_{\mathbb{C}} = \mathfrak{so}(n, \mathbb{R})_{\mathbb{C}} \cong \mathfrak{o}(p, q)_{\mathbb{C}} = \mathfrak{so}(p, q)_{\mathbb{C}} \\
\mathfrak{o}(2m, \mathbb{C}) &= \mathfrak{so}(2m, \mathbb{C}) \cong \mathfrak{o}^*(2m)_{\mathbb{C}} = \mathfrak{so}^*(2m)_{\mathbb{C}} \\
\mathfrak{sp}(2n, \mathbb{C}) &\cong \mathfrak{sp}(n)_{\mathbb{C}} \cong \mathfrak{sp}(2n, \mathbb{R})_{\mathbb{C}} \cong \mathfrak{sp}(p, q)_{\mathbb{C}}.
\end{aligned} \tag{3.25}$$

Although the real Lie algebras $\mathfrak{su}(n)_{\mathbb{C}}, \mathfrak{sl}(n, \mathbb{R})_{\mathbb{C}}$, and $\mathfrak{su}(p, q)_{\mathbb{C}}$ are not isomorphic to each other as real Lie algebras, their complexifications are isomorphic to each other, as explained in the above table. Since $\mathfrak{su}(n)_{\mathbb{C}}$ is a compact real Lie algebra and $\mathfrak{sl}(n, \mathbb{R})_{\mathbb{C}}$ and $\mathfrak{su}(p, q)_{\mathbb{C}}$ are non-compact real Lie algebras, $\mathfrak{sl}(n, \mathbb{R})_{\mathbb{C}}$ and $\mathfrak{su}(p, q)_{\mathbb{C}}$ are not isomorphic to $\mathfrak{su}(n)_{\mathbb{C}}$ as real Lie algebras. Even though the complexifications of given Lie algebras are isomorphic to each other, the original real Lie algebras are not necessarily isomorphic to each other. This fact indicates that a real Lie algebra has a more complicated structure than a complex Lie algebra.

The following isomorphic relation for real Lie algebras holds in the lower dimensional case.

$$\mathfrak{sp}(2, \mathbb{R}) = \mathfrak{sl}(2, \mathbb{R}) \cong \mathfrak{su}(1, 1) \tag{3.26}$$

$$\mathfrak{so}(3, \mathbb{R}) \cong \mathfrak{su}(2), \quad \mathfrak{so}(4, \mathbb{R}) \cong \mathfrak{su}(2) \oplus \mathfrak{su}(2) \tag{3.27}$$

$$\begin{aligned}
&\mathfrak{so}(5, \mathbb{R}) \cong \mathfrak{sp}(4, \mathbb{R}), \quad \mathfrak{so}(6, \mathbb{R}) \cong \mathfrak{su}(4) \\
&\mathfrak{sl}(2, \mathbb{C}) \cong \mathfrak{so}(1, 3).
\end{aligned} \tag{3.28}$$

In particular, the second isomorphic relation in (3.27) is discussed precisely in Sect. 5.2.2. Hence, the following isomorphic relation for complex Lie algebras holds from (3.25).

$$\begin{aligned}
&\mathfrak{sp}(2, \mathbb{C}) \cong \mathfrak{sl}(2, \mathbb{C}), \quad \mathfrak{so}(3, \mathbb{C}) \cong \mathfrak{sl}(2, \mathbb{C}) \\
&\mathfrak{so}(4, \mathbb{C}) \cong \mathfrak{sl}(2, \mathbb{C}) \oplus \mathfrak{sl}(2, \mathbb{C}), \quad \mathfrak{so}(5, \mathbb{C}) \cong \mathfrak{sp}(4, \mathbb{C}) \\
&\mathfrak{so}(6, \mathbb{C}) \cong \mathfrak{sl}(2, \mathbb{C}).
\end{aligned} \tag{3.29}$$

As mentioned later, it is known that complex simple Lie algebras are completely classified. That is, they are limited to classical types $\mathfrak{sl}(n, \mathbb{C})$, $\mathfrak{so}(n, \mathbb{C})$, $\mathfrak{sp}(n, \mathbb{C})$ and exceptional types E_6, E_7, E_8, F_4, G_2. In these examples, as explained in (3.29), only $\mathfrak{so}(4, \mathbb{C})$ is not simple.

Similarly, as mentioned later, since any compact simple Lie algebra uniquely corresponds to a complex simple Lie algebra, compact simple Lie algebras are completely classified. Although the classification of non-compact simple Lie algebras is not easy, they has been completely classified [66, Theorem 6.105, Sect. 10, Chap. VI].

3.3.3 Central Extension of Real Lie Algebra

We have discussed the central extension by a real Lie group in Sect. 3.1.2. Similarly, we can consider the central extension by a real Lie algebra.

For a real Lie algebra $\mathfrak{g}$, we consider the direct sum space $\mathbb{R} + \mathfrak{g}$ as a linear space. Then, we denote the generator of the subspace $\mathbb{R}$ by Z. So, any element of $\mathbb{R} + \mathfrak{g}$ can be written as $aZ + X$ by using $a \in R$ and $X \in \mathfrak{g}$. We define the commutator for two elements $aZ + X, a'Z + X' \in \mathbb{R} + \mathfrak{g}$ as

$$[aZ + X, a'Z + X'] = c(X, X')Z + [X, X'], \tag{3.30}$$

where c is a bilinear form satisfying

$$c(X, X') = -c(X', X). \tag{3.31}$$

That is, using a bilinear form c on $\mathfrak{g}$ satisfying (3.31), we can define the above commutator on the direct sum space $\mathbb{R} + \mathfrak{g}$. The constructed Lie algebra $\mathfrak{g}_c$ is called the central extension of $\mathbb{R}$ by $\mathfrak{g}$. Unless c is 0, the center $\mathfrak{c}(\mathfrak{g}_c)$ of $\mathfrak{g}_c$ is $\mathbb{R}$.

Next, we will show that the central extension by a Lie algebra is the Lie algebra corresponding to the central extension by a Lie group. Consider the central extension of U(1) by Lie group G. Let Z be the generator of the normal subgroup U(1) of the central extension. Since Z is commutative with any elements of the central extension, the relations

$$\begin{aligned}&\lim_{t\to 0} \frac{1}{t}(e^{t(aZ+X)}(a'Z + X')e^{-t(aZ+X)} - (a'Z + X'))\\ =&\lim_{t\to 0} \frac{1}{t}(e^{tX}X'e^{-tX} - X') = c(X, X')Z + [X, X']\end{aligned} \tag{3.32}$$

hold for $X, X' \in \mathfrak{g}(G)$. The property of Lie algebra guarantees the bilinearity of c and the condition (3.31).

For example, the bilinear form c characterizing the central extension by the commutative Lie algebra $\mathbb{R}^2$ is given as $\lambda\langle((p, q), (p', q'))\rangle$ by using a symplectic inner product and a real number λ. In the central extension, the generators $Q := (1, 0)$ and $P := (0, 1)$ of the commutative Lie algebra $\mathbb{R}^2$ satisfy the following relations

$$[Q, P] = -\lambda Z, [Q, Z] = [P, Z] = 0. \tag{3.33}$$

Especially, the above Lie algebra with $\lambda = 1$ is called the Heisenberg algebra with degree 1 and is written as $\mathfrak{h}(2, \mathbb{R})$. This fact shows that (3.12) gives the product of the central extension of U(1) by the commutative Lie group $\mathbb{R}^2$. An element of Heisenberg group $H(2, \mathbb{R})$ is written as $g_h(\vec{z}, \theta) := e^{pQ-qP}e^{\theta Z}$ by using real numbers $\vec{z} := (p, q)$ and θ.

Similarly, any central extension by the commutative Lie algebra $\mathbb{R}^{2r}$ is given by a bilinear form c, which is given as $\langle\langle \lambda(\boldsymbol{p}, \boldsymbol{q}), \lambda(\boldsymbol{p}, \boldsymbol{q})\rangle\rangle$ with a matrix λ and the symplectic inner product $\langle\langle\ ,\ \rangle\rangle$. In particular, when λ is the identity matrix, the central extension is called the Heisenberg algebra of degree r and is written as $\mathfrak{h}(2r, \mathbb{R})$. Let e_j be the element of the commutative Lie algebra $\mathbb{R}^2$ that has the j-th entry 1. The Lie algebra $\mathfrak{h}(2r, \mathbb{R})$ is generated by $Q := e_j$, $P_j := e_{r+j}$, and the generator Z of the center. The relations

$$[P_j, Q_{j'}] = \delta_{j,j'}Z,\ [P_j, Z] = [Q_j, Z] = [P_j, P_{j'}] = [Q_j, Q_{j'}] = 0$$

hold. Hence, (3.15) gives the product of the central extension of U(1) by the commutative Lie group $\mathbb{R}^{2r}$. Any element of Heisenberg group $H(2, \mathbb{R})$ is written as $g_h(\vec{z}, \theta) := \exp(\sum_{j=1}^{r} p_j Q_j - q_j P_j)e^{\theta Z}$ by using a real vector $\vec{z} := (q_1, \ldots, q_r, p_1, \ldots, p_r)^T$ and θ. Since the Lie algebra $\mathfrak{h}(2r, \mathbb{R})$ is nilpotent, it requires a different treatment from semi simple Lie algebras.

3.4 Representation of Lie Algebra

3.4.1 Representation of Real Lie Algebra

The structure of a Lie group and its representations can be easily understood via the structure of a Lie algebra and its representations. Hence, we focus on a representation of a Lie algebra.

Since this subsection and the next subsection address only representations of a real Lie algebra $\mathfrak{g}$, we abbreviate real Lie algebra to Lie algebra. Given a representation f of the Lie group G corresponding to a Lie algebra $\mathfrak{g}$, we define the element of Lie algebra $\mathfrak{gl}(\mathcal{H})$ by

$$\mathsf{f}(X) := \lim_{t \to 0} \frac{\mathsf{f}(e^{tX}) - I}{t} \tag{3.34}$$

for an element $X \in \mathfrak{g}$. Theorem 3.1 guarantees that f is a homomorphism from the Lie algebra $\mathfrak{g}$ to the Lie algebra $\mathfrak{gl}(\mathcal{H})$ on the complex Hilbert space $\mathcal{H}$. Generally, a homomorphism f from the Lie algebra $\mathfrak{g}$ to the Lie algebra $\mathfrak{gl}(\mathcal{H})$ on the complex Hilbert space $\mathcal{H}$ is called a **representation** of the Lie algebra $\mathfrak{g}$ on $\mathcal{H}$. As mentioned after Theorem 3.1, a representation of the Lie group G uniquely determines a representation of the Lie algebra $\mathfrak{g}$, however, the converse does not hold in general.

A representation f of a Lie algebra $\mathfrak{g}$ is called **faithful** when it is injective. It is called a **skew-Hermitian representation** when all elements of the image of f are skew-Hermitian matrices, i.e., the image of f is included in the Lie algebra $\mathfrak{u}(\mathcal{H})$. Let f be a unitary representation of a Lie group G. As mentioned in Sect. 3.6, the corresponding representation of the Lie algebra $\mathfrak{g}$ is a skew-Hermitian representation. Hence, the skew-Hermitian representation of a Lie algebra is directly related to the quantum theory.

The final aim is the characterization of unitary representations of real Lie groups. The procedure of this book is the following.

Characterization of skew-Hermitian representation of real Lie algebra	$\Longrightarrow$	Characterization of unitary representation of real Lie group

There exists another reason why a skew-Hermitian representation of a real Lie algebra is important for quantum theory. In quantum theory, an important physical quantity is often written as $i\mathsf{f}(X)$ by using an element X of a real Lie algebra $\mathfrak{g}$ and a skew-Hermitian representation f of the real Lie algebra $\mathfrak{g}$. This is a reason for the importance of skew-Hermitian representations of real Lie algebras.

Given a representation f of a Lie algebra $\mathfrak{g}$ on $\mathcal{H}$, a linear subspace $\mathcal{K}$ of $\mathcal{H}$ is called an **invariant subspace** when it satisfies the invariance condition given below;

Invariance: $\mathsf{f}(X)u \in \mathcal{K}, \quad \forall u \in \mathcal{K}, \forall X \in \mathfrak{g}.$

The representation f of a Lie algebra $\mathfrak{g}$ is called **irreducible** when its invariant subspace is limited to $\{0\}$ or $\mathcal{H}$. On the other hand, f is called **reducible** when there exists an invariant subspace different from $\{0\}$ and $\mathcal{H}$. Then, employing the above defined invariance, for a representation of a Lie algebra $\mathfrak{g}$, we can define the concepts of **decomposable**, **indecomposable**, **completely reducible**, **complex conjugate representation** and **self-adjoint representation** in the same way as the case of a representation of a group.

When $\mathcal{H}$ is an infinite-dimensional Hilbert space, f is defined to be a skew-Hermitian representation if $i\mathsf{f}(X)$ is a self-adjoint operator for any $X \in \mathfrak{g}$. Then, we have the Schur's lemma as follows.

Lemma 3.2 *Let f be an irreducible representation of a Lie algebra $\mathfrak{g}$ on a finite-dimensional Hilbert space $\mathcal{H}$. When the relation*

$$A\mathsf{f}(X) = \mathsf{f}(X)A, \quad \forall X \in \mathfrak{g} \tag{3.35}$$

holds for a matrix A on $\mathcal{H}$, A is a constant matrix.

Proof Let c be an eigenvalue of A. This lemma can be shown by considering the Kernel of $A - cI$ in the same way as Schur's lemma for a representation of a group. ■

In fact, Lemma 3.2 can be extended to the case of an irreducible skew-Hermitian representation on an infinite-dimensional space $\mathcal{H}$. The extension can be shown by the following steps. Firstly, we show the equivalence between $A \exp(\mathfrak{f}(X)) = \exp(\mathfrak{f}(X))A$ and (3.35). Secondly, we focus on the spectral decompositions of A and $AA^\dagger$ in the same way as Lemma 2.4.

Given two representations $\mathfrak{f}_1$ and $\mathfrak{f}_2$ of a Lie algebra $\mathfrak{g}$ on two different spaces $\mathcal{H}_1$ and $\mathcal{H}_2$, we can define the **direct sum representation** $\mathfrak{f}_1 \oplus \mathfrak{f}_2$ of the Lie algebra $\mathfrak{g}$ on the direct sum space $\mathcal{H}_1 \oplus \mathcal{H}_2$ and the **tensor product representation** $\mathfrak{f}_1 \otimes \mathfrak{f}_2$ of the Lie algebra $\mathfrak{g}$ on the tensor product space $\mathcal{H}_1 \otimes \mathcal{H}_2$ as follows.

$$\mathfrak{f}_1 \oplus \mathfrak{f}_2(X) := \mathfrak{f}_1(X) \oplus \mathfrak{f}_2(X), \quad \mathfrak{f}_1 \otimes \mathfrak{f}_2(X) := \mathfrak{f}_1(X) \otimes I + I \otimes \mathfrak{f}_2(X).$$

On the other hand, given two representations $\mathfrak{f}_1$ and $\mathfrak{f}_2$ of the Lie algebras $\mathfrak{g}_1$ and $\mathfrak{g}_2$ on $\mathcal{H}_1$ and $\mathcal{H}_2$, respectively, we can define the representation $\mathfrak{f}_1 \bar{\otimes} \mathfrak{f}_2$ of the Lie algebra $\mathfrak{g}_1 \oplus \mathfrak{g}_2$ on the tensor product space $\mathcal{H}_1 \otimes \mathcal{H}_2$ as follows.

$$(\mathfrak{f}_1 \bar{\otimes} \mathfrak{f}_2)(X_1 \oplus X_2) := \mathfrak{f}_1(X_1) \otimes I + I \otimes \mathfrak{f}_2(X_2), \quad \forall X_1 \oplus X_2 \in \mathfrak{g}_1 \oplus \mathfrak{g}_2 .$$

Then, we can show the following lemma similar to Lemma 2.8.

Lemma 3.3 *Given two irreducible representations* $\mathfrak{f}_1$ *and* $\mathfrak{f}_2$ *of the Lie algebras* $\mathfrak{g}_1$ *and* $\mathfrak{g}_2$, *the representation* $\mathfrak{f}_1 \bar{\otimes} \mathfrak{f}_2$ *is also irreducible. Conversely, any irreducible representation of the Lie algebra* $\mathfrak{g}_1 \oplus \mathfrak{g}_2$ *can be written as* $\mathfrak{f}_1 \bar{\otimes} \mathfrak{f}_2$ *by using two irreducible representations* $\mathfrak{f}_1$ *and* $\mathfrak{f}_2$ *of* $\mathfrak{g}_1$ *and* $\mathfrak{g}_2$.

Therefore, when a Lie algebra $\mathfrak{g}$ is decomposed as a direct sum $\mathfrak{g}_1 \oplus \cdots \oplus \mathfrak{g}_l$, an irreducible representation of the Lie algebra $\mathfrak{g}$ is given as $\mathfrak{f}_1 \bar{\otimes} \cdots \bar{\otimes} \mathfrak{f}_l$ by using irreducible representations $\mathfrak{f}_1, \ldots, \mathfrak{f}_l$ of $\mathfrak{g}_1, \ldots, \mathfrak{g}_l$.

3.4.2 Real Representation

On the other hand, a homomorphism from a Lie algebra $\mathfrak{g}$ to the Lie algebra $\mathfrak{gl}(V)$ on a real linear space V is called a **real representation** of Lie algebra $\mathfrak{g}$ on V. A representation with a complex Hilbert space is called a complex representation when it is needed to distinguish it from a real representation. A real representation of a Lie algebra $\mathfrak{g}$ corresponds to a real representation of a real Lie group G. A real representation $\mathfrak{f}$ of a Lie algebra $\mathfrak{g}$ is called an **alternative representation** when the image of $\mathfrak{f}$ is included in the Lie algebra $\mathfrak{o}(V)$, which is the set of alternative matrices on V. It corresponds to an orthogonal representation of a Lie group G. For a real representation, we can define the concepts of **irreducible**, **faithful**, **decomposable**,

indecomposable, and **completely reducible** in a way similar to those for a complex representation. For a finite-dimensional complex representation of a Lie algebra, we define **complex conjugate representation** in the same way as (2.40). Also, we define a representation **underlying real form** and a **complexifiable** representation of a Lie algebra in the same way as the case of a group. Schur's lemma does not hold for a real representation of a Lie algebra $\mathfrak{g}$ due to the same reason as the case of a real representation of a group. In order to employ Schur's lemma, the irreducibility of a real representation f on a real linear space V means the irreducibility of the complex representation on the complex representation space $V \otimes \mathbb{C}$.

3.4.3 *Representation of Complex Lie Algebra*

Next, we focus on a complex Lie algebra $\mathfrak{g}$, a homomorphism for complex Lie algebras from $\mathfrak{g}$ to the Lie algebra $\mathfrak{gl}(\mathcal{H})$ on a complex linear space $\mathcal{H}$ is called a **representation for complex Lie algebras** of $\mathfrak{g}$ on $\mathcal{H}$. For simplicity, we simplify it to a $\mathbb{C}$-representation of $\mathfrak{g}$ on $\mathcal{H}$. On the other hand, a representation of a complex Lie algebra $\mathfrak{g}$ on $\mathcal{H}$ means a homomorphism for real Lie algebras from $\mathfrak{g}$ to $\mathfrak{gl}(\mathcal{H})$. So, it is not necessarily a $\mathbb{C}$-representation. For a $\mathbb{C}$-representation, we define the concept of **irreducible**, **faithful**, **decomposable**, **indecomposable**, **completely reducible**. Schur's lemma holds in the same way as the case of a real Lie algebra.

3.4.4 *Adjoint Representation*

As an example of a representation, given a real (complex) Lie algebra $\mathfrak{g}$, we define the real representation ad of a real linear space $\mathfrak{g}$ (the representation ad of a complex Lie algebra $\mathfrak{g}$ for complex Lie algebras) on the real (complex) linear space $\mathfrak{g}$ by $\mathsf{ad}(X)(Y) := [X, Y]$, which is called the **adjoint representation** ad of $\mathfrak{g}$. Hence, an ideal can be defined as an invariant subspace with respect to the adjoint representation. Since

$$\begin{aligned}\mathsf{ad}(X)\mathsf{ad}(Y)(Z) = [X, [Y, Z]] = [Y, [X, Z]] + [[X, Y], Z] \\ =(\mathsf{ad}(Y)\mathsf{ad}(X) + \mathsf{ad}([X, Y]))(Z),\end{aligned}$$

we have

$$\mathsf{ad}([X, Y]) = [\mathsf{ad}(X), \mathsf{ad}(Y)]. \tag{3.36}$$

In particular, if and only if a real (complex) Lie algebra $\mathfrak{g}$ is simple, the adjoint representation is irreducible. For example, when $\mathfrak{g}$ is a subspace of $\mathfrak{u}(\mathcal{H})$ and the inner product of $\mathfrak{g}$ is defined as $\langle X, Y\rangle := \operatorname{Tr} X^{\dagger} Y$, the adjoint representation is skew-

Hermitian. Since elements of Lie algebras $\mathfrak{gl}(r)$, $\mathfrak{sl}(r)$, $\mathfrak{u}(r)$, and $\mathfrak{su}(r)$ are written as matrices on $\mathbb{C}^r$, we can define natural representations of these Lie algebras on $\mathbb{C}^r$. These representations on $\mathbb{C}^r$ are called the **fundamental representation** and are irreducible. Then, we can naturally define their representations on the n-fold symmetric tensor product space of $\mathbb{C}^r$, which are irreducible representations and are called n-fold **symmetric tensor representations**. Similarly, their representations on the n-fold alternative tensor product space of $\mathbb{C}^r$ are irreducible representations and are called the n-fold **alternative tensor representations**.

Next, we define the **adjoint representation** ad of the Lie group G corresponding to the Lie algebra $\mathfrak{g}$ on the vector space $\mathfrak{g}$ by $\mathsf{ad}(g) := \exp(\mathsf{ad}(tX))$ for an arbitrary element $g = \exp(tX) \in G$. Although this definition seems to depend on the choice of t and X, the representation is independent of this choice, as shown in Sect. 3.6. When $\mathfrak{g}$ is given as a linear subspace of $\mathfrak{gl}(\mathcal{H})$, the relation $\mathsf{ad}(g)Y = gYg^{-1}$ holds for $Y \in \mathfrak{g}$.

3.4.5 *Projective Representation*

We can consider a projective representation for a real Lie algebra. A linear map f is called **projective representation** of the Lie algebra $\mathfrak{g}$ when the relation

$$[\mathsf{f}(X), \mathsf{f}(Y)] = \mathsf{f}([X, Y]) + c(X, Y)iI. \tag{3.37}$$

holds, where $c(X, Y)$ is a real-valued bilinear map satisfying (3.31). In particular, the linear map f is called a **projective skew-Hermitian representation** of the Lie algebra $\mathfrak{g}$ when all elements of the image of f are skew-Hermitian matrices. (skew-self-adjoint operators in the infinite-dimensional case)

Then, we have the following lemma.

Lemma 3.4 *The following conditions are equivalent to each other for a projective representation* f *of* $\mathfrak{g}$ *with a real-valued bilinear map* $c(X, Y)$.

(1) *The real number* $c(X, Y)$ *depends only on* $[X, Y] \in \mathfrak{g}$.
(2) *There exists a representation* f' *such that*

$$\mathsf{f}'(X) = \mathsf{f}(X) + \tilde{c}(X)iI, \tag{3.38}$$

where $\tilde{c}$ *is a real-valued linear function on* $\mathfrak{g}$.

Proof The direction **(2)** $\Rightarrow$ **(1)** can be easily shown by a simple calculation. The direction **(1)** $\Rightarrow$ **(2)** can be shown as follows. Condition **(1)** guarantees the existence of a real-valued linear function $\tilde{c}$ on $\mathfrak{g}$ such that $c(X, Y) = \tilde{c}([X, Y])$ for $X, Y \in \mathfrak{g}$. When we define the projective representation f' of $\mathfrak{g}$ by (3.38), f' is a representation of $\mathfrak{g}$. ∎

Therefore, when the condition of Lemma 3.4 holds, the projective representation f of $\mathfrak{g}$ can be replaced by the representation f' of $\mathfrak{g}$. Hence, to avoid the above situation, we call a projective representation f a **proper projective representation** when it does not satisfy the above condition.

Lemma 3.5 *A projective representation of a Lie algebra $\mathfrak{g}$ on a finite-dimensional space is not a proper projective representation.*

Proof We consider a projective representation f of the Lie algebra $\mathfrak{g}$ on the finite-dimensional space $\mathcal{H}$. Then, the map

$$\mathsf{f}'(X) := \mathsf{f}(X) - \frac{1}{\dim \mathcal{H}}(\operatorname{Tr} \mathsf{f}(X))I \tag{3.39}$$

is a projective representation of $\mathfrak{g}$. We give a real-valued bilinear function c according to (3.37) for the projective representation f'. The relation

$$[\mathsf{f}'(X), \mathsf{f}'(Y)] = \mathsf{f}'([X, Y]) + c(X, Y)iI \tag{3.40}$$

holds. Since the traces of the left hand side and $\mathsf{f}'([X, Y])$ are 0, c is also 0. So, f' is a representation. That is, f is not a proper projective representation. ■

Since a proper projective skew-Hermitian representation is limited to a projective representation on an infinite-dimensional space, it is sufficient to treat skew-Hermitian representations for a finite-dimensional quantum system.

3.4.6 Semi Direct Product Lie Algebra and Representation

Given an automorphic action T of a Lie group G_2 on a Lie group G_1, the semi direct product $G_1 \rtimes G_2$ can be defined according to (2.2). When $\mathfrak{g}_1$ and $\mathfrak{g}_2$ are the Lie algebras corresponding to G_1 and G_2, respectively, we define the action T of G_2 on $\mathfrak{g}_1$ as

$$T(g)(X_1) := \lim_{t\to 0} \frac{T(g)(e^{tX_1}) - T(g)(I)}{t}, \quad \forall g \in G_2, \forall X_1 \in \mathfrak{g}_1 .$$

Since $\mathfrak{g}_1$ is a linear space, T is a real representation of G_2 on $\mathfrak{g}_1$. Then, the map

$$T(X_2)(X_1) := \lim_{t\to 0} \frac{T(e^{tX_2})(X_1) - X_1}{t}, \quad \forall X_2 \in \mathfrak{g}_2, \forall X_1 \in \mathfrak{g}_1 .$$

is a real representation of Lie algebra $\mathfrak{g}_2$ on $\mathfrak{g}_1$.

Given a real representation T of the Lie algebra $\mathfrak{g}_2$ on $\mathfrak{g}_1$, we define the bilinear map for the direct product set $\mathfrak{g}_1 \cdot \mathfrak{g}_2$ of $\mathfrak{g}_1$ and $\mathfrak{g}_2$.

$$[(X_1, X_2), (Y_1, Y_2)] := (T(X_2)Y_1 - T(Y_2)X_1 + [X_1, Y_1], [X_2, Y_2]). \tag{3.41}$$

The above bilinear map satisfies the conditions for a Lie algebra, and the Lie algebra is called the **semi direct product Lie algebra** $\mathfrak{g}_1 \rtimes \mathfrak{g}_2$ of Lie algebras $\mathfrak{g}_1$ and $\mathfrak{g}_2$. In particular, when T is 0, the semi direct product Lie algebra $\mathfrak{g}_1 \rtimes \mathfrak{g}_2$ is the direct sum Lie algebra $\mathfrak{g}_1 \oplus \mathfrak{g}_2$. For example, the Lie algebra $\mathfrak{aff}(n, \mathbb{R})$ is given as the semi direct product Lie algebra $\mathbb{R}^n \rtimes \mathfrak{gl}(n, \mathbb{R})$ of the Lie algebras $\mathbb{R}^n$ and $\mathfrak{gl}(n, \mathbb{R})$.

When representations f_1 and f_2 of Lie algebras $\mathfrak{g}_1$ and $\mathfrak{g}_2$ on $\mathcal{H}$ satisfy

$$[\mathsf{f}_2(X_2), \mathsf{f}_1(X_1)] = \mathsf{f}_1(T(X_2)X_1), \tag{3.42}$$

we can define the semi direct product representation $\mathsf{f}_1 \rtimes \mathsf{f}_2$ of the semi direct product Lie algebra $\mathfrak{g}_1 \rtimes \mathfrak{g}_2$ on $\mathcal{H}$ as

$$\mathsf{f}_1 \rtimes \mathsf{f}_2((X_1, X_2)) := \mathsf{f}_1(X_1) + \mathsf{f}_2(X_2). \tag{3.43}$$

Finally, we remark that the semi direct product representation of the semi direct product Lie algebra can be derived from the semi direct product representation of the semi direct product Lie group.

Exercise 3.6 Given a real representation T of a Lie algebra $\mathfrak{g}_2$ on another Lie algebra $\mathfrak{g}_1$, we define the bilinear map on the direct product set $\mathfrak{g}_1 \cdot \mathfrak{g}_2$ by (3.41). Show that the bilinear map satisfies the conditions for a Lie algebra.

Exercise 3.7 Assume that the condition (3.42) holds. Show that the map $\mathsf{f}_1 \rtimes \mathsf{f}_2$ defined in (3.43) is a representation of the semi direct product Lie algebra $\mathfrak{g}_1 \rtimes \mathfrak{g}_2$ on $\mathcal{H}$.

3.5 Killing Form and Compactness

3.5.1 Killing Form

We define a bilinear form on a real (complex) Lie algebra $\mathfrak{g}$ that is invariant for Lie algebra. That is, we define the **Killing form** $(X, Y)_{\mathfrak{g}} := \operatorname{Tr} \mathsf{ad}(X)\mathsf{ad}(Y)$ for arbitrary $X, Y \in \mathfrak{g}$. Then, we have

$$\begin{aligned}\operatorname{Tr} \mathsf{ad}([X, Y])\mathsf{ad}(Z) &= \operatorname{Tr}[\mathsf{ad}(X), \mathsf{ad}(Y)]\mathsf{ad}(Z) \\ &= \operatorname{Tr} \mathsf{ad}(X)[\mathsf{ad}(Y), \mathsf{ad}(Z)] = \operatorname{Tr} \mathsf{ad}(X)\mathsf{ad}([Y, Z]).\end{aligned}$$

The Killing form satisfies the following associative law.

Associative law: $([X, Y], Z)_{\mathfrak{g}} = (X, [Y, Z])_{\mathfrak{g}}$.

This fact implies the equation

$$(-\mathsf{ad}(Y)X, Z)_{\mathfrak{g}} = (X, \mathsf{ad}(Y)Z)_{\mathfrak{g}}. \tag{3.44}$$

Applying this property to the Lie group, we find that the action $\mathsf{ad}(g)$ of the Lie group preserves the Killing form. That is, the equation $(\mathsf{ad}(g)X, \mathsf{ad}(g)Z)_{\mathfrak{g}} = (X, Z)_{\mathfrak{g}}$ holds.

The following lemmas hold with respect to the Killing form.

Lemma 3.6 *The Killing form is non-degenerate for a real (complex) Lie algebra $\mathfrak{g}$ if and only if $\mathfrak{g}$ is semi simple.*

Let $A^{\diamond}$ be the dual map of a linear map A on a semi simple Lie algebra $\mathfrak{g}$ with respect to Killing form. Then, (3.44) implies that

$$\mathsf{ad}(Y)^{\diamond} = -\mathsf{ad}(Y). \tag{3.45}$$

That is, $\mathsf{ad}(Y)$ is an alternative map with respect to the Killing form.

Lemma 3.7 *When a real (complex) Lie algebra $\mathfrak{g}$ is simple, any inner product satisfying the associative law is a constant times of the Killing form.*

Lemma 3.8 *When a real (complex) Lie algebra $\mathfrak{g}$ is nilpotent, its Killing form satisfies $(X, Y)_{\mathfrak{g}} = 0$. In this case, any inner product satisfying the associative law is degenerate. That is, there exists a non-zero element $X \in \mathfrak{g}$ such that $(X, X)_{\mathfrak{g}} = 0$.*

When a complex Lie algebra is semi simple, its Killing form is non-degenerate. When a real Lie algebra is semi simple, we have another information for its Killing form in addition to the non-degeneracy. Even though an inner product of a real linear space is non-degenerate, there are many possibilities for the signs of the inner product. That is, the matrix corresponding to the inner product might have positive and negative eigenvalues. When all of eigenvalues are positive, the inner product is called **positive definite**. When all of eigenvalues are negative, it is called **negative definite**. When the matrix has both positive and negative eigenvalues, it is called **non-definite**.

3.5.2 Compactness of Real Lie Algebra $\mathfrak{g}$

In this subsection, we address the compactness of a real Lie algebra $\mathfrak{g}$, and abbreviate real Lie algebra and real Lie group to Lie algebra and Lie group, respectively. Firstly, we discuss the compactness of an element X of the Lie algebra $\mathfrak{g}$. When $\{\exp(tX)\}_t$ is a compact subgroup of the corresponding Lie group G, the one-dimensional Lie group $\{\mathsf{ad}(\exp(tX))\}_t = \{\exp(\mathsf{ad}(tX))\}_t$ is compact. There is no one-to-one correspondence between a Lie algebra $\mathfrak{g}$ and a Lie group G. When a group $\{\exp(\mathsf{ad}(tX))\}_t$ is included in a commutative compact subgroup of $\mathsf{ad}(G)$, X is called **compact**.

Lemma 3.9 *The following conditions are equivalent to each other for an element X of a Lie algebra $\mathfrak{g}$.*

(1) *X is compact.*
(2) *$\mathsf{ad}(X)$ is diagonalizable, and its eigenvalues are pure imaginary numbers.*
(3) *By choosing a suitable basis of $\mathfrak{g}$, $\mathsf{ad}(X)$ can be diagonalized as*

$$\mathsf{ad}(X) \cong \begin{pmatrix} 0 & c_1 & & & & & & \\ -c_1 & 0 & & & & & & \\ & & \ddots & & & & & \\ & & & 0 & c_l & & & \\ & & & -c_l & 0 & & & \\ & & & & & 0 & & \\ & & & & & & \ddots & \\ & & & & & & & 0 \end{pmatrix}. \tag{3.46}$$

Proof **(2)** $\Rightarrow$ **(1)**: Trivial.
(2) $\Leftrightarrow$ **(3)**: Trivial.
(1) $\Rightarrow$ **(2)**: Let $\mathsf{ad}(X_1), \ldots, \mathsf{ad}(X_l)$ be the generator of the Lie algebra corresponding to the commutative compact subgroup including $\{\exp(\mathsf{ad}(tX))\}_t$. Due to the compactness of the subgroup, there exist real numbers $t_1, \ldots, t_l$ such that $\exp(t_j \mathsf{ad}(X_j))$ is the identity matrix. Thus, $\mathsf{ad}(X_1), \ldots, \mathsf{ad}(X_l)$ are diagonalizable and their eigenvalues are pure imaginary numbers. Since $\mathsf{ad}(X)$ is written as a linear sum of $\mathsf{ad}(X_1), \ldots, \mathsf{ad}(X_l)$, which are commutative with each other, Condition **(2)** holds. ∎

Hence, when $X \in \mathfrak{g}$ is compact, the relation $(X, X)_{\mathfrak{g}} < 0$ or $\mathsf{ad}(X) = 0$ holds. Then, we have the following relation between compactness and the Killing form for a semi simple Lie algebra $\mathfrak{g}$.

Theorem 3.2 *The following conditions are equivalent to each other for a semi simple Lie algebra $\mathfrak{g}$.*

(1) *The Killing form is negative definite.*
(2) *The Killing form is not non-definite.*
(3) *Any element of $\mathfrak{g}$ is compact.*

Proof **(1)** $\Rightarrow$ **(2)**: Trivial.
(2) $\Rightarrow$ **(3)**: Since any element $X \in \mathfrak{g}$ satisfies (3.45) under the non-degenerate inner product that is not non-definite, Condition **(3)** holds.
(3) $\Rightarrow$ **(1)**: Since any eigenvalue of $\mathsf{ad}(X)$ is a pure imaginary number and its Killing form is non-degenerate, $\mathrm{Tr}\, \mathsf{ad}(X)$ is a negative number. So, Condition **(1)** holds. ∎

Then, we obtain the following characterization for the compactness of an element of the Lie algebra $\mathfrak{g}$.

Theorem 3.3 *The following three conditions are equivalent to each other for a Lie algebra $\mathfrak{g}$.*

(1) *Any element of $\mathfrak{g}$ is compact.*
(2) *There exists a compact Lie group G corresponding to the Lie algebra $\mathfrak{g}$.*
(3) *There exists a faithful skew-Hermitian representation f of $\mathfrak{g}$ on a finite-dimensional Hilbert space $\mathcal{H}$.*

Proof **(1)** $\Rightarrow$ **(2)**: Since the Lie group $\{\exp(\mathsf{ad}(X))\}_{\mathfrak{g}}$ is compact, Condition **(2)** holds.
(2) $\Rightarrow$ **(1)**: This relation follows from the definition of compactness of X.
(1) $\Rightarrow$ **(3)**: The adjoint representation satisfies Condition **(3)**.
(3) $\Rightarrow$ **(1)**: Condition **(1)** guarantees that $\mathsf{f}(\mathfrak{g})$ is a Lie subalgebra of $\mathfrak{u}(\mathcal{H})$. For any element $X \in \mathfrak{g}$, the matrix $\mathsf{ad}(\mathsf{f}(X))$ on $\mathfrak{u}(\mathcal{H})$ is a diagonalizable and has pure imaginary eigenvalues. The subspace $\mathsf{f}(\mathfrak{g})$ is an invariant subspace with respect to the action $\mathsf{ad}(\mathsf{f}(X)) = \mathsf{f}(\mathsf{ad}(X))$. When the action $\mathsf{ad}(\mathsf{f}(X)) = \mathsf{f}(\mathsf{ad}(X))$ is limited to $\mathsf{f}(\mathfrak{g})$, the matrix $\mathsf{f}(\mathsf{ad}(X))$ on $\mathsf{f}(\mathfrak{g})$ is diagonalizable and all of its eigenvalues are pure imaginary. This fact guarantees that $\mathsf{ad}(X)$ is diagonalizable and all of its eigenvalues are pure imaginary because f is faithful. Thus, Condition **(1)** holds. ∎

Now, a Lie algebra $\mathfrak{g}$ is called **compact** when the condition of Theorem 3.3 holds. All of Lie algebras of Table 3.5 are compact under this definition, and all of Lie algebras of Table 3.6 are non-compact. Then, the Killing form of the quotient Lie algebra $\mathfrak{g} / \mathfrak{c}(\mathfrak{g})$ with respect to the center $\mathfrak{c}(\mathfrak{g})$ of $\mathfrak{g}$ is negative definite. That is, $\mathfrak{g} / \mathfrak{c}(\mathfrak{g})$ is semi simple. Thus, $\mathfrak{g} / \mathfrak{c}(\mathfrak{g}) = [\mathfrak{g} / \mathfrak{c}(\mathfrak{g}), \mathfrak{g} / \mathfrak{c}(\mathfrak{g})]$, which implies that

$$\mathfrak{g} = [\mathfrak{g}, \mathfrak{g}] + \mathfrak{c}(\mathfrak{g}). \tag{3.47}$$

In other words, the compact Lie algebra $\mathfrak{g}$ is decomposed to the direct sum of the commutative part $\mathfrak{c}(\mathfrak{g})$ and the semi simple part $[\mathfrak{g}, \mathfrak{g}]$.

Due to Theorem 3.3, there exists a faithful skew-Hermitian representation of the Lie algebra $\mathfrak{g}$ on a finite-dimensional Hilbert space $\mathcal{H}$ if and only if $\mathfrak{g}$ is compact. Hence, to realize a faithful skew-Hermitian representation of a non-compact Lie algebra $\mathfrak{g}$, we need to consider a representation on an infinite-dimensional Hilbert space $\mathcal{H}$. However, we need some additional cares to discuss a representation on an infinite-dimensional Hilbert space. When we focus on a unitary representation of a Lie group, the appearing operator on $\mathcal{H}$ is bounded so that no serious problem occurs. However, when we focus on a skew-Hermitian representation of a Lie algebra, the appearing operator on $\mathcal{H}$ is unbounded so that we need to care about the domain of the operator. To consider this problem, we need to employ self-adjoint operators [88] because this concept covers the difficulty of the domain. That is, a skew-Hermitian representation f of a Lie algebra $\mathfrak{g}$ on an infinite-dimensional Hilbert space $\mathcal{H}$ is defined as a homomorphism for Lie algebras satisfying that $i\mathsf{f}(X)$ is a self-adjoint operator for any element $X \in \mathfrak{g}$. In particular, when the Killing form of a semi simple Lie algebra $\mathfrak{g}$ is negative definite, the representation space of its irreducible skew-Hermitian representation is limited to finite-dimensional. When the Killing form of a semi simple Lie algebra $\mathfrak{g}$ is non-definite, the representation space of its irreducible skew-Hermitian representation is limited to infinite-dimensional.

In general, even though the physical system is given as a representation space of a representation f of a non-compact group, it seems that the Hamiltonian of our interest is usually given as the operator $i\mathsf{f}(X)$ with a compact element X of the Lie algebra $\mathfrak{g}$. In this book, even for a non-compact Lie algebra, we treat its representation based on its compact elements due to the above physical reason.

Exercise 3.8 Show that $(\mathsf{ad}(g)X, \mathsf{ad}(g)Z)_{\mathfrak{g}} = (X, Z)_{\mathfrak{g}}$.

3.5.3 Casimir Operator

Next, we define an invariant operator called the **Casimir operator**. We choose a basis $\{e_1, \ldots, e_d\}$ of a semi simple real (complex) Lie algebra $\mathfrak{g}$ and its dual basis $\{e'_1, \ldots, e'_d\}$ with respect to its Killing form. That is, we choose the dual basis $\{e'_1, \ldots, e'_d\}$ such that $(e_k, e'_j)_{\mathfrak{g}} = \delta_{k,j}$. Then, identifying the dual space $\mathfrak{g}^*$ of $\mathfrak{g}$ with $\mathfrak{g}$ via the Killing form, we define the Casimir operator $C := \sum_{j=1}^{d} e_j \otimes e'_j$ as an element of $\mathfrak{g} \otimes \mathfrak{g}^*$. Then, C is the identity operator on $\mathfrak{g}$. When A is a linear map preserving the Killing form, we have $C = \sum_{j=1}^{d} (Ae_j) \otimes (Ae'_j)$. Since the action of the Lie group G on $\mathfrak{g}$ preserves the Killing form, C is invariant for the action of the Lie group G. For a representation f on the Hilbert space $\mathcal{H}$, we define the map $\mathsf{f}_* : \sum_t X_t \otimes X'_t \mapsto \sum_t \mathsf{f}(X_t)\mathsf{f}(X'_t)$ from $\mathfrak{g} \otimes \mathfrak{g}^*$ to the set of linear maps on $\mathcal{H}$. Then, defining the Casimir operator C_{f} on the Hilbert space $\mathcal{H}$ as the image of C with respect to the map f_*, we have the relation $C_{\mathsf{f}} = \sum_{j=1}^{d} \mathsf{f}(e_j)\mathsf{f}(e'_j)$. C_{f} is commutative with the action of the Lie group $G(\mathfrak{g})$ corresponding to the Lie algebra $\mathfrak{g}$. That is, the relations

$$
\begin{aligned}
\mathsf{f}(g)C_{\mathsf{f}}\mathsf{f}(g)^{-1} &= \sum_{j=1}^{d} \mathsf{f}(g)\mathsf{f}(e_j)\mathsf{f}(g)^{-1}\mathsf{f}(g)\mathsf{f}(e'_j)\mathsf{f}(g)^{-1} \\
&= \sum_{j=1}^{d} \mathsf{f}(\mathsf{ad}(g)e_j)\mathsf{f}(\mathsf{ad}(g)e'_j) \\
&= \mathsf{f}\left(\sum_{j=1}^{d} (\mathsf{ad}(g)e_j) \otimes (\mathsf{ad}(g)e'_j)\right) = \mathsf{f}\left(\sum_{j=1}^{d} e_j \otimes e'_j\right) = C_{\mathsf{f}}
\end{aligned}
$$

holds for an element $g \in G(\mathfrak{g})$. Considering the derivative of $g = \exp(tX)$ with respect to t, we have the following relation for any element $X \in \mathfrak{g}$

$$\mathfrak{f}(X)C_{\mathfrak{f}} = C_{\mathfrak{f}}\mathfrak{f}(X). \tag{3.48}$$

As mentioned in the previous section, many physical quantities are given as $i\mathfrak{f}(X)$ by using a representation $\mathfrak{f}$ of a Lie algebra. Some of other physical quantities are written as the Casimir operator $C_{\mathfrak{f}}$. When the Casimir operator $C_{\mathfrak{f}}$ takes different values dependently of irreducible representations, we can realize the measurement corresponding to the irreducible decomposition by measuring the physical quantity $C_{\mathfrak{f}}$.

3.6 Relation Between Lie Group and Lie Algebra II

3.6.1 *Universal Covering Group*

This section discusses the relation between a representation of a Lie group and that of a Lie algebra. From a representation (unitary representation) $\mathfrak{f}$ of the real (complex) Lie group G on $\mathcal{H}$, we can define a representation (skew-Hermitian representation) $\mathfrak{f}(X) := \lim_{t\to 0} \frac{\mathfrak{f}(\exp(tX))-I}{t}$ of the corresponding real (complex) Lie algebra $\mathfrak{g}$ on $\mathcal{H}$. However, the converse is not true in general. That is, we cannot necessarily define a representation (unitary representation) of real (complex) Lie group G from a representation (skew-Hermitian representation) of a real (complex) Lie algebra $\mathfrak{g}$. The aim of this section is clarifying this relation.

For simplicity, we assume that the group is connected. To address this problem, we consider a differentiable curve $c(t)$ (a path) in G. Two paths $c(t)$ and $\tilde{c}(t)$ with the same initial and final points are called equivalent to each other and their relation is written as $g \sim \tilde{g}$ when they are continuously converted to each other under the constraint that the initial and final points are fixed.

When the final point of $c(t)$ is the initial point of $\tilde{c}(t)$, we define the product between $c(t)$ and $\tilde{c}(t)$ as

$$(c \cdot \tilde{c})(t) := \begin{cases} \tilde{c}(2t) & 0 \le t \le \frac{1}{2} \\ c(2t-1)) & \frac{1}{2} \le t \le 1. \end{cases} \tag{3.49}$$

For a path $c(t)$, we also define the inverse path $c^{-1}(t) := c(1-t)$. Then, the set of paths with the initial and final points g_0 forms a group whose identity element is a path $g(t) = g_0, \forall t \in [0, 1]$. Especially, the set of paths that are equivalent to the identity element forms a normal subgroup. Hence, we can define the quotient group with respect to the normal subgroup. The quotient group has the product $[g] \cdot [g'] := [g \cdot g']$, is called the **fundamental group** of G, and is written as $\pi_1(G)$. When G is connected, the fundamental group does not depend on the choice of g_0. Since G is a group, we define another product for any two paths $c(t)$ and $\tilde{c}(t)$ as follows.

$$(c\tilde{c})(t) := c(t)\tilde{c}(t). \tag{3.50}$$

These two products have the following relation.

$$\begin{aligned}(c \cdot \tilde{c})(c' \cdot \tilde{c}')(t) = (c \cdot \tilde{c})(t)(c' \cdot \tilde{c}')(t) &= \begin{cases} c(2t)c'(2t) & 0 \le t \le \frac{1}{2} \\ \tilde{c}(2t-1)\tilde{c}'(2t-1) & \frac{1}{2} \le t \le 1 \end{cases} \\ &= (cc') \cdot (\tilde{c}\tilde{c}')(t).\end{aligned}$$

Hence, we have $[(c \cdot \tilde{c})(c' \cdot \tilde{c}')] = [(cc') \cdot (\tilde{c}\tilde{c}')]$. Thus, denoting the path that constantly maps the element e by e, we have

$$\begin{aligned}[cc'] &= [(c \cdot e)(e \cdot c')] = [(ce) \cdot (ec')] = [c \cdot c'] \\ [cc'] &= [(e \cdot c)(c' \cdot e)] = [(ec') \cdot (ce)] = [c' \cdot c].\end{aligned}$$

Since $[c \cdot c'] = [c' \cdot c]$, we have the following lemma.

Lemma 3.10 *The fundamental group $\pi_1(G)$ of a real (complex) Lie group G is a commutative group.*

When the fundamental group consists only of the identity element, G is called **simply connected**. Then, when the Lie group G is simply connected, the representation f of the Lie group G can be uniquely defined from a representation f of the corresponding Lie algebra, as shown in the following way.

Since the neighborhood of the identity element e of the Lie group G can be identified with the corresponding Lie algebra $\mathfrak{g}$, the representation of the element $e^{t_1 X_1}$ can be given as $\mathsf{f}(e^{t_1 X_1}) = e^{t_1 \mathsf{f}(X_1)}$. In general, given a path $c(t)$ in G, we can define the representation of the Lie group G along the path. That is, we can define $\mathsf{f}(c)(t)$ when the initial point of c is e. Hence, we can define $\mathsf{f}(g)$ for an element $g \in G$ as follows. First, we choose a path $c(t)$ such that the initial point is e and the final point is g. Then, we define $\mathsf{f}(g)$ to be $\mathsf{f}(c)(1)$.

Since this definition depends on the choice of the path $c(t)$, we need to show that it does not depend on the choice. Now, we consider another path $\tilde{c}(t)$ such that the initial point is e and the final point is g. Then, the initial and final points of the path $(c \cdot \tilde{c}^{-1})(t)$ are g. Then, the initial and final points of the path $(\mathsf{f}(c) \cdot \mathsf{f}(\tilde{c})^{-1})(t)$ in $\mathrm{GL}(\mathcal{H})$ are $\mathsf{f}(\tilde{c})(1)$ and $\mathsf{f}(c)(1)$. Since G is simply connected, this path can be continuously converted to the path that always maps to g. Since the image of f is also simply connected, the path $(\mathsf{f}(c) \cdot \mathsf{f}(\tilde{c})^{-1})(t)$ in $\mathrm{GL}(\mathcal{H})$ can be continuously converted to a path that maps to a fixed point in $\mathrm{GL}(\mathcal{H})$, where the initial and final points are fixed during this continuous conversion. Hence, $\mathsf{f}(\tilde{c})(1)$ is $\mathsf{f}(c)(1)$. Next, we show that the map f is a representation of the Lie group G. For two elements $g_1, g_2 \in G$, we choose three paths $c_1(t)$, $c_2(t)$, and $c_3(t)$ such that their initial points are e and their final points are g_1, g_2, and $g_1 g_2$, respectively. Then, it is sufficient to show that $\mathsf{f}(c_1)(1)\mathsf{f}(c_2)(1) = c_3(1)$. Since $\mathsf{f}(c_1)(t)\mathsf{f}(c_2)(t)$ is the image along the path $c_1 c_2(t)$, it is $\mathsf{f}(c_1 c_2)(t)$. Since the path $c_1 c_2(t)$ has the initial point e and the final point $g_1 g_2$, we can show the relation $\mathsf{f}(c_1 c_2)(1) = \mathsf{f}(c_3)(1)$ in the same way as the above discus-

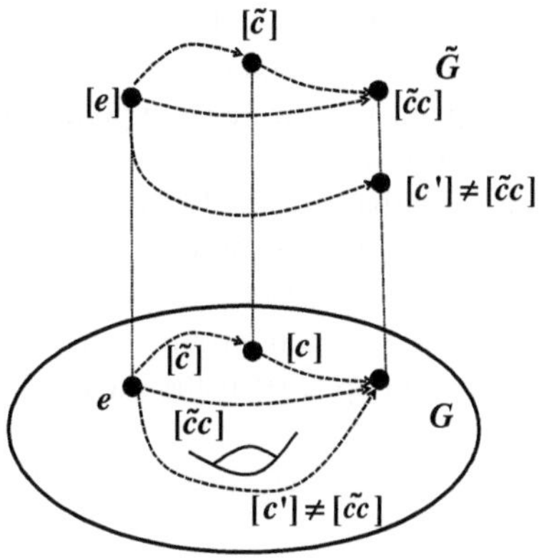

Fig. 3.3 Universal covering group

sion. Therefore, we find that the above defined map f is a representation of the Lie group G.

On the other hand, when the connected Lie group G is not simply connected, there is a possibility that no representation of the connected Lie group G corresponds to the representation f of the Lie algebra $\mathfrak{g}$. However, even in such a case, there exists a projective representation of the connected Lie group G corresponding to the representation f of the Lie algebra $\mathfrak{g}$, as explained later. That is, there exists a quantum mechanical symmetry of the connected Lie group G on $\mathcal{H}$ if and only if there exists a skew-Hermitian representation of the Lie algebra $\mathfrak{g}$ on $\mathcal{H}$. To see this fact, we make a central extension of the fundamental group $\pi_1(G)$ by the real (complex) Lie group G as follows. We consider the group of paths in G of the initial point e with the product defined in (3.50). The identity element of this group is the path that constantly maps to e. The existence of the inverse element can be easily checked. The quotient group of this group with respect to the equivalent relation $\sim$ is called the **universal covering group** of G and is written as $\tilde{G}$ (see Fig. 3.3). Now, we define the map from the universal covering group $\tilde{G}$ to G by $\pi : c(t) \to c(1)$. We have $\pi^{-1}(e) = \pi_1(G)$. Thus, the universal covering group $\tilde{G}$ is a central extension of a fundamental group $\pi(G)$ by G. For example, SU(2) is the universal covering group of the Lie group SO(3, $\mathbb{R}$) that expresses the rotations in the 3-dimensional space. In this case, since the center of SU(2) is $\{\pm e\}$ isomorphic to $\mathbb{Z}_2$ and SU(2)/$\mathbb{Z}_2 \cong$ SO(3, $\mathbb{R}$), and SU(2) is a central extension of SO(3, $\mathbb{R}$).

When a central extension of the fundamental group $\pi_1(G)$ by G is connected, it must be the universal covering group $\tilde{G}$. Further, we focus on a subgroup H of the fundamental group $\pi_1(G)$. Then, when a central extension of a quotient group $\pi_1(G)/H$ by G is connected, it also must be $\tilde{G}/H$. Let $[\phi(g_1, g_2)]$ be the residue class containing $\phi(g_1, g_2)$ with respect to H. Then, the quotient group $\tilde{G}/H$ is the central extension of the factor system $\{[\phi(g_1, g_2)]\}$. In particular, when $\pi_1(G)/H$ is $\mathbb{Z}_2$, the quotient group $\check{G} := \tilde{G}/H$ satisfies the relation $\check{G}/\mathbb{Z}_2 \cong G$, and is a central extension of $\mathbb{Z}_2$ by G. Then, $\check{G}$ is called the **double covering group** of G.

For example, we focus on the 2-dimensional Lorentz group that describes the relativistic symmetry between the 2-dimensional space and time. The fundamental group of its connected component $SO(2, 1, \mathbb{R})_0$ is $\mathbb{Z}$. Here, we choose the subgroup $2\mathbb{Z}$ consists of even integers as subgroup H of $\mathbb{Z}$. Then, we have $\mathbb{Z}/2\mathbb{Z} = \mathbb{Z}_2$. Especially, we have $\widetilde{SO(2, 1, \mathbb{R})_0}/2\mathbb{Z} \cong SU(1, 1)$, which implies that $SU(1, 1)$ is a central extension of $\mathbb{Z}_2$ by $SO(2, 1, \mathbb{R})_0$.

In general, when we choose a discrete subgroup H of the center $C(\tilde{G})$ of the universal covering group $\tilde{G}$ of the Lie algebra $\mathfrak{g}$, the Lie algebra corresponding to the Lie group $\tilde{G}/H$ is the Lie algebra $\mathfrak{g}$. Thus, the relation

$$C(\tilde{G}) = \mathsf{ad}^{-1}(I) \tag{3.51}$$

holds for an adjoint representation ad of $\tilde{G}$. The relation $\subset$ of (3.51) is trivial from the definition. The relation $\supset$ of (3.51) follows from the fact that any element of $\tilde{G}$ can be written as $\exp(X_1)\ldots\exp(X_l)$ with a suitable choice of $X_1, \ldots, X_l \in \mathfrak{g}$.

Hence, we can define the adjoint representation ad for the Lie group $\tilde{G}/H$. In particular, when $\mathfrak{g}$ is semi simple, since the center $C(\tilde{G})$ is a discrete subgroup of $\tilde{G}$, the Lie algebra corresponding to the Lie group $\tilde{G}/C(\tilde{G})$ is the Lie algebra $\mathfrak{g}$, and hence, (3.51) guarantees that the center of the Lie group $\tilde{G}/C(\tilde{G})$ is trivial.[2] When the representation is irreducible, the action on the set of density matrices is the Lie group $\tilde{G}/C(\tilde{G})$ because any element in $C(\tilde{G})$ does not change any density matrix. When $\mathfrak{g}$ is simple, the Lie group $\tilde{G}/C(\tilde{G})$ is a simple group (as a group). That is, the action of the center $C(\tilde{G})$ does not change any density matrix. Hence, in these cases, not the applications of $SU(2)$ and $SU(1, 1)$ but those of $SO(3, \mathbb{R})$ and $SO(1, 2, \mathbb{R})_0$ are directly reflected to the action on the set of density matrices.

3.6.2 Relation to Representation

Since an infinitesimal change in the universal covering group $\tilde{G}$ equals an infinitesimal change in the Lie group G, the Lie algebra corresponding to $\tilde{G}$ equals the Lie algebra $\mathfrak{g}$ corresponding to the Lie group G. Then, we can define the representation $\tilde{\mathsf{f}}$ of the universal covering group $\tilde{G}$ from the representation f of the Lie algebra $\mathfrak{g}$ as follows. We choose a path $c(t)$ corresponding to an element of the universal covering group $\tilde{G}$, and define $\tilde{\mathsf{f}}([c])$ to be $\mathsf{f}(c)(1)$ by considering the image $\mathsf{f}(c)(t)$ along the path $c(t)$. The independence of the choice of c in this definition can be shown in the same way as the simply connected case. In particular, when a representation f of $\mathfrak{g}$ is an irreducible representation, the representation $\tilde{\mathsf{f}}$ of the universal covering group $\tilde{G}$ is also an irreducible representation. Summarizing the above discussion, we have the following theorem.

---

[2]When $\mathfrak{g}$ is not semi simple, the center $C(\tilde{G})$ might be a continuous group. In particular, when $\mathfrak{g}$ is a nilpotent Lie algebra, the center $C(\tilde{G})$ is a continuous group, and hence, the Lie algebra of the Lie group $\tilde{G}/C(\tilde{G})$ is different from the Lie algebra $\mathfrak{g}$.

Theorem 3.4 *Given a connected Lie group G, there is a one-to-one correspondence between a (unitary) representation $\tilde{\mathsf{f}}$ of the universal covering group $\tilde{G}$ and a (skew-Hermitian) representation f of the Lie algebra $\mathfrak{g}$, and the relation*

$$\tilde{\mathsf{f}}(e^{tX}) = \exp(t\mathsf{f}(X)), \quad \mathsf{f}(X) = \lim_{t\to 0} \frac{\tilde{\mathsf{f}}(e^{tX}) - I}{t}$$

holds.

Since the universal covering group $\tilde{G}$ is a central extension by G, the discussion in Sect. 2.4 guarantees that an irreducible representation $\tilde{\mathsf{f}}$ of $\tilde{G}$ uniquely yields a projective irreducible representation of G. The following theorem holds for this relation.

Theorem 3.5 *Let G be a connected Lie group. Assume that its fundamental group $\pi_1(G)$ is a finite group. Let k be the maximum value of the orders of elements of $\pi_1(G)$. Then, given a finite-dimensional projective representation f, there exists a factor system $\{e^{i\theta'(g,g')}\}_{g,g'}$ taking values in U_k that is equivalent to the factor system $\{e^{i\theta(g,g')}\}_{g,g'}$ of the projective representation f.*

Hence, to discuss a factor system of a finite-dimensional projective representation, it is enough to treat the group $H(G : \mathbb{Z}_k)$. So, given a projective representation f with a factor system $\mathcal{E} \in H(G : \mathbb{Z}_k)$, there exists a representation $\tilde{\mathsf{f}}$ of the universal covering group $\tilde{G}$ such that for an element $g \in G = \tilde{G}/\pi_1(G)$, there exists a representative $\tilde{g} \in \tilde{G}$ of g satisfying that $\mathsf{f}(g) = \tilde{\mathsf{f}}(\tilde{g})$. This fact can be shown from the one-to-one correspondence given in Theorem 3.4.

To express this fact from another viewpoint, we introduce notations as follows. Let $\hat{\mathfrak{g}}$ be the set of labels of irreducible skew-Hermitian representations of the Lie algebra $\mathfrak{g}$, and $\hat{\mathfrak{g}}_{\mathrm{f}}$ be the subset of labels corresponding to finite-dimensional representations. Applying this fact to irreducible projective unitary representations, we have

$$\bigcup_{\mathcal{E}\in H(G:\mathbb{Z}_k)} \hat{G}_{\mathrm{f}}(\mathcal{E}) = \hat{\tilde{G}}_{\mathrm{f}} = \hat{\mathfrak{g}}_{\mathrm{f}}. \tag{3.52}$$

When the dimension k of the representation space of the irreducible representation f of the Lie algebra $\mathfrak{g}$ is coprime to the maximum order k of elements of $\pi_1(G)$, due to Corollary 2.2, any projective representation f of G is a representation.

Proof Let $\{e^{i\theta(g,g')}\}_{g,g'}$ be the factor system of a given d-dimensional projective representation f of G. We define a projective representation $\bar{\mathsf{f}}$ equivalent to f by $\bar{\mathsf{f}}(g) := \mathsf{f}(g)(\frac{\det \mathsf{f}(g)}{|\det \mathsf{f}(g)|})^{-1/d} \in \mathrm{SL}(\mathcal{H})$. Then, the factor system $\{e^{i\theta'(g,g')}\}_{g,g'}$ of $\bar{\mathsf{f}}$ takes values in U_d. Let G' be the central extension of the group U_d by the Lie group G based on the factor system, and G'_0 be the connected component of G' containing e. There exists a subgroup U_l of U_d such that U_l is included in the center of G'_0 and $G'_0/\mathrm{U}_l = G$. Due to this fact and Lemma 3.1, the factor system $\{e^{i\theta'(g,g')}\}_{g,g'}$ takes values in U_l.

Since the Lie algebra corresponding to G'_0 is the Lie algebra $\mathfrak{g}$ corresponding to G, the universal covering group of G'_0 is the universal covering group $\tilde{G}$ of G, which implies that $\tilde{G}/\pi_1(G'_0) = G'_0$. Since $\tilde{G}/\pi_1(G) = G$, U_l is a subgroup of $\pi_1(G)$. This fact indicates that l is a divisor of the maximum order k of elements of $\pi_1(G)$. Hence, the factor system $\{e^{i\theta'(g,g')}\}_{g,g'}$ takes values in U_k. ∎

The following corollary holds.

Corollary 3.1 *The compact Lie group G and the maximum order k of elements of* $\pi_1(G)$ *satisfies that* $\bigcup_{\mathcal{E}\in H(G:\mathbb{Z}_k)} \hat{G}(\mathcal{E}) = \hat{\tilde{G}} = \hat{\mathfrak{g}}$.

Proof Theorem 3.7 in the next section guarantees that any irreducible projective unitary representation of a compact Lie group G has a finite-dimensional representation space. Then, the discussion after Theorem 3.5 yields the argument of this corollary. ∎

3.6.3 Projective Representation

As mentioned above, when a Lie group G is a universal covering group, there is one-to-one correspondence between a (skew-Hermitian) representation of the corresponding Lie algebra and a (unitary) representation of the Lie group G. Even though the Lie group G is not a universal covering group, as mentioned before Theorem 3.5, we can make a projective (unitary) representation of the Lie group G from a (skew-Hermitian) representation of the corresponding Lie algebra. Applying the discussion in Sect. 3.6.2 to a projective (unitary) representation, we can make a projective (skew-Hermitian) representation of the corresponding Lie algebra from a projective (unitary) representation of the Lie group G. Conversely, when G is a universal covering group, we can make a projective (unitary) representation of the Lie group G from a projective (skew-Hermitian) representation of the corresponding Lie algebra. Further, due to the discussion in Sect. 2.4, from a projective (unitary) representation of a universal covering group, we can make a projective (skew-Hermitian) representation of another Lie group that corresponds to the same Lie algebra but is not a universal covering group. Hence, even when the Lie group is not necessarily a universal covering group, we can make a projective (unitary) representation of the Lie group from a projective (skew-Hermitian) representation of the corresponding Lie algebra.

However, Lemma 3.5 shows that there does not exist a proper projective representation on a finite-dimensional representation space. Due to Theorem 3.7 in the next section, any irreducible projective unitary representation of a compact Lie group has a finite-dimensional representation space. Hence, whenever the Lie algebra is compact, it is sufficient to treat only a (skew-Hermitian) representation of the Lie algebra. On the other hand, a non-compact Lie algebra has a representation whose representation space is infinite-dimensional. Hence, we need to treat a skew-Hermitian representation of a non-compact Lie algebra.

3.6.4 Representation for Complex Lie Groups

We can define the universal covering group $\tilde{G}$ for a complex Lie algebra $\mathfrak{g}$. For this definition, we give the definition of a representation for complex Lie groups. A homomorphism for groups from a complex Lie group G_1 to a complex Lie group G_2 is called a homomorphism for complex Lie groups when the homomorphism is a holomorphic function for both coordinates. In particular, a homomorphism for complex Lie groups from a complex Lie group G to the complex Lie group $\mathrm{GL}(\mathcal{H})$ is called a **representation for complex Lie groups** of a complex Lie group G. For simplicity, it is simplified to a $\mathbb{C}$-representation of a complex Lie group G.

Then, we have the same relation between a $\mathbb{C}$-representation of a complex Lie group G and a $\mathbb{C}$-representation of the corresponding complex Lie algebra $\mathfrak{g}$ as that between a representation of a real Lie group and a representation of real Lie algebra. We should remark that a representation (for real Lie groups) of a complex Lie group G is not necessarily a $\mathbb{C}$-representation of a complex Lie group G. Since a unitary group is not a complex Lie group, a unitary representation of a complex Lie group is not a $\mathbb{C}$-representation.

3.7 Invariant Measures on Group and Homogeneous Space

This section addresses an integral (a measure) $\nu(dg)$ on a group G. The measure ν is defined by determining the values $\nu(B)$ with every Borel set B of G. The measure ν is called **left-invariant** when the relation $\nu(gB) = \nu(B)$ holds for an element $g \in G$ and a Borel set B of G. Similarly, it is called **right-invariant** when $\nu(Bg) = \nu(B)$. Both left-invariant and right-invariant measures are uniquely determined except for a constant factor [66, Sect. 2, Chap. VIII]. When they are the same, it is called the **invariant measure** (**Haar measure**) of G. When the group G consists of finite elements, the left-invariant and right-invariant measures are the same, and given as $\nu(B) = \#B$. Then, the following theorem is known [66, Sect. 2, Chap. VIII].

Theorem 3.6 *Assume that a Lie group G consists of finite number of connected components. The left-invariant and right-invariant measures of the Lie group G are the same if and only if the corresponding Lie algebra $\mathfrak{g}$ satisfies that*

$$\mathrm{Tr}\,\mathsf{ad}(X) = 0, \forall X \in \mathfrak{g}\,. \tag{3.53}$$

The group G is called **unimodular** when it satisfies the above condition. For example, when the corresponding Lie algebra $\mathfrak{g}$ is simple, any element $X \in \mathfrak{g}$ can be written as $[Y, Z]$ with two elements $Y, Z \in \mathfrak{g}$. Since (3.36) implies that $\mathsf{ad}(X) = [\mathsf{ad}(Y), \mathsf{ad}(Z)]$, we have (3.53), which guarantees the existence of the invariant measure of G. Since a semi simple Lie algebra can be written as a direct sum of simple Lie algebras, it satisfies (3.53) and thus, the Lie group G has the invariant measure. Further, since any element of the Lie algebra corresponding to a compact Lie group is compact, it satisfies (3.53). So, any compact Lie group has the invariant measure. On the other hand, the affine group $\mathrm{Aff}(n)$ satisfies the relation $\mathsf{ad}((X,y))(X',y') = (\mathsf{ad}(X)X', Xy' - X'y)$ for $(X,y), (X',y') \in \mathrm{Aff}(n) \cong \mathfrak{gl}(n) \oplus \mathbb{R}^n$, which implies $\mathrm{Tr}\,\mathsf{ad}((X,y)) = \mathrm{Tr}\,X$. So, the affine group $\mathrm{Aff}(n)$ does not satisfy (3.53).

A Lie group G (that might be a discrete group) is compact if and only if the total measure $\mu(G)$ of the invariant measure μ of G is finite. Hence, a compact Lie group (that might be a finite group) has the **normalized** invariant measure μ_G, whose total measure is normalized to 1. Thus, when G is a compact Lie group, an irreducible unitary representation f of G on $\mathcal{H}$ satisfies

$$\begin{aligned} &\mathsf{f}(g') \int_G \mathsf{f}(g)|v\rangle\langle v|\mathsf{f}(g)^\dagger \mu_G(dg)\mathsf{f}(g')^\dagger \\ =& \int_G \mathsf{f}(g'g)|v\rangle\langle v|\mathsf{f}(g'g)^\dagger \mu_G(dg) = \int_G \mathsf{f}(g)|v\rangle\langle v|\mathsf{f}(g)^\dagger \mu_G(dg) \end{aligned}$$

for $v \in \mathcal{H}$ and $g' \in G$. Thus, Lemma 2.4 guarantees that $\mathsf{f}(g)|v\rangle\langle v|\mathsf{f}(g)^\dagger \mu_G(dg)$ is a constant times of I. Also, the relation $\int_G \langle u|\mathsf{f}(g)|v\rangle\langle v|\mathsf{f}(g)^\dagger|u\rangle \mu_G(dg) > 0$ holds for $u \in \mathcal{H}$. Letting $\frac{1}{d_\mathsf{f}}$ be the above integral, we have $d_\mathsf{f} \int_G \mathsf{f}(g)|v\rangle\langle v|\mathsf{f}(g)^\dagger \mu_G(dg) = I$. Comparing the traces of both sides, we find that $\mathcal{H}$ is a finite-dimensional space and its dimension is d_f. Summarizing this discussion, we have the following theorem.

Theorem 3.7 *When G is a compact Lie group, an irreducible unitary representation of G is a finite-dimensional representation. Hence, letting d_λ be the dimension of the representation space of f_λ for $\lambda \in \hat{G}$, we obtain*

$$d_\lambda \int_G \mathsf{f}_\lambda(g)|v\rangle\langle v|\mathsf{f}_\lambda(g)^\dagger \mu_G(dg) = I. \tag{3.54}$$

The same fact can be shown for an irreducible projective unitary representation in the same way. Notice that the integral (3.54) is not only a mathematical formula but also a POVM, which gives a measurement in the physical system because any decomposition of the identity matrix (the identity operator corresponds to a measurement, as explained in Sect. 1.1.

Next, we consider the case when G is non-compact. In this case, while G has the invariant measure μ_G, the total measure μ_G is not necessarily finite. Also, an irreducible representation space is not necessarily finite-dimensional. In the following, we assume that there exist a vector v_0 with norm 1 and a real number d_λ such that

$$d_\lambda^{-1} = \int_G |\langle v_0|\mathsf{f}_\lambda(g)|v_0\rangle|^2 \mu_G(dg) < \infty. \tag{3.55}$$

Then, we have the following theorem.

Theorem 3.8 *Assume that G is a Lie group satisfying (3.53). G is not necessarily compact. For a representation f_λ, we choose a real number d_λ satisfying (3.55). The* *relation (3.54) holds for any vector v of norm* 1*. The value d_λ does not depend on the choice of v_0.*

In the following, the real number d_λ satisfying (3.55) is called the **formal dimension** of the representation f_λ with respect to the measure μ_G.

Proof Schur's lemma guarantees that $d_\lambda \int_G \mathsf{f}_\lambda(g)|v_0\rangle\langle v_0|\mathsf{f}_\lambda(g)^\dagger \mu_G(dg)$ is a constant times of I. By considering the trace of the product with the projection $|v_0\rangle\langle v_0|$, the relation (3.55) guarantees that the operator is I. Hence, the relation $1 = d_\lambda \int_G |\langle v_0|\mathsf{f}_\lambda(g)|v\rangle|^2 \mu_G(dg)$ holds for any vector v of norm 1. Similarly, Schur's lemma guarantees that the operator $d_\lambda \int_G \mathsf{f}_\lambda(g)|v\rangle\langle v|\mathsf{f}_\lambda(g)^\dagger \mu_G(dg)$ is also a constant times of I. Using the same discussion, we find that this operator is also I. ∎

A subgroup K of G is called a **maximal compact subgroup** of G when there is no compact subgroup K' of G such that $K \subsetneq K'$. When a Lie group G has a non-trivial compact subgroup, it is known that there exist a maximal compact subgroup K of G and a subgroup P such that $G = KP$ and any element $g \in G$ uniquely has two elements $k \in K$ and $p \in P$ such that $g = kp$. Such a decomposition of G is uniquely determined for a given Lie group G, and is called the **Cartan decomposition** of G [67, p.3]. Hence, the invariant measure μ_G of G can be written as $\mu_G(dg) = \mu_K(dk)\mu_P(dp)$ by using the invariant measures μ_K and μ_P of K and P. When μ_K is the normalized invariant measure of K, the measure $\mu_G(dg) = \mu_K(dk)\mu_P(dp)$ is called the **normalized** invariant measure of G. When a Lie group G has a non-trivial compact subgroup, we denote the invariant measure of G by μ_G.

Next, we focus on the homogeneous space $\Theta := G/H$ of Lie group G with a stabilizer H. A measure μ_Θ on Θ is called the **invariant measure** of Θ when the relation $\mu_\Theta(B) = \mu_\Theta(gB)$ holds for any Borel subset B of Θ. The invariant measure is uniquely determined except for constant factor [66, Theorem 8.36, Sect. 3, Chap. VIII].

When G has the invariant measure μ, we define a measure μ_Θ on Θ by $\mu_\Theta(B) := \mu(\pi_H^{-1}(B))$, and find that it is the invariant measure of Θ, where π_H is a projection from G to G/H. When the state $|v\rangle\langle v|$ is invariant with respect to the action of H, the action onto the state $|v\rangle\langle v|$ can be described by an element of G/H. Hence, we can choose the representative $g(\theta)$ of the residue class θ for an element $\theta \in G/H$ so that the relation (3.54) can be rewritten as

$$d_\lambda \int_\Theta \mathsf{f}_\lambda(g(\theta))|v\rangle\langle v|\mathsf{f}_\lambda(g(\theta))^\dagger \mu_\Theta(d\theta) = I. \tag{3.56}$$

Especially, when G has a non-trivial compact subgroup, we denote the normalized invariant measure of Θ by μ_Θ as we do that of G by μ_G. Then, d_λ in (3.56) is the formal dimension (the dimension when G is compact).

On the other hand, when a non-compact group G does not have a non-trivial compact subgroup, we cannot define the normalized invariant measure in this method. For example, when G is the group $\mathbb{R}$ with respect to addition, which is a non-compact group, we choose the invariant measure $\mu_{\mathbb{R}}([a,b]) = \int_a^b \frac{1}{\sqrt{2\pi}} dx$ as the normalized invariant measure of $\mathbb{R}$.

3.8 Fourier Transform on Lie Group

3.8.1 Commutative Case

In the above discussion, we have seen that irreducible representations of a Lie group G can be derived from a representation of the corresponding Lie algebra. However, a part of properties of the irreducible representations can be characterized without through the corresponding Lie algebra. For the characterization, Fourier transform plays a key role via the invariant measure of the Lie group G. The purpose of this section to characterize an irreducible representation of a Lie group G by using Fourier transform. For this purpose, we need to generalize Fourier transform to a general Lie group.

In this subsection, we discuss the commutative case. Let us consider the unit circle $\mathrm{U}(1)$ as the simplest example of Lie group. Since $\mathrm{U}(1)$ is a commutative group, the representation space of an irreducible representation f of $\mathrm{U}(1)$ is one-dimensional and f has the form $\mathsf{f}(\omega) = \omega^{-n}$, $(\forall \omega \in \mathrm{U}(1))$ with use of an integer k. Hence, the set $\hat{\mathrm{U}(1)}$ of its irreducible representation can be identified with the set $\mathbb{Z}$ of integers. Then, the set $L^2(\mathrm{U}(1)) := \{\phi | \int_0^{2\pi} |\phi(e^{i\theta})|^2 d\theta < \infty\}$ of square-integrable functions on $\mathrm{U}(1)$ is isomorphic to the set $L^2(\hat{\mathrm{U}(1)}) := \{\{a_n\}_{n\in\mathbb{Z}} | \sum_{n=-\infty}^{\infty} |a_n|^2 < \infty\}$ of square-integrable functions on $\hat{\mathrm{U}(1)} = \mathbb{Z}$ as linear spaces. That is, we define the Fourier transform $\mathcal{F}_{\mathrm{U}(1)}$ and the inverse Fourier transform $\mathcal{F}_{\mathrm{U}(1)}^{-1}$ as the linear map $\mathcal{F}_{\mathrm{U}(1)}$ from $L^2(\mathrm{U}(1))$ to $L^2(\hat{\mathrm{U}(1)})$ and the linear map $\mathcal{F}_{\mathrm{U}(1)}^{-1}$ from $L^2(\hat{\mathrm{U}(1)})$ to $L^2(\mathrm{U}(1))$

$$\mathcal{F}_{\mathrm{U}(1)}[\phi](n) := \frac{1}{2\pi} \int_0^{2\pi} \phi(e^{i\theta}) e^{-in\theta} d\theta \tag{3.57}$$

$$\mathcal{F}_{\mathrm{U}(1)}^{-1}[\{a_n\}](\omega) := \sum_{n\in\mathbb{Z}} a_n \omega^n. \tag{3.58}$$

Then, the maps $\mathcal{F}_{\mathrm{U}(1)}$ and $\mathcal{F}_{\mathrm{U}(1)}^{-1}$ are unitary maps, and the maps $\mathcal{F}_{\mathrm{U}(1)} \circ \mathcal{F}_{\mathrm{U}(1)}^{-1}$ and $\mathcal{F}_{\mathrm{U}(1)}^{-1} \circ \mathcal{F}_{\mathrm{U}(1)}$ are the identity map. (Exercises 3.9 and 3.10)

Next, we proceed to the case of $G = \mathbb{R}$, in which, any irreducible representation of $\mathbb{R}$ is one-dimensional because $\mathbb{R}$ is a commutative group. Then, the irreducible unitary representation corresponding to $\lambda \in \hat{\mathbb{R}}$ is given as $\mathbb{R} \ni x \mapsto \mathsf{f}_\lambda(x) := e^{ix\lambda}$. Then, the Fourier transform and the inverse Fourier transform are as

$$\mathcal{F}[\phi](\lambda) := \frac{1}{\sqrt{2\pi}} \int_{\mathbb{R}} \phi(x) e^{-ix\lambda} dx, \tag{3.59}$$

$$\mathcal{F}^{-1}[\psi](x) := \frac{1}{\sqrt{2\pi}} \int_{\mathbb{R}} \psi(x) e^{ix\lambda} d\lambda = \mathcal{F}[\psi](-x). \tag{3.60}$$

These transformations are isomorphic transformations preserving the L^2 norm, and the maps $\mathcal{F} \circ \mathcal{F}^{-1}$ and $\mathcal{F}^{-1} \circ \mathcal{F}$ are the identity map on $L^2(\mathbb{R})$ (Exercise 3.16). Here, we list useful properties as

$$\mathcal{F}[\mathsf{P}\psi](\lambda) = \lambda \mathcal{F}[\psi](\lambda) \tag{3.61}$$

That is, we have

$$\mathcal{F}\mathsf{P} = \mathsf{Q}\mathcal{F}. \tag{3.62}$$

Applying $\mathcal{F}^{-1}$ twice, we have $\mathsf{P}\mathcal{F}^{-1} = \mathcal{F}^{-1}\mathsf{Q}$. Thus,

$$\mathcal{F}[\mathsf{Q}\psi](-\lambda) = -i\frac{d}{d\lambda}\mathcal{F}[\psi](-\lambda). \tag{3.63}$$

As an integral version, we have

$$\mathcal{F}[e^{ip\mathsf{Q}}\phi](\lambda) = \mathcal{F}[\phi](\lambda - p) \tag{3.64}$$

$$e^{-iq\mathsf{P}}\phi(x) = \phi(x-q). \tag{3.65}$$

When we define the convolution $\phi_1 * \phi_2$ for two functions $\phi_1, \phi_2 \in \mathcal{S}(\mathbb{R})$ as

$$\phi_1 * \phi_2(x) := \frac{1}{\sqrt{2\pi}}\int_{-\infty}^{\infty} \phi_1(y)\phi_2(x-y)dy, \tag{3.66}$$

we have

$$\mathcal{F}[\phi_1 \cdot \phi_2] = \mathcal{F}[\phi_1] * \mathcal{F}[\phi_2]. \tag{3.67}$$

In the d-dimensional space $\mathbb{R}^d$, the Fourier transform and the inverse Fourier transform are as

$$\mathcal{F}_d[\phi](\lambda) := \frac{1}{(2\pi)^{\frac{d}{2}}}\int_{\mathbb{R}^d} \phi(x)e^{-ix\cdot\lambda}dx, \tag{3.68}$$

$$\mathcal{F}_d^{-1}[\phi](\lambda) := \mathcal{F}_d[\phi](-\lambda). \tag{3.69}$$

Similarly, these transformations are isomorphic transformations preserving the L^2 norm, and the maps $\mathcal{F}_d \circ \mathcal{F}_d^{-1}$ and $\mathcal{F}_d^{-1} \circ \mathcal{F}_d$ are the identity map on $L^2(\mathbb{R}^d)$ (Exercise 3.16). The d-dimensional Fourier transform $\mathcal{F}_d$ has similar properties as

$$\mathcal{F}_d[\mathsf{P}_j\psi](\lambda) = \lambda_j \mathcal{F}_d[\psi](\lambda) \tag{3.70}$$

$$\mathcal{F}_d[\mathsf{Q}_j\psi](\lambda) = -i\frac{\partial}{\partial\lambda_j}\mathcal{F}_d[\psi](\lambda). \tag{3.71}$$

Similarly, for two functions $\phi_1, \phi_2 \in \mathcal{S}(\mathbb{R}^d)$, the convolution $\phi_1 * \phi_2$ is defined as

$$\phi_1 * \phi_2(x) := \frac{1}{(2\pi)^{\frac{d}{2}}} \int_{-\infty}^{\infty} \phi_1(y)\phi_2(x-y)dy, \tag{3.72}$$

we have

$$\mathcal{F}_d[\phi_1 \cdot \phi_2] = \mathcal{F}_d[\phi_1] * \mathcal{F}_d[\phi_2]. \tag{3.73}$$

Remark 3.1 The precise derivations of the above relations require slightly difficult mathematics. That is, we need to discuss the set $\mathscr{S}(\mathbb{R})$ of rapidly decreasing functions, which was introduced in Sect. 1.6. These derivations are given in Exercises in this subsection. In these derivations, we firstly derive required relations on $\mathscr{S}(\mathbb{R})$. Then, since $\mathscr{S}(\mathbb{R})$ is a dense subspace of $L^2(\mathbb{R})$, we extend these properties to the whole of $L^2(\mathbb{R})$. Since Exercises 3.11–3.22 are too difficult, a reader who is not interested in mathematical details can skip them.

Exercise 3.9 Show that the maps $\mathcal{F}_{\mathrm{U}(1)} \circ \mathcal{F}_{\mathrm{U}(1)}^{-1}$ and $\mathcal{F}_{\mathrm{U}(1)}^{-1} \circ \mathcal{F}_{\mathrm{U}(1)}$ are the identity maps.

Exercise 3.10 Show that the maps $\mathcal{F}_{\mathrm{U}(1)}$ and $\mathcal{F}_{\mathrm{U}(1)}^{-1}$ are unitary maps.

Exercise 3.11 Consider a function f in $\mathscr{S}(\mathbb{R}^d)$. Show that $-i\mathcal{F}_d[\frac{df}{dx_j}](\lambda)=\lambda_j\mathcal{F}_d[f](\lambda)$, i.e., $\mathcal{F}_d[\mathsf{P}_j f](\lambda) = \lambda_j \mathcal{F}_d[f](\lambda)$.

Exercise 3.12 Show the equation $\overline{\mathcal{F}_d^{-1}[\phi]} = \mathcal{F}_d[\overline{\phi}]$, where ϕ^* is the complex conjugate function of ϕ.

Exercise 3.13 Show the equation $\mathcal{F}_d[\phi](-\lambda) = \mathcal{F}_d^{-1}[\phi]$.

Exercise 3.14 Consider a function ϕ in $\mathscr{S}(\mathbb{R}^d)$. Show that $\mathcal{F}_d[\phi]$ also belongs to $\mathscr{S}(\mathbb{R}^d)$.

Exercise 3.15 Consider a function ϕ_1, ϕ_2 in $\mathscr{S}(\mathbb{R}^d)$. Show that

$$\langle \phi_1 | \mathcal{F}_d[\phi_2] \rangle = \langle \mathcal{F}_d^{-1}[\phi_1] | \phi_2 \rangle.$$

Exercise 3.16 Define the function $\varphi_\epsilon(x) := \frac{1}{(2\pi\epsilon)^{\frac{1}{4}}} e^{-\frac{x^2}{2\epsilon}}$. Show that the maps $\mathcal{F}_d \circ \mathcal{F}_d^{-1}$ and $\mathcal{F}_d^{-1} \circ \mathcal{F}_d$ are the identity map on $L^2(\mathbb{R}^d)$ by the following steps below.

a: Show that $\mathcal{F}[\varphi_{1/\epsilon}] = \varphi_\epsilon$.
b: Show that $(\frac{1}{2\pi\epsilon})^{\frac{1}{4}} \int_{-\infty}^{\infty} \phi(x-y)\varphi_\epsilon(y)dy = \frac{1}{\sqrt{2\pi}} \int_{-\infty}^{\infty} \mathcal{F}[\phi](y) e^{ixy} e^{-\epsilon y^2/2} dy$ for $\phi \in \mathscr{S}(\mathbb{R})$ by using Exercise 3.15.
c: Show the desired statement with $d = 1$ on $\mathscr{S}(\mathbb{R})$.
d: Show the desired statement with $d = 1$.

e: Show the desired statement with the general case.

Exercise 3.17 Show that $\mathcal{F}_d$ and $\mathcal{F}_d^{-1}$ are unitary maps on $L^2(\mathbb{R}^d)$ by using Exercises 3.15 and 3.16

Exercise 3.18 Show (3.61) and (3.70) by using Exercise 3.11

Exercise 3.19 Show (3.64).

Exercise 3.20 Show (3.65) by using (3.64)

Exercise 3.21 Show the convolution $\phi_1 * \phi_2$ belongs to $\mathscr{S}(\mathbb{R}^d)$ for two functions $\phi_1, \phi_2 \in \mathscr{S}(\mathbb{R}^d)$.

Exercise 3.22 Show the following relation for two functions $\phi_1, \phi_2 \in \mathscr{S}(\mathbb{R}^d)$.

$$\mathcal{F}_d[\phi_1 \cdot \phi_2](\lambda) = \frac{1}{(2\pi)^{\frac{d}{2}}} \int_{\mathbb{R}^d} \mathcal{F}_d[\phi_1](\lambda')\mathcal{F}_d[\phi_2](\lambda - \lambda')d\lambda'. \tag{3.74}$$

3.8.2 Non-commutative Case

The aim of this section is generalizing this discussion to compact Lie groups. Firstly, any (projective) representation of a compact group G can be converted to a (projective) unitary representation in the same way as the case of a finite group via the inner product (2.52) with replacement of $\sum_g$ by the Haar measure μ_G. Due to the same reason as the finite group case, to investigate the orbit of a vector in a representation of a compact Lie group, it is sufficient to treat the representation space $\bigoplus_{\lambda\in\hat{G}} \mathcal{U}_\lambda \otimes \mathbb{C}^{d_\lambda}$, which is isomorphic to the direct sum space $\bigoplus_{\lambda\in\hat{G}} \mathfrak{gl}(\mathcal{U}_\lambda)$ with respect to the left side action by the Lie group G. We define the inner product in the direct sum space $\bigoplus_{\lambda\in\hat{G}} \mathfrak{gl}(\mathcal{U}_\lambda)$ as $\langle A, B\rangle_{\oplus_\lambda \mathfrak{gl}(\mathcal{U}_\lambda)} := \sum_{\lambda\in\hat{G}}^k \operatorname{Tr} A_\lambda^\dagger B_\lambda$ for two elements $A = (A_1, \ldots)$ and $B = (B_1, \ldots)$ in the direct sum space $\bigoplus_{\lambda\in\hat{G}} \mathfrak{gl}(\mathcal{U}_\lambda)$, which is the inner product of the representation space $\bigoplus_{\lambda\in\hat{G}} \mathcal{U}_\lambda \otimes \mathbb{C}^{d_\lambda}$. We denote the complete space with respect to the inner product by $L^2(\hat{G})$. Then, we study how to treat the space $L^2(\hat{G})$ under the action of the Lie group G. As defined in (2.41), the space $\mathfrak{gl}(\mathcal{U}_\lambda)$ also has the unitary representation $\mathsf{D}(\mathsf{f}_\lambda)$ of $G \times G$, that is, the space $\mathfrak{gl}(\mathcal{U}_\lambda)$ is written as $\mathcal{U}_\lambda \otimes \mathcal{U}_{\lambda^*}$. Hence, the space $L^2(\hat{G})$ also has the unitary representation of $G \times G$ as well as that of G and can be regarded as the complete space of $\bigoplus_{\lambda\in\hat{G}} \mathcal{U}_\lambda \otimes \mathcal{U}_{\lambda^*}$.

For this purpose, we make a similar discussion for representations of a compact Lie group G as that in Sect. 2.8. Since it is a compact Lie group G is usually not a finite group, we have several technical difficulties to make a parallel discussion as that in Sect. 2.8. So, we omit the detail of the derivation in the following discussion. Firstly, we focus on the set $L^2(G) := \{\phi | \int_G |\phi(g)|^2 \mu_G(dg) < \infty\}$ of square-integrable functions on the normalized invariant measure μ_G of G.

Theorem 3.7 guarantees that any irreducible unitary representation of a compact Lie group G has a finite-dimensional representation space. The number of its irreducible unitary representations is countable at most. We denote these irreducible unitary representations by f_λ, and their representation spaces by $\mathcal{U}_\lambda$. Letting $|\lambda, m\rangle$ be a CONS of $\mathcal{U}_\lambda$, we define the function $R(\lambda, m, m') := \langle \lambda, m'|\mathsf{f}_\lambda(g)|\lambda, m\rangle$. Then, the orthogonal relation

$$\int_G R_{\lambda,m,m'}(g)R_{\bar{\lambda},\bar{m}',\bar{m}}(g)\mu_G(dg) = \frac{1}{d_\lambda}\delta_{\lambda,\bar{\lambda}}\delta_{m,\bar{m}}\delta_{m',\bar{m}'} \tag{3.75}$$

holds.

Similar to (2.60), the relation (3.75) implies the orthogonal relation of characters

$$\int_G \chi_{\lambda'}(g)\overline{\chi_\lambda(g)}\mu_G(dg) = \int_G \chi_{\lambda'}(g)\chi_{\lambda^*}(g)\mu_G(dg) = \delta_{\lambda',\lambda} \tag{3.76}$$

and the relation

$$\int_G \overline{\chi_\lambda(g)}\mathsf{f}_{\lambda'}(g)\mu_G(dg) = \frac{\delta_{\lambda',\lambda}}{d_\lambda}I_{\lambda'}. \tag{3.77}$$

Especially, when the irreducible representations f_λ and f_{λ^*} are still irreducible under the representation of a subgroup G' of G, an orthogonal relation similar to (3.76) holds even though the integral is replaced by the integral with the invariant measure of G'. Hence, similar to (2.64), (3.76) implies the relation

$$C^{\lambda''}_{\lambda',\lambda} = C^{\lambda''}_{\lambda,\lambda'} = C^{\lambda'^*}_{\lambda,\lambda''^*} = C^{\lambda^*}_{\lambda',\lambda''^*}. \tag{3.78}$$

Here, the first equation follows from the Definition 2.45 of the multiplicity coefficient $C^{\lambda''}_{\lambda,\lambda'}$.

We define **Fourier transform** $\mathcal{F} : L^2(G) \to L^2(\hat{G})$ and **Inverse Fourier transform** $\mathcal{F}^{-1} : L^2(\hat{G}) \to L^2(G)$ as

$$(\mathcal{F}[\phi])_\lambda := \sqrt{d_\lambda}\int_G \mathsf{f}_\lambda(g)^\dagger\phi(g)\mu_G(dg) \tag{3.79}$$

$$\mathcal{F}^{-1}[A](g) := \sum_\lambda \sqrt{d_\lambda}\,\mathrm{Tr}\,\mathsf{f}_\lambda(g)A_\lambda. \tag{3.80}$$

Then, the relation (3.77) implies that $\mathcal{F}\mathcal{F}^{-1}[A] = A$. As shown later, the kernel of $\mathcal{F}$ is $\{0\}$. Hence, the domain $L^2(G)$ of $\mathcal{F}$ is isomorphic to the image of $\mathcal{F}^{-1}$, which implies that $\mathcal{F}^{-1}\mathcal{F}[\phi] = \phi$. Thus, the set of functions $\{\frac{1}{\sqrt{d_\lambda}}R_{\lambda,m,m'}\}_{\lambda,m,m'}$ forms a CONS of $L^2(G)$. Here, we should remark that the function $\mathcal{F}^{-1}[A] \in L^2(G)$ contains all information for the entangled state $A = \sum_{\lambda\in\hat{G}} |A_\lambda\rangle\rangle \in L^2(\hat{G})$ with respect to the representation.

Proof of Ker $\mathcal{F} = \{0\}$. Section 2.8 employs the relation $\mu_G(g) = \frac{1}{|G|}$ in the proof of the relation $\mathcal{F}^{-1}\mathcal{F}[\phi] = \phi$. However, in the case of Lie group, we cannot employ the same method because the denominator is infinity. Assume that Ker $\mathcal{F} \neq \{0\}$. Then, there exists a continuous function ϕ such that $\mathcal{F}[\phi] = 0$. Any representation f_λ satisfies that $\int_G \phi(g)\mathsf{f}_\lambda(g)^\dagger \mu_G(dg) = 0$.

Now, we focus on the left hand side action of G to $\psi \in L^2(G)$, which is given as $\psi(g') \mapsto \psi(g^{-1}g')$ for an element g of G. We denote the representation of G by L. Since the representation L is given as a direct sum representation of irreducible representations, the above relation implies that $\int_G \phi(g)L(g)\mu_G(dg) = 0$. Since $L(g)\psi(g') = \psi(g^{-1}g')$, we have $\int_G \phi(g)\psi(g^{-1}g')\mu_G(dg) = 0$. We choose $\psi \in L^2(G)$ satisfying that $\psi(g^{-1}) = \overline{\phi(g)}$ and substitute e into g'. Then, we have $\int_G |\phi(g)|^2 \mu_G(dg) = 0$, which implies that $\phi(g) = 0$. So, we obtain Ker $\mathcal{F} = \{0\}$. ∎

Proof of (3.75). We choose an inner product of the space $\oplus_\lambda \mathcal{H}_\lambda$ such that the spaces $\mathcal{U}_\lambda$ and $\mathcal{U}_{\tilde{\lambda}}$ are orthogonal to each other for distinct representations λ and $\tilde{\lambda}$. Letting $\tilde{\mathsf{f}}$ be the representation of G on $\oplus_\lambda \mathcal{U}_\lambda$, we have

$$\begin{aligned}
&\int_G R_{\lambda,m,m'}(g) R_{\tilde{\lambda},\tilde{m}',\tilde{m}}(g)\mu_G(dg) \\
=&\int_G \langle \lambda, m'|\mathsf{f}_\lambda(g)|\lambda, m\rangle\langle\tilde{\lambda}, \tilde{m}|\mathsf{f}_{\tilde{\lambda}}(g)|\tilde{\lambda}, \tilde{m}'\rangle \mu_G(dg) \\
=&\langle \lambda, m'| \left(\int_G \tilde{\mathsf{f}}(g)|\lambda, m\rangle\langle\tilde{\lambda}, \tilde{m}|\tilde{\mathsf{f}}(g)\mu_G(dg)\right) |\tilde{\lambda}, \tilde{m}'\rangle.
\end{aligned}$$

The choice of the inner product guarantees that this value is 0 when $\lambda \neq \tilde{\lambda}$.

On the other hand, when $\lambda = \tilde{\lambda}$, Schur's lemma and the equality for the traces on both sides guarantee that $\int_G \tilde{\mathsf{f}}(g)|\lambda, m\rangle\langle\tilde{\lambda}, \tilde{m}|\tilde{\mathsf{f}}(g)^\dagger \mu_G(dg) = \frac{\langle\tilde{\lambda},\tilde{m}|\lambda,m\rangle}{d_\lambda} I_\lambda$, where I_λ is the identity matrix on $\mathcal{U}_\lambda$. Thus, we obtain (3.75). ∎

The above discussion can be extended to the case with projective representations with factor system $\mathcal{E}$ as follows. Using the set $\hat{G}[\mathcal{E}]$ of irreducible projective representations corresponding to the factor system $\mathcal{E} := [\{e^{i\theta(g,g')}\}_{g,g'}] \in H(G : \mathrm{U}(1))$, we define the linear space $L^2(\hat{G}[\mathcal{E}]) := \bigoplus_{\lambda\in\hat{G}[\mathcal{E}]} \mathfrak{gl}(\mathcal{U}_\lambda)$. Defining Fourier transform and Inverse Fourier transform based on (3.79) and (3.80), we can show that they are isomorphisms between $L^2(G)$ and $L^2(\hat{G}[\mathcal{E}])$ in the same way. Hence, we obtain (3.75)–(3.77) by replacing the set $\hat{G}$ by the set $\hat{G}[\mathcal{E}]$. In this case, the Fourier transform is defined for a function $L^2(G)$ as well as in the representation case. To clarify the dependence of the factor system $\mathcal{F}$, we denote the Fourier transform by $\mathcal{F}_{\mathcal{E}}$.

Next, we consider the case when G is non-compact. A unimodular group G is called **Type I** when the following condition holds with the following definition [32];

$$\mathsf{f}(G)' := \{A|\mathsf{f}(g)A = A\mathsf{f}(g),\ \forall g \in G\} \tag{3.81}$$

$$\mathsf{f}(G)'' := \{A|BA = AB,\ \forall B \in \mathsf{f}(G)'\}. \tag{3.82}$$

When any element of the set $\mathfrak{f}(G)' \cap \mathfrak{f}(G)''$ is a constant times of I, the representation $\mathfrak{f}$ is limited to an irreducible representation with multiplicity. In this case, since the representation space $\mathcal{U}_\lambda$ might be infinite-dimensional for $\lambda \in \hat{G}[\mathcal{E}]$, in general, we define the linear space $\mathfrak{gl}(\mathcal{U}_\lambda)$ as the set of Hilbert–Schmidt operators on $\mathcal{U}_\lambda$, and define the space $L^2(\hat{G}[\mathcal{E}]) := \bigoplus_{\lambda \in \hat{G}[\mathcal{E}]} \mathfrak{gl}(\mathcal{U}_\lambda)$. When the cardinality of $\hat{G}[\mathcal{E}]$ is countable or finite and the formal dimension d_λ of $\mathcal{U}_\lambda$ is finite for any $\lambda \in \hat{G}[\mathcal{E}]$, we can define the Fourier transform and the Inverse Fourier transform based on (3.79) and (3.80). These are isomorphisms between $L^2(G)$ and $L^2(\hat{G}[\mathcal{E}])$. Similarly, the relations (3.75)–(3.77) hold. Even though $\hat{G}[\mathcal{E}]$ does not satisfy these conditions, when the formal dimension d_λ is finite for an element λ of a countable or finite subset S of $\hat{G}[\mathcal{E}]$, the inverse Fourier transform can be defined on the linear space $L^2(S) := \bigoplus_{\lambda \in S} \mathfrak{gl}(\mathcal{U}_\lambda)$.

Even when $\hat{G}[\mathcal{E}]$ has cardinality of the continuum, when Lie group G is Type I, the Fourier transform can be defined as follows [34, 32]. We define

$$(\mathcal{F}[\phi])_\lambda := \int_G \mathfrak{f}_\lambda(g)^\dagger \phi(g) \mu_G(dg) \tag{3.83}$$

for $\lambda \in \hat{G}[\mathcal{E}]$. Then, it is known in [34, 32] that there uniquely exists a measure $\mu_{\hat{G}[\mathcal{E}]}$ (called the **Plancherel measure**) on $\hat{G}[\mathcal{E}]$ such that $\mathcal{F}$ is a unitary under the norm defined by

$$\|\mathcal{F}[\phi]\|^2 := \int_{\hat{G}[\mathcal{E}]} \mathrm{Tr}(\mathcal{F}[\phi])_\lambda (\mathcal{F}[\phi])_\lambda^\dagger \mu_{\hat{G}[\mathcal{E}]}(d\lambda). \tag{3.84}$$

Then, the inverse Fourier transform is given as follows.

$$\mathcal{F}^{-1}[A](g) := \int_{\hat{G}[\mathcal{E}]} \mathrm{Tr}\, \mathfrak{f}_\lambda(g) A_\lambda \mu_{\hat{G}[\mathcal{E}]}(d\lambda). \tag{3.85}$$

Finally, we see how to recover the Fourier transform (3.59) and the inverse Fourier transform (3.60) in the case of $G = \mathbb{R}$. Since any irreducible representation of $\mathbb{R}$ is one-dimensional due to the communitativity of $\mathbb{R}$, the irreducible unitary representation corresponding to $\lambda \in \hat{\mathbb{R}}$ is given as $\mathbb{R} \ni x \mapsto \mathfrak{f}_\lambda(x) := e^{ix\lambda}$. That is, the set $\hat{\mathbb{R}}$ of irreducible unitary representations of $\mathbb{R}$ is $\mathbb{R}$. Hence, $L^2(\hat{\mathbb{R}})$ can be identified with the square-integrable space $L^2(\mathbb{R})$. Choosing the invariant measure μ_G as $\frac{dx}{\sqrt{2\pi}}$, as mentioned in the end of Sect. 3.7, the Plancherel measure is given as $\frac{d\lambda}{\sqrt{2\pi}}$. Therefore, the Fourier transform is given as $\mathcal{F}[\phi](\lambda) = \frac{1}{\sqrt{2\pi}} \int_{\mathbb{R}} \phi(x) e^{-ix\lambda} dx$, and the inverse Fourier transform is as $\mathcal{F}^{-1}[\psi](x) = \frac{1}{\sqrt{2\pi}} \int_{\mathbb{R}} \psi(x) e^{ix\lambda} d\lambda$. The above general discussion also shows that these transformations are isomorphic transformations preserving the L^2 norm, and the maps $\mathcal{F} \circ \mathcal{F}^{-1}$ and $\mathcal{F}^{-1} \circ \mathcal{F}$ are the identity map.

Finally, we consider the physical meaning of the Inverse Fourier transform $\mathcal{F}^{-1}$. For this purpose, we prepare the operator T on the representation space $\bigoplus_{\lambda\in\hat{G}} \mathcal{U}_\lambda \otimes \mathbb{C}^{d_\lambda}$ as

$$T := \left(\sum_{\lambda\in\hat{G}} \sqrt{d_\lambda}|I_\lambda\rangle\rangle\right)\left(\sum_{\lambda\in\hat{G}} \sqrt{d_\lambda}\langle\langle I_\lambda|\right).$$

When the state is a pure entangled state $\sum_\lambda |A_\lambda\rangle$ and we apply the measurement corresponding to the POVM $\{\mathsf{f}(g)T\mathsf{f}(g)^\dagger\}_{g\in G}$, the measurement outcome g obeys the distribution $|\mathcal{F}^{-1}[A](g)|^2\mu_G(dg)$. For example, for $G=\mathbb{R}$, we consider the representation $p \mapsto e^{ipx}$. The above measurement is the measurement of the momentum P. This measurement relates the optimal measurement of the group action, which is discussed in [50, Sect. 4.3]. In the projective representation case with the group $\mathbb{R}^2$, the Inverse Fourier transform $\mathcal{F}^{-1}$ is closely related to Wigner function. That is, the discussion of this section gives the bridge between group representation and Fourier analysis.

Finally, we consider the physical meaning of the inverse Fourier transform [illegible]. For this purpose, we prepare the operator [illegible] on the representation space [illegible]

[illegible]

When [illegible] compressed [illegible] we [illegible] corresponding to [illegible] the [illegible]

[illegible]

Chapter 4
Representations of Typical Lie Groups and Typical Lie Algebras

Abstract This chapter deals with representations of special Lie groups and special Lie algebras. Since a representation of a Lie algebra can be classified with the highest weight, those of a Lie group can be easily treated through those of the corresponding Lie algebra. Also, Lie algebras provide several concepts important for quantum theory. Hence, this chapter is organized so that it constructs a representation of a Lie group via that of the corresponding Lie algebra. This chapter starts with representations of Lie algebras $\mathfrak{su}(2)$ and $\mathfrak{su}(1, 1)$. Since the Lie algebra $\mathfrak{su}(2)$ is compact and the Lie algebra $\mathfrak{su}(1, 1)$ is not compact, they require different treatment caused in this difference. As they have unexpected common features, we handle both in a unified way. Then, we proceed to representation of the Lie algebra $\mathfrak{su}(r)$ by using Young diagrams. Especially, the representation of the Lie algebra $\mathfrak{su}(r)$ on the tensor product space is closely related to that of the permutation group on the same tensor product space. The relation is called Schur duality. We also consider what a finite subgroup of a Lie group can replace the Lie group when its representation is given. Such a problem is called design, and is discussed in this chapter.

4.1 SL(2, $\mathbb{C}$) and Its Subgroup

The groups SO(3, $\mathbb{R}$) and SO(1, 2, $\mathbb{R}$)$_0$ play a very important role in physics. The previous chapter clarified that their Lie algebras are isomorphic to $\mathfrak{su}(2)$ and $\mathfrak{su}(1, 1)$, respectively. Important physical quantities are expressed by the skew–Hermitian representations of their Lie algebras. In the following, we investigate the structure of the group SL(2, $\mathbb{C}$) firstly. Then, based on this investigation, we study the structures of the groups SU(2) and SU(1, 1), which are required to discuss the skew–Hermitian representations of $\mathfrak{su}(2)$ and $\mathfrak{su}(1, 1)$, respectively.

M. Hayashi, *Group Representation for Quantum Theory*,
DOI 10.1007/978-3-319-44906-7_4

4.1.1 *Gauss Decomposition of* SL(2, ℂ)

Since the complex Lie group SL(2, ℂ) has the simplest structure among non-commutative complex Lie groups, this section studies typical properties of SL(2, ℂ), its subgroup, and its complex Lie algebra, which are the foundation of the latter discussion. We choose the following three elements in the complex Lie algebra $\mathfrak{sl}(2, \mathbb{C})$ of SL(2, ℂ) as its basis.

$$E_{0,1} := \frac{1}{2}\begin{pmatrix} -1 & 0 \\ 0 & 1 \end{pmatrix}, \quad K_{+,1} := \begin{pmatrix} 0 & 0 \\ 1 & 0 \end{pmatrix}, \quad K_{-,1} := \begin{pmatrix} 0 & 1 \\ 0 & 0 \end{pmatrix}, \tag{4.1}$$

where the lower subscript 1 indicates the Lie group SU(2), i.e., this basis is deeply related the real Lie group SU(2) because these matrices produce a Gauss decomposition of SU(2) as (4.5). This basis satisfies the commutation relation $[E_{0,1}, K_{\pm,1}] = \pm K_{\pm,1}, \quad [K_{+,1}, K_{-,1}] = 2E_{0,1}$, which plays an important role in the representation theory of a Lie algebra. Any element of $\mathfrak{sl}(2, \mathbb{C})$ can be written as a linear combination of this basis. The same fact can be shown for almost all the elements of SL(2, ℂ). That is, any matrix g ∈ SL(2, ℂ) can be written as

$$\begin{aligned} g &= \exp(\zeta K_{-,1}) \exp(\eta E_{0,1}) \exp(\zeta' K_{+,1}) \\ &= \begin{pmatrix} 1 & \zeta \\ 0 & 1 \end{pmatrix}\begin{pmatrix} e^{-\eta/2} & 0 \\ 0 & e^{\eta/2} \end{pmatrix}\begin{pmatrix} 1 & 0 \\ \zeta' & 1 \end{pmatrix} = \begin{pmatrix} e^{-\eta/2} + \zeta\zeta' e^{\eta/2} & \zeta e^{\eta/2} \\ \zeta' e^{\eta/2} & e^{\eta/2} \end{pmatrix} \end{aligned} \tag{4.2}$$

with a suitable choice of $x, y, z \in \mathbb{C}$ unless it is written as $\begin{pmatrix} b & -a^{-1} \\ a & 0 \end{pmatrix}$. In other words, by letting $M_\gamma := \begin{pmatrix} \gamma^{-1} & 0 \\ 0 & \gamma \end{pmatrix}$, $P_\zeta := \begin{pmatrix} 1 & \zeta \\ 0 & 1 \end{pmatrix}$, g is written as $g = P_\zeta M_\gamma J P_{\zeta'} J^{-1}$ unless g is written as $\begin{pmatrix} b & -a^{-1} \\ a & 0 \end{pmatrix}$.

On the other hand, when g is written as $\begin{pmatrix} b & -a^{-1} \\ a & 0 \end{pmatrix}$, we have $g = M_a J P_{-ba}$. However, the set of such elements has measure 0 under the proper measure of SL(2, ℂ). That is, except for a subset with measure 0, an element of SL(2, ℂ) has the decomposition (4.2), which is called the **Gauss decomposition** of g and plays an important role for studying the structures of SL(2, ℂ) and its subgroup.

4.1.2 *Structure of* SU(2)

The basis

$$F_1^x := \frac{1}{2}\begin{pmatrix} 0 & -i \\ -i & 0 \end{pmatrix}, \quad F_1^y := \frac{1}{2}\begin{pmatrix} 0 & -1 \\ 1 & 0 \end{pmatrix} = \frac{-1}{2}J, \quad F_1^z := \frac{1}{2}\begin{pmatrix} -i & 0 \\ 0 & i \end{pmatrix}$$

is very useful for discussing the real Lie algebra $\mathfrak{su}(2)$ of the subgroup SU(2) of SL(2, ℂ), and satisfies the commutation relation $[F_1^z, F_1^x] = F_1^y$, $[F_1^z, F_1^y] = -F_1^x$, $[F_1^x, F_1^y] = F_1^z$. We replace the super subscripts x, y, z by 1, 2, 3 and define $\epsilon_{j,k,l}$ as

$$\epsilon_{j,k,l} = \begin{cases} 1 & \text{when } (j,k,l) = (1,2,3), (2,3,1), (3,1,2) \\ -1 & \text{when } (j,k,l) = (3,2,1), (2,1,3), (1,3,2) \\ 0 & \text{otherwise.} \end{cases} \tag{4.3}$$

Then, the commutation relation can be rewritten as

$$[F_1^j, F_1^k] = \sum_{l=1}^{3} \epsilon_{j,k,l} F_1^l. \tag{4.4}$$

Combining these basis, we can recover the above basis of $\mathfrak{sl}(2, \mathbb{C})$ as $E_{0,1} = -iF_1^z$, $K_{\pm,1} = iF_1^x \pm F_1^y$. The matrices $E_{0,1}$ and $K_{\pm,1}$ are not elements of $\mathfrak{su}(2)$, however, they are useful for discussing representations of $\mathfrak{su}(2)$ later. We focus on the subgroup $H = \{\exp(tF_1^z)\}$ of SU(2) isomorphic to U(1). By using two complex numbers μ and ν satisfying the condition $|\mu|^2 + |\nu|^2 = 1$, any element of SU(2) can be written as $\begin{pmatrix} \mu & \nu \\ -\overline{\nu} & \overline{\mu} \end{pmatrix}$, which is converted to $\begin{pmatrix} \mu e^{-it/2} & \nu e^{it/2} \\ -\overline{\nu} e^{-it/2} & \overline{\mu} e^{it/2} \end{pmatrix}$ by applying the element $\exp(tF_1^z)$ of H from the right hand side. Hence, an element of the homogeneous space SU(2)/H with the stabilizer H can be written as the ratio ν and $\overline{\mu}$. Hence, SU(2)/H is **Riemann sphere** $\hat{\mathbb{C}}$, which is the addition of ∞ to ℂ, where ∞ corresponds to $\overline{\mu} = 0$ (see Fig. 4.1). When $\overline{\mu} \neq 0$, an element of SU(2)/H is given by $\nu/\overline{\mu} \in \mathbb{C}$. So, we choose the representative of the element SU(2)/H with the condition $\overline{\mu} \neq 0$ such that $\mu \in \mathbb{R}$. The representative can be written as $\begin{pmatrix} \cos\tau & e^{i\phi}\sin\tau \\ -e^{-i\phi}\sin\tau & \cos\tau \end{pmatrix}$ by using $\tau \in \mathbb{R}, \phi \in \mathbb{R}$. Using $\zeta := e^{i\phi}\tan\tau$ and $\eta := -\log(1 + |\zeta|^2)$, we obtain its Gauss

Fig. 4.1 Riemann sphere

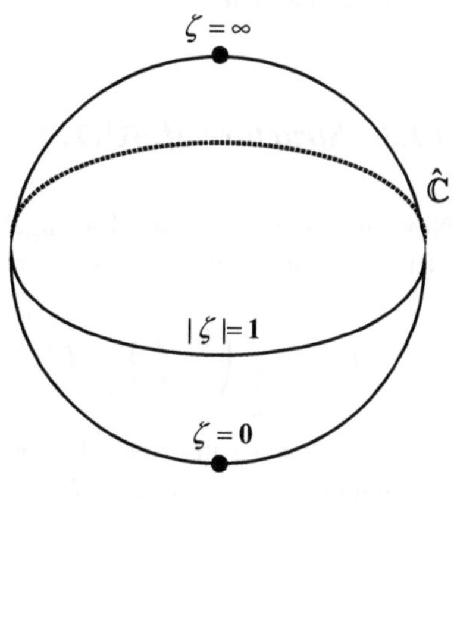

decomposition given in (4.2) as follows.

$$\begin{pmatrix} \cos\tau & e^{i\phi}\sin\tau \\ -e^{-i\phi}\sin\tau & \cos\tau \end{pmatrix} = \exp(\zeta K_{-,1})\exp(\eta E_{0,1})\exp(-\overline{\zeta}K_{+,1}).$$

Letting $\xi := e^{i\phi}\tau$, we also have the relations

$$\begin{pmatrix} \cos\tau & e^{i\phi}\sin\tau \\ -e^{-i\phi}\sin\tau & \cos\tau \end{pmatrix} = \exp\begin{pmatrix} 0 & e^{i\phi}\tau \\ -e^{-i\phi}\tau & 0 \end{pmatrix} = \exp(\xi K_{-,1} - \overline{\xi}K_{+,1}), \quad (4.5)$$

which play an important role in the latter sections. Since the matrix is uniquely determined by $\zeta = e^{i\phi}\tan\tau$ and $\xi = e^{i\phi}\tau$, the above matrix is denoted by $g_{\zeta,1}$ and $\hat{g}_{\xi,1}$.

The set of traceless 2×2 Hermitian matrices is closed with respect to the action of SU(2). For an element $g \in \mathrm{SU}(2)$, we define the map on the set of traceless matrices as $\begin{pmatrix} z & x-yi \\ x+yi & -z \end{pmatrix} \mapsto g\begin{pmatrix} z & x-yi \\ x+yi & -z \end{pmatrix}g^{-1}$. Since $-(x^2+y^2+z^2) = \det\begin{pmatrix} z & x-yi \\ x+yi & -z \end{pmatrix} = \det g\begin{pmatrix} z & x-yi \\ x+yi & -z \end{pmatrix}g^{-1}$, So, any element $g \in \mathrm{SU}(2)$ gives a 3×3 orthogonal matrix $\iota(g)$ based on the coordinate $(x, y, z) \in \mathbb{R}^3$, which is a representation of SU(2) on $\mathbb{R}^3$. Thus, the relations

$$\iota(F_1^x) = \begin{pmatrix} 0 & 0 & 0 \\ 0 & 0 & 1 \\ 0 & -1 & 0 \end{pmatrix}, \quad \iota(F_1^y) = \begin{pmatrix} 0 & 0 & -1 \\ 0 & 0 & 0 \\ 1 & 0 & 0 \end{pmatrix}, \quad \iota(F_1^z) = \begin{pmatrix} 0 & -1 & 0 \\ 1 & 0 & 0 \\ 0 & 0 & 0 \end{pmatrix}$$

hold. Hence, the Lie algebras $\mathfrak{su}(2)$ and $\mathfrak{so}(3,\mathbb{R})$ are isomorphic to each other. However, the homomorphism ι from SU(2) to $\mathrm{SO}(3,\mathbb{R})$ is not an isomorphism because the matrices $I, -I \in \mathrm{SU}(2)$ are mapped to the unit matrix on $\mathbb{R}^3$. The fundamental group of the Lie group SU(2) is trivial, i.e., $\{e\}$ and that of the Lie group $\mathrm{SO}(3,\mathbb{R})$ is $\mathbb{Z}_2$. The central extension of $\mathbb{Z}_2$ by $\mathrm{SO}(3,\mathbb{R})$ is isomorphic to SU(2). So, we obtain the first equation in (3.8).

4.1.3 *Structure of* SU(1, 1)

Next, let us consider the Lie algebra $\mathfrak{su}(1,1)$ of the Lie subgroup SU(1, 1) of $\mathrm{SL}(2,\mathbb{C})$. For this purpose, we choose the following basis

$$F_{-1}^x := \frac{1}{2}\begin{pmatrix} 0 & -i \\ i & 0 \end{pmatrix}, \quad F_{-1}^y := \frac{1}{2}\begin{pmatrix} 0 & -1 \\ -1 & 0 \end{pmatrix}, \quad F_{-1}^z := \frac{1}{2}\begin{pmatrix} -i & 0 \\ 0 & i \end{pmatrix},$$

where the lower subscript -1 expresses the Lie group SU(1, 1). They satisfy the commutation relations $[F_{-1}^z, F_{-1}^x] = F_{-1}^y$, $[F_{-1}^z, F_{-1}^y] = -F_{-1}^x$, and $[F_{-1}^x, F_{-1}^y] =$

$-F^z_{-1}$. Combining these basis, we can recover the basis of $\mathfrak{sl}(2,\mathbb{C})$ as follows

$$E_{0,-1} := -iF^z_{-1} = \frac{1}{2}\begin{pmatrix} -1 & 0 \\ 0 & 1 \end{pmatrix},$$
$$K_{+,-1} := iF^x_{-1} + F^y_{-1} = \begin{pmatrix} 0 & 0 \\ -1 & 0 \end{pmatrix},$$
$$K_{-,-1} := iF^x_{-1} - F^y_{-1} = \begin{pmatrix} 0 & 1 \\ 0 & 0 \end{pmatrix}.$$

They satisfy the commutation relation $[E_{0,-1}, K_{\pm,-1}] = \pm K_{\pm,-1}$, $[K_{+,-1}, K_{-,-1}] = -2E_{0,-1}$. The matrices $E_{0,-1}$ and $K_{\pm,-1}$ are not elements of $\mathfrak{su}(1,1)$, but are useful for discussing representations of $\mathfrak{su}(1,1)$ later. The group H defined above is a subgroup of SU(1, 1), which is isomorphic to U(1). By using two complex numbers μ and ν satisfying that $|\mu|^2 - |\nu|^2 = 1$, any element of SU(1, 1) can be written as $\begin{pmatrix} \mu & \nu \\ \overline{\nu} & \overline{\mu} \end{pmatrix}$, which is converted to $\begin{pmatrix} \mu e^{-it/2} & \nu e^{it/2} \\ \overline{\nu} e^{-it/2} & \overline{\mu} e^{it/2} \end{pmatrix}$ by applying the element $\exp(tF^z_1)$ of H from the right hand side. Hence, an element of the homogeneous space $\mathrm{SU}(1,1)/H$ with the stabilizer H is given as the ratio between ν and $\overline{\mu}$. Hence, the homogeneous space $\mathrm{SU}(1,1)/H$ is the **unit disk** $D := \{\zeta \in \mathbb{C} | |\zeta| < 1\}$ (Fig. 4.2), and its elements are given as $\nu/\overline{\mu} \in \mathbb{C}$. So, we choose the representative of the element $\mathrm{SU}(1,1)/H$ such that $\mu \in \mathbb{R}$. Then, by using $\tau \in \mathbb{R}$, $\phi \in \mathbb{R}$, the representative is written as $\begin{pmatrix} \cosh\tau & e^{i\phi}\sinh\tau \\ e^{-i\phi}\sinh\tau & \cosh\tau \end{pmatrix}$. Letting $\zeta := e^{i\phi}\tan\tau$ and $\eta := -\log(1-|\zeta|^2)$, and using (4.2), we have its Gauss decomposition based on $K_{\pm,-1}$, $E_{0,-1}$ as follows

$$\begin{pmatrix} \cosh\tau & e^{i\phi}\sinh\tau \\ e^{-i\phi}\sinh\tau & \cosh\tau \end{pmatrix} = \exp(\zeta K_{-,-1})\exp(\eta E_{0,-1})\exp(-\overline{\zeta}K_{+,-1}). \tag{4.6}$$

Hence, letting $\xi := e^{i\phi}\tau$, we obtain

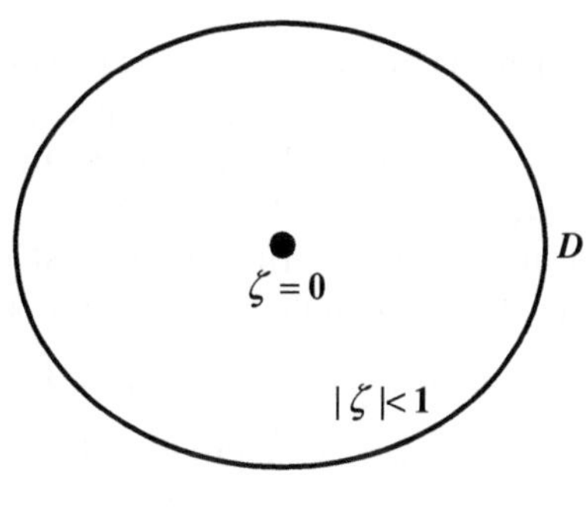

Fig. 4.2 Unit diskUnit disk

$$\begin{pmatrix} \cosh \tau & e^{i\phi} \sinh \tau \\ e^{-i\phi} \sinh \tau & \cosh \tau \end{pmatrix} = \exp \begin{pmatrix} 0 & e^{i\phi}\tau \\ e^{-i\phi}\tau & 0 \end{pmatrix} = \exp(\xi K_{-,-1} - \overline{\xi} K_{+,-1}), \quad (4.7)$$

which will play an important role later. Since the matrix is uniquely determined by $\zeta = e^{i\phi} \tan \tau$ and $\xi = e^{i\phi}\tau$, the above matrix is denoted by $g_{\zeta,-1}$ and $\hat{g}_{\xi,-1}$.

As mentioned in (3.18), the Lie group SU(1, 1) is isomorphic to the Lie groups $SL(2, \mathbb{R})$ and $Sp(2, \mathbb{R})$. In particular, the Lie groups $Sp(2, \mathbb{R})$ and $SL(2, \mathbb{R})$ are the subgroup of $SL(2, \mathbb{C})$. Now, we study their relation with SU(1, 1). By using two complex numbers $\mu = a - ib$ and $\nu = c + id$ satisfying that $|\mu|^2 - |\nu|^2 = 1$, an element g of SU(1, 1) can be described as

$$g = \begin{pmatrix} \mu & \nu \\ \overline{\nu} & \overline{\mu} \end{pmatrix} = \begin{pmatrix} a - ib & c + id \\ c - id & a + ib \end{pmatrix}.$$

Then, the map $\kappa : SU(1, 1) \to SL(2, \mathbb{R})$ is defined as

$$\kappa(g) := W \begin{pmatrix} a - ib & c + id \\ c - id & a + ib \end{pmatrix} W^{-1} = \begin{pmatrix} a + c & b + d \\ -b + d & a - c \end{pmatrix}, \quad (4.8)$$

$$W := \frac{1}{\sqrt{2}} \begin{pmatrix} 1 & 1 \\ -i & i \end{pmatrix}. \quad (4.9)$$

Here, we consider the action of g to the pair of complex numbers $z = x + iy$ and $\overline{z} = x - iy$ (defined by two real numbers x and y) as

$$\begin{pmatrix} x + iy \\ x - iy \end{pmatrix} \mapsto \begin{pmatrix} gz \\ g\overline{z} \end{pmatrix} := g \begin{pmatrix} x + iy \\ x - iy \end{pmatrix}. \quad (4.10)$$

When we focus on the real numbers x and y, by using κ defined the above, the transformation (4.10) can be written as

$$\begin{pmatrix} x \\ y \end{pmatrix} \mapsto \kappa(g) \begin{pmatrix} x \\ y \end{pmatrix}.$$

Then, the matrix $\kappa(g)$ is contained in $SL(2, \mathbb{R})$. Since the map $g \mapsto \kappa(g)$ preserves the product of matrices and is one-to-one, SU(1, 1) and $SL(2, \mathbb{R}) = Sp(2, \mathbb{R})$ are isomorphic to each other. Their fundamental group is $\mathbb{Z}$. We denote the universal covering group of SU(1, 1) by $SU(\tilde{1}, 1)$. The central extension of $\mathbb{Z}_2$ by SU(1, 1) is the double covering group of SU(1, 1) and is denoted by $SU(\check{1}, 1)$. Then, taking the derivative in the group isomorphic relation given the above, we obtain the map κ from $\mathfrak{su}(1, 1)$ to $\mathfrak{sl}(2, \mathbb{R})$ as $\kappa \begin{pmatrix} -zi & y + xi \\ y - xi & zi \end{pmatrix} = \begin{pmatrix} y & z + x \\ -z + x & -y \end{pmatrix}$. Since the map κ preserves the product of Lie algebras, the map κ gives the isomorphic relation between $\mathfrak{su}(1, 1)$ and $\mathfrak{sl}(2, \mathbb{R})$, which is described in (3.26). Hence, the basis of $\mathfrak{sl}(2,\mathbb{R}) = \mathfrak{sp}(2,\mathbb{R})$ is given as $\kappa(F^x_{-1}) = \frac{1}{2}\begin{pmatrix} 0 & 1 \\ 1 & 0 \end{pmatrix}$, $\kappa(F^y_{-1}) = \frac{1}{2}\begin{pmatrix} 1 & 0 \\ 0 & -1 \end{pmatrix}$, and $\kappa(F^z_{-1}) = \frac{1}{2}\begin{pmatrix} 0 & 1 \\ -1 & 0 \end{pmatrix} = \frac{-1}{2} J$.

Exercise 4.1 Show the relation $SO(1,2)_0 \cong SU(1,1)/\mathbb{Z}_2$ by steps below.

a: Show that the set $\mathcal{M}_{1,1} := \left\{ \begin{pmatrix} z & x - yi \\ x + yi & z \end{pmatrix} | (x, y, z) \in \mathbb{R}^3 \right\}$ is invariant under the map $X \mapsto gXg^\dagger$ for any matrix $g \in SU(1,1)$.

b: Show that $\det X = \det gXg^\dagger$ for $X \in \mathcal{M}_{1,1}$ and any matrix $g \in SU(1,1)$.

c: Show the relation $SO(1,2)_0 \cong SU(1,1)/\mathbb{Z}_2$.

4.2 Irreducible Skew–Hermitian Representation of $\mathfrak{su}(2)$ and $\mathfrak{su}(1,1)$

4.2.1 Construction of Irreducible Skew–Hermitian Representation

This section treats irreducible unitary representations of the Lie groups SU(2) and SU(1, 1), which have the simplest structure among non-commutative Lie groups. For this purpose, we deal with irreducible skew–Hermitian representations of the corresponding Lie algebras $\mathfrak{su}(2)$ and $\mathfrak{su}(1,1)$. The complexifications of these real Lie algebras are $\mathfrak{sl}(2,\mathbb{C})$. However, the structures of their irreducible skew–Hermitian representations are completely different to each other.[1]

In the following, we deal with irreducible skew–Hermitian representations of $\mathfrak{su}(2)$ and $\mathfrak{su}(1,1)$ based on a unified approach, which will clarify their definitive difference. For this purpose, we choose the parameter s taking the value 1 or -1, where 1 indicates the Lie algebra $\mathfrak{su}(2)$ and -1 does the Lie algebra $\mathfrak{su}(1,1)$. Then, we obtain the commutation relation

$$[F^z_s, F^x_s] = F^y_s,\ [F^z_s, F^y_s] = -F^x_s,\ [F^x_s, F^y_s] = sF^z_s \tag{4.11}$$

with respect to the basis. We also denote the images of the generators with respect to the skew–Hermitian representation f by

$$\mathsf{F}^a_s := \mathsf{f}(F^a_s), \quad a = x, y, z, \ s = 1, -1.$$

Then, we obtain

$$\mathsf{E}_{0,s} := -i\mathsf{F}^z_s, \quad \mathsf{K}_{\pm,s} := i\mathsf{F}^x_s \pm \mathsf{F}^y_s, \ s = 1, -1. \tag{4.12}$$

[1]For the details of skew–Hermitian representations of $\mathfrak{su}(1,1)$, see the reference [59].

Since the Hamiltonian is often given by $\mathsf{E}_{0,s}$ in a quantum system, it is important to study the Hermitian matrix (the self-adjoint operator) $\mathsf{E}_{0,s}$.

By extending the representation f to the representation of the complex Lie algebra $\mathfrak{sl}(2, \mathbb{C})$, the matrices/ the operators $\mathsf{E}_{0,s}$ and $\mathsf{K}_{\pm,s}$ can be formally described by using $E_{0,s}$, $K_{\pm,s}$ given in Sect. 4.1 as follows

$$\mathsf{E}_{0,s} := \mathsf{f}(E_{0,s}), \quad \mathsf{K}_{\pm,s} := \mathsf{f}(K_{\pm,s}).$$

These operators satisfy the following relations

$$\begin{aligned} [\mathsf{E}_{0,s}, \mathsf{K}_{\pm,s}] = \pm\mathsf{K}_{\pm,s}, \quad [\mathsf{K}_{+,s}, \mathsf{K}_{-,s}] = 2s\mathsf{E}_{0,s} \\ \mathsf{E}_{0,s}^{\dagger} = \mathsf{E}_{0,s}, \quad \mathsf{K}_{+,s}^{\dagger} = \mathsf{K}_{-,s}. \end{aligned} \tag{4.13}$$

Since we can conversely recover F_s^x, F_s^y, and F_s^z from $\mathsf{E}_{0,s}$, $\mathsf{K}_{\pm,s}$ as

$$\mathsf{F}_s^z = i\mathsf{E}_{0,s}, \ \mathsf{F}_s^x = \frac{i}{2}(\mathsf{K}_{-,s} + \mathsf{K}_{+,s}), \ \mathsf{F}_s^y = \frac{1}{2}(\mathsf{K}_{-,s} - \mathsf{K}_{+,s}),$$

it is sufficient to discuss the actions of $\mathsf{E}_{0,s}$ and $\mathsf{K}_{\pm,s}$. Further, the two relations

$$[\mathsf{E}_{0,s}, \mathsf{K}_{-,s}^n] = -n\mathsf{K}_{-,s}^n, \quad [\mathsf{K}_{+,s}, \mathsf{K}_{-,s}^n] = \mathsf{K}_{-,s}^{n-1} ns(2\mathsf{E}_{0,s} - (n-1)I) \tag{4.14}$$

hold (see Exercise 4.3).

In the following, we assume that there exists an eigenvector v_0 corresponding to the eigenvalue λ of the Hermitian matrix /the self-adjoint operator $\mathsf{E}_{0,s}$ such that the eigenvector v_0 is contained in the Kernel of $\mathsf{K}_{+,s}$ and $\|v_0\| = 1$. It is known that this assumption holds for an irreducible skew–Hermitian representation of $\mathfrak{su}(2)$, but does not necessarily hold for an irreducible skew–Hermitian representation of $\mathfrak{su}(1, 1)$. In fact, there exists a skew–Hermitian representation of $\mathfrak{su}(1, 1)$ that does not satisfy this assumption.

When $\mathsf{K}_{-,-1}$ has a non-trivial Kernel instead of $\mathsf{K}_{+,-1}$, the following discussion is valid by exchanging the roles of $\mathsf{K}_{+,-1}$ and $\mathsf{K}_{-,-1}$. However, it is known that there exist irreducible skew–Hermitian representations of $\mathfrak{su}(1, 1)$ such that $\mathsf{K}_{+,-1}$ nor $\mathsf{K}_{-,-1}$ does not have a non-trivial Kernel. A part of such representations are called **principal series representations** and another part of them are called **complementary series representations** [59]. Since this book does not deal with such an irreducible skew–Hermitian representation, we assume the above assumption. The relation (4.14) guarantees that the vector $v_n := \mathsf{K}_{-,s}^n v_0$ satisfies

$$\mathsf{E}_{0,s} v_n = (\mathsf{K}_{-,s}^n \mathsf{E}_{0,s} + [\mathsf{E}_{0,s}, \mathsf{K}_{-,s}^n])v_0 = (\mathsf{K}_{-,s}^n \lambda - n\mathsf{K}_{-,s}^n)v_0 = (\lambda - n)v_n.$$

Hence, the vector v_n is the eigenvector of the Hermitian matrix /the self-adjoint operator $\mathsf{E}_{0,s}$ corresponding to distinct eigenvalues. So, the vector v_n is orthogonal to the vector v_m with $n \neq m$. Since

$$\mathsf{K}_{+,s}\mathsf{K}_{-,s}v_{n-1} = ns(2\lambda - (n-1))v_{n-1}, \tag{4.15}$$

(see Exercise 4.2) we have

$$\langle v_{n-1}|\mathsf{K}_{-,s}^{\dagger}\mathsf{K}_{-,s}|v_{n-1}\rangle = ns(2\lambda - (n-1))\langle v_{n-1}|v_{n-1}\rangle. \tag{4.16}$$

The property of the inner product of the representation space of the skew–Hermitian representation guarantees the relation $ns(2\lambda - (n-1)) \geq 0$ for any integer $n \geq 0$. So, the condition with $n = 1$ implies that $\lambda s \geq 0$.

When $\lambda = 0$, we have $\mathsf{K}_{-,s}v_0 = 0$. Thus, we have $\mathsf{F}_s^y = \mathsf{F}_s^x = 0$, which implies that the skew–Hermitian representation is the trivial one-dimensional skew–Hermitian representation. Hence, we discuss only the case with $\lambda s > 0$. Assume that $s = 1$. We have $\lambda > 0$. When 2λ is an integer, we have $\mathsf{K}_{-,s}v_{2\lambda} = 0$, which implies that the skew–Hermitian representation is a $2\lambda + 1$-dimensional irreducible skew–Hermitian representation. When 2λ is not an integer, the vector v_n does not vanish for any integer $n \geq 0$. So, there exists a positive integer n such that $(2\lambda - (n-1)s)$ is a negative value, which yields the contradiction. Thus, we find that 2λ is an integer. Now, we assume that $s = -1$. So, we have $\lambda < 0$, which implies the inequality $ns(2\lambda - (n-1)) > 0$ for any integer $n \geq 1$. Hence, the space spanned by $\{v_n|n \geq 0\}$ is an irreducible representation space.

The eigenvalue λ of $\mathsf{E}_{0,s}$ corresponding to the vector in the kernel $\operatorname{Ker}\mathsf{K}_{-,s}$ uniquely determines the skew–Hermitian representation. So, in the following, we regard the eigenvalue λ as the element of the set $\hat{G}$ labeling the irreducible skew–Hermitian representation so that the representation space is written as $\mathcal{U}_\lambda$. In particular, we denote it by $\mathcal{U}_\lambda(\mathfrak{g})$ when we need to clarify the Lie algebra $\mathfrak{g}$ of our interest.

Further, since $\langle v_n|v_n\rangle = \langle v_{n-1}|\mathsf{K}_{-,s}^{\dagger}\mathsf{K}_{-,s}|v_{n-1}\rangle = ns(2\lambda - (n-1))\langle v_{n-1}|v_{n-1}\rangle$, we have $\langle v_n|v_n\rangle = n!s(2\lambda - (n-1))s(2\lambda - (n-2))\cdots s(2\lambda)$. Hence, the normalized eigenvector $|\lambda; \lambda - n\rangle := \frac{1}{\|v_n\|}v_n$ of $\mathsf{E}_{0,s}$ corresponding to the eigenvalue $\lambda - n$ satisfies $\mathsf{K}_{-,s}|\lambda; m+1\rangle = \mathsf{K}_{-,s}\frac{1}{\|v_{n-1}\|}v_{n-1} = \sqrt{ns(2\lambda - (n-1))}\frac{1}{\|v_n\|}v_n = \sqrt{s(\lambda - m)(\lambda + m + 1)}|\lambda; m\rangle$ when $m + 1 = \lambda - (n-1)$. The eigenvalue $\lambda - n$ of $\mathsf{E}_{0,s}$ is called a **weight**, and the normalized eigenvector $|\lambda; \lambda - n\rangle$ is called the **weight vector**. So, we obtain

$$\mathsf{E}_{0,s}|\lambda; m\rangle = m|\lambda; m\rangle \tag{4.17}$$

$$\mathsf{K}_{+,s}|\lambda; m\rangle = \sqrt{s(\lambda - m)(\lambda + m + 1)}|\lambda; m+1\rangle \tag{4.18}$$

$$\mathsf{K}_{-,s}|\lambda; m\rangle = \sqrt{s(\lambda - m + 1)(\lambda + m)}|\lambda; m-1\rangle. \tag{4.19}$$

Since the real number λ is the minimum value among weights in the representation space $\mathcal{U}_\lambda$, λ is called the **highest weight**. Hence, we denote the skew–Hermitian representation with the highest weight λ by f_λ.

In the quantum theory, the Hamiltonian is often given by $-\mathsf{E}_{0,s}$. In this case, the value $-m$ is the energy of the eigenstate $|\lambda; m\rangle$. In particular, the highest weight state has the minimum energy and is called the **ground state**.

Remark 4.1 In the case of $\mathrm{U}(1,1)$, the dimension of $\mathcal{H}$ is infinite. Since the operators F^x_s, F^y_s, and F^z_s are bounded, we need careful treatment for them. Since $i\mathsf{F}^x_s$, $i\mathsf{F}^y_s$, and $i\mathsf{F}^z_s$ are Hermitian band diagonal with respect to the base $\{|\lambda; m\rangle\}$, Lemmas 1.4 and 1.5 guarantee that they are essentially self-adjoint.

Remark 4.2 The skew–Hermitian representation of $\mathfrak{su}(2)$ on $\mathcal{U}_\lambda$ can be extended to an irreducible $\mathbb{C}$-representation of the complex Lie algebra $\mathfrak{sl}(2,\mathbb{C})$. However, the same discussion is not true for the skew–Hermitian representation of $\mathfrak{su}(1,1)$ on $\mathcal{U}_\lambda$ due to the infinite-dimensionality of the representation space. This is because there is no method to define a non-skew–Hermitian representation on an infinite-dimensional space.

Exercise 4.2 Show (4.15) by using (4.13).

Exercise 4.3 Show the equations in (4.14) by using (4.13).

4.2.2 Unitary Representation and Real Representation of Lie Group

We assume that $\lambda > 0$. Based on a skew–Hermitian representation of $\mathfrak{su}(2)$ on $\mathcal{U}_\lambda$, we can define a unitary representation of simply connected Lie group SU(2) by considering the map exp. Hence, when $\lambda > 0$ is an integer, the dimension of the representation space $\mathcal{U}_\lambda$ is an odd number and coprime to the order of the fundamental group of $\mathrm{SO}(3,\mathbb{R})$, which is 2. Since SU(2) is a central extension of the fundamental group of $\mathrm{SO}(3,\mathbb{R})$ by $\mathrm{SO}(3,\mathbb{R})$, Lemma 2.2 guarantees that the unitary representation of $\mathrm{SO}(3,\mathbb{R})$ on the representation space $\mathcal{U}_\lambda$ is an irreducible unitary representation. On the other hand, when λ is a half-integer, the skew–Hermitian representation of $\mathfrak{su}(2)$ cannot be extended to a unitary representation of $\mathrm{SO}(3,\mathbb{R})$. However, as mentioned in the end of Sect. 2.4, it can be extended to a unitary representation of SU(2), which is a central extension of U_2 by $\mathrm{SO}(3,\mathbb{R})$. So, there exists a common factor system for the projective representations whose highest weight is a half-integer λ. Hence, when $\lambda_1, \ldots$, and λ_k are odd numbers, the direct sum representation $\mathsf{f}_{\lambda_1} \oplus \ldots \oplus \mathsf{f}_{\lambda_k}$ can be extended to a projective representation of $\mathrm{SO}(3,\mathbb{R})$.

We assume that $\lambda < 0$. Based on a skew–Hermitian representation of $\mathfrak{su}(1,1)$ on $\mathcal{U}_\lambda$, we can define a unitary representation of universal covering group $\widetilde{\mathrm{SU}(1,1)}$ by considering the map exp. The skew–Hermitian representation of $\mathfrak{su}(1,1)$ can be extended to a unitary representation of $\mathrm{SU}(1,1)$ if and only if 2λ is an integer. Otherwise, it can be extended only to a projective unitary representation of $\mathrm{SU}(1,1)$. When 4λ is an odd number, it cannot be extended to a unitary representation of $\mathrm{SU}(1,1)$, but can be to a unitary representation of a double covering group $\check{\mathrm{SU}}(1,1)$.

Due to the same discussion, when $c_1, \ldots$, and c_k are odd numbers, the direct sum representation $\mathsf{f}_{c_1/4} \oplus \ldots \oplus \mathsf{f}_{c_k/4}$ can be extended to a projective representation of $\check{\mathrm{SU}}(1, 1)$.

Here, we consider the complex conjugate representation of an irreducible representation f_λ. We can make the complex conjugate representation by replacing $\mathsf{E}_{0,s}, \mathsf{K}_{\pm,s}$ by $-\mathsf{E}_{0,s}, -\mathsf{K}_{\mp,s}$ in (4.17)–(4.19). Hence, when $s = 1$, the complex conjugate representation is isomorphic to f_λ via the isomorphism $|\lambda; m\rangle \mapsto |\lambda; -m\rangle$, i.e., it is a self-adjoint representation. When λ is an integer, it is underlying real form. However, when λ is a half-integer, it is not underlying real form. On the other hand, when $s = -1$, in the complex conjugate representation, the operator $\mathsf{K}_{-,-1}$ has a non-trivial kernel instead of $\mathsf{K}_{+,-1}$. That is, while the representation space of the representation f_λ is generated by the highest weight vector, the representation space of the complex conjugate representation is generated by the minimum weight vector. Hence, f_λ is not a self-adjoint representation.

4.2.3 Casimir Operator

Next, we calculate the Casimir operator of these representations. For this purpose, we calculate the Killing form. Let $x = 1, y = 2, z = 3$. Based on the basis F_s^x, F_s^y, F_s^z, we make the matrix expression of $\mathsf{ad}(F_s^x)$, $\mathsf{ad}(F_s^y)$, and $\mathsf{ad}(F_s^z)$ as follows.

$$\mathsf{ad}(F_s^x) = \begin{pmatrix} 0 & 0 & 0 \\ 0 & 0 & -1 \\ 0 & s & 0 \end{pmatrix}, \quad \mathsf{ad}(F_s^y) = \begin{pmatrix} 0 & 0 & 1 \\ 0 & 0 & 0 \\ -s & 0 & 0 \end{pmatrix}, \quad \mathsf{ad}(F_s^z) = \begin{pmatrix} 0 & -1 & 0 \\ 1 & 0 & 0 \\ 0 & 0 & 0 \end{pmatrix}.$$

Then, the matrix expression of the Killing form is given as follows.

$$[(F_s^a, F_s^b)_{\mathfrak{g}}]_{1 \le a,b \le 3} = \begin{pmatrix} -2s & 0 & 0 \\ 0 & -2s & 0 \\ 0 & 0 & -2 \end{pmatrix}.$$

Then, the Casimir operator C_λ of f_λ can be calculated as

$$\begin{aligned} C_\lambda &= -\frac{1}{2}(s(\mathsf{F}_s^x)^2 + s(\mathsf{F}_s^y)^2 + (\mathsf{F}_s^z)^2) \\ &= \frac{1}{2}(\mathsf{E}_{0,s}^2 + \frac{s}{2}(\mathsf{K}_{+,s}\mathsf{K}_{-,s} + \mathsf{K}_{-,s}\mathsf{K}_{+,s})). \end{aligned} \tag{4.20}$$

Hence, since $C_\lambda v_0 = \frac{1}{2}(\lambda^2 + \lambda)v_0$, we have $C_\lambda = \frac{1}{2}\lambda(\lambda + 1)$. Considering the matrix-valued vector $\vec{\mathsf{F}}_1 := (\mathsf{F}_1^x, \mathsf{F}_1^y, \mathsf{F}_1^z)$, we have

$$\vec{\mathsf{F}}_1 \cdot \vec{\mathsf{F}}_1 = -\lambda(\lambda + 1). \tag{4.21}$$

When $s = 1(\mathfrak{su}(2))$, since there is a one-to-one correspondence between λ and $\frac{1}{2}\lambda(\lambda+1)$ under the condition $\lambda > 0$, the representation is uniquely determined by the Casimir operator. However, when $s = -1(\mathfrak{su}(1,1))$, there is no one-to-one correspondence between λ and $\frac{1}{2}\lambda(\lambda+1)$ under the condition $-1 < \lambda < 0$ and $\lambda \neq -\frac{1}{2}$. This fact is so crucial that the negative Casimir operator cannot uniquely determine the representation. Hence, to determine the representation, we need a more careful discussion. In other words, in the case of $\mathfrak{su}(2)$, we can realize the measurement corresponding to the irreducible decomposition by measuring the physical quantity corresponding the Casimir operator $C_{\mathfrak{f}}$. However, in the case of $\mathfrak{su}(1,1)$, even though we measure the physical quantity corresponding the Casimir operator $C_{\mathfrak{f}}$, we cannot necessarily realize the measurement corresponding to the irreducible decomposition. That is, this method cannot distinguish two distinct irreducible components realizing the same value as the Casimir operator.

4.2.4 Tensor Product

Given two quantum systems, their composite system is the tensor product space. Hence, when two quantum systems have transformations by the same group $SU(2)$ or $SU(1,1)$, the tensor product space also has transformations by the same group. To investigate what transformation is caused by the group in the tensor product space, we need the irreducible decomposition of the representation on the tensor product space. For this purpose, we assume that the representation of the generator F^a_s on $\mathcal{U}_{\lambda_1} \otimes \mathcal{U}_{\lambda_2}$ is given by $\mathsf{F}^a_{s;1} + \mathsf{F}^a_{s;2}$, where $\mathsf{F}^a_{s;1} := \mathsf{F}^a_s \otimes I$ and $\mathsf{F}^a_{s;2} := I \otimes \mathsf{F}^a_s$. We denote the tensor product $|\lambda_1; m_1\rangle \otimes |\lambda_2; m_2\rangle$ by $|\lambda_1, \lambda_2; m_1, m_2\rangle$.

In the case of $\mathfrak{su}(2)$, we have

$$\mathcal{U}_{\lambda_1}(\mathfrak{su}(2)) \otimes \mathcal{U}_{\lambda_2}(\mathfrak{su}(2)) = \bigoplus_{|\lambda_1-\lambda_2| \le \lambda \le \lambda_1+\lambda_2} \mathcal{U}_{\lambda}(\mathfrak{su}(2)), \tag{4.22}$$

where λ takes values in $\{|\lambda_1 - \lambda_2|, |\lambda_1 - \lambda_2| + 1, \ldots, \lambda_1 + \lambda_2\}$. (4.22) can be checked as follows. In the space $\mathcal{U}_{\lambda_1} \otimes \mathcal{U}_{\lambda_2}(\mathfrak{su}(2))$, the eigenspace of $\mathsf{E}_{0,1}$ with the eigenvalue $\lambda_1 + \lambda_2$ is the one-dimensional space spanned by $|\lambda_1, \lambda_2, \lambda_1, \lambda_2\rangle$, which is the generator of $\mathcal{U}_{\lambda_1+\lambda_2}(\mathfrak{su}(2))$.

The eigenspace of $\mathsf{E}_{0,1}$ with the eigenvalue $\lambda_1 + \lambda_2 - 1$ is the two-dimensional space spanned by $|\lambda_1, \lambda_2; \lambda_1 - 1, \lambda_2\rangle$ and $|\lambda_1, \lambda_2; \lambda_1, \lambda_2 - 2\rangle$. Then, the vector in the eigenspace orthogonal to $\mathsf{K}_{-,1}|\lambda_1, \lambda_2; \lambda_1, \lambda_2\rangle$ is the generators of the different representation space $\mathcal{U}_{\lambda_1+\lambda_2-1}(\mathfrak{su}(2))$. The eigenspace of $\mathsf{E}_{0,1}$ with the eigenvalue $\lambda_1 + \lambda_2 - 2$ is three-dimensional. From the eigenspace, we choose the generator of the representation space $\mathcal{U}_{\lambda_1+\lambda_2-2}(\mathfrak{su}(2))$. We repeat this process up to the eigenvalue $|\lambda_1 - \lambda_2|$. Then, when the eigenvalue is larger than $|\lambda_1 - \lambda_2|$, the dimension of the eigenspace does not increase. This discussion shows (4.22).

In the case of $\mathfrak{su}(1,1)$, we can apply the same discussion with the order decreasing the eigenvalues. Then, the dimension of the eigenspace increases up to infinity. Thus, we obtain

$$\mathcal{U}_{\lambda_1} \otimes \mathcal{U}_{\lambda_2}(\mathfrak{su}(1,1)) = \bigoplus_{n=0}^{\infty} \mathcal{U}_{\lambda_1+\lambda_2-n}(\mathfrak{su}(1,1)).$$

Notice that the linear space of right hand side of (4.22) is generated by the basis $\{|\lambda; m\rangle\}_{|\lambda_1-\lambda_2|\le\lambda\le\lambda_1+\lambda_2, |m|\le\lambda}$. Then, an element $|\lambda_1, \lambda_2; m_1, m_2\rangle$ of $\mathcal{U}_{\lambda_1}(\mathfrak{su}(1,1)) \otimes \mathcal{U}_{\lambda_2}(\mathfrak{su}(1,1))$ is expressed by using the above basis as

$$|\lambda_1, \lambda_2; m_1, m_2\rangle = \sum_{\lambda,m} |\lambda; m\rangle\langle\lambda; m|\lambda_1, \lambda_2; m_1, m_2\rangle.$$

Since the coefficient $\langle\lambda; m|\lambda_1, \lambda_2; m_1, m_2\rangle$ has the ambiguity for the phase factor, we fix the phase factors of the basis such that $\langle\lambda; \lambda|\lambda_1, \lambda_2; \lambda_1, \lambda - \lambda_1\rangle > 0$. When our algebra is restricted to $\mathfrak{su}(2)$, the coefficient $\langle\lambda; m|\lambda_1, \lambda_2; m_1, m_2\rangle$ is called **Clebsh-Gordan coefficient**. The coefficient is always a real number, and the following relation is known [78, Chap. 13 and Appendix C].

$$\begin{aligned}
\langle\lambda; m|\lambda_1, \lambda_2; m_1, m_2\rangle &= (-1)^{\lambda_1+\lambda_2-\lambda}\langle\lambda; m|\lambda_2, \lambda_1; m_2, m_1\rangle \\
&= (-1)^{\lambda_1+\lambda+m_2}\sqrt{\frac{2\lambda+1}{2\lambda_1+1}}\langle\lambda_1; m_1|\lambda, \lambda_2; m, -m_2\rangle \\
&= (-1)^{\lambda_2-\lambda-m_1}\sqrt{\frac{2\lambda+1}{2\lambda_2+1}}\langle\lambda_2; m_2|\lambda_1, \lambda; -m_1, m\rangle \\
&= (-1)^{\lambda_1+\lambda_2-\lambda}\langle\lambda; -m|\lambda_1, \lambda_2; -m_1, -m_2\rangle. \qquad (4.23)
\end{aligned}$$

To characterize the coefficient, we focus on the relation

$$\begin{aligned}
&-\vec{\mathsf{F}}_s \cdot \vec{\mathsf{F}}_s + \vec{\mathsf{F}}_{s;1} \cdot \vec{\mathsf{F}}_{s;1} + \vec{\mathsf{F}}_{s;2} \cdot \vec{\mathsf{F}}_{s;2} = -2\vec{\mathsf{F}}_{s;1} \cdot \vec{\mathsf{F}}_{s;2} \\
&= 2\mathsf{E}_{0,1;1}\mathsf{E}_{0,1;2} + \mathsf{K}_{+,1;1}\mathsf{K}_{-,1;2} + \mathsf{K}_{-,1;1}\mathsf{K}_{+,1;2},
\end{aligned}$$

where $\mathsf{K}_{\pm,1;x}$ and $\mathsf{E}_{0,1;x}$ are defined as the same way as $\mathsf{F}^a_{1;x}$. Thus, multiplying $|\lambda; m\rangle$ and $\langle\lambda; m|$ from both sides, we have

$$\begin{aligned}
&\lambda(\lambda+1) - \lambda_1(\lambda_1+1) - \lambda_2(\lambda_2+1) \\
=& \sum_{m_2,m_2'} \langle\lambda_1, \lambda_2; m_1, m_2|(2\mathsf{E}_{0,1;1}\mathsf{E}_{0,1;2} + \mathsf{K}_{+,1;1}\mathsf{K}_{-,1;2} + \mathsf{K}_{-,1;1}\mathsf{K}_{+,1;2}) \\
&|\lambda_1, \lambda_2; m_1', m_2'\rangle \cdot \langle\lambda, m|\lambda_1, \lambda_2; m_1, m_2\rangle\langle\lambda, m|\lambda_1, \lambda_2; m_1', m_2'\rangle \\
=& \sum_{m_2} m_1 m_2 \langle\lambda, m|\lambda_1, \lambda_2; m_1, m_2\rangle^2
\end{aligned}$$

$$+\sum_{m_2}\sqrt{(\lambda_1-m_1)(\lambda_1+m_1+1)(\lambda_2-m_2+1)(\lambda_2+m_2)}$$
$$\cdot\langle\lambda,m|\lambda_1,\lambda_2;m_1,m_2\rangle\langle\lambda,m|\lambda_1,\lambda_2;m_1+1,m_2-1\rangle$$
$$+\sum_{m_2}\sqrt{(\lambda_1-m_1+1)(\lambda_1+m_1)(\lambda_2-m_2)(\lambda_2+m_2+1)}$$
$$\cdot\langle\lambda,m|\lambda_1,\lambda_2;m_1,m_2\rangle\langle\lambda,m|\lambda_1,\lambda_2;m_1-1,m_2+1\rangle, \tag{4.24}$$

where m_1 and m_1' are given as $m-m_2$ and $m-m_2'$, respectively.

For example, it can be calculated as follows.

Example 4.1 We consider the case when $\lambda_2=\frac{1}{2}$. We put $\alpha:=\langle\lambda,m|\lambda_1,\frac{1}{2},m-\frac{1}{2},\frac{1}{2}\rangle$ and $\beta:=\langle\lambda,m|\lambda_1,\frac{1}{2},m+\frac{1}{2},-\frac{1}{2}\rangle$. Equation (4.24) implies that

$$\lambda(\lambda+1)-\lambda_1(\lambda_1+1)-\frac{3}{4}$$
$$=(m-\frac{1}{2})\alpha^2-(m+\frac{1}{2})\beta^2+2\sqrt{(\lambda_1+\frac{1}{2})^2-m^2}\alpha\beta. \tag{4.25}$$

Employing the new number $t:=\frac{m}{\lambda+1}$ and the relation $\alpha^2+\beta^2=1$, we have

$$t(\alpha^2-\beta^2)+2\sqrt{1-t^2}\alpha\beta=\begin{cases}1 & \text{when } \lambda=\lambda_1+\frac{1}{2}\\ -1 & \text{when } \lambda=\lambda_1-\frac{1}{2}.\end{cases} \tag{4.26}$$

So, when $\lambda=\lambda_1+\frac{1}{2}$, we have

$$\alpha=\sqrt{\frac{1+t}{2}}=\sqrt{\frac{\lambda_1+m+\frac{1}{2}}{2\lambda_1+1}},\quad \beta=\sqrt{\frac{1-t}{2}}=\sqrt{\frac{\lambda_1-m+\frac{1}{2}}{2\lambda_1+1}}. \tag{4.27}$$

When $\lambda=\lambda_1-\frac{1}{2}$, we have

$$\alpha=\sqrt{\frac{1-t}{2}}=\sqrt{\frac{\lambda_1-m+\frac{1}{2}}{2\lambda_1+1}},\quad \beta=-\sqrt{\frac{1+t}{2}}=-\sqrt{\frac{\lambda_1+m+\frac{1}{2}}{2\lambda_1+1}}. \tag{4.28}$$

Example 4.2 When $m=\pm\lambda$, the following relation is known [78, Appendix C].

$$\langle\lambda;\lambda|\lambda_1,\lambda_2;m_1,m_2\rangle=\langle\lambda;-\lambda|\lambda_2,\lambda_1;-m_2,-m_1\rangle$$
$$=(-1)^{\lambda_1-m_1}\sqrt{\frac{(2\lambda+1)!(\lambda_1+\lambda_2-\lambda)!(\lambda_1+m_1)!(\lambda_2+m_2)!}{(\lambda_1+\lambda_2+\lambda+1)!(\lambda+\lambda_1-\lambda_2)!(\lambda+\lambda_2-\lambda_1)!}}$$
$$\cdot\sqrt{\frac{1}{(\lambda_1-m_1)!(\lambda_2-m_2)!}}, \tag{4.29}$$

where $\lambda = m_1 + m_2$. Hence, using (4.23), we can show the following relation when $m_1 = \pm\lambda_1$

$$\begin{aligned}\langle\lambda; m|\lambda_1, \lambda_2; \lambda_1, m_2\rangle &= (-1)^{\lambda_1+\lambda_2-\lambda}\langle\lambda; -m|\lambda_1, \lambda_2; -\lambda_1, -m_2\rangle \\ &= \sqrt{\frac{(2\lambda+1)(2\lambda_1)!(\lambda+\lambda_2-\lambda_1)!(\lambda+m)!(\lambda_2-m_2)!}{(\lambda+\lambda_2+\lambda_1+1)!(\lambda_1+\lambda-\lambda_2)!(\lambda_1+\lambda_2-\lambda)!(\lambda-m)!(\lambda_2+m_2)!}},\end{aligned} \tag{4.30}$$

where $\lambda_1 = m + m_2$.

Example 4.3 When $\lambda = \lambda_1 + \lambda_2$, the following relation is known [78, Appendix C].

$$\begin{aligned}&\langle\lambda; m|\lambda_1, \lambda_2; \lambda_1, m_2\rangle \\ &= \sqrt{\frac{(2\lambda_1)!(2\lambda_2)!(\lambda+m)!(\lambda-m)!}{(2\lambda)!(\lambda_1+m_1)!(\lambda_1-m_1)!(\lambda_2+m_2)!(\lambda_2-m_2)!}}.\end{aligned}$$

Exercise 4.4 Show (4.26) by using the relations (4.25) and $\alpha^2 + \beta^2 = 1$.

4.2.5 *Coherent Vector*

Next, when a representation of the group SU(2) or SU(1, 1) is given on a quantum system, we investigate the orbit of the ground state by the action of the group. Based on discussion in Sect. 4.1, we define the **coherent vector** based on an element ζ of the homogeneous space, the Riemann sphere $\hat{\mathbb{C}}$ or the unit disk D with the stabilizer $H = \{e^{i\theta \mathsf{E}_{0,s}}\}$ [52].[2] The coherent vector is the orbit of the ground state. For a given ζ, we choose $\xi \in \mathbb{C}, \eta \in \mathbb{R}$ according to (4.5) and (4.7) in Sect. 4.1. Then, $\exp(\xi K_{+,s} - \bar{\xi} K_{-,s})) = g_{\zeta,s} = \hat{g}_{\xi,s}$ is an element of the Lie group SU(2) or SU(1, 1). We obtain

$$\begin{aligned}\mathsf{U}_{\zeta,s} &:= \mathsf{f}_\lambda(g_{\zeta,s}) = \mathsf{f}_\lambda(\exp(\xi K_{+,s} - \bar{\xi} K_{-,s})) = \exp(\xi \mathsf{K}_{+,s} - \bar{\xi}\mathsf{K}_{-,s}) \\ &= \exp(\zeta \mathsf{K}_{-,s}) \exp(\eta \mathsf{E}_{0,s}) \exp(-\bar{\zeta}\mathsf{K}_{+,s}),\end{aligned} \tag{4.31}$$

where $\mathsf{U}_{\infty,1}$ is $\exp(\frac{\pi}{2}(\mathsf{K}_{+,s} - \mathsf{K}_{-,s}))$. Using $\mathsf{U}_{\zeta,s}$ and the highest weight vector $|\lambda;\lambda\rangle$, we define the coherent vector $|\lambda : \zeta\rangle := \mathsf{U}_{\zeta,s}|\lambda;\lambda\rangle$.

Since $\exp(\eta\lambda) = (1 + s|\zeta|^2)^{-\lambda}$, the coherent vector $|\lambda : \zeta\rangle$ can be written as

$$|\lambda : \zeta\rangle = (1 + s|\zeta|^2)^{-\lambda} \exp(\zeta \mathsf{K}_{-,s})|\lambda;\lambda\rangle$$

[2] When the group is SU(1, 1), this complex parametrization is different from that by Perelomov [82]. Lisiecki [74, Sect. 4] defined the same coherent vector with the same parametrization as a holomorphic function via the one-to-one correspondence between a square integrable function and a holomorphic function.

$$= (1 + s|\zeta|^2)^{-\lambda} \sum_{n=0}^{*} \frac{(\zeta \mathsf{K}_{-,s})^n}{n!} |\lambda; \lambda\rangle,$$

where $*$ is 2λ for $s = 1$ and $*$ is ∞ for $s = -1$.

The normalized invariant measure $\mu_{\hat{\mathbb{C}}}$ on the compact homogeneous space $\hat{\mathbb{C}}$, is written as $\mu_{\hat{\mathbb{C}}}(d\zeta) = \frac{dxdy}{2\pi(1+|\zeta|^2)^2}$, where $\zeta = \frac{x+iy}{\sqrt{2}}$. When $\lambda > 0$, the dimension of $\mathcal{U}_\lambda$ is $2\lambda + 1$, since the vector $|\lambda; \lambda\rangle\langle\lambda; \lambda|$ is invariant with respect to the action of the stabilizer H_1, the relation (3.56) yields (see Exercise 4.5)

$$(2\lambda + 1) \int_{\hat{\mathbb{C}}} |\lambda : \zeta\rangle\langle\lambda : \zeta| \frac{dxdy}{2\pi(1 + |\zeta|^2)^2} = I. \tag{4.32}$$

On the other hand, the invariant measure μ_D on the other homogeneous space D is also written as $\mu_D(d\zeta) = \frac{dxdy}{2\pi(1-|\zeta|^2)^2}$. When $\lambda < -\frac{1}{2}$, since the vector $|\lambda; \lambda\rangle\langle\lambda; \lambda|$ is invariant with respect to the action of the stabilizer H_{-1}, the relation (3.56) yields

$$-(2\lambda + 1) \int_D |\lambda : \zeta\rangle\langle\lambda : \zeta| \frac{dxdy}{2\pi(1 - |\zeta|^2)^2} = I, \tag{4.33}$$

where the constant factor $-(2\lambda + 1)$ will be given in Exercise 4.6. Notice that the above integral diverges when $-\frac{1}{2} \le \lambda < 0$. When $-\frac{1}{2} > \lambda$, the formal dimension of $\mathcal{U}_\lambda$ is $-(2\lambda + 1)$.

Hence, when $\lambda > 0$ or $\lambda < -\frac{1}{2}$, the invariant measure on Θ_s is unifiedly written as $\mu_{\Theta_s}(d\zeta) = \frac{dxdy}{2\pi(1+s|\zeta|^2)^2}$ and the relations (4.32) and (4.33) are unified to

$$s(2\lambda + 1) \int_{\Theta_s} |\lambda : \zeta\rangle\langle\lambda : \zeta| \mu_{\Theta_s}(d\zeta) = I. \tag{4.34}$$

As long as $s(2\lambda + 1)$ is positive, this equation shows that the coherent states give a measurement (POVM). Since the POVM has a good symmetry, it is useful for information processing with a symmetry.

Exercise 4.5 Show (4.32) by using only the invariance of the LHS of (4.32) without use of (3.56).

Exercise 4.6 Show (4.33) by using only the invariance of the LHS of (4.33).

Exercise 4.7 Calculate the outcome of $\zeta = \frac{x+iy}{\sqrt{2}}$ when the measurement corresponding to the POVM given in (4.32) is performed to the system whose state is $|\lambda; \lambda\rangle$.

Exercise 4.8 Calculate the outcome of $\zeta = \frac{x+iy}{\sqrt{2}}$ when the measurement corresponding to the POVM given in (4.33) is performed to the system whose state is $|\lambda; \lambda\rangle$.

4.3 Irreducible Unitary Representations of SU(r) and U(r)

4.3.1 Irreducible Unitary Representation of SU(r)

The unitary group U(r) and the special unitary group SU(r) with a general dimension r often appear in a quantum system. To deal with an irreducible unitary representation of SU(r), this section addresses a skew–Hermitian representation of $\mathfrak{su}(r)$. Since the Lie group SU(r) is simply connected, any skew–Hermitian representation of the real Lie algebra $\mathfrak{su}(r)$ can be directly converted to a unitary representation of the Lie group SU(r). Hence, when it is irreducible, it can be regarded as an element of $\hat{\mathrm{SU}}(r)$. An irreducible $\mathbb{C}$-representation of the complex Lie algebra $\mathfrak{sl}(r, \mathbb{C})$ can be obtained by extending a representation of the real Lie algebra $\mathfrak{su}(r)$ to complex coefficients. Then, since the complex Lie group SL$(r, \mathbb{C})$ is simply connected, a $\mathbb{C}$-representation of SL$(r, \mathbb{C})$ is also obtained from the $\mathbb{C}$-representation discussed here.

To treat a representation of the Lie algebra $\mathfrak{su}(r)$, we denote the Lie subalgebra consisting of the diagonal elements of $\mathfrak{su}(r)$ by $\mathfrak{sd}(r)$, $\mathfrak{d}$ is the Fraktur of d. The Lie subalgebra $\mathfrak{sd}(r)$ is the maximal Cartan subalgebra, which will be defined later. As extensions of F_1^x, F_1^y, F_1^z defined in Sect. 4.2, we introduce

$$F_{j,l}^x := \frac{i}{2}(|l\rangle\langle j| + |j\rangle\langle l|), \quad 1 \le j < l \le r \tag{4.35}$$

$$F_{j,l}^y := \frac{1}{2}(|j\rangle\langle l| - |l\rangle\langle j|), \quad 1 \le j < l \le r \tag{4.36}$$

$$F_{j,l}^z := -i(|j\rangle\langle j| - |l\rangle\langle l|), \quad j = 1, \dots, r-1. \tag{4.37}$$

The set $\{F_{j,j+1}^z\}_{j=1}^{r-1}$ forms a basis of $\mathfrak{sd}(r)$. Adding $\{F_{j,l}^x, F_{j,l}^y\}_{1\le j<l\le r}$ to the basis of $\mathfrak{sd}(r)$, we obtain a basis of $\mathfrak{su}(r)$. Given an irreducible skew–Hermitian representation f of $\mathfrak{su}(r)$ on $\mathcal{H}$, we define $\mathsf{E}_j := -i\mathsf{f}(F_{j,j+1}^z)$ and $\mathsf{K}_{j,l;\pm} := -i\mathsf{f}(F_{j,l}^x) \mp \mathsf{f}(F_{j,l}^y)$ in the same way as that of $\mathfrak{su}(2)$. Defining $E_j := -|j\rangle\langle j| + |j+1\rangle\langle j+1|$, $K_{j,l;+} := |l\rangle\langle j|$, and $K_{j,l;-} := |j\rangle\langle l|$, we can define $\mathsf{E}_j := \mathsf{f}(E_j)$ and $\mathsf{K}_{j,l;\pm} := \mathsf{f}(K_{j,l;\pm})$ formally.

Notice that $\mathfrak{sd}(r)$ is the set of $-i$ times of traceless diagonal Hermitian matrices. Given $1 \le j < l \le r$, we define the element $\alpha_{j,l}$ of the dual space $\mathfrak{sd}(r)^*$ of $\mathfrak{sd}(r)$ as follows

$$\alpha_{j,l}(X) := i(\langle j|X|j\rangle - \langle l|X|l\rangle). \tag{4.38}$$

Then, we have

$$[\mathsf{E}_{j'}, \mathsf{E}_j] = 0 \tag{4.39}$$

$$[\mathsf{E}_{j'}, \mathsf{K}_{j,l;\pm}] = \pm\alpha_{j,l}(F_{j',j'+1}^z)\mathsf{K}_{j,l;\pm} \tag{4.40}$$

$$[\mathsf{K}_{j,l;+}, \mathsf{K}_{j,l;-}] = \mathsf{E}_j + \cdots + \mathsf{E}_{l-1}.$$

Since the representation is skew–Hermitian, we find that $(\mathsf{K}_{j,l;+})^\dagger = \mathsf{K}_{j,l;-}$ and E_j are Hermitian. Hence, the relation (4.39) guarantees that the representation space $\mathcal{H}$ can be decomposed to the direct sum of the simultaneous eigenspace with respect to $\mathfrak{sd}(r)$. Hence, the simultaneous eigenvalues of elements of $\mathfrak{sd}(r)$ can be described by an element $\boldsymbol{m}$ of the dual space $\mathfrak{sd}(r)^*$ of $\mathfrak{sd}(r)$. The simultaneous eigenspace corresponding to $\boldsymbol{m}$ is denoted by $\mathcal{U}^w_{\boldsymbol{m}}$, i.e.,

$$\mathcal{U}^w_m := \{v \in \mathcal{H} | \mathsf{f}(X)v = im(X)v, \quad \forall X \in \mathfrak{sd}(r)\}.$$

Then, an element $m \in \mathfrak{sd}(r)^*$ is called a **weight** when $\mathcal{U}^w_m \neq \{0\}$. Using the basis $F^z_{1,2}, \ldots, F^z_{r-1,r}$ of $\mathfrak{sd}(r)$, we can make a vector representation of $\boldsymbol{m}$ as $[m_1, \ldots, m_{r-1}]$ $:=[\boldsymbol{m}(F^z_{1,2}), \ldots, \boldsymbol{m}(F^z_{r-1,r})]$, which is called the **Dynkin index**. Then, the set $\Phi := \{\boldsymbol{\alpha}_{j,l}, -\boldsymbol{\alpha}_{j,l}\}_{1\le j<l\le r}$ is called a **root system**, and its element is called a **root**. We abbreviate the root $\boldsymbol{\alpha}_{j,j+1}$ to $\boldsymbol{\alpha}_j$. The set $\Delta := \{\boldsymbol{\alpha}_j\}_{1\le j\le r-1}$ is called the **fundamental system**, and its element is called **simple root**. An element of $\Phi^+ := \{\boldsymbol{\alpha}_{j,l}\}_{1\le j<l\le r}$ ($\Phi^- := \{-\boldsymbol{\alpha}_{j,l}\}_{1\le j<l\le r}$) are called a **positive root** (a **negative root**). Then, any element of $\Phi^+ = \{\boldsymbol{\alpha}_{j,l}\}_{1\le j<l\le r}$ is given as a sum of simple roots with positive integer coefficients.

The following array of the Dynkin indexes of all of simple roots $\boldsymbol{\alpha}_1, \ldots, \boldsymbol{\alpha}_{r-1}$ is called **Cartan matrix**

$$\begin{bmatrix} \boldsymbol{\alpha}_1 \\ \boldsymbol{\alpha}_2 \\ \boldsymbol{\alpha}_3 \\ \cdots \\ \boldsymbol{\alpha}_{r-1} \end{bmatrix} = \begin{bmatrix} 2 & -1 & 0 & 0 & \cdots & 0 \\ -1 & 2 & -1 & 0 & \cdots & 0 \\ 0 & -1 & 2 & -1 & \cdots & 0 \\ \vdots & \vdots & \vdots & \vdots & \ddots & \vdots \\ 0 & 0 & 0 & 0 & -1 & 2 \end{bmatrix}. \tag{4.41}$$

Due to the relation (4.40), for $v \in \mathcal{U}^w_{\boldsymbol{m}}$, $\mathsf{K}_{j,l;\pm}v$ is an element of $\mathcal{U}^w_{\boldsymbol{m}\pm\boldsymbol{\alpha}_{j,l}}$ unless $\mathsf{K}_{j,l;\pm}v$ is 0. For an arbitrary irreducible skew–Hermitian representation space $\mathcal{H}$, there uniquely exists a weight $\boldsymbol{\lambda}$ such that

(1) $\mathcal{H}_{\boldsymbol{\lambda}}$ is a one-dimensional space.
(2) An element v_0 of $\mathcal{H}_{\boldsymbol{\lambda}}$ satisfies $\mathsf{K}_{j,l;+}v_{\boldsymbol{\lambda}} = 0$.

Such a weight is called the **highest weight**. This representation space is generated by $\left\{\prod_{1\le j<l\le r} \mathsf{K}^{n_{j,l}}_{j,l;-} v_0\right\}_{\{n_{j,l}\}}$, and is uniquely determined by the highest weight $\boldsymbol{\lambda}$. There exists a skew–Hermitian representation whose highest weight is $\boldsymbol{\lambda}$ if and only if all entries of Dynkin index of $\boldsymbol{\lambda}$ are non-negative integer. This equivalence relation holds for both representations of the Lie group $\mathrm{SU}(r)$ and the Lie algebra $\mathfrak{su}(r)$.

As mentioned before, since $\mathrm{SU}(r)$ is a universal covering group, an irreducible skew–Hermitian representation $\mathsf{f}_{\boldsymbol{\lambda}}$ of $\mathfrak{su}(r)$ uniquely gives an irreducible unitary representation of $\mathrm{SU}(r)$. Hence, we regard a highest weight $\boldsymbol{\lambda}$ as an element of the set $\widehat{\mathrm{SU}}(r)$, i.e., a label indicating an irreducible representation, and denote the representation space by $\mathcal{U}_{\boldsymbol{\lambda}}(\mathrm{SU}(r))$.

An element of $\mathfrak{su}(r)$ is an $r \times r$ matrix, which directly gives a skew–Hermitian representation on the representation space $\mathbb{C}^r$. This representation is called the **fundamental representation**. The highest weight of this representation is $[1, 0, \ldots, 0]$. Given a representation $\boldsymbol{\lambda} = [\lambda_1, \ldots, \lambda_{r-1}]$, its complex conjugate representation is $\boldsymbol{\lambda}^* = [\lambda_{r-1}, \ldots, \lambda_1]$. For example, the representation with the highest weight $[1, \ldots, 1]$ is a self-adjoint representation. The adjoint representation is a self-adjoint representation, and has the highest weight $\boldsymbol{\lambda} = [1, 0, \ldots, 0, 1]$. The n-fold **symmetric tensor representation** (the n-fold **alternative tensor representation**) has the highest weight $[n, 0, \ldots, 0, 0]$ ($[0, 0, \ldots, 0, n]$). When we employ Dynkin index, we have no possibility to confuse another group. So, we simplify $\mathcal{U}_{[\lambda_1,\ldots,\lambda_{r-1}]}(\mathrm{SU}(r))$ to $\mathcal{U}_{[\lambda_1,\ldots,\lambda_{r-1}]}(r)$. When the space $\mathbb{C}^r$ is written as $\mathcal{H}$, we may denote it by $\mathcal{U}_{[\lambda_1,\ldots,\lambda_{r-1}]}(\mathcal{H})$.

When the quantum system has a representation of the group SU(d), the Hamiltonian is often given as a sum of $-\mathsf{E}_i$ with non-negative coefficients. In this case, the ground state, i.e., the minimum energy state, is given as the highest weight vector $|\boldsymbol{\lambda}; \boldsymbol{\lambda}\rangle$. Hence, we are often interested in the orbit of the ground state with respect to the representation. The subgroup H of SU(r) that stabilizes the state $|\boldsymbol{\lambda}; \boldsymbol{\lambda}\rangle\langle\boldsymbol{\lambda}; \boldsymbol{\lambda}|$ plays an important role in the analysis of the orbit. In this case, the orbit of the ground state is characterized not by SU(r) but by the homogeneous space $\Theta := \mathrm{SU}(r)/H$. That is, given an element $\theta \in \Theta$, the coherent state $|\boldsymbol{\lambda}, \theta\rangle$ is given with its representative g_θ as

$$|\boldsymbol{\lambda}, \theta\rangle := \mathsf{f}_{\boldsymbol{\lambda}}(g_\theta)|\boldsymbol{\lambda}; \boldsymbol{\lambda}\rangle. \tag{4.42}$$

Further, the normalized invariant measure μ_Θ on the homogeneous space Θ satisfies

$$d_{\boldsymbol{\lambda}} \int_\Theta |\boldsymbol{\lambda}, \theta\rangle\langle\boldsymbol{\lambda}, \theta|\mu_\Theta(d\theta) = I, \tag{4.43}$$

which can be shown by Schur's lemma (Lemma 2.4) because the left hand side is commutative with $\mathsf{f}_{\boldsymbol{\lambda}}(g)$ for an arbitrary element g of SU(r). Hence, we can construct a measurement with symmetric property by using of coherent states. Similar to the case with SU(2), the LHS of (4.43) gives a POVM, which describes the statistical behavior of the measurement outcome.

4.3.2 *Irreducible Unitary Representation of* U(r)

Next, we deal with an irreducible representation of the Lie group U(r). Similar to the case of $\mathfrak{su}(r)$, we focus on the Lie subalgebra composed of diagonal elements of the Lie algebra $\mathfrak{u}(r)$. The Lie subalgebra is given by adding the one-dimensional space spanned by $-iI$ to $\mathfrak{sd}(r)$, and is written as $\mathfrak{d}(r)$. Then, the basis of $\mathfrak{d}(r)$ is given by

$$F_j^{z,u} := -i|j\rangle\langle j|, \quad j = 1, \ldots, r. \tag{4.44}$$

Since all of elements of $\mathfrak{d}(r)$ are commutative with each other, we can define the simultaneous eigenspace $\mathcal{U}_m^w := \{v \in \mathcal{H} | \mathfrak{f}(X)v = im(X)v, \forall X \in \mathfrak{d}(r)\}$. An element $m \in \mathfrak{d}(r)^*$ is called the **weight** of the representation of $\mathfrak{u}(r)$ when $\mathcal{U}_m^w \neq \{0\}$. Using the basis $F_1^{z,u}, \ldots, F_r^{z,u}$ of $\mathfrak{d}(r)$, we can express the weight $\boldsymbol{m}$ by the vector form $(m_1, \ldots, m_{r-1}) := (\boldsymbol{m}(F_1^{z,u}), \ldots, \boldsymbol{m}(F_r^{z,u}))$, which is called the **Young index**. The entries of Young index are expressed by ( ) and those of Dynkin index are by [ ]. We can check that the entries of Young index of the weight are integers by considering the representation of the subgroup $\{e^{\theta F_j^{z,u}}\}$ isomorphic to U(1). Now, we focus on the dual map $in^* : \mathfrak{d}(r)^* \to \mathfrak{sd}(r)^*$ of the immbeding $in : \mathfrak{sd}(r) \hookrightarrow \mathfrak{d}(r)$. Then, we find that the vector $m = (m_1, \ldots, m_r) \in \mathfrak{d}(r)^*$ satisfies $in^*(m) = [m_1 - m_2, \ldots, m_{r-1} - m_r] \in \mathfrak{sd}(r)^*$.

Since a representation of U(r) is also that of the subgroup SU(r), similar to a representation of $\mathfrak{su}(r)$, an irreducible unitary representation of the Lie group $\mathfrak{u}(r)$ is given by an element $\boldsymbol{\lambda} \in \mathfrak{d}(r)^*$ such that $in^*(\boldsymbol{\lambda}) \in \mathfrak{sd}(r)^*$ satisfies Conditions **(1)** and **(2)** given above, which is called the **highest weight** of U(r).

Since the Lie algebra $\mathfrak{u}(r)$ is isomorphic to $\mathfrak{su}(r) \oplus \mathfrak{u}(1)$, the highest weight $\boldsymbol{\lambda} \in \mathfrak{d}(r)^*$ generates an irreducible skew–Hermitian representation of the Lie algebra $\mathfrak{u}(r)$ if and only if each entry $\lambda_j - \lambda_{j+1}$ of Dynkin index of $in^*(\boldsymbol{\lambda})$ is a non-negative integer. Then, the highest weight $\boldsymbol{\lambda} \in \mathfrak{d}(r)^*$ generates an irreducible unitary representation of U(r) if and only if each entry λ_j of Young index is an integer as well as the above condition holds.

In the following, we regard the highest weight $\boldsymbol{\lambda}$ as an element of the set $\widehat{\mathrm{U}(r)}$ labeling an irreducible representation, and denote its representation space by $\mathcal{U}_{\boldsymbol{\lambda}}(\mathrm{U}(r))$. Especially, when all of λ_j are non-negative integer, the highest weight $\boldsymbol{\lambda}$ of an irreducible unitary representation of U(r) is given as a Young diagram of depth r. Each basis of the representation space $\mathcal{U}_{\boldsymbol{\lambda}}(\mathrm{U}(r))$ is given as a Semistandard Young tableau of Young diagram $\boldsymbol{\lambda}$. The eigenvalue of $\mathfrak{f}_{\boldsymbol{\lambda}}(F_j^{z,u})$ corresponding to the respective base is given as the number of j contained in the corresponding Semistandard Young tableau.

Next, using the contents of Sect. 2.4, we consider the relation between irreducible unitary representations of U(r) and SU(r). The centralizer $C(\mathrm{SU}(r))$ of the Lie group SU(r) is U_r. Defining $\omega := e^{i2\pi/r}I \in C(\mathrm{SU}(r))$ and using the notation in Sect. 2.4, we find that the Lie group U(r) is $\overline{\mathrm{SU}(r)}_\omega$. That is, an element of the Lie group U(r) can be written as $(g, e^{i\theta})$ by using $g \in \mathrm{SU}(r)$ and $\theta \in [0, 2\pi/r)$. Given an irreducible unitary representation $\mathfrak{f}_{\boldsymbol{\lambda}}$ of the Lie group U(r) whose highest weight is $\boldsymbol{\lambda} \in \mathfrak{d}(r)^*$, we choose an integer l such that $\mathfrak{f}_{\boldsymbol{\lambda}}(\omega) = e^{i2\pi l/k}I$. Then, the representation $\mathfrak{f}_{\boldsymbol{\lambda}}$ can be written by using an integer $n \in \mathbb{Z}$ as

$$\mathfrak{f}_{\boldsymbol{\lambda}}(g, e^{i\theta}) = \mathfrak{f}_{in^*(\boldsymbol{\lambda})}(g)e^{i\theta(nr+l)}. \tag{4.45}$$

Then, we have $nr + l = \sum_{j=1}^{r} \lambda_j$.

4.3.3 Dimension of Irreducible Representation Space

The dimension of $\mathcal{U}_{\boldsymbol{\lambda}}(\mathrm{SU}(r))$ is the number $|Y_{U,r}(\boldsymbol{n})|$ of Semistandard Young tableau, which is given as

$$\dim \mathcal{U}_{\boldsymbol{n}}(\mathrm{U}(r)) = |\mathcal{T}_{u,r}(\boldsymbol{n})| = \prod_{1\le j<l\le r} \frac{l-j+n_l-n_j}{l-j}, \tag{4.46}$$

where n_j is Young index and the second formula is known as **Weyl's dimension formula** [37, Theorem 7.1.7]. Applying this formula to the representation of $\mathrm{SU}(r)$, we have

$$\dim \mathcal{U}_{\boldsymbol{\lambda}}(\mathrm{SU}(r)) = \prod_{1\le j<l\le r} \frac{l-j+\sum_{t=j}^{l-1}\lambda_t}{l-j}, \tag{4.47}$$

where λ_t is Dynkin index. We should remark that the eigenspace $\mathcal{H}_{\boldsymbol{m}}$ corresponding to a non-highest weight $\boldsymbol{m}$ is not one-dimensional. For example, for $r=3$, we have

$$\dim \mathcal{U}_{\boldsymbol{\lambda}}(\mathrm{SU}(3)) = \frac{(\lambda_1+1)(\lambda_1+\lambda_2+2)(\lambda_2+1)}{2}.$$

When $\lambda_1=1$ and $\lambda_2=1$, the dimension is 8 while the number of possible weights is 7. This typical example appears in Baryon octet in Sect. 5.5.3. In particular, the vectors Λ^0 and Σ^0 given in Sect. 5.5.3 have the same weight, but are orthogonal to each other.

For $r=4$, we have

$$\begin{aligned}&\dim \mathcal{U}_{\boldsymbol{\lambda}}(\mathrm{SU}(4))\\ =&\frac{(\lambda_1+1)(\lambda_1+\lambda_2+2)(\lambda_1+\lambda_2+\lambda_3+3)(\lambda_2+1)(\lambda_2+\lambda_3+2)(\lambda_3+1)}{2!3!}.\end{aligned}$$

Even for a general number r, when all entries of Dynkin index λ_j except for the first and second entries are zero, we can calculate the dimension as follows.

$$\dim \mathcal{U}_{[\lambda_1,\lambda_2,0,\ldots,0]}(\mathrm{SU}(r)) = \frac{(\lambda_1+\lambda_2+r-1)!(\lambda_1+r-2)!(\lambda_1+1)}{(r-1)!(r-2)!(\lambda_1+\lambda_2+1)!\lambda_2!}. \tag{4.48}$$

When we employ Young index n_j, this relation can be rewritten as

$$\dim \mathcal{U}_{(n_1,n_2,0,\ldots,0,0)}(\mathrm{U}(r)) = \frac{(n_1+r-1)!(n_2+r-2)!(n_1-n_2+1)}{(r-1)!(r-2)!(n_1+1)!n_2!}. \tag{4.49}$$

Exercise 4.9 Show the following relation

$$\dim \mathcal{U}_{[n,0,\ldots,0]}(\mathrm{U}(r)) = \dim \mathcal{U}_{(n,0,\ldots,0)}(\mathrm{U}(r)) = \frac{(n+r-1)!}{(r-1)!n!}. \tag{4.50}$$

Exercise 4.10 Show the following relation

$$\dim \mathcal{U}_{[1,1,0,\ldots,0]}(\mathrm{U}(r)) = \frac{(r+1)r(r-1)}{3}. \tag{4.51}$$

4.3.4 *Tensor Product*

Next, we consider the irreducible decomposition of the tensor product representation of two irreducible skew–Hermitian representations $\mathfrak{f}_{\boldsymbol{\lambda}}$ and $\mathfrak{f}_{\boldsymbol{\lambda}'}$ of $\mathrm{U}(r)$. For simplicity, we assume that all of entries of $\boldsymbol{\lambda}$ and $\boldsymbol{\lambda}'$ are non-negative integers.

The multiplicity coefficient $C^{\boldsymbol{n}}_{\boldsymbol{\lambda},\boldsymbol{\lambda}'}(\mathrm{U}(r))$ given in (2.45) satisfies the relation

$$\mathcal{U}_{\boldsymbol{\lambda}}(\mathrm{U}(r)) \otimes \mathcal{U}_{\boldsymbol{\lambda}'}(\mathrm{U}(r)) = \bigoplus_{\boldsymbol{n} \in Y^r_n} \mathbb{C}^{C^{\boldsymbol{n}}_{\boldsymbol{\lambda},\boldsymbol{\lambda}'}(\mathrm{U}(r))} \otimes \mathcal{U}_{\boldsymbol{n}}(\mathrm{U}(r)), \tag{4.52}$$

and is called the **Littlewood–Richardson coefficient**, where $n := |\boldsymbol{\lambda}| + |\boldsymbol{\lambda}'|$. The coefficient can be calculated from the relation among three Young diagrams $\boldsymbol{\lambda}$, $\boldsymbol{\lambda}'$, and $\boldsymbol{n}$ as follows. Given $\boldsymbol{n}$ and $\boldsymbol{\lambda}$, we define the **skew-diagram** $\boldsymbol{n}\backslash\boldsymbol{\lambda}$ as a diagram given as the Young diagram $\boldsymbol{n}$ with substraction of the Young diagram $\boldsymbol{\lambda}$. For example, the skew-diagram $\boldsymbol{n}\backslash\boldsymbol{\lambda}$ is defined as the diagram 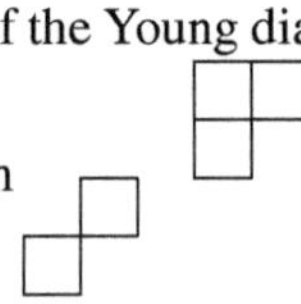 that is 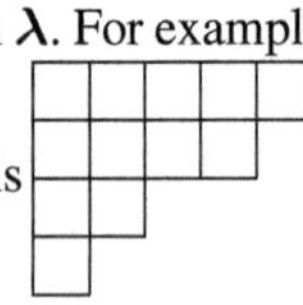 with substraction of 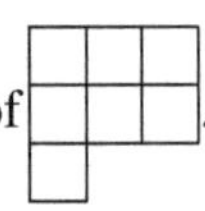. Then, we insert the letters $1, \ldots, r$ in the skew-diagram $\boldsymbol{n}\backslash\boldsymbol{\lambda}$ according the following rule. The Littlewood–Richardson coefficient $C^{\boldsymbol{n}}_{\boldsymbol{\lambda},\boldsymbol{\lambda}'}(\mathrm{U}(r))$ is given as the number of possible types of the insertions of the letters $1, \ldots, r$. This fact is called the **Littlewood–Richardson rule** [75, p.68].

(1) The letter j is written in λ'_j boxes among the skew-diagram $\boldsymbol{n}\backslash\boldsymbol{\lambda}$.
(2) The letter is strictly increasing from up to down in each column of the skew-diagram $\boldsymbol{n}\backslash\boldsymbol{\lambda}$.
(3) The letter is increasing from left to right in each low of the skew-diagram $\boldsymbol{n}\backslash\boldsymbol{\lambda}$.
(4) We make the order for the boxes in the skew-diagram $\boldsymbol{n}\backslash\boldsymbol{\lambda}$ as follows. The first to $n_1 - \lambda_1$-th boxes are given in the first low from left to right. (This order is opposite to Condition (3).) Next, we make the order of the boxes in the second low in the same way. We repeat this numbering method up to the $|\boldsymbol{n}| - |\boldsymbol{\lambda}|$-th box. Then, for an arbitrary integer $0 \le m \le |\boldsymbol{n}| - |\boldsymbol{\lambda}|$, the number of boxes with j is larger than that with $j+1$ among the first to m-th boxes.

Example 4.4 Assume that $\boldsymbol{\lambda}' = (1, 0, \ldots, 0)$. The Littlewood–Richardson coefficient $C^{\boldsymbol{n}}_{\boldsymbol{\lambda},\boldsymbol{\lambda}'}(\mathrm{U}(r))$ is 1 when $\boldsymbol{n} \in Y^r$ satisfies the following two conditions. It is 0 otherwise.

(1) Only one entry n_i of $\boldsymbol{n}$ satisfies that $n_i = \lambda_i + 1$.
(2) The remaining entries n_j of $\boldsymbol{n}$ satisfy that $n_j = \lambda_j$.

Example 4.5 Assume that $\boldsymbol{\lambda}'$ is $(1, \ldots, 1, 0)$. The Littlewood–Richardson coefficient $C^{\boldsymbol{n}}_{\boldsymbol{\lambda},\boldsymbol{\lambda}'}(\mathrm{U}(r))$ is 1 when $\boldsymbol{n} \in Y^r$ satisfies the following two conditions. It is 0 otherwise.

(1) Only one entry n_i of $\boldsymbol{n}$ satisfies that $n_i = \lambda_i$.
(2) The remaining entries n_j of $\boldsymbol{n}$ satisfy that $n_j = \lambda_j + 1$.

Examples 4.4 and 4.5 yield that

$$\mathcal{U}_{(1,0,\ldots,0)} \otimes \mathcal{U}_{(0,\ldots,0,1)} = \mathcal{U}_{(0,\ldots,0)} \oplus \mathcal{U}_{(2,1,\ldots,1,0)}. \tag{4.53}$$

When we employ Dynkin index, we have

$$\mathcal{U}_{[1,0,\ldots,0]}(\mathrm{U}(r)) \otimes \mathcal{U}_{[0,\ldots,0,1]}(\mathrm{U}(r)) = \mathcal{U}_{[1,0,\ldots,0,1]}(\mathrm{U}(r)) \oplus \mathcal{U}_{[0,\ldots,0]}(\mathrm{U}(r)). \tag{4.54}$$

Example 4.6 Assume that $\boldsymbol{\lambda}{=}(m, 0, \ldots, 0)$ and $\boldsymbol{\lambda}'{=}(n, 0, \ldots, 0)$ $(m \geq n)$. The Littlewood–Richardson coefficient $C^{\boldsymbol{n}}_{\boldsymbol{\lambda},\boldsymbol{\lambda}'}(\mathrm{U}(r))$ is 1 when $\boldsymbol{n}{=}(n + m - k, k, 0, \ldots, 0)$ and $k{=}0, 1, \ldots,$ or n. Otherwise, it is 0. That is,

$$\mathcal{U}_{(m,0,\ldots,0)}(\mathrm{U}(r)) \otimes \mathcal{U}_{(n,0,\ldots,0)}(\mathrm{U}(r)) = \bigoplus_{k=0}^{n} (\mathcal{U}_{(m+n-k,k,0,\ldots,0)}(\mathrm{U}(r)). \tag{4.55}$$

Example 4.7 Assume that $\boldsymbol{\lambda}'$ is the adjoint representation $(2, 1, \ldots, 1, 0)$. We have the following three cases.

(1) When all entries n_i of $\boldsymbol{n}$ satisfies that $n_i = \lambda_i + 1$, the Littlewood–Richardson coefficient $C^{\boldsymbol{n}}_{\boldsymbol{\lambda},\boldsymbol{\lambda}'}(\mathrm{U}(r))$ is the number of i satisfying the condition $n_i > n_{i+1}$.
(2) Assume that one entry n_i of $\boldsymbol{n}$ satisfies that $n_i = \lambda_i + 2$, another entry n_j of $\boldsymbol{n}$ satisfies that $n_j = \lambda_j$, and the remaining entries n_l of $\boldsymbol{n}$ satisfy that $n_l = \lambda_l + 1$. Then, the Littlewood–Richardson coefficient $C^{\boldsymbol{n}}_{\boldsymbol{\lambda},\boldsymbol{\lambda}'}(\mathrm{U}(r))$ is 1.
(3) Otherwise, the Littlewood–Richardson coefficient $C^{\boldsymbol{n}}_{\boldsymbol{\lambda},\boldsymbol{\lambda}'}(\mathrm{U}(r))$ is 0.

Example 4.8 Assume that $\boldsymbol{\lambda} = (m, 0, \ldots, 0)$, $\boldsymbol{\lambda}'{=}(n, 0, \ldots, 0)$, $\boldsymbol{\lambda}''{=}(1, 0, \ldots, 0)$, and $m \geq n$. In the following, we omit $\mathrm{U}(r)$.

(1) When $r = 2$ and $m = n$, we have

$$\begin{aligned} &\mathcal{U}_{(n,0)} \otimes \mathcal{U}_{(n,0)} \otimes \mathcal{U}_{(1,0)} \\ &= \mathcal{U}_{(2n+1,0)} \oplus \bigoplus_{k=1}^{n} (\mathcal{U}_{(2n+1-k,k)} \otimes \mathbb{C}^2). \end{aligned} \tag{4.56}$$

(2) When $r = 2$ and $m > n$, we have

$$\begin{aligned}&\mathcal{U}_{(m,0)} \otimes \mathcal{U}_{(n,0)} \otimes \mathcal{U}_{(1,0)}\\ &= \mathcal{U}_{(m+n+1,0)} \oplus \mathcal{U}_{(m,n+1)} \oplus \bigoplus_{k=1}^{n} (\mathcal{U}_{(m+n+1-k,k)} \otimes \mathbb{C}^2).\end{aligned}$$

(3) When $r \geq 3$ and $m = n$, we have

$$\begin{aligned}&\mathcal{U}_{(n,0,\ldots,0)} \otimes \mathcal{U}_{(n,0,\ldots,0)} \otimes \mathcal{U}_{(1,0,\ldots,0)}\\ &= \mathcal{U}_{(2n+1,0,\ldots,0)} \oplus \bigoplus_{k=1}^{n} (\mathcal{U}_{(2n+1-k,k,0,\ldots,0)} \otimes \mathbb{C}^2) \oplus \bigoplus_{k=1}^{n} \mathcal{U}_{(2n-k,k,1,\ldots,0)}.\end{aligned} \tag{4.57}$$

(4) When $r \geq 3$ and $m > n$, we have

$$\begin{aligned}&\mathcal{U}_{(m,0,\ldots,0)} \otimes \mathcal{U}_{(n,0,\ldots,0)} \otimes \mathcal{U}_{(1,0,\ldots,0)}\\ &= \mathcal{U}_{(m+n+1,0,\ldots,0)} \oplus \mathcal{U}_{(m,n+1,\ldots,0)} \oplus \bigoplus_{k=1}^{n} (\mathcal{U}_{(m+n+1-k,k,0,\ldots,0)} \otimes \mathbb{C}^2)\\ &\oplus \bigoplus_{k=1}^{n} \mathcal{U}_{(m+n-k,k,1,0,\ldots,0)}.\end{aligned}$$

Exercise 4.11 Show the following relation

$$\dim \mathcal{U}_{[1,0,\ldots,0,1]}(\mathrm{U}(r)) = r^2 - 1. \tag{4.58}$$

4.3.5 *Irreducible Representation of* SL(r, $\mathbb{C}$), GL(r, $\mathbb{C}$)

An irreducible unitary representation of SU(r) can be extended to an irreducible $\mathbb{C}$-representation of SL$(r, \mathbb{C})$, and an irreducible unitary representation of U(r) can be extended to an irreducible $\mathbb{C}$-representation of GL$(r, \mathbb{C})$. In this case, a non-unitary element of GL$(r, \mathbb{C})$ does not directly give a physical transformation, however, the following discussion plays an important role for analysis of a density matrix. For simplicity, we consider the case with $r = 2$. So, Young index $\boldsymbol{n}$ can be written as $(\frac{n+k}{2}, \frac{n-k}{2})$ with an integer k between n and $n/2$ so that $n - k$ is an even number. Assume that g has the eigenvalues a and b. The character $\chi_{\frac{n+k}{2},\frac{n-k}{2}}(g)$ of the representation $\mathsf{f}_{\boldsymbol{n}}$ is given as

$$\chi_{\frac{n+k}{2},\frac{n-k}{2}}(g) = \sum_{m=-k/2}^{k/2} a^{\frac{n}{2}+m} b^{\frac{n}{2}-m} = \frac{a^{\frac{n-k}{2}} b^{\frac{n+k}{2}} - a^{\frac{n+k}{2}+1} b^{\frac{n-k}{2}-1}}{1 - ab}.$$

In particular, when $a = 1 - p$ and $b = p$ with $p \in (1/2, 1)$,

$$\chi_{\frac{n+k}{2}, \frac{n-k}{2}}(g) = \frac{p^{\frac{n+k}{2}+1}(1-p)^{\frac{n-k}{2}} - p^{\frac{n-k}{2}}(1-p)^{\frac{n+k}{2}+1}}{(2p-1)}. \tag{4.59}$$

Next, we consider the case with GL($r, \mathbb{C}$). When an element $g \in \mathrm{GL}(r, \mathbb{C})$ is diagonalizable, and its eigenvalues $a_1, \ldots, a_r$ are distinct with each other, by using $\delta := (r-1, r-2, \ldots, 0)$, the character of the representation $\mathfrak{f}_{\boldsymbol{n}}$ is given as the **Weyl's dimension formula** [37, Theorem 7.1.1].

$$\chi_{\boldsymbol{n}}(g) = \frac{\sum_{\sigma \in S_r} \mathrm{sgn}(\sigma) \prod_{l=1}^{r} a_l^{n_{\sigma(l)}+\delta_{\sigma(l)}}}{\prod_{j<l}(a_j - a_l)}. \tag{4.60}$$

The right hand side is a function of the eigenvalues $a_1, \ldots, a_r$, and is called the **Schur function** $\chi_{\boldsymbol{n}}(a_1, \ldots, a_r)$.

4.4 Many-Particle System: Composite System of n Quantum Systems

4.4.1 Distinguishable Particles: Schur Duality

Given an element g of the Lie group SU(r) or U(r), we consider its n-fold tensor product $g^{\otimes n} := \underbrace{g \otimes \cdots \otimes g}_{n}$, which gives a unitary representation of SU(r) on the composite system $(\mathbb{C}^r)^{\otimes n}$. This unitary representation is called the n-fold **tensor product representation** of SU(r), and describes the situation that n quantum systems (n particles) with the representation space $\mathbb{C}^r$ are independently prepared and the same transformation is applied on all of them. Further, this description assumes that these systems or these particles are distinguishable. When such a transformation is applied on n quantum systems, this representation is helpful. However, even when such a transformation is not given, this representation is still helpful for the composite system $(\mathbb{C}^r)^{\otimes n}$ due to the following symmetry.

In the previous section, given $\boldsymbol{n} \in Y^{r-1}$, we gave the irreducible unitary representation $\mathcal{U}_{\boldsymbol{n}}(\mathrm{U}(r))$ of U(r) and the irreducible unitary representation $\mathcal{U}_{in^*(\boldsymbol{n})}(\mathrm{U}(r))$ of SU(r). On the other hand, we consider the following unitary representation $\mathfrak{f}_n$ of the permutation group S_n on $(\mathbb{C}^r)^{\otimes n}$.

$$\mathfrak{f}_n(g)(v_1 \otimes \cdots \otimes v_n) = v_{g(1)} \otimes \cdots \otimes v_{g(n)}, \quad \forall g \in S_n. \tag{4.61}$$

Hence, combining these two unitary representations, we obtain the unitary representation of the direct product group $\mathrm{SU}(r) \times S_n$ or $\mathrm{U}(r) \times S_n$ on $(\mathbb{C}^r)^{\otimes n}$, which is called **Schur duality**.

When $r = 2$, we rearrange 2^n basis of $(\mathbb{C}^2)^{\otimes n}$ dependently of the weight of $\mathsf{E}_{0,+1}$. In this case, the number of basis with weight m is $\binom{n}{m}$. When we additionally assume that $n = 4$, we can classify the basis as the following figure, where m is the weight and λ is the highest weight. In this figure, the action of $\mathfrak{su}(2)$ moves the basis in the vertical direction. $\mathsf{K}_{+,+1}$ moves the basis upward, and $\mathsf{K}_{-,+1}$ does them downward. On the other hand, the permutation group group S_4 moves the basis in the horizontal direction. Hence, the vertical arrows indicate irreducible representation spaces of $\mathfrak{su}(2)$, and the horizontal arrows do irreducible representation spaces of S_4. That is, irreducible representation spaces can be classified by λ (Fig. 4.3).

Now, we concretely construct each representation space as follows. Let $|\uparrow\rangle$ be the vector of weight $m = 1/2$ in $\mathbb{C}^2$, and $|\downarrow\rangle$ be the vector of weight $m = -1/2$. Then, the leftmost representation space of $\mathfrak{su}(2)$ (i.e., the representation space with the highest weight $\lambda = 2$) is a symmetric tensor product space, the vector of weight $m = 1$ in this space is $|\uparrow,\uparrow,\uparrow,\downarrow\rangle + |\downarrow,\uparrow,\uparrow,\uparrow\rangle + |\uparrow,\downarrow,\uparrow,\uparrow\rangle + |\uparrow,\uparrow,\downarrow,\uparrow\rangle$. Hence, the vector of weight $m = 1$ in the representation space with the highest weight $\lambda = 1$ is the vector of weight $m = 1$ orthogonal to the above vector. The vector of weight $m = 0$ in the representation space with the highest weight $\lambda = 1$ is given by applying $\mathsf{K}_{-,+1}$ to this vector. The representation space with the highest weight $\lambda = 0$ contains only the vector of weight $m = 0$. This vector is the vector of weight $m = 0$ orthogonal to the vector of weight $m = 0$ in the representation space with the highest weight $\lambda = 2$ and the vector of weight $m = 0$ in the representation space with the highest weight $\lambda = 1$. For a general integer n, we can make the basis of each representation in the same way.

For a general integer r, we can make the following irreducible decomposition of $(\mathbb{C}^r)^{\otimes n}$ with respect to the unitary representation of the direct product group $\mathrm{SU}(r) \times S_n$.

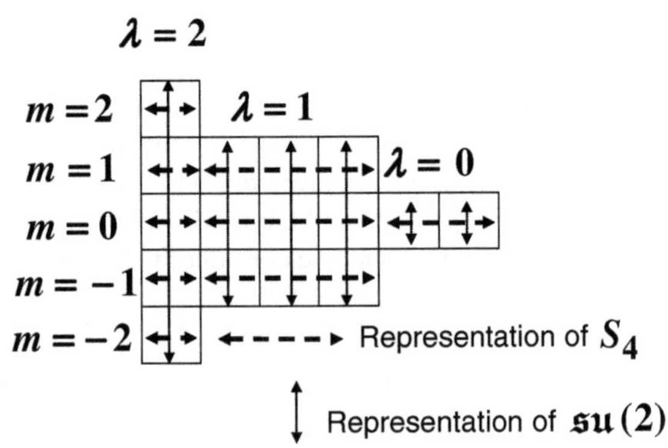

Fig. 4.3 Irreducible decomposition of $(\mathbb{C}^2)^{\otimes 4}$

$$(\mathbb{C}^r)^{\otimes n} = \bigoplus_{\boldsymbol{n}\in Y_n^r} \mathcal{U}_{in^*(\boldsymbol{n})}(\mathrm{SU}(r)) \otimes \mathcal{U}_{\boldsymbol{n}}(S_n) \tag{4.62}$$

$$= \bigoplus_{\boldsymbol{n}\in Y_n^r} \mathcal{U}_{\boldsymbol{n}}(\mathrm{SU}(r) \times S_n). \tag{4.63}$$

Then, given $\boldsymbol{n} \in Y_n^r$, the space $\mathcal{U}_{\boldsymbol{n}}(\mathrm{SU}(r) \times S_n)$ is given as the irreducible representation space $\mathcal{U}_{in^*(\boldsymbol{n})}(\mathrm{SU}(r)) \otimes \mathcal{U}_{\boldsymbol{n}}(S_n)$ of the direct product group $\mathrm{SU}(r) \times S_n$. In the irreducible space, the permutation group S_n acts only on $\mathcal{U}_{\boldsymbol{n}}(S_n)$, and its action on $\mathcal{U}_{in^*(\boldsymbol{n})}(\mathrm{SU}(r))$ is given as the identity matrix. Similarly, the Lie group $\mathrm{SU}(r)$ acts only on the space $\mathcal{U}_{in^*(\boldsymbol{n})}(\mathrm{SU}(r))$, and its action on the space $\mathcal{U}_{\boldsymbol{n}}(S_n)$ is given as the identity matrix. Hence, in the irreducible space, the multiplicity of the unitary representation $\mathcal{U}_{\boldsymbol{n}}(S_n)$ of S_n is the dimension of $\mathcal{U}_{in^*(\boldsymbol{n})}(\mathrm{SU}(r))$, and the multiplicity of the unitary representation $\mathcal{U}_{in^*(\boldsymbol{n})}(\mathrm{SU}(r))$ of $\mathrm{SU}(r)$ is the dimension of $\mathcal{U}_{\boldsymbol{n}}(S_n)$. These dimensions are given in (4.47) and (2.72). Especially, when $\boldsymbol{n} = (n, 0, \ldots, 0)$, $\mathcal{U}_{\boldsymbol{n}}(S_n)$ is a one-dimensional space, and the subspace $\mathcal{U}_{in^*(\boldsymbol{n})}(\mathrm{SU}(r)) \otimes \mathcal{U}_{\boldsymbol{n}}(S_n) \cong \mathcal{U}_{in^*(\boldsymbol{n})}(\mathrm{SU}(r))$ of $(\mathbb{C}^r)^{\otimes n}$ is the n-fold **symmetric tensor product space**.

The tensor product representation can be naturally extended to a $\mathbb{C}$-representation of $\mathrm{GL}(r, \mathbb{C})$. In this case, the relation (4.62) also gives the irreducible decomposition with respect to the direct product group $\mathrm{GL}(r, \mathbb{C}) \times S_n$ of $(\mathbb{C}^r)^{\otimes n}$ (Table 4.1).

When we denote the projection to irreducible component corresponding to $\boldsymbol{n}$ by $P_{\boldsymbol{n}}$, the irreducible decomposition (4.63) gives the measurement $\{P_{\boldsymbol{n}}\}$. When the densities of all systems are given as the same density matrix ρ and these systems are independent of each other, the density matrix of the composite system is given as $\rho^{\otimes n}$. When ρ does not have eigenvalue 0, it can be regarded as an element of $\mathrm{GL}(r, \mathbb{C})$. Hence, when the above measurement is applied, the probability to obtain the outcome $\boldsymbol{n}$ is

$$\mathrm{Tr}\, P_{\boldsymbol{n}} \rho^{\otimes n} = \mathrm{Tr}\, \mathfrak{f}_{\boldsymbol{n}}(\rho) \cdot \mathrm{Tr}\, \mathcal{U}_{\boldsymbol{n}}(S_n). \tag{4.64}$$

Since $\mathrm{Tr}\, \mathfrak{f}_{\boldsymbol{n}}(\rho)$ is the character of ρ, its calculation formula is available in (4.60). On the other hand, the calculation formula of $\mathrm{Tr}\, \mathcal{U}_{\boldsymbol{n}}(S_n)$ is given in (2.72). Several formulas given in group representation theory are useful for the calculations of physically important values in this way.

4.4.2 Indistinguishable Particles: Boson and Fermion

In Sect. 1.1, we have explained that the composite system is given as $\mathcal{H}^{\otimes n}$ when every particle is distinguishable from each other and its representation space is $\mathcal{H}$. However, even when the quantum system of the internal degree is $\mathcal{H}$, when the particles are in the same place as another particle, they cannot be distinguished from each other. In this case, the composite system is not $\mathcal{H}^{\otimes n}$. Instead of this description, we have other two types of descriptions as follows. One is the case when more than

Table 4.1 Young diagram and representation for $r \geq n$

Name	Young diagram	Young index of $\mathcal{U}(\mathfrak{su}(r))$	Dynkin index of $\mathcal{U}(\mathfrak{su}(r))$	Dimension of $\mathcal{U}(\mathfrak{su}(r))$	Dimension of $\mathcal{U}(S_n)$
Fundamental representation		$(1, 0, \ldots, 0)$	$[1, 0, \ldots, 0]$	r	1
2-fold symmetric tensor product representation		$(1, 1, 0, \ldots, 0)$	$[0, 1, 0, \ldots, 0]$	$\frac{r(r-1)}{2}$	1
n-fold symmetric tensor product representation	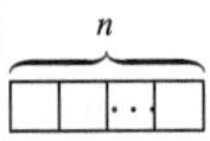	$(n, 0, \ldots, 0)$	$[n, 0, \ldots, 0]$	$\binom{n+r-1}{r-1}$	1
n-fold alternative tensor product representation	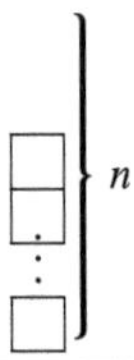	$(\overbrace{1, \ldots, 1}^{n}, 0, \ldots, 0)$	$[\overbrace{0, \ldots, 0}^{n-1}, 1, 0, \ldots, 0]$	$\binom{r}{n}$	1
Adjoint representation (only when $n = r - 1$, no name in other cases.)	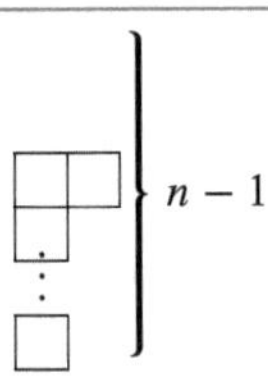	$(2, \overbrace{1, \ldots, 1}^{n-1}, 0, \ldots, 0)$	$[1, \overbrace{0, \ldots, 0}^{n-2}, 1, 0, \ldots, 0]$	$n^2 - 1$	$n-1$

one particles can simultaneously take the same state. The other is the case when more than one particles cannot. The choice depends on the type of the particle. The particle choosing the former is called a **boson**, and the particle choosing the latter is called a **fermion**. A typical example of a boson is a photon and examples of a fermion particle are an electron and a quark.

When there are n boson particles in the quantum system $\mathcal{H}$, the representation system of the composite system is the n-fold symmetric tensor product representation of $\mathcal{H}$. As Fig. 4.4, the space is the subspace of vectors invariant for the permutation in the n-fold tensor product space $\mathcal{H}^{\otimes n}$, which has the representation of $\mathrm{SU}(r) \times S_n$. That is, the quantum system is given as the irreducible space $\mathcal{U}_{[n,0,\ldots,0]}(r)$ when $\mathcal{H} = \mathbb{C}^r$.

As a typical example, we consider the case when $\mathcal{H} = \mathbb{C}^r \otimes \mathbb{C}^2$. the two-dimensional system $\mathbb{C}^2$ is called spin and is spanned by $\{|\uparrow\rangle, |\downarrow\rangle\}$.Hence, t Its physical meaning is explained in Sect. 5.4. Then, the symmetric subspace

Fig. 4.4 boson and fermion

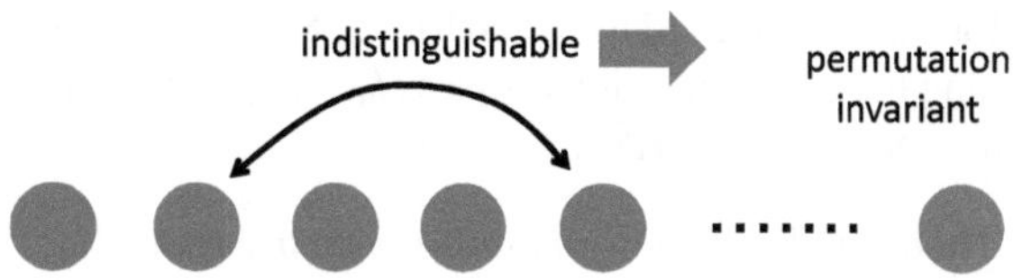

$\mathcal{U}_{[3,0,\dots,0]}(2r)$ is irreducible for $\mathrm{SU}(2r)$, but is not necessarily irreducible for the subgroup $\mathrm{SU}(r) \times \mathrm{SU}(2) \subset \mathrm{SU}(2r)$ as follows. To discuss this issue, we define the normalized symmetrized vector of $|k_1 s_1, k_2 s_2, k_3 s_3\rangle$ with respect to the permutation by $\mathcal{S}(|k_1 s_1, k_2 s_2, k_3 s_3\rangle)$.

Theorem 4.1 *The space* $\mathcal{U}_{[3,0,\dots,0]}(2r)$ *is irreducibly decomposed for the representation of the subgroup* $\mathrm{SU}(r) \times \mathrm{SU}(2)$.

$$\begin{aligned}&\mathcal{U}_{[3,0,\dots,0]}(2r)\\&=\mathcal{U}_{[3,0,\dots,0]}(r)\otimes\mathcal{U}_{[3]}(2)\oplus\mathcal{U}_{[1,1,0,\dots,0]}(r)\otimes\mathcal{U}_{[1]}(2).\end{aligned} \tag{4.65}$$

The space $\mathcal{U}_{[3,0,\dots,0]}(r)\otimes\mathcal{U}_{[3]}(2)$ *can be generated by the element* $|1,1,1,\uparrow,\uparrow,\uparrow\rangle$ *with the action of* $\mathrm{SU}(r)\times\mathrm{SU}(2)$. *The other space* $\mathcal{U}_{[1,1,0,\dots,0]}(r)\otimes\mathcal{U}_{[1]}(2)$ *can be generated by the element* $\sqrt{\frac{2}{3}}\mathcal{S}(|2\downarrow,1\uparrow,1\uparrow\rangle)-\sqrt{\frac{1}{3}}\mathcal{S}(|2\uparrow,1\downarrow,1\uparrow\rangle)$ *with the action of* $\mathrm{SU}(r)\times\mathrm{SU}(2)$.

Proof Any element of the subspace $\mathcal{U}_{[3,0,\dots,0]}(r)\otimes\mathcal{U}_{[3]}(2)$ is permutation-symmetric. So, the subspace is included in $\mathcal{U}_{[3,0,\dots,0]}(2r)$. Next, we focus on the subspace spanned by $\mathcal{S}(|2\downarrow,1\uparrow,1\uparrow\rangle)$ and $\mathcal{S}(|2\uparrow,1\downarrow,1\uparrow\rangle)$. The subspace $\mathcal{U}_{[3,0,\dots,0]}(r)\otimes\mathcal{U}_{[3]}(2)$ contains the vector $\sqrt{\frac{1}{3}}\mathcal{S}(|2\downarrow,1\uparrow,1\uparrow\rangle)+\sqrt{\frac{2}{3}}\mathcal{S}(|2\uparrow,1\downarrow,1\uparrow\rangle)$. Under the action of $\mathrm{SU}(r)\times\mathrm{SU}(2)$, the orthogonal vector $\sqrt{\frac{2}{3}}\mathcal{S}(|2\downarrow,1\uparrow,1\uparrow\rangle)-\sqrt{\frac{1}{3}}\mathcal{S}(|2\uparrow,1\downarrow,1\uparrow\rangle)$ generates the space $\mathcal{U}_{[1,1,0,\dots,0]}(r)\otimes\mathcal{U}_{[1]}(2)$. Hence, we obtain the relation $\supset$ in (4.65). Exercises 4.9 and 4.10 guarantee that the dimension of LHS of (4.65) is $\frac{2r(r+1)(2r+1)}{3}$ and that the dimension of RHS of (4.65) is $4\times\frac{(r+2)(r+1)r}{6}+2\times\frac{(r+1)r(r-1)}{3}=\frac{2r(r+1)(2r+1)}{3}$. Hence, we obtain the equality of (4.65). ∎

On the other hand, when there are n fermionic particles in the quantum system $\mathcal{H}$, the representation system of the composite system is the n-fold alternative tensor product representation of $\mathcal{H}$. That is, the quantum system is given as $\mathcal{U}_{[\underbrace{0,\dots,0}_{n-1},1,0,\dots,0]}(r)$. Only the sign is changed under the action of permutation in this space. So, a density matrix in the space is invariant for the action of permutation. Therefore, to discuss bosons or fermions, we need to discuss the special unitary group $\mathrm{SU}(r)$ in the above way.

As a typical example, we consider the case when $\mathcal{H}=\mathbb{C}^r\otimes\mathbb{C}^n$, which is spanned by $\{|k,l\rangle\}$ with $k=1,\dots,r$ and $l=1,\dots,n$. Then, the n-fold anti-symmetric subspace $\mathcal{U}_{[\underbrace{0,\dots,0}_{n-1},1,0,\dots,0]}(nr)$ is irreducible for $\mathrm{SU}(nr)$, but is not necessarily irreducible for the subgroup $\mathrm{SU}(r) \times \mathrm{SU}(n) \subset \mathrm{SU}(nr)$ as follows. Let $|\psi\rangle$ be the vector spanning the one-dimensional space $\mathcal{U}_{[0,\ldots,0]}(n)$ of $(\mathbb{C}^n)^{\otimes n}$. Then, $|\psi\rangle\langle\psi|$ can be regarded as a projection in $\mathcal{H}^{\otimes n}$.

Theorem 4.2 *The n-fold anti-symmetric space* $\mathcal{U}_{[\underbrace{0,\ldots,0}_{n-1},1,0,\ldots,0]}(\mathcal{H})$ *satisfies*

$$|\psi\rangle\langle\psi|\mathcal{U}_{[\underbrace{0,\ldots,0}_{n-1},1,0,\ldots,0]}(\mathcal{H}) = \mathcal{U}_{[n,0,\ldots,0]}(r) \otimes \mathcal{U}_{[\underbrace{0,\ldots,0}_{n-1}]}(n), \tag{4.66}$$

which is an irreducible subspace of the subgroup $\mathrm{SU}(r) \times \mathrm{SU}(n)$.

This theorem shows that the subspace of the anti-symmetric space can be regarded as the symmetric space of the smaller system. This statement plays an important role in Sect. 5.5.

Proof Any vector in $|\psi\rangle\langle\psi|\mathcal{U}_{[\underbrace{0,\ldots,0}_{n-1},1,0,\ldots,0]}(\mathcal{H})$ has a from $|\phi\rangle|\psi\rangle$ with an element of $(\mathbb{C}^r)^{\otimes n}$. Since any permutation g satisfies $\mathsf{f}(g)|\psi\rangle = \mathrm{sgn}(g)|\psi\rangle$, the other vector $|\phi\rangle$ is invariant for any permutation. Hence, $|\phi\rangle$ belongs to the h-fold symmetric space $\mathcal{U}_{[n,0,\ldots,0]}(r)$. ∎

To discuss another typical example, we define the spaces $\mathcal{U}_{[1,0,\ldots,0]}(r), \mathcal{U}_{[0,\ldots,0,1]}(r)$, $\mathcal{U}_{[1,0,\ldots,0]}(d)$, and $\mathcal{U}_{[0,\ldots,0,1]}(d)$ spanned by $\{|q\rangle\}_{q=1}^r$, $\{|\bar{q}\rangle\}_{q=1}^r$, $\{|k\rangle\}_{k=1}^d$, and $\{|\bar{k}\rangle\}_{k=1}^d$, respectively. Then, we focus on the $2rd$-dimensional space $\mathcal{H} := \mathcal{U}_{[1,0,\ldots,0]}(r) \otimes \mathcal{U}_{[1,0,\ldots,0]}(d) \oplus \mathcal{U}_{[0,\ldots,0,1]}(r) \otimes \mathcal{U}_{[0,\ldots,0,1]}(d)$. Hence, we can naturally define the representation of the subgroup $\mathrm{SU}(r) \times \mathrm{SU}(d) \subset \mathrm{SU}(2rd)$ on $\mathcal{H}$. The vector $|\psi'\rangle := \sum_{k=1}^d \sqrt{\frac{1}{2d}}(|k,\bar{k}\rangle - |\bar{k},k\rangle)$ spans the one-dimensional space $\mathcal{U}_{[0,\ldots,0]}(d)$ of $\mathcal{U}_{[1,0,\ldots,0]}(d) \otimes \mathcal{U}_{[0,\ldots,0,1]}(d)$. Hence, $|\psi'\rangle\langle\psi'|$ can be regarded as a projection in $\mathcal{H}^{\otimes 2}$.

Theorem 4.3 *The two-fold alternative tensor product space* $\mathcal{U}_{[0,1,0,\ldots,0]}(\mathcal{H})$ *satisfies*

$$|\psi'\rangle\langle\psi'|\mathcal{U}_{[0,1,0,\ldots,0]}(\mathcal{H}) \cong \mathcal{U}_{[1,0,\ldots,0]}(r) \otimes \mathcal{U}_{[0,\ldots,0,1]}(r) \otimes \mathcal{U}_{[0,\ldots,0]}(d), \tag{4.67}$$

which is closed with respect to the subgroup $\mathrm{SU}(r) \times \mathrm{SU}(d) \subset \mathrm{SU}(2rd)$.

Proof Any vector in $\mathcal{H}^{\otimes 2}$ can be written as a liner combination of the following vectors; $|q,k,q',k'\rangle$, $|\bar{q},\bar{k},\bar{q}',\bar{k}'\rangle$ $|q,k,\bar{q}',\bar{k}'$, $|\bar{q},\bar{k},q',k'\rangle$, and $|\bar{q},\bar{k},\bar{q}',\bar{k}'\rangle$. The vectors $|q,k,q',k'\rangle$ and $|\bar{q},\bar{k},\bar{q}',\bar{k}'\rangle$ vanishes by the application of the projection $|\psi'\rangle\langle\psi'|$. We also have

$$\begin{aligned} |\psi'\rangle\langle\psi'||q',k',\bar{q},\bar{k}\rangle &= -|\psi'\rangle\langle\psi'||\bar{q},\bar{k},q',k'\rangle \\ &= \delta_{k,k'}(|q',\bar{q}\rangle + |\bar{q},q'\rangle)\sum_{k=1}^d \sqrt{\frac{1}{2d}}(|k,\bar{k}\rangle - |\bar{k},k\rangle). \end{aligned}$$

Hence, the space $|\psi'\rangle\langle\psi'|\mathcal{U}_{[0,1,0,\ldots,0]}(\mathcal{H})$ is given as linear combination of the vectors $(|q',\bar{q}\rangle + |\bar{q},q'\rangle)\sum_{k=1}^{d}\sqrt{\frac{1}{2d}}(|k,\bar{k}\rangle - |\bar{k},k\rangle)$. We can make one-to-one correspondence the basis and the basis of $\mathcal{U}_{[1,0,\ldots,0]}(r)\otimes\mathcal{U}_{[0,\ldots,0,1]}(r)$. So, we obtain (4.67). ∎

Exercise 4.12 Show

$$\mathcal{U}_{[0,0,\ldots,3]}(2r) = \mathcal{U}_{[0,0,\ldots,3]}(r)\otimes\mathcal{U}_{[3]}(2)\oplus\mathcal{U}_{[0,\ldots,0,1,1]}(r)\otimes\mathcal{U}_{[1]}(2). \tag{4.68}$$

4.5 Discretization of Lie Group and Homogeneous Space

4.5.1 Design on Group

Section 3.7 deals with the invariant measure μ_G on the Lie group G. Although many quantum information processings employ the invariant measure μ_G, it is not easy to realize such quantum information processings because the support of G is so large. Hence, to resolve this problem, instead of the measure μ_G, we often employ another measure μ' whose support is much smaller than G. To discuss how another measure μ' works as an alternative of the measure μ_G under a quantum information processing, we need to clarify the required property of μ' for the quantum information processing. If we need to realize the completely same performance in any situation, we cannot replace the invariant measure μ_G by another measure. The invariant measure μ_G can be used in various purposes, however, it is used for a specific purpose in practice. Hence, for the specific purpose, it is sufficient to employ a measure that satisfies a part of properties of the invariant measure μ_G. Thus, to decide whether another measure μ' realizes the same performance under the specific purpose as the invariant measure μ_G, we need to clarify the required property in terms of the group representation. However, in the current situation, we have not discussed its concrete application, we give only a mathematical definition of the required property for the measure μ' while the usefulness of the definition is discussed in [20, 27] and [50, Chap. 4]. Since the invariant measure does not necessarily exist for a non-compact Lie group G, we discuss a Lie group G whose invariant measure exists.

Now, we prepare the following theorem.

Theorem 4.4 *The following conditions are equivalent to each other for a measure ν on G and a representation f of G on $\mathcal{H}$, where we assume that any irreducible component of $\mathcal{H}$ has the formal dimension when the representation space $\mathcal{H}$ is infinite-dimensional.*

(1) $\int_G \mathsf{f}(g)\otimes\mathsf{f}(g)^\dagger\mu_G(dg) = \int_G \mathsf{f}(g)\otimes\mathsf{f}(g)^\dagger\nu(dg)$.

(2) *Any matrix X on $\mathcal{H}$ satisfies $\int_G \mathfrak{f}(g)X\mathfrak{f}(g)^\dagger \mu_G(dg) = \int_G \mathfrak{f}(g)X\mathfrak{f}(g)^\dagger \nu(dg)$. (When the representation space $\mathcal{H}$ is infinite-dimensional, any trace class operator X satisfies the above equation).*

(3) *Any Hermitian matrix X satisfies $\int_G \mathfrak{f}(g)X\mathfrak{f}(g)^\dagger \mu_G(dg) = \int_G \mathfrak{f}(g)X\mathfrak{f}(g)^\dagger \nu(dg)$. (When the representation space $\mathcal{H}$ is infinite-dimensional, any trace class self-adjoint operator X satisfies the above equation).*

Proof **(2)**⇔**(3)**: Any matrix X can be written as $X_1 + X_2 i$ by using two Hermitian matrices X_1 and X_2. Hence, Conditions **(2)** and **(3)** are equivalent to each other.
(1)⇒**(2)**: Assume that v_1, v_2, v_3, and v_4 are arbitrary vectors in $\mathcal{U}_\lambda$. Considering the matrix $|v_1\rangle\langle v_2| \otimes |v_3\rangle\langle v_4|$, we have

$$\begin{aligned}
&\mathrm{Tr} \int_G \mathfrak{f}(g) \otimes \mathfrak{f}(g)^\dagger \mu_G(dg) |v_1\rangle\langle v_2| \otimes |v_3\rangle\langle v_4| \\
=& \int_G \langle v_2|\mathfrak{f}(g)|v_1\rangle\langle v_4|\mathfrak{f}(g)^\dagger|v_3\rangle \mu_G(dg) \\
=& \langle v_2| \mathrm{Tr} \int_G \mathfrak{f}(g)|v_1\rangle\langle v_4|\mathfrak{f}(g)^\dagger \mu_G(dg)|v_3\rangle.
\end{aligned} \tag{4.69}$$

Hence,

$$\langle v_2| \int_G \mathfrak{f}(g)|v_1\rangle\langle v_4|\mathfrak{f}(g)^\dagger \mu_G(dg)|v_3\rangle = \langle v_2| \int_G \mathfrak{f}(g)|v_1\rangle\langle v_4|\mathfrak{f}(g)^\dagger \nu(dg)|v_3\rangle. \tag{4.70}$$

Since v_3 and v_2 are arbitrary, we have

$$\int_G \mathfrak{f}(g)|v_1\rangle\langle v_4|\mathfrak{f}(g)^\dagger \mu_G(dg) = \int_G \mathfrak{f}(g)|v_1\rangle\langle v_4|\mathfrak{f}(g)^\dagger \nu(dg). \tag{4.71}$$

Since an arbitrary matrix X can be written as a linear combination of matrices with the form $|v_1\rangle\langle v_4|$, we obtain Condition **(2)**.
(2)⇒**(1)**: Assume that v_1, v_2, v_3, and v_4 are arbitrary vector in $\mathcal{U}_\lambda$. To show **(1)**, it is sufficient to show that the traces of the products with the matrix $|v_1\rangle\langle v_2| \otimes |v_3\rangle\langle v_4|$ take the same value in both side in the equation in **(1)**. Condition **(2)** yields (4.71). Hence, we have (4.70). Thus, using (4.69), we find that the traces of the products with the matrix $|v_1\rangle\langle v_2| \otimes |v_3\rangle\langle v_4|$ take the same value in both side in the equation in **(1)**. ■

When the measure ν on G satisfies the condition of this theorem under the representation $\mathfrak{f}$ of G, the measure ν on G is called an **$\mathfrak{f}$-design**. Using Condition **(1)**, we obtain the following lemma.

Lemma 4.1 *For a measure ν on the group G and an element g of G, we define the measure ν_g on the group G as $\nu_g(B) := \nu(gB)$. The following conditions are equivalent to each other.*

(1) *ν is an $\mathfrak{f}$-design.*

(2) *ν_g is an $\mathfrak{f}$-design.*

The discretization of the invariant measure μ_G can be realized by choosing a measure whose support is a finite subset G' of G. So, we focus on a finite subset G' of G. In this case, we define the measure $\mu_{G'}$ as the $\frac{1}{|G'|}$-times of the counting measure on G'. Then, when the measure $\mu_{G'}$ is an $\mathfrak{f}$-design G' is called an $\mathfrak{f}$-**design**. We abbreviate an $\mathfrak{f}_{\boldsymbol{\lambda}}$-design to a $\boldsymbol{\lambda}$-**design**. When a subset $G' \subset G$ is a $\boldsymbol{\lambda}$-design for any element $\boldsymbol{\lambda}$ of a subset $S \subset \hat{G}$, the subset G' is called an S-**design**.

Lemma 4.2 *When a probability measure ν is an S design, the relation*

$$\int_G \chi_{\boldsymbol{\lambda}}(g)\overline{\chi_{\boldsymbol{\lambda}'}(g)}\nu(dg) = \delta_{\boldsymbol{\lambda},\boldsymbol{\lambda}'}$$

holds for $\boldsymbol{\lambda}, \boldsymbol{\lambda}' \in S$. Then, when $\boldsymbol{\lambda}' \in S$ and $\{\boldsymbol{\lambda}''' \in \hat{G} | C_{\boldsymbol{\lambda}'''}^{\boldsymbol{\lambda},\boldsymbol{\lambda}''} \neq 0\} \subset S$, the relation

$$\int_G \chi_{\boldsymbol{\lambda}}(g)\chi_{\boldsymbol{\lambda}''}(g)\overline{\chi_{\boldsymbol{\lambda}'}(g)}\nu(dg) = C_{\boldsymbol{\lambda},\boldsymbol{\lambda}''}^{\boldsymbol{\lambda}'}$$

holds.

Lemma 4.3 *Assume that G is a compact Lie group and $\mathfrak{f}$ is its finite-dimensional representation. A finite subgroup G' of G is an $\mathfrak{f}$-design if and only if the irreducible decomposition of the restriction to G' of the representation $\mathfrak{f}$ is the irreducible decomposition of the representation $\mathfrak{f}$ of G*

Proof Let $\mathcal{K}$ be an arbitrary irreducible non-trivial subspace of the restriction to G' of the representation $\mathfrak{f}$. For a vector $u \in \mathcal{K}$, we have

$$\int_{G'} \mathfrak{f}_{\boldsymbol{\lambda}}(g)|u\rangle\langle u|\mathfrak{f}_{\boldsymbol{\lambda}}(g)^\dagger \mu_{G'}(dg) = \frac{\|u\|^2}{d_{\mathcal{K}}} I_{\mathcal{K}}, \tag{4.72}$$

where $d_{\mathcal{K}}$ is the dimension of $\mathcal{K}$. Let $\mathcal{K}'$ be the irreducible space with respect to the representation of G including u, and $d_{\mathcal{K}'}$ be its dimension. Then, we have

$$\int_{G} \mathfrak{f}_{\boldsymbol{\lambda}}(g)|u\rangle\langle u|\mathfrak{f}_{\boldsymbol{\lambda}}(g)^\dagger \mu_{G}(dg) = \frac{\|u\|^2}{d_{\mathcal{K}'}} I_{\mathcal{K}'}. \tag{4.73}$$

Hence, when G' is an $\mathfrak{f}$-design, $\mathcal{K}$ and $\mathcal{K}'$ are the same space. So, the irreducible decomposition of the restriction to G' of the representation $\mathfrak{f}$ is the irreducible decomposition of the representation $\mathfrak{f}$ of G.

Conversely, we assume that any irreducible component of $\mathfrak{f}$ is still irreducible for the restriction to G'. Since any Hermitian matrix is written with a linear combination of matrices $|u\rangle\langle u|$, it is sufficient to show that any vector $u \in \mathcal{H}$ and any matrix $|u\rangle\langle u|$ satisfy

$$\int_G \mathsf{f}_\lambda(g)|u\rangle\langle u|\mathsf{f}_\lambda(g)^\dagger \mu_G(dg) = \int_G \mathsf{f}_\lambda(g)|u\rangle\langle u|\mathsf{f}_\lambda(g)^\dagger \nu(dg). \tag{4.74}$$

Now, we decompose the vector u into vectors of irreducible representation spaces as $u = \sum_j u_j$. Let $\mathcal{K}_j$ be the irreducible space containing u_j and d_j be its dimension. Then,

$$\int_G \mathsf{f}_\lambda(g)|u\rangle\langle u|\mathsf{f}_\lambda(g)^\dagger \mu_G(dg) = \sum_j \frac{\|u_j\|^2}{d_j} I_{\mathcal{K}_j}.$$

Since the right hand side depends only on the irreducible decomposition, we obtain (4.74). ∎

Notice that the map $X \mapsto \mathsf{f}(g)X\mathsf{f}(g)^\dagger$ is a completely positive map on $\mathcal{H}$ for a representation f on a d-dimensional space $\mathcal{H}$. Since the dimension of the convex set of completely positives map on $\mathcal{H}$ has the dimension $d^4 - d^2$, applying the following Caratheodory's theorem (Theorem 4.5) to this convex set, we find that there exists a probability distribution p such that its support has cardinality $d^4 - d^2 + 1$ and p is an f-design. A similar conclusion has been be shown by using a different method [53].

Theorem 4.5 (Caratheodory's Theorem) *[92, Sect. 17] Assume that X is an m-dimensional convex set and the convex set X is given as the convex closure of the set* Ext(X) *of its extremal points. For an element $x \in X$, there exist $m+1$ extremal points $x_1, \ldots, x_{m+1} \in \mathrm{Ext}(X)$ and a probability distribution p on $\{x_1, \ldots, x_{m+1}\}$ such that $x = \sum_{i=1}^{m+1} p(i)x_i$.*

Generalizing the proof of Lemma 4.3, we can show the following corollary.

Corollary 4.1 *Assume that G is a compact Lie group and f is its finite-dimensional representation. We also assume that $\tilde{\mathsf{f}}$ is a map from a finite group G'' to a Lie group G and $\mathsf{f} \circ \tilde{\mathsf{f}}$ is a projective representation. Then, a subset $\tilde{\mathsf{f}}(G'')$ of G is an f-design if and only if the irreducible decomposition of the projective representation $\mathsf{f} \circ \tilde{\mathsf{f}}$ of G'' is the irreducible decomposition of the representation f of G.*

Next, we concentrate the case when G is $\mathrm{SU}(d)$. A measure ν on G or a subgroup G' of G is called a (t_1, t_2)-**design** when it is an f_{t_1,t_2}-design for the representation $\mathsf{f}_{t_1,t_2}(g) := g^{\otimes t_1} \otimes (g^{\otimes t_2})^\dagger$. Then, a measure ν on $\mathrm{SU}(d)$ is a (t_1, t_2)-design if and only if any matrices $X_1, \ldots, X_{2t_1+2_2}$ on $\mathbb{C}^d$ satisfy

$$\begin{aligned}
&\int_G \mathrm{Tr}\, X_1 g X_2 g \cdots X_{t_1} g X_{t_1+1} g^\dagger \cdots X_{t_1+t_2} g^\dagger X_{t_1+t_2+1} g^\dagger \\
&\qquad \cdots X_{2t_1+t_2} g^\dagger X_{2t_1+t_2+1} g \cdots X_{2t_1+2t_2} g \mu_G(dg) \\
=& \int_G \mathrm{Tr}\, X_1 g X_2 g \cdots X_{t_1} g X_{t_1+1} g^\dagger \cdots X_{t_1+t_2} g^\dagger X_{t_1+t_2+1} g^\dagger \\
&\qquad \cdots X_{2t_1+t_2} g^\dagger X_{2t_1+t_2+1} g \cdots X_{2t_1+2t_2} g \nu(dg).
\end{aligned}$$

This fact can be shown in the same way as Theorem 4.4 by replacing X_i by a linear sum of matrices $|u_i\rangle\langle v_i|$. Hence, a measure ν on SU(d) is a (t_1, t_2)-design if and only if it is a $(t_1 + t_2, 0)$-design. Since it is enough to discuss a $(t, 0)$-design due to this relation, we simplify it to a t-design.[3] When we need to emphasize the design for set of unitaries, we call it a unitary t-design. The properties were discussed in details in [21, 24, 39, 45, 94, 96]. Recently, it was shown that Clifford group is a 3-design when the dimension is a power of 2 (including 2) [106, 117]. (For the definition of Clifford group, see Sect. 8.2). Deep connection with discrete Wigner functions was revealed in [117, 250].

By using Dynkin index, the representation $\mathfrak{f}_{1,1}$ can be decomposed into the direct sum representation of the $d^2 - 1$-dimensional representation with the highest weight $[1, 0, \ldots, 0, 1]$ and the one-dimensional representation with the highest weight $[0, \ldots, 0]$. On the other hand, the representation $\mathfrak{f}_{2,0}$ can be decomposed into the direct sum representation of the $d(d+1)/2$-dimensional representation with the highest weight $[2, 0, \ldots, 0]$ and the $d(d-1)/2$-dimensional representation with the highest weight $[1, 1, 0, \ldots, 0, 0]$. Hence, a measure ν on SU(d) is a $[1, 0, \ldots, 0, 1]$-design if and only if it is a $[2, 0, \ldots, 0]$-design and is a $[1, 1, 0, \ldots, 0]$-design. That is, for a subgroup G' of SU(d) or U(d), the restriction to G' of the representation $\mathfrak{f}_{[1,0,\ldots,0,1]}$ is still irreducible if and only if both restrictions to G' of the representations $\mathfrak{f}_{[2,0,\ldots,0]}$ and $\mathfrak{f}_{[1,1,0,\ldots,0]}$ are still irreducible.

4.5.2 Design on Homogeneous Space

The previous subsection is the discretization of the group G. We can consider the discretization of the homogeneous space Θ in a similar way as follows. In this case, we employ the invariant measure μ_Θ of Θ instead of a measure ν of Θ. A measure ν of Θ is called a $\boldsymbol{\lambda}$-design for $\boldsymbol{\lambda} \in \hat{G}$ when the irreducible representation $\mathfrak{f}_{\boldsymbol{\lambda}}$ of the Lie group G satisfies

$$d_{\boldsymbol{\lambda}} \int_\Theta |\boldsymbol{\lambda} : \theta\rangle\langle\boldsymbol{\lambda} : \theta|\nu(d\theta) = I_{\boldsymbol{\lambda}}. \tag{4.75}$$

As mentioned in Sect. 4.3.1, $|\boldsymbol{\lambda} : \theta\rangle\langle\boldsymbol{\lambda} : \theta|$ gives a POVM. In the above definition, the existence of the formal dimension is assumed when the representation $\mathfrak{f}_{\boldsymbol{\lambda}}$ is an infinite-dimensional representation. Especially, a finite subset Θ' of Θ is called a **$\boldsymbol{\lambda}$-design** when its counting measure is a $\boldsymbol{\lambda}$-design.

For example, when a given finite subgroup G' of the Lie group G is a $\boldsymbol{\lambda}$-design, the orbit $\{g\theta_0 | g \in G'\}$ of $\theta_0 \in \Theta$ with respect to the action of G' is a $\boldsymbol{\lambda}$-design as a subset of the homogeneous space Θ. Let Θ be the set of coherent vectors in the fundamental representation of SU(d), i.e., the set of pure states on $\mathcal{H} = \mathbb{C}^d$

[3] Originally, the name of t-design was given in [23].

When a measure ν on Θ satisfies

$$\int_{\Theta} |u\rangle\langle u|^{\otimes t_1} \otimes |\overline{u}\rangle\langle\overline{u}|^{\otimes t_2} \nu(d\theta) = \int_{\Theta} |u\rangle\langle u|^{\otimes t_1} \otimes |\overline{u}\rangle\langle\overline{u}|^{\otimes t_2} \mu_{\Theta}(d\theta), \tag{4.76}$$

the measure ν is called a (t_1, t_2)-**design**. Additionally, a finite subset Θ' of Θ is called a (t_1, t_2)-**design** when its counting measure is a $\boldsymbol{\lambda}$-design.

When G is a compact group, the dimension $d_{\boldsymbol{\lambda}}$ of the representation space of the representation $\mathsf{f}_{\boldsymbol{\lambda}}$ is finite. Then, (6.40) guarantees that the matrix I is contained in the convex set generated by $\{|\boldsymbol{\lambda} : \theta\rangle\langle\boldsymbol{\lambda} : \theta|\}_{\theta}$, whose dimension is $d_{\boldsymbol{\lambda}}^2 - 1$ at most. Hence, applying Theorem 4.5 to the pair of the convex set and the matrix I, we find that there exists a probability distribution p on Θ such that the cardinality of the support is $d_{\boldsymbol{\lambda}}^2$ and it is a $\boldsymbol{\lambda}$-design.

Then, letting Γ be the operation to take transpose only in the space $(\mathbb{C}^d)^{\otimes t_2}$ at (4.76), we have $\Gamma(|u\rangle\langle u|^{\otimes t_1} \otimes |\overline{u}\rangle\langle\overline{u}|^{\otimes t_2}) = |u\rangle\langle u|^{\otimes t_1+t_2}$. Since the condition for (t_1, t_2)-design and that for $(t_1 + t_2, 0)$-design can be converted to each other via the map Γ, this fact yields the following theorem.

Theorem 4.6 *A measure ν on Θ is a (t_1, t_2)-design if and only if it is a $(t_1 + t_2, 0)$-design.*

Moreover, the following theorem is known [16, 57, 73, 72, 95].

Theorem 4.7 *When a measure ν on Θ is a $(t, 0)$-design, the relation $|\operatorname{supp}(\nu)| \ge \binom{d+\lceil t/2\rceil-1}{\lceil t/2\rceil}^2$ holds. When the equality holds, the relation $\nu(|u\rangle) = 1/|\operatorname{supp}(\nu)|$ holds for any vector $|u\rangle \in \operatorname{supp}(\nu)$, i.e., ν is a distribution on $\operatorname{supp}(\nu)$ with equal weight.*

Since it is enough to discuss a $(t, 0)$-design due to this relation, we simplify it to a t-design. Sections 8.3 and 8.4 give several $(2, 0)$-designs. A more efficient $(2, 0)$-design was given in [21] when the dimension is a power of 2. In this case, since Clifford group is a unitary 3-design [106, 117], the set of stabilizer states forms a 3-design as a set of pure states [69, 117]. (For the definition of stabilizer state, see [50, Sect. 3.3]). Also, t-design is related to several topics. For example, the relation with spherical codes is discussed in [15, 26]. The connection with SIC vectors defined in Sect. 8.4 is discussed in [7, 90, 89, 95]. Since t-design guarantees the symmetry, it can be applied to quantum state estimation [53, 95, 121, 114, 116], quantum state discrimination [2, 76], and detecting maximally entangled state [47]. Further, using t-design, we can construct phase retrieval with the phase lift algorithm [40]. Also, we can discuss the designs in projective spaces [56].

4.5.3 Approximating Design

In general, it is not easy to find an f-design for a representation f. Even though we could find an f-design, its structure might be too complicated for a practical use. Hence, as an alternative idea, we often seek to approximately realize it. As measures of the quality of the approximation, we introduce two criteria, one is based on the matrix (operator) norm, and the other is based on the trace norm. Given a representation f of G, a measure ν on G is called a trace norm ϵ approximate f-design when

$$\left\| \int_G \mathsf{f}(g) \otimes \mathsf{f}(g)^\dagger \mu_G(dg) - \int_G \mathsf{f}(g) \otimes \mathsf{f}(g)^\dagger \nu(dg) \right\|_{\mathrm{Tr}} \le \epsilon .$$

Similarly, it is called a matrix (operator) norm ϵ approximate f-design when

$$\left\| \int_G \mathsf{f}(g) \otimes \mathsf{f}(g)^\dagger \mu_G(dg) - \int_G \mathsf{f}(g) \otimes \mathsf{f}(g)^\dagger \nu(dg) \right\| \le \epsilon .$$

A measure ν on Θ is called a trace norm ϵ approximate $\boldsymbol{\lambda}$-design when the following relation (4.77) holds, and it is called a matrix (operator) norm ϵ approximate $\boldsymbol{\lambda}$-design when the following relation (4.78) holds.

$$\left\| d_{\boldsymbol{\lambda}} \int_{\Theta'} |\boldsymbol{\lambda} : \theta\rangle\langle \boldsymbol{\lambda} : \theta| \nu(d\theta) - I_{\boldsymbol{\lambda}} \right\|_{\mathrm{Tr}} \le \epsilon , \tag{4.77}$$

$$\left\| d_{\boldsymbol{\lambda}} \int_{\Theta'} |\boldsymbol{\lambda} : \theta\rangle\langle \boldsymbol{\lambda} : \theta| \nu(d\theta) - I_{\boldsymbol{\lambda}} \right\| \le \epsilon . \tag{4.78}$$

Chapter 5
Application to Physical Systems

Abstract We apply representation theory to several real quantum systems. A real microscopic system has the symmetry of three-dimensional rotation as well as the classical macroscopic system. We firstly discuss this rotation symmetry on the function space $L^2(\mathbb{R}^3)$ by using the algebra $\mathfrak{su}(2)$. Since this symmetry is related to the three-dimensional rotation, it is called orbital angular momentum. Then, we discuss the Schrödinger equation with central potential. As a typical example, we address the case of hydrogen-type atom and explain the degeneracy of Hamiltonian in this case by using SO(4, $\mathbb{R}$). We also extend the discussion of the rotation symmetry to the d-dimensional case. Then, we introduce the spin system and discuss an interaction between the spin system and the orbital angular momentum system. Also, we discuss quarks that constitutes elements of atoms based on several groups. Since a quark is a fermion and our discussion is limited to the cases with two or three particles, the discussion for quarks in this chapter is much simpler than that of the photonic system because we need to discuss unlimited number of bosons in the photonic system. Finally, we discuss uncertainty relation for wake packets on various spaces, i.e., not only on the set of real numbers $\mathbb{R}$ but also on the one-dimensional and three-dimensional spheres S^1 and S^3.

5.1 3-Dimensional Physical System

5.1.1 Orbital Angular Momentum

This chapter discusses spectral decompositions of physically important operators. As a typical example of the group representation, we focus on the physical system $L^2(\mathbb{R}^3)$ for a particle. For this analysis, we focus on two self-adjoint operators Q_k and $\hat{\mathsf{P}}_k$ on $L^2(\mathbb{R}^3)$ defined as[1]

[1]In this book, P_k expresses the operator $-i\frac{\partial}{\partial x_k}$, which equals $\hat{\mathsf{P}}_k$ when $\hbar = 1$. For the precise definition of the operators in this chapter, see Sect. 1.6.

M. Hayashi, *Group Representation for Quantum Theory*,
DOI 10.1007/978-3-319-44906-7_5

$$\mathsf{Q}_k f(x) := x_k f(x), \quad \hat{\mathsf{P}}_k f(x) := -i\hbar \frac{\partial}{\partial x_k} f(x) \tag{5.1}$$

for $x \in \mathbb{R}^3$, where $\hbar$ is the Planck constant. Here, Q_k is called the **position operator** of the k-th coordinate, and P_k is called the **momentum operator** of the k-th coordinate. These operators satisfy the commutation relation as

$$[\mathsf{Q}_l, \hat{\mathsf{P}}_k] = i\hbar\delta_{l,k} I. \tag{5.2}$$

Then, we define the **orbital angular momentum operator** L_t on $L^2(\mathbb{R}^3)$ as

$$L_t := \sum_{k,l} \epsilon_{t,k,l} \mathsf{Q}_k \hat{\mathsf{P}}_l \tag{5.3}$$

for $x \in \mathbb{R}^3$. Due to the commutation relation (5.2), the self-adjoint operator L_k satisfies the commutation relation

$$[L_t, L_k] = i\hbar \sum_l \epsilon_{t,k,l} L_l. \tag{5.4}$$

Hence, $\frac{-i}{\hbar} L_k$ satisfies the commutation relation (4.11) for $\mathfrak{su}(2)$, which is equivalent to $\mathfrak{so}(3)$. This is natural because the operator L_k generates the rotation with k-axis in $L^2(\mathbb{R}^3)$.

We also define the operators $L_\pm$ and the **total orbital angular momentum** $\boldsymbol{L}^2$ as

$$L_\pm := L_1 \pm i L_2, \quad \boldsymbol{L}^2 := \sum_{k=1}^{3} L_k^2. \tag{5.5}$$

In this chapter, we denote the vector composed of operators by the bold capital letter as $\boldsymbol{L}$, and the sum of the squares of components by the bold capital letter with superscript 2 as $\boldsymbol{L}^2$. We also denote the sum of product of components of two vectors composed of operators by the inner product of the two bold capital letter as $\boldsymbol{P} \cdot \boldsymbol{Q} := \sum_{k=1}^{3} \hat{\mathsf{P}}_k \mathsf{Q}_k$.

So, we have

$$L_+^\dagger = L_-, \quad \boldsymbol{L}^2 = L_3^2 + \frac{1}{2}(L_+ L_- + L_- L_+). \tag{5.6}$$

Now, we introduce the **spherical coordinate** (r, θ_1, θ_2) as (Fig. 5.1)

$$x_1 = r \sin\theta_1 \cos\theta_2, \quad x_2 = r \sin\theta_1 \sin\theta_2, \quad x_3 = r\cos\theta_1, \tag{5.7}$$

where $r \in [0, \infty)$, $\theta_1 \in [0, 2\pi)$, and $\theta_2 \in [0, \pi)$. That is, (θ_1, θ_2) gives a coordinate on the two-dimensional sphere S^2.

Fig. 5.1 Spherical coordinate

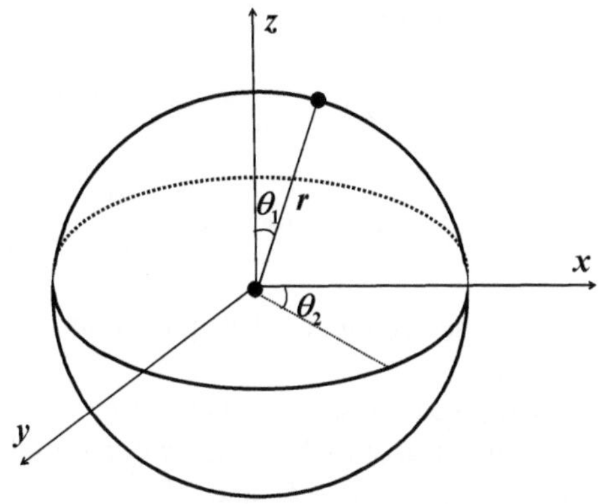

We define a linear map $\mathcal{I}$ from $L^2(\mathbb{R}_+) \otimes L^2(S^2) = L^2(\mathbb{R}_+ \times S^2)$ to $L^2(\mathbb{R}^3)$ as follows, where $\mathbb{R}_+$ is the set of positive real numbers. For $\phi \in L^2(\mathbb{R}_+ \times S^2)$, we define

$$\mathcal{I}(\phi)(x) := \frac{1}{\sqrt{4\pi} r}\phi(r, \theta_1, \theta_2) \tag{5.8}$$

with the condition (5.7). Since we have

$$\begin{aligned} &\int_{\mathbb{R}^3} |\mathcal{I}(\phi)(x)|^2 d^3x \\ =& \int_0^\infty \int_{S^2} 4\pi r^2 \Big| \frac{1}{\sqrt{4\pi} r}\phi(r, \theta_1, \theta_2)\Big|^2 dr \mu_2(d\theta_1, d\theta_2) \\ =& \int_0^\infty \int_{S^2} \Big|\phi(r, \theta_1, \theta_2)\Big|^2 dr \mu_2(d\theta_1, d\theta_2) \end{aligned} \tag{5.9}$$

with the normalized invariant measure μ_2 on S^2, the map $\mathcal{I}$ is an isometry. So, we can identity the system $L^2(\mathbb{R}^3)$ with the composite system $L^2(\mathbb{R}_+) \otimes L^2(S^2)$ via the isometry $\mathcal{I}$ defined in (5.8).

Then, we have

$$L_1 = i\hbar\left(\sin\theta_2 \frac{\partial}{\partial\theta_1} + \frac{\cos\theta_1 \cos\theta_2}{\sin\theta_1}\frac{\partial}{\partial\theta_2}\right) \tag{5.10}$$

$$L_2 = -i\hbar\left(\cos\theta_2 \frac{\partial}{\partial\theta_1} - \frac{\cos\theta_1 \sin\theta_2}{\sin\theta_1}\frac{\partial}{\partial\theta_2}\right) \tag{5.11}$$

$$L_3 = -i\hbar\frac{\partial}{\partial\theta_2}. \tag{5.12}$$

Hence, we have

$$\boldsymbol{L}^2 = -\hbar^2\left(\frac{1}{\sin\theta_1}\frac{\partial}{\partial\theta_1}\left(\sin\theta_1\frac{\partial}{\partial\theta_1}\right)+\frac{1}{\sin^2\theta_1}\frac{\partial^2}{\partial\theta_2^2}\right) \tag{5.13}$$

$$L_\pm = \hbar e^{\pm i\theta_2}\left(\pm\frac{\partial}{\partial\theta_1}+i\frac{\cos\theta_1}{\sin\theta_1}\frac{\partial}{\partial\theta_2}\right). \tag{5.14}$$

That is, the orbital angular momentum operator L_j can be given as an operator on $L^2(S^2)$.

Now, we choose the generator $F_1^k := \frac{-i}{\hbar}L_k$ to satisfy the commutation relation (4.11) because of (5.4). Hence, we can employ the discussion in Sect. 4.2. In this case, we have $\boldsymbol{F_1}\cdot\boldsymbol{F_1} = \frac{1}{\hbar^2}\boldsymbol{L}^2$ and $F_1^\pm = \frac{1}{\hbar}L_\mp$. Also, F_1^3 and $\boldsymbol{F_1}\cdot\boldsymbol{F_1}$ are commutative. The operator $\boldsymbol{F_1}\cdot\boldsymbol{F_1}$ is a constant times of the Casimir operator, as explained in (4.20). So, it takes only the value $\ell(\ell+1)$ with a non-negative half integer ℓ, as explained in (4.21). This constant ℓ determines the type of the irreducible representation as the maximum weight. Under the irreducible representation with the maximum weight ℓ, the operator $-iF_1^3$ takes only the value $\ell, \ell-1, \ldots, -\ell$. So, m is a (half-) integer when ℓ is a (half-) integer.

Now, we denote the eigenvector with eigenvalues $\ell(\ell+1)$ and m of $\boldsymbol{F_1}\cdot\boldsymbol{F_1}$ and $-iF_1^3$ by the function $Y_\ell^m \in L^2(\mathbb{R}^3)$. Due to the eigen equations for $\boldsymbol{F_1}\cdot\boldsymbol{F_1}$ and $-iF_1^3$, we have

$$-i\hbar\frac{\partial}{\partial\theta_2}Y_\ell^m = m\hbar Y_\ell^m \tag{5.15}$$

$$-\hbar^2\left(\frac{1}{\sin\theta_1}\frac{\partial}{\partial\theta_1}\left(\sin\theta_1\frac{\partial}{\partial\theta_1}\right)+\frac{1}{\sin^2\theta_1}\frac{\partial^2}{\partial\theta_2^2}\right)Y_\ell^m = \ell(\ell+1)\hbar^2 Y_\ell^m. \tag{5.16}$$

Since these partial differential equations depend on θ_1 and θ_2, Y_ℓ^m can be regarded as a function of θ_1 and θ_2. Now, we define another function $\Theta_\ell^m := Y_\ell^m e^{-im\theta_2}$. Then, (5.15) and (5.16) are converted to

$$\frac{\partial}{\partial\theta_2}\Theta_\ell^m = 0 \tag{5.17}$$

$$\left(\frac{1}{\sin\theta_1}\frac{\partial}{\partial\theta_1}\left(\sin\theta_1\frac{\partial}{\partial\theta_1}\right)+\frac{m^2}{\sin^2\theta_1}+\ell(\ell+1)\right)\Theta_\ell^m = 0. \tag{5.18}$$

That is, Θ_ℓ^m is a function only of θ_1.

Firstly, we discuss these equations only when $m = \ell$. Since the relation (4.18) with $m = \ell$ shows that the operator L_+ annihilates the function Y_ℓ^ℓ, the relation (5.14) implies

$$e^{i\theta_2}\left(\frac{\partial}{\partial\theta_1}+i\frac{\cos\theta_1}{\sin\theta_1}\frac{\partial}{\partial\theta_2}\right)Y_\ell^\ell = 0, \tag{5.19}$$

which is equivalent to

$$\left(\frac{\partial}{\partial \theta_1} - \ell \frac{\cos\theta_1}{\sin\theta_1}\right)\Theta_\ell^\ell = 0, \tag{5.20}$$

i.e., we have $\frac{\partial}{\partial\theta_1}\log\Theta_\ell^\ell = \ell\frac{\cos\theta_1}{\sin\theta_1}$. Since $\log\sin^\ell\theta_1$ is a solution of $\log\Theta_\ell^\ell$, the general solution of Θ_ℓ^ℓ is written as $c_\ell \sin^\ell\theta_1$ with a constant c_ℓ. That is, the eigenfunction Y_ℓ^ℓ is $c_\ell e^{i\ell\theta_2}\sin^\ell\theta_1$. However, since θ_1 takes values in $[0, 2\pi)$, the eigenfunction has to be smooth at the neighborhood of $\theta_1 = \pi$, i.e., $\sin\theta_1 = 0$. To satisfy this requirement, the maximum weight ℓ needs to be an integer. In this way, the maximum weight ℓ is restricted to be an integer. Due to this restriction, our representation of $\mathfrak{su}(2)$ can be extended to the unitary representation of $\mathrm{SO}(3,\mathbb{R})$. It is natural because the function space $L^2(S^2)$ has the symmetry of $\mathrm{SO}(3,\mathbb{R})$ naturally.

Now, we introduce another variable $\xi := \cos\theta_1$. Then, the eigenfunction $P_\ell^m(\xi) := \Theta_\ell^m(\theta_1)$ satisfies **Legendre equation**;

$$(1-\xi^2)\frac{d^2}{d\xi^2}P_\ell^m - 2\xi\frac{d}{d\xi}P_\ell^m\left(1 - \frac{m^2}{1-\xi^2}\right)P_\ell^m = 0. \tag{5.21}$$

The solution is given as $P_\ell^m(\xi) = \frac{(-1)^\ell}{2^\ell \ell}\sqrt{\frac{(2\ell+1)(\ell+m)!}{4\pi(\ell-m)!}}\frac{1}{(1-\xi^2)}\frac{d^{\ell-m}}{d\xi^{\ell-m}}(1-\xi^2)^\ell$, which is called **Legendre polynomial**. Therefore, the eigenfunction Y_ℓ^m is given as

$$Y_\ell^m(\theta_1,\theta_2) = P_\ell^m(\cos\theta_1)e^{im\theta_2}. \tag{5.22}$$

The function Y_ℓ^m is called the spherical harmonic function. Indeed, while $P_\ell^m(\xi)$ has a singularity at $\xi = 1$, the singularity is resolved with the parameter θ_1. So, $Y_\ell^m(\theta_1,\theta_2)$ has no singularity even at $\theta_1 = \pi$. When we treat the function Y_ℓ^m with an element of $L^2(S^2)$, we denote it by $|\ell, m\rangle$. It is known that $\{|\ell,m\rangle\}_{\ell,m}$ forms a CONS of $L^2(S^2)$, which implies that $\{|\ell,m\rangle\langle\ell,m|\}_{\ell,m}$ is a PVM, i.e., gives a simultaneous measurement of $\boldsymbol{L}^2$ and L_3.

The number ℓ is called the **azimuthal quantum number** and the number m is called the **magnetic quantum number**. The azimuthal quantum number ℓ does not depend on the choice of z-axis, but the magnetic quantum number depends on the choice of z-axis. That is, the choice of the wave function Y_ℓ^m depends on the choice of z-axis. The azimuthal quantum number ℓ corresponds to the subshell of the orbit of electron as Table 5.1.

Finally, we give the concrete forms of $Y_\ell^m(\theta_1,\theta_2)$ with $\ell = 0, 1, 2$ as follows.

Table 5.1 Structures of subshell

Azimuthal quantum number (ℓ)	Name of subshell	Shape	Picture
0	s (sharp)	Sphere	
1	p (principal)	Dumbbel	
2	d (diffuse)	Double dumbbel	
3	f (fundamental)	Complex	?

$$Y_0^0(\theta_1, \theta_2) = \frac{1}{\sqrt{4\pi}} \tag{5.23}$$

$$Y_1^0(\theta_1, \theta_2) = \sqrt{\frac{3}{4\pi}} \cos\theta_1 \tag{5.24}$$

$$Y_1^{\pm 1}(\theta_1, \theta_2) = \mp\sqrt{\frac{3}{8\pi}} \sin\theta_1 \, e^{\pm i\theta_2} \tag{5.25}$$

$$Y_2^0(\theta_1, \theta_2) = \sqrt{\frac{5}{16\pi}} (3\cos^2\theta_1 - 1) \tag{5.26}$$

$$Y_2^{\pm 1}(\theta_1, \theta_2) = \mp\sqrt{\frac{15}{8\pi}} \sin\theta_1 \cos\theta_1 \, e^{\pm i\theta_2} \tag{5.27}$$

$$Y_2^{\pm 2}(\theta_1, \theta_2) = \sqrt{\frac{15}{32\pi}} \sin^2\theta_1 \, e^{\pm 2i\theta_2}. \tag{5.28}$$

Exercise 5.1 Show (5.4).

5.1.2 Hamiltonian with Central Potential

Next, we consider a particle in the three-dimensional system, which has the potential $V(r)$ as a function of the radius $r = \sqrt{\sum_{j=1}^3 x_j^2}$. When there is only a electric charge only at the charge, the potential satisfies this condition. In this case, the total **Hamiltonian** is given as the sum of the potential and the kinetic energy $\frac{1}{2\mu}\boldsymbol{P}^2$ as

$$H := \frac{1}{2\mu}\boldsymbol{P}^2 + V(r), \tag{5.29}$$

where μ is the constant describing the reduced mass of the system consisting of the electron and the nucleus, which is given by $\frac{m_e M}{m_e+M}$ with the mass of the electron m_e and the mass of the atom M.

Now, we discuss the eigenequation of the Hamiltonian H as

$$\left(\frac{1}{2\mu}\boldsymbol{P}^2 + V(r)\right)f(x) = Ef(x). \tag{5.30}$$

For this purpose, we consider the relation among three operators $\boldsymbol{L}^2$, L_3, and the Hamiltonian H. From the definition of $\boldsymbol{L}^2$ and L_j, we have

$$\boldsymbol{L}^2 = \boldsymbol{Q}^2\boldsymbol{P}^2 - (\boldsymbol{Q}\cdot\boldsymbol{P})^2 + i\hbar\boldsymbol{Q}\cdot\boldsymbol{P}. \tag{5.31}$$

So, we have

$$\boldsymbol{L}^2 f(x) = r^2\boldsymbol{P}^2 f(x) + \hbar^2\left(r^2\frac{\partial^2}{\partial r^2}f(x) + 2r\frac{\partial}{\partial r}f(x)\right). \tag{5.32}$$

Thus,

$$\frac{1}{2\mu}\boldsymbol{P}^2 f(x) = -\frac{\hbar^2}{2\mu}\left(\frac{\partial^2}{\partial r^2}f(x) + \frac{2}{r}\frac{\partial}{\partial r}f(x)\right) + \frac{1}{2\mu r^2}\boldsymbol{L}^2 f(x). \tag{5.33}$$

Hence, the eigenequation (5.30) can be simplified as

$$\left(-\frac{\hbar^2}{2\mu r^2}\frac{\partial}{\partial r}\left(r^2\frac{\partial}{\partial r}\right) + \frac{\ell(\ell+1)\hbar^2}{2\mu r^2} + V(r)\right)f = Ef. \tag{5.34}$$

However, to discuss the Hamiltonian on $L^2(\mathbb{R}_+ \times S^2) = L^2(\mathbb{R}_+)\otimes L^2(S^2)$, we need the unitary transformation $\mathcal{I}^{-1}$, i.e., we focus on $\hat{f}(r,\theta_1,\theta_2) := \sqrt{2\pi}rf(r,\theta_1,\theta_2)$. Then, (5.34) is converted to

$$H\hat{f} = \left(-\frac{\hbar^2}{2\mu}\frac{\partial}{\partial r^2} + \frac{1\hbar^2\ell(\ell+1)}{2\mu r^2} + V(r)\right)\hat{f} = E\hat{f}. \tag{5.35}$$

Hence, we find that the operators $\frac{1}{2\mu}\boldsymbol{P}^2$ and $V(r)$ are commutative with the operators $\boldsymbol{L}^2$ and L_3, i.e., the Hamiltonian H is commutative with the operators $\boldsymbol{L}^2$ and L_3 defined on $L^2(S^2)$.

Now, we apply the discussion in Sect. 1.4.2. The simultaneous measurement of $\boldsymbol{L}^2$ and L_3 on $L^2(S^2)$ is given as $\{|\ell,m\rangle\langle\ell,m|\}_{l,m}$. Then, we have

$$H_{\ell,m} = -\frac{\hbar^2}{2\mu}\frac{\partial}{\partial r^2} + \frac{1}{2\mu r^2}\ell(\ell+1) + V(r) \tag{5.36}$$

as an operator on $L^2(\mathbb{R}_+)$. Hence, we obtain the following **Schrodinger equation** dependently of ℓ;

$$-\frac{\hbar^2}{2\mu}\frac{\partial}{\partial r^2}\chi_{E,\ell}+\left(\frac{\ell(\ell+1)\hbar^2}{2\mu r^2}+V(r)\right)\chi_{E,\ell}=E\chi_{E,\ell}. \tag{5.37}$$

That is, when the function $\chi_{E,\ell}$ in $L^2(\mathbb{R}_+)$ is a solution of the above equation, $f_{E,\ell,m}(r,\theta_1,\theta_2):=Y_\ell^m(\theta_1,\theta_2)\frac{\chi_{E,\ell}(r)}{\sqrt{2\pi r}}$ is a solution of the original (5.30).

The eigen equation (5.37) can be regarded as a Schrödinger equation in the one-dimensional system $L^2(\mathbb{R}_+)$ with the potential term $(\frac{\ell(\ell+1)\hbar^2}{2\mu r^2}+V(r))$. That is, $\frac{\ell(\ell+1)\hbar^2}{2\mu r^2}$ can be regarded as the centrifugal force.

In some case, the main term H_0 of the Hamiltonian is given in (5.29), and the fluctuation ΔH is given as a small constant ϵ times of L_3. For example, when there is an interaction with the external magnetic field. Then, $\{f_{E,\ell,m}\}E,\ell,m$ gives the eigenvectors of the Hamiltonian, as discussed in the end of Sect. 1.4. In this case, we need to choose the z-axis dependently of the direction of the external magnetic field.

5.2 Laplacian on Sphere

5.2.1 *General Case*

Now, we generalize the above discussion to general d-dimensional case. For this analysis, we focus on two self-adjoint operators Q_k and $\hat{\mathsf{P}}_k$ on the function space $L^2(\mathbb{R}^d)$ on the d-dimensional space $\mathbb{R}^d$, which is defined as (5.1). Then, the total kinetic energy is given as

$$\frac{1}{2m}\sum_{j=1}^{d}\hat{\mathsf{P}}_j^2=-\frac{\hbar}{2m}\sum_{k=1}^{d}\partial_k^2, \tag{5.38}$$

where $\partial_k:=\frac{\partial}{\partial x_k}$. To discuss the total kinetic energy, we introduce the **Laplacian** $\Delta_{\mathbb{R}^d}:=\sum_{k=1}^{d}\partial_k^2$ on the function space $L^2(\mathbb{R}^d)$ on the d-dimensional space.

As another important self-adjoint operator, we introduce the **angular momentum** operator

$$L_{k,j}:=\mathsf{Q}_k\hat{\mathsf{P}}_j-\mathsf{Q}_j\hat{\mathsf{P}}_k=-i\hbar(\mathsf{Q}_k\partial_j-\mathsf{Q}_j\partial_k) \tag{5.39}$$

for $k\neq j$. For mathematical convenience, we employ another operator $F_{k,l}:=\mathsf{Q}_k\partial_j-\mathsf{Q}_j\partial_k$. Now, we investigate the relation between the operators $F_{k,l}$ and the Laplacian $\Delta_{\mathbb{R}^d}$. Since

$$\sum_{k>l} F_{k,l}^2 = \sum_{k\neq l} Q_k^2 \partial_l^2 - \sum_{k\neq l} Q_k Q_l \partial_k \partial_l - (d-1)\sum_k Q_k \partial_k \tag{5.40}$$

$$\Big(\sum_k Q_k \partial_k\Big)^2 = \sum_k Q_k^2 \partial_k^2 + \sum_k Q_k \partial_k + \sum_{k\neq l} Q_k Q_l \partial_k \partial_l \tag{5.41}$$

$$\Big(\sum_k Q_k^2\Big)\Big(\sum_l \partial_l^2\Big) = \sum_k Q_k^2 \partial_k^2 + \sum_{k\neq l} Q_k^2 \partial_l^2, \tag{5.42}$$

we have

$$\begin{aligned}\sum_{k>l} L_{k,l}^2 + \Big(\sum_k Q_k \partial_k\Big)^2 &= \sum_{k\neq l} Q_k^2 \partial_l^2 + \sum_k Q_k^2 \partial_k^2 - (d-2)\sum_k Q_k \partial_k \\ &= \Big(\sum_k Q_k^2\Big)\Big(\sum_l \partial_l^2\Big) - (d-2)\sum_k Q_k \partial_k .\end{aligned} \tag{5.43}$$

Now, in the d-dimensional space $\mathbb{R}^d$, we introduce the spherical coordinate r, w with $r > 0$ and $w \in S^{d-1}$ as $x_k = r w_k$, where S^{d-1} is the $d-1$-dimensional sphere. More precisely, we employ **spherical coordinate** r and $\theta = (\theta_1, \ldots, \theta_{d-1})$ as

$$r = \sqrt{\sum_k x_k^2} \tag{5.44}$$

$$w_k = \frac{x_k}{r} = \sin\theta_1 \cdots \sin\theta_{k-1}\cos\theta_k \tag{5.45}$$

$$w_d = \frac{x_d}{r} = \sin\theta_1 \cdots \sin\theta_d . \tag{5.46}$$

for $k = 1, \ldots, d-1$. Then, we have $\frac{\partial}{\partial r} = \sum_k w_k \partial_k$, i.e., $\sum_k x_k \partial_k = r\frac{\partial}{\partial r}$. To analyze the Laplacian $\Delta_{\mathbb{R}^d}$, we introduce the **spherical Laplacian** $\Delta_{S^{d-1}} := \sum_{k>l} F_{k,l}^2$. The relation (5.43) is simplified to

$$\Delta_{S^{d-1}} + \Big(r\frac{\partial}{\partial r}\Big)^2 = r^2 \Delta_{\mathbb{R}^d} - (d-2) r\frac{\partial}{\partial r}, \tag{5.47}$$

which implies that

$$\Delta_{\mathbb{R}^d} = \frac{1}{r^2}\Delta_{S^{d-1}} + \frac{\partial^2}{\partial r^2} + \frac{d-1}{r}\frac{\partial}{\partial r} \tag{5.48}$$

because $(r\frac{\partial}{\partial r})^2 = r^2\frac{\partial^2}{\partial r^2} + r\frac{\partial}{\partial r}$. The relation (5.48) is the generalization of (5.33).

Now, we show that the operator $F_{k,l}$ and the spherical Laplacian $\Delta_{S^{d-1}}$ can be regarded as operators on the function space $L^2(S^{d-1})$. First, we notice that

$$r = \sqrt{\sum_{k=1}^{d} x_k^2}, \quad \theta_k = \cos^{-1} \frac{x_k}{\sqrt{\sum_{k'=k}^{d} x_{k'}^2}} \tag{5.49}$$

for $k = 1, \ldots, d-1$. Since

$$\frac{\partial r}{\partial x_k} = \frac{w_k}{r} \tag{5.50}$$

$$\frac{\partial \theta_k}{\partial x_k} = \frac{\sqrt{\sum_{k'=k+1}^{d} w_{k'}^2}}{r \sum_{k'=k}^{d} w_{k'}^2} \tag{5.51}$$

$$\frac{\partial \theta_k}{\partial x_j} = \frac{w_k w_j}{r\left(\sum_{k'=k}^{d} w_{k'}^2\right)\sqrt{\sum_{k'=k+1}^{d} w_{k'}^2}} \tag{5.52}$$

for $k = 1, \ldots, d-1$, $j > k$.

$\frac{d}{dx} \cos^{-1} f(x) = \frac{-1}{\sqrt{1-f(x)^2}} \frac{df}{dx}$. Hence, we have

$$\begin{aligned} \frac{\partial}{\partial x_j} =& \sum_{k=1}^{d-1} \frac{\partial \theta_k}{\partial x_j} \frac{\partial}{\partial \theta_k} + \frac{\partial r}{\partial x_j} \frac{\partial}{\partial r} \\ =& \sum_{k=1}^{j-1} \frac{w_k w_j}{r\left(\sum_{k'=k}^{d} w_{k'}^2\right)\sqrt{\sum_{k'=k+1}^{d} w_{k'}^2}} \frac{\partial}{\partial \theta_k} \\ &+ \frac{\sqrt{\sum_{k'=j+1}^{d} w_{k'}^2}}{r \sum_{k'=j}^{d} w_{k'}^2} \frac{\partial}{\partial \theta_j} + \frac{w_j}{r} \frac{\partial}{\partial r} \end{aligned} \tag{5.53}$$

for $j = 1, \ldots, d-1$, and

$$\begin{aligned} \frac{\partial}{\partial x_d} =& \sum_{k=1}^{d-1} \frac{\partial \theta_k}{\partial x_d} \frac{\partial}{\partial \theta_k} + \frac{\partial r}{\partial x_d} \frac{\partial}{\partial r} \\ =& \sum_{k=1}^{d-1} \frac{w_k w_d}{r\left(\sum_{k'=k}^{d} w_{k'}^2\right)\sqrt{\sum_{k'=k+1}^{d} w_{k'}^2}} \frac{\partial}{\partial \theta_k} + \frac{w_d}{r} \frac{\partial}{\partial r}. \end{aligned} \tag{5.54}$$

Thus, for $j > l$, we have

$$F_{j,l} = \sum_{k=j}^{l-1} \frac{w_j w_l w_k}{\left(\sum_{k'=k}^{d} w_{k'}^2\right)\sqrt{\sum_{k'=k+1}^{d} w_{k'}^2}} \frac{\partial}{\partial \theta_k} + w_j \frac{\sqrt{\sum_{k'=l+1}^{d} w_{k'}^2}}{\sum_{k'=l}^{d} w_{k'}^2} \frac{\partial}{\partial \theta_l} - w_l \frac{\sqrt{\sum_{k'=j+1}^{d} w_{k'}^2}}{\sum_{k'=j}^{d} w_{k'}^2} \frac{\partial}{\partial \theta_j} \tag{5.55}$$

$$F_{j,d} = \sum_{k=j}^{d-2} \frac{w_k w_j w_d}{\left(\sum_{k'=k}^{d} w_{k'}^2\right)\sqrt{\sum_{k'=k+1}^{d} w_{k'}^2}} \frac{\partial}{\partial \theta_k} + w_j \frac{w_{d-1}}{w_{d-1}^2 + w_d^2} \frac{\partial}{\partial \theta_{d-1}} - w_d \frac{\sqrt{\sum_{k'=j+1}^{d} w_{k'}^2}}{\sum_{k'=j}^{d} w_{k'}^2} \frac{\partial}{\partial \theta_j} \tag{5.56}$$

for $k > l > d - 1$. Therefore, the operator $F_{j,l}$ can be written as an operator on $L^2(S^{d-1})$. The spherical Laplacian $\Delta_{S^{d-1}} = \sum_{k>l} F_{k,l}^2$ is also written as an operator on $L^2(S^{d-1})$.

In the following, we consider the function space $L^2(S^{d-1})$, and discuss the eigen equation with eigenvalue E of the spherical Laplacian $\Delta_{S^{d-1}}$ as

$$\Delta_{S^{d-1}} f(\theta) = E f(\theta). \tag{5.57}$$

We define the set $P_n(d)$ of polynomials with degree n as a subspace of $L^2(\mathbb{R}^d)$. Dividing all elements of $P_n(d)$ by r^n, the set $P_n(d)$ can be regarded as a subspace $L^2(S^{d-1})$. The Laplacian $\Delta_{\mathbb{R}^d}$ can be regarded as a linear map from $P_n(d)$ to $P_{n-1}(d)$. For any element f of the kernel Ker $\Delta_{\mathbb{R}^d}$ in $P_n(d)$, (5.48) implies that

$$\begin{aligned} 0 &= \frac{1}{r^2}\Delta_{S^{d-1}} f + \frac{\partial^2}{\partial r^2} f + \frac{d-1}{r}\frac{\partial}{\partial r} f \\ &= \frac{1}{r^2}(\Delta_{S^{d-1}} f + n(n-1) f + n(d-1) f), \end{aligned} \tag{5.58}$$

which yields that

$$\Delta_{S^{d-1}} f = -n(n+d-2) f. \tag{5.59}$$

The kernel of the Laplacian $\Delta_{\mathbb{R}^d}$ on the subspace $P_n(d)$ is the eigenspace of the spherical Laplacian $\Delta_{S^{d-1}}$ with the eigenvalue $-n(n+d-2)$. In the following, we denote the subspace by $K_n(d)$. Since the dimension of $P_n(d)$ is $\binom{n+d-1}{n}$, the dimension of $K_n(d)$ is calculated as

$$\dim K_n(d) = \binom{n+d-1}{n} - \binom{n+d-3}{n-2}$$
$$= \frac{(2n+d-2)((n+d-3)\cdots(n+1))}{(d-2)!}. \tag{5.60}$$

Indeed, the spherical Laplacian $\Delta_{S^{d-1}}$ is closely related to the Lie algebra $\mathfrak{so}(d)$ as follows. Define the element $X_{k,l}$ of $\mathfrak{so}(d)$ as $|k\rangle\langle l| - |l\rangle\langle k|$. Since we have commutation relations

$$[X_{k,l}, X_{k',l'}] = X_{k,l'}\delta_{l,k'} - X_{l,l'}\delta_{k,k'} - X_{k,k'}\delta_{j,j'} + X_{l,k'}\delta_{k,l'} \tag{5.61}$$
$$[L_{k,l}, L_{k',l'}] = L_{k,l'}\delta_{l,k'} - L_{l,l'}\delta_{k,k'} - L_{k,k'}\delta_{l,l'} + L_{l,k'}\delta_{k,l'}, \tag{5.62}$$

$L_{k,l}$ can be regarded as the representation of $X_{k,l}$ on $L^2(S^{d-1})$. Since the killing form is given as $(X_{k,l}, X_{k,l})_{\mathfrak{so}(d)} = \operatorname{Tr} \mathrm{ad}(X_{k,l})\mathrm{ad}(X_{k',l'}) = -(d-1)\delta_{k,k'}\delta_{l,l'}$ for $k > l, k' > l'$, the Casimir operator C is given as $-\frac{1}{d-1}\Delta_{S^{d-1}}$.

Exercise 5.2 Show (5.40) and (5.41).

Exercise 5.3 Show (5.50)–(5.52), (5.55), and (5.56).

Exercise 5.4 Show (5.60).

Exercise 5.5 Show (5.62).

Exercise 5.6 Show that the function $(x_2 + i x_1)^n$ belongs to $K_n(d)$.

Exercise 5.7 Show that the function $\sum_{k=0}^{\lfloor \frac{n}{2} \rfloor}(-1)^k \binom{n}{2k} x_1^{n-2k} x_2^{2k}$ belongs to $K_n(d)$.

5.2.2 *Three-Dimensional Sphere*

Since we have discussed the case with the two-dimensional sphere, we discuss the detail with the three-dimensional sphere, i.e., $d = 4$. The case $d = 4$ has the symmetry by the 4-dimensional **special orthogonal group** $\mathrm{SO}(4, \mathbb{R})$. The Lie group $\mathrm{SO}(4, \mathbb{R})$ is isometric to $\mathrm{SU}(2) \times \mathrm{SU}(2)/\mathbb{Z}_2$. To see this relation, we express the real vector space $\mathbb{R}^4$ as the matrix space $\hat{R}^4 := \left\{ \begin{pmatrix} a+bi & ci+d \\ ci-d & a-bi \end{pmatrix} \right\}$, whose element is a constant times of an element of SU(2). For an element $(g_1, g_2) \in \mathrm{SU}(2) \times \mathrm{SU}(2)$, we give an element of $\mathrm{SO}(4, \mathbb{R})$ that maps $\begin{pmatrix} a+bi & ci+d \\ ci-d & a-bi \end{pmatrix} \mapsto g_1 \begin{pmatrix} a+bi & ci+d \\ ci-d & a-bi \end{pmatrix} g_2^{-1}$ in the sense of coordinate $(a, b, c, d) \in \mathbb{R}^4$. That is, this correspondence gives a homomorphism from $\mathrm{SU}(2) \times \mathrm{SU}(2)$ to $\mathrm{SO}(4, \mathbb{R})$. The kernel is $\{(I, I), (-I, -I)\}$. Hence, we find that $\mathrm{SO}(4, \mathbb{R})$ is isometric to $\mathrm{SU}(2) \times \mathrm{SU}(2)/\mathbb{Z}_2$, which implies that $\mathfrak{so}(4) \cong \mathfrak{su}(2) \oplus \mathfrak{su}(2)$.

The set of elements of $\hat{R}^4$ with $a^2+b^2+c^2+d^2=1$ is SU(2). Hence, we can regard $L^2(S^3)$ as $L^2(\mathrm{U}(2))$. Due to Fourier transform on the group SU(2) defined in Sect. 3.8, we find that

$$L^2(\mathrm{SU}(2)) \cong L^2(\hat{\mathrm{SU}(2)}) = \oplus_{\lambda\in\hat{\mathrm{SU}(2)}} \mathcal{U}_\lambda \otimes \mathcal{U}_\lambda. \tag{5.63}$$

Hence, the space $\mathcal{U}_\lambda \otimes \mathcal{U}_\lambda$ each $\lambda \in \hat{\mathrm{SU}(2)}$ is the irreducible component of $L^2(\mathrm{SU}(2))$ with respect to the representation of $\mathfrak{so}(4)$. The dimension of $\mathcal{U}_\lambda \otimes \mathcal{U}_\lambda$ is $(2\lambda+1)^2$. Due to (5.60), dim $K_n(4)=(n+1)^2$. Since the space $K_n(4)$ is closed with respect to the representation of $\mathfrak{so}(4)$, we have $\mathcal{U}_{\frac{n}{2}} \otimes \mathcal{U}_{\frac{n}{2}} = K_n(4)$.

Now, we consider the subgroup $\mathrm{SO}(3,\mathbb{R}) \subset \mathrm{SO}(4,\mathbb{R})$ in the sense of the subset $\{(g,g)|g\in \mathrm{SU}(2)\}$ of $\mathrm{SU}(2)\times\mathrm{SU}(2)$. This subset does not change a and $b^2+c^2+d^2$. That is, this action of $\mathrm{SO}(3,\mathbb{R})$ can be regarded as rotation for the three parameters b,c,d. In this subrepresentation of $\mathfrak{so}(3)=\mathfrak{su}(2)$, due to (4.22), the space $\mathcal{U}_{\frac{n}{2}} \otimes \mathcal{U}_{\frac{n}{2}}$ is irreducibly decomposed to

$$\mathcal{U}_{\frac{n}{2}} \otimes \mathcal{U}_{\frac{n}{2}} = \mathcal{U}_n \oplus \mathcal{U}_{n-1} \oplus \cdots \oplus \mathcal{U}_0. \tag{5.64}$$

Now, we consider an integral operator T on $L^2(S^3)$ as

$$T(f)(w) = \int_{S^3} \frac{f(v)}{|v-w|^2}\mu_3(dv), \tag{5.65}$$

where μ_3 is the invariant measure on S^3 whose full measure is $2\pi^2$, i.e., the measure of the volume of 3-dimensional unit sphere in $\mathbb{R}^4$. Since the operator T is invariant with respect to the action of $\mathrm{SO}(4,\mathbb{R})$, the operator T is a constant on the irreducible component $K_n(4)$. Hence, we have the following theorem.

Theorem 5.1 *The operator has the following spectral decomposition;*

$$T = \sum_{n=0}^{\infty} \frac{2\pi^2}{n+1} P_n, \tag{5.66}$$

where P_n is the projection to $K_n(4)$.

Proof Now, we choose a function $f(v) = (v_2 + iv_1)^n$ in $K_n(4)$. When $w = (1,0,0,0)$, we have $|v-w|^2 = (2\sin\frac{\theta_1}{2})^2 = 2(1-\cos\theta_1)$ as Fig. 5.2.

Hence, when $n \geq 1$,

$$\begin{aligned} T(f)(w) &= \int_0^{2\pi} d\theta_1 \int_0^{\pi} d\theta_2 \int_0^{\pi} d\theta_3 \frac{(\sin\theta_1\cos\theta_2 + i\cos\theta_1)^n}{2(1-\cos\theta_1)} \sin^2\theta_1 \sin\theta_2 \\ &\overset{(a)}{=} \frac{2\pi i^n}{n+1}\int_0^{2\pi} d\theta_1 \frac{\sin\theta_1}{2(1-\cos\theta_1)}\sin(n\theta_1) \overset{(b)}{=} \frac{2\pi^2 i^n}{n+1}, \end{aligned} \tag{5.67}$$

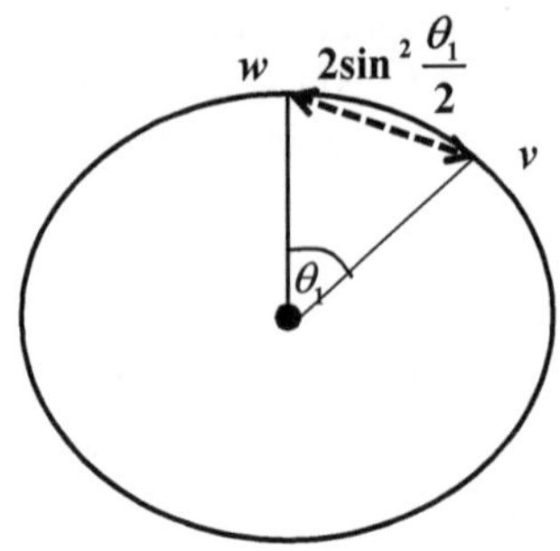

Fig. 5.2 Difference between v and w

where the steps (a) and (b) follow from Exercises 5.8 and 5.9. Since $f(w) = i^n$, the eigenvalue is $\frac{2\pi^2}{n+1}$. Also, the case when $n = 0$ can be shown as (Exercise 5.10)

$$T(f)(w) = \int_0^{2\pi} d\theta_1 \int_0^{\pi} d\theta_2 \int_0^{\pi} d\theta_3 \frac{\sin^2\theta_1 \sin\theta_2}{2(1-\cos\theta_1)} = 2\pi^2. \tag{5.68}$$

■

Exercise 5.8 Show the step (a) in (5.67).

Exercise 5.9 Show the equation $\int_0^{2\pi} \frac{\sin\theta \sin n\theta}{1-\cos\theta} d\theta = 2\pi$ for any integer $n \geq 1$ by following steps below.

a: Show that $\int_0^{2\pi} \sin\theta \sin n\theta d\theta = \int_0^{2\pi} \cos\theta \cos n\theta d\theta = 0$ for $n \geq 2$.
b: Show the equation $\int_0^{2\pi} \frac{\sin^2\theta}{1-\cos\theta} d\theta = 2\pi$.
c: Show the equation $\int_0^{2\pi} \frac{\sin^2\theta \sin 2\theta}{1-\cos\theta} d\theta = 2\pi$.
d: Show the equation $\int_0^{2\pi} \frac{\sin\theta \sin(n+1)\theta}{1-\cos\theta} d\theta = \int_0^{2\pi} \frac{\sin\theta \sin n\theta}{1-\cos\theta} d\theta$ for $n \geq 2$.

Exercise 5.10 Show (5.68).

5.3 Hydrogen-Type Hamiltonian and SO(4, $\mathbb{R}$)

5.3.1 *Puzzle for Degeneracy*

Next, we discuss the orbital Hamiltonian of an electron around hydrogen-type atom. In this case, the Hamiltonian H is given in (5.29) with the potential $V(r) = -\frac{Ze^2}{4\pi\epsilon_0 r}$, where ϵ_0 is the permittivity of the vacuum, Z is the atomic number (charge of the nucleus in unit e; number of protons in the nucleus), e is the elementary charge (charge of an electron), μ is the reduced mass of the system consisting of the electron and the nucleus. Then, it is known [77] that **Schrödinger equation** is given as[2]

[2]The Hamiltonian $\frac{1}{2\mu}\boldsymbol{P}^2 - \frac{Ze^2}{4\pi\epsilon_0 r}$ is defined as a self-adjoint operator on the second Sobolov space, i.e., the domain of $\boldsymbol{P}^2$.

Table 5.2 Summary of quantum numbers

Eigen equation	Range of quantum number	Name of quantum number
$H\ \psi_{n\ell m} = \frac{Z^2 E_h}{2n^2} \psi_{n\ell m}$	$n = 1, 2, \ldots, \infty$	Principal quantum number
$L^2\ \psi_{n\ell m} = \hbar^2 \ell(\ell+1)\ \psi_{n\ell m}$	$\ell = 0, 1, \ldots, n-1$	Azimuthal quantum number
$L_z\ \psi_{n\ell m} = \hbar m\ \psi_{n\ell m}$	$m = -\ell, 1-\ell, \ldots, \ell$	Magnetic quantum number

$$\left(\frac{1}{2\mu}\boldsymbol{P}^2 - \frac{Ze^2}{4\pi\epsilon_0 r}\right)\phi = E\phi. \tag{5.69}$$

In this case, the solution $\chi_{n,\ell}$ of the equation (5.37) is given as

$$\chi_{n,\ell}(r) = \left(\frac{2Z}{na_\mu}\right)^{3/2} \left[\frac{(n-\ell-1)!}{2n[(n+\ell)!]}\right]^{1/2} e^{-Zr/na_\mu} \left(\frac{2Zr}{na_\mu}\right)^{l} L_{n-\ell-1}^{2\ell+1}\left(\frac{2Zr}{na_\mu}\right). \tag{5.70}$$

with $a_\mu = \frac{4\pi\epsilon_0\hbar^2}{\mu e^2}$ and a new quantum number $n = 1, 2, \ldots, \infty$, which is called the **principal quantum number**. Here, the azimuthal quantum number ℓ takes values $n-1, n-2, \ldots, 0$. The eigenvalue is $-\frac{Z^2 E_h}{2n^2}$ with $E_h := \mu(\frac{e^2}{4\pi\epsilon_0\hbar})^2$. Hence, the total wave function on $L^2(\mathbb{R}_+ \times S^2)$ is given as

$$\tilde{\psi}_{n,\ell,m}(r, \theta_1, \theta_2) := \chi_{n,\ell}(r)\, Y_{\ell,m}(\theta_1, \theta_2). \tag{5.71}$$

That is, the wave function on $L^2(\mathbb{R}^3)$ is given as

$$\psi_{n,\ell,m}(x) := \frac{1}{\sqrt{4\pi} r}\chi_{n,\ell}(r)\, Y_{\ell,m}(\theta_1, \theta_2) \tag{5.72}$$

with (5.7). Further, the LHS of (5.37) also has continuous spectrum on $(0, \infty)$ [100, (10.6) p. 222]. These spectrums correspond to states that are not bounded by the atom. Such a solution is called a free electron. Therefore, the quantum numbers of states bounded by the atom are summarized as Tables 5.2 and 5.3. That is, the eigenvalue of the Hamiltonian depends only on the principal quantum number n, which decides the shell of orbit, which is illustrated in Fig. 5.3. So, this equation has a large degeneracy n^2 of the Hamiltonian. Since the above discussion explain the symmetry with respect to 2-dimensional sphere, it can explain the degeneracy $2\ell+1$ in each subshell, cannot explain the large degeneracy n^2 in each shell. In the next subsection, we solve this puzzle by using the larger symmetry based on SO(4, ℝ).

Table 5.3 Relation among principal, azimuthal, and magnetic quantum numbers

Shell n	Azimuthal quantum number	Subshell ℓ	Magnetic quantum numbers	Number orbitals of subshell	Number of orbitals of shell
K-shell ($n = 1$)	0	1s	0	1	1
L-shell ($n = 2$)	0	2s	0	1	4
	1	2p	$1, 0, -1$	3	
M-shell ($n = 3$)	0	3s	0	1	9
	1	3p	$1, 0, -1$	3	
	2	3d	$2, 1, 0, -1, -2$	5	
N-shell ($n = 3$)	0	4s	0	1	16
	1	4p	$1, 0, -1$	3	
	2	4d	$2, 1, 0, -1, -2$	5	
	3	4f	$3, 2, 1, 0, -1, -2, -3$	7	

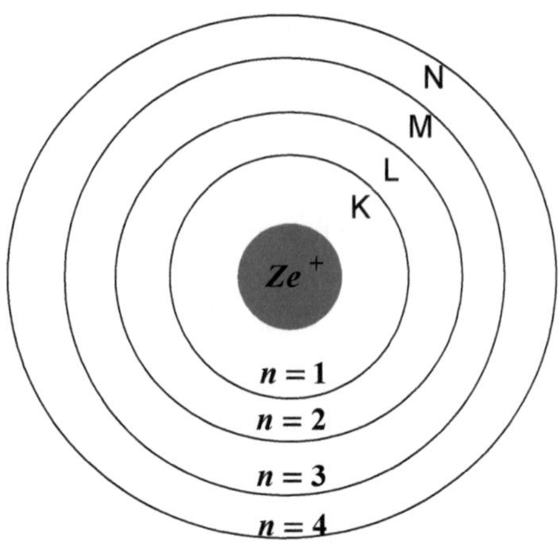

Fig. 5.3 Shells of hydrogen type atom

5.3.2 *Symmetry Based on* SO(4, $\mathbb{R}$)

When the eigenvalue E is negative, the eigenspace $\mathcal{L}_E$ associated with the eigenvalue E has symmetry based on the group SO(4, $\mathbb{R}$) behind [30]. When the eigenvalue E is positive, the eigenspace $\mathcal{L}_E$ associated with the eigenvalue E is irreducible representation (principal series) of the group SO(3, 1, $\mathbb{R}$) [83]. We discuss only the hidden symmetry of SO(4, $\mathbb{R}$) according to [13], [22, Chap. 9] because the symmetry by the group SO(3, 1) is too complicated. To see the symmetry, we apply Fourier transform $\tilde{\phi}(p) := \mathcal{F}_3[\phi](p)$ given in (3.68) to a function ϕ in $L^2(\mathbb{R}^2)$. Then, due to the discussion in the next subsection, the eigenequation (5.69) with eigenvalue E is converted to

$$\frac{\hbar^2 p^2}{2\mu}\tilde{\phi}(p) - \frac{Ze^2}{8\pi^3\epsilon_0}\int_{\mathbb{R}^3}\frac{\tilde{\phi}(p')}{|p-p'|^2}dp' = E\tilde{\phi}(p). \tag{5.73}$$

Using $\mu_h := \frac{\mu}{\hbar^2}$, we have

$$\frac{p^2}{2}\tilde{\phi}(p) - \frac{Z\sqrt{E_h\mu_h}}{2\pi^2}\int_{\mathbb{R}^3}\frac{\tilde{\phi}(p')}{|p-p'|^2}dp' = E\mu_h\tilde{\phi}(p). \tag{5.74}$$

Dependently of the eigenvalue E, we introduce the coordinate of S^3 based on the momentum coordinate $\mathbb{R}^3$ by the inverse of the **stereographic projection** from S^3 to $\mathbb{R}^3$ as

$$w(p)_j := \frac{2p_0 p_j}{p^2 + p_0^2}, \quad j = 1, 2, 3 \tag{5.75}$$

$$w(p)_4 := \frac{p^2 - p_0^2}{p^2 + p_0^2} \tag{5.76}$$

where $p^2 = \sum_{j=1}^3 p_j^2$ and p_0 is set to be $\sqrt{-2E\mu_h}$. That is, using the vector $n = (0, 0, 0, 1)$, $w(p)$ is given as

$$w(p) = \frac{p^2 - p_0^2}{p^2 + p_0^2}n + \frac{2p_0}{p^2 + p_0^2}p. \tag{5.77}$$

Then, $p_i(w)$ is given as

$$p(w)_i := \frac{p_0 w_i}{1 - w_4} \tag{5.78}$$

for $i = 1, 2, 3$. That is, by using the vector $n = (0, 0, 0, 1)^T$, the relation between w and p is illustrated in Fig. 5.4 and is given as

$$w = \frac{p^2 - p_0^2}{p^2 + p_0^2}n + \frac{2p_0}{p^2 + p_0^2}p. \tag{5.79}$$

Now, we have the following theorem.

Theorem 5.2 *When the points w and w' in S^3 correspond to the points p and p' in $\mathbb{R}^3$ in the above way, we have*

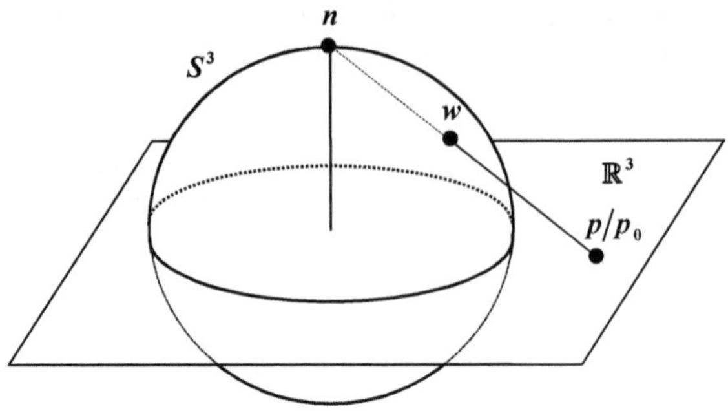

Fig. 5.4 Stereographic projection from S^3 to $\mathbb{R}^3$

$$|p-p'|^2 = \frac{(p^2+p_0^2)(p'^2+p_0^2)}{(2p_0)^2}|w-w'|^2. \tag{5.80}$$

Proof It is sufficient to show the special case $p_0 = 1$ because the special case with replacing p and p' by $\frac{p}{p_0}$ and $\frac{p'}{p_0}$ yields the general case. Our proof consists of five steps.

Step 1: We show that

$$|p-n| \cdot |w-n| = 2 \tag{5.81}$$

when p is outside the unit sphere S^3. We choose the point s as the south pole point as Fig. 5.5. Since the angles $\angle nws$ and $\angle nop$ are $\frac{\pi}{2}$, the triangle snw is homothetic to the triangle pno. So, we have

$$\frac{|w-n|}{|o-n|} = \frac{|s-n|}{|p-n|}. \tag{5.82}$$

Thus,

$$|w-n| \cdot |p-n| = |o-n| \cdot |s-n| = 2. \tag{5.83}$$

Step 2: We show that (5.81) when p is on the unit sphere S^3. Since $p = w = a$, we have (5.83).

Step 3: We show that (5.81) when p is inside of the unit sphere S^3. We can show (5.83) in the same way as Step 1 (See Fig. 5.6).

Step 4: We show that the triangle $\Delta nww'$ is similar to the triangle $\Delta np'p$. We can show (5.81) for p' and w'. Thus, $|p-n| \cdot |w-n| = |p'-n| \cdot |w'-n|$, i.e., $\frac{|p-n|}{|w'-n|} = \frac{|p'-n|}{|w-n|}$. Also, $\angle nww' = \angle nww'$ (common). Therefore, the triangle $\Delta nww'$ is similar to the triangle $\Delta np'p$.

Step 5: We show (5.80). The similarity relationship between the triangles $\Delta nww'$ and $\Delta np'p$ guarantees that

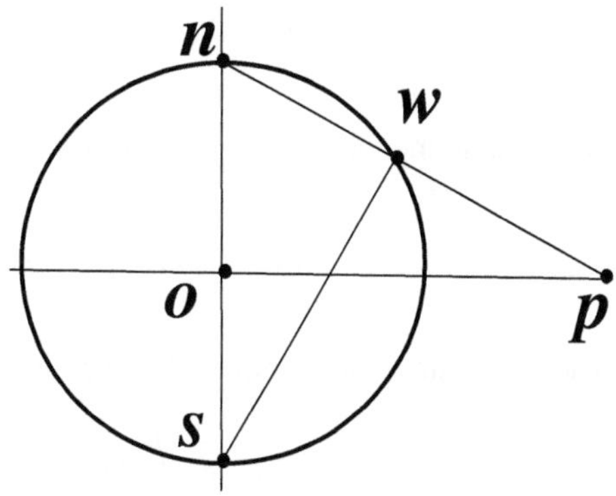

Fig. 5.5 Cross section with *onwp* when p is inside of the unit *sphere* S^3

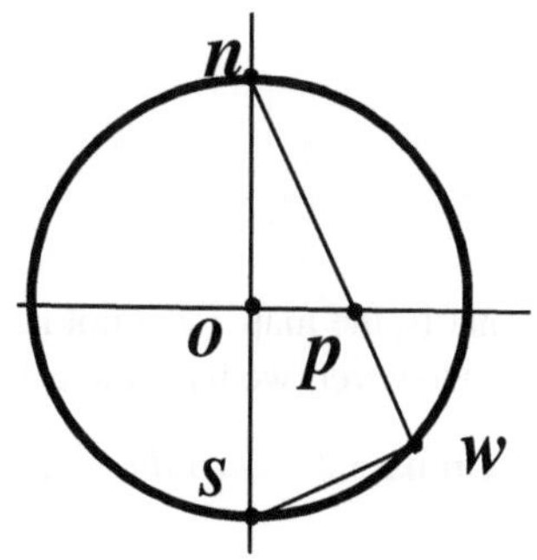

Fig. 5.6 Cross section with *onwp* when p is outside of the unit *sphere* S^3

$$|p - p'| = |w - w'|\frac{|p' - n|}{|w - n|} \overset{(a)}{=} |w - w'|\frac{|p' - n| \cdot |p - n|}{2}$$
$$\overset{(b)}{=} |w - w'|\frac{\sqrt{p^2 + 1} \cdot \sqrt{p'^2 + 1}}{2}, \tag{5.84}$$

where (a) follows from (5.81) and (b) follows from the relations $|p - n| = \sqrt{p^2 + 1}$ and $|p' - n| = \sqrt{p'^2 + 1}$. Taking the squares, we obtain (5.80). ∎

Now, we derive the description $\hat{\mu}_3(dp)$ of the volume integral $\mu_3(dw)$ in the 3-dimensional unit sphere S^3 in terms of the coordinate $p = (p_1, p_2, p_3)$, whose full measure is $2\pi^2$. The metric is given as

$$g(p)_{k,l} := \sum_{j=1}^{4} \frac{\partial w(p)_j}{\partial p_k} \frac{\partial w(p)_j}{\partial p_l}. \tag{5.85}$$

As shown later, we have the following lemma.

Lemma 5.1 *The determinant of* $g(p)_{k,l}$ *is* $(\frac{2p_0}{p^2+p_0^2})^6$.

Hence, the normalized invariant volume integral is

$$\hat{\mu}_3(dp) = \sqrt{\det g(p)} d^3 p = \left(\frac{2p_0}{p^2 + p_0^2} \right)^3 d^3 p. \tag{5.86}$$

Now, we define the map F_E from $L^2(\mathbb{R}^3)$ to $L^2(S^3)$ as

$$F_E(f)(w) := \left(\frac{p(w)^2 + p_0^2}{2p_0} \right)^2 f\Big(p(w) \Big). \tag{5.87}$$

Then, the square of the norm of $F_E(f)$ is given as

$$\begin{aligned} \| F_E(f)(w) \|_{S^3} &:= \int_{\mathbb{R}^3} |F_E(f)(w(p))|^2 \hat{\mu}_3(dp) \\ &= \int_{\mathbb{R}^3} \left(\frac{p^2 + p_0^2}{2p_0} \right)^4 |f(p)|^2 \left(\frac{2p_0}{p^2 + p_0^2} \right)^3 d^3 p \\ &= \int_{\mathbb{R}^3} \frac{p^2 + p_0^2}{2p_0} |f(p)|^2 d^3 p. \end{aligned} \tag{5.88}$$

That is, the map F_E is not necessarily a unitary map, in general.

However, we have the following lemma.

Lemma 5.2 *When the map F_E is restricted to the subspace $\mathcal{L}_E$, the map F_E is an isometry.*

Proof We define the operator $A := \frac{1}{2\hbar} \sum_{k=1}^3 (\hat{\mathsf{P}}_k \mathsf{Q}_k + \mathsf{Q}_k \hat{\mathsf{P}}_k)$. Then, using the commutation relation (5.2), we have the commutation relation

$$i[H, A] = \frac{1}{2\mu} \boldsymbol{P}^2 + H. \tag{5.89}$$

For any normalized wave function $f \in \mathcal{L}_E$, we have $\langle f, HAf \rangle = E \langle f, Af \rangle = \langle f, AHf \rangle$, which implies that

$$\begin{aligned} 0 = i \langle f, [H, A] f \rangle &= \Big\langle f, (\frac{1}{2\mu} \sum_{j=1}^3 \hat{\mathsf{P}}_j^2 + H) f \Big\rangle \\ = \Big\langle f, (\frac{1}{2\mu} \sum_{j=1}^3 \hat{\mathsf{P}}_j^2) f \Big\rangle + E &= \int_{\mathbb{R}^3} \frac{\hbar^2 p^2}{2\mu} |\tilde{f}(p)|^2 d^3 p + E. \end{aligned}$$

Thus,

$$\int_{\mathbb{R}^3} p^2|\tilde{f}(p)|^2 d^3p = -2E\mu_h = p_0^2. \tag{5.90}$$

Combining (5.88) and (5.90), we have

$$\|F_E(f)(w)\|_{S^3} = \frac{p_0^2 + p_0^2}{2p_0} = 1, \tag{5.91}$$

which implies that the map F_E preserves the norm for any element in $\mathcal{L}_E$. ∎

Lemma 5.3 *Any element Φ of $F_E(\mathcal{L}_E)$ satisfies*

$$\int_{S^3} \frac{\Phi(w')}{|w - w'|^2} \Phi \mu_3(dw') = \frac{\pi p_0}{Z\sqrt{E_h \mu_h}} \Phi(w). \tag{5.92}$$

Proof The equation (5.74) is written as

$$\left(\frac{p^2}{2} - E\mu_h\right) F_E^{-1}(\Phi)(p) = \frac{Z\sqrt{E_h\mu_h}}{2\pi^2} \int_{\mathbb{R}^3} \frac{F_E^{-1}(\Phi)(p')}{|p - p'|^2} dp'. \tag{5.93}$$

The LHS of (5.93) is calculated to

$$\left(\frac{p^2}{2} - E\mu_h\right) F_E^{-1}(\Phi)(p) = \frac{p^2 + p_0^2}{2} F_E^{-1}(\Phi)(p)$$
$$= \frac{p^2 + p_0^2}{2} \left(\frac{2p_0}{p^2 + p_0^2}\right)^2 \Phi(w(p)) = \frac{2p_0^2}{p^2 + p_0^2} \Phi(w(p)). \tag{5.94}$$

Let w and w' be $w(p)$ and $w(p')$. Then, Theorem 5.2 guarantees that

$$|p - p'|^2 = \frac{(p^2 + p_0^2)(p'^2 + p_0^2)}{(2p_0)^2} |w - w'|^2. \tag{5.95}$$

The LHS of (5.93) is calculated to

$$\frac{Z\sqrt{E_h\mu_h}}{2\pi^2} \int_{\mathbb{R}^3} \frac{F_E^{-1}(\Phi)(p')}{|p - p'|^2} d^3p'$$
$$= \frac{Z\sqrt{E_h\mu_h}}{2\pi^2} \int_{\mathbb{R}^3} \frac{\left(\frac{2p_0}{p(w')^2 + p_0^2}\right)^2 (\Phi)(w')}{\frac{(p^2+p_0^2)(p(w')^2+p_0^2)}{(2p_0)^2} |w(p) - w'|^2} \left(\frac{p(w')^2 + p_0^2}{2p_0}\right)^3 \mu(dw')$$
$$= \frac{2p_0\sqrt{E_h\mu_h}}{2\pi^2(p^2 + p_0^2)} \int_{\mathbb{R}^3} \frac{(\Phi)(w')}{|w(p) - w'|^2} \mu_3(dw'). \tag{5.96}$$

Combining (5.93), (5.94), and (5.96), we have

$$\frac{2p_0^2}{p^2+p_0^2}\Phi(w) = \frac{2p_0\sqrt{E_h\mu_h}}{2\pi^2(p^2+p_0^2)}\int_{\mathbb{R}^3}\frac{(\Phi)(w')}{|w-w'|^2}\mu_3(dw'), \tag{5.97}$$

which implies the desired equation. ■

Now, we recall Theorem 5.1. From Lemma 5.3, we see that the eigenvalue $\frac{2\pi^2 p_0}{Z\sqrt{E_h\mu_h}}$ is restricted to $\frac{2\pi^2}{n}$ with a positive integer n. That is, p_0 is $\frac{2\pi^2 Z\sqrt{E_h\mu_h}}{2\pi^2 n} = \frac{Z\sqrt{E_h\mu_h}}{n}$. So, the energy E is $-\frac{p_0^2}{2\mu_h} = -\frac{Z^2E_h}{2n^2}$.

In summary, we obtain the following theorem.

Theorem 5.3 *The negative eigenvalue E is restricted to $-\frac{Z^2E_h}{2n^2}$ with a positive integer n. Then, the map $F_E|_{\mathcal{L}_E}$ is a unitary map from $\mathcal{L}_E$ to $\mathcal{K}_{n-1}(4)$.*

Now, we consider the representation of $\mathrm{SO}(3,\mathbb{R})$ on $L^2(\mathbb{R}^3)$. When we consider the $L^2(\mathbb{R}^3)$ with the momentum coordinate, the irreducible decomposition of $\mathrm{SO}(3,\mathbb{R})$ is the same as that with the position coordinate. When $E = -\frac{Z^2E_h}{2n^2}$, due to (5.64), the eigenspace $F_E|_{\mathcal{L}_E}(\mathcal{L}_E)$ is irreducibly decomposed to

$$\mathcal{U}_{n-1}\oplus\mathcal{U}_{n-2}\oplus\cdots\oplus\mathcal{U}_0, \tag{5.98}$$

which clarifies the degeneracy of the Hamiltonian for the azimuthal quantum number ℓ.

Proof of Lemma 5.1 As shown later, the diagonal element $g(p)_{k,k}$ is $(\frac{2p_0}{p^2+p_0^2})^2$ and the off diagonal element $g(p)_{k,l}$ is 0 with $k\neq l$. Hence, the determinant of $g(p)_{k,l}$ is $(\frac{2p_0}{p^2+p_0^2})^6$.

To obtain the diagonal element $g(p)_{k,k}$, we calculate $\frac{\partial w_j}{\partial p_k}$ as

$$\frac{\partial w_j}{\partial p_k} = \begin{cases} \frac{-4p_0p_kp_j}{(p^2+p_0^2)^2} & \text{when } k\neq j,\ j=1,2,3 \\ \frac{-4p_0(p^2-2p_j^2+p_0^2)}{(p^2+p_0^2)^2} & \text{when } k=j,\ j=1,2,3 \\ \frac{4p_0^2p_k}{(p^2+p_0^2)^2} & \text{when } j=4. \end{cases} \tag{5.99}$$

So, we have

$$g(p)_{k,k} = \sum_{j=1}^4\left(\frac{\partial w_j}{\partial p_k}\right)^2 = \left(\frac{2p_0}{p^2+p_0^2}\right)^2. \tag{5.100}$$

Similarly, we have

$$g(p)_{k,l} = \sum_{j=1}^4\frac{\partial w_j}{\partial p_k}\frac{\partial w_j}{\partial p_l} = 0. \tag{5.101}$$

■

Exercise 5.11 Show the commutation relation (5.89).

Exercise 5.12 Show (5.100) and (5.101).

5.3.3 *Fourier Transform of Coulomb Potential**

This subsection discusses Fourier transform of Coulomb potential. That is, we show the following lemma.

Lemma 5.4 *Any element* $\phi \in D((\sum_{j=1}^3 \mathsf{Q}_j^2)^{-\frac{1}{2}})$ *satisfies*

$$\mathcal{F}_3\left[\left(\sum_{j=1}^{3} \mathsf{Q}_j^2\right)^{-\frac{1}{2}} \phi\right](\lambda) = \frac{1}{2\pi^2} \int_{\mathbb{R}^3} \frac{\mathcal{F}_3[\phi](\lambda')}{|\lambda - \lambda'|^2} d\lambda'. \tag{5.102}$$

Proof For this proof, we recall that the Coulomb potential is defined as the closed extension of the operator defined in $\mathscr{S}(\mathbb{R}^3)$. in Sect. 1.6. Hence, it is sufficient to show (5.102) when ϕ belongs to $\mathscr{S}(\mathbb{R}^3)$. Now, we define the convolution $\phi * \phi'$ for two functions ϕ and ϕ' as

$$\phi * \phi'(x) := \frac{1}{(2\pi)^{\frac{3}{2}}} \int_{\mathbb{R}^3} \phi(x')\phi'(\lambda - x')dx'. \tag{5.103}$$

We define two functions $\phi_3(x) := \frac{1}{|x|}$ and $\phi_4(x) := \frac{\sqrt{2}}{\sqrt{\pi}|x|^2}$. Indeed, these functions ϕ_3 and ϕ_4 are not square integrable. However, when a function ϕ_5 is a rapidly decreasing function, we can define $\langle \phi_5 | \phi_3 \rangle$ and $\langle \phi_5 | \phi_4 \rangle$ as

$$\langle \phi_5 | \phi_j \rangle := \int_{\mathbb{R}^3} \phi_5(x)^* \phi_j(x) d^3x \tag{5.104}$$

for $j = 3, 4$. Since $\mathscr{S}(\mathbb{R}^3)$ is a dense subset, it is enough to show that

$$\langle \phi_1 | \mathcal{F}_3(\phi_3\phi_2) \rangle = \langle \phi_1 | \phi_4 * \mathcal{F}_3(\phi_2) \rangle \tag{5.105}$$

for any $\phi_1 \in \mathscr{S}(\mathbb{R}^3)$.

Thus, we have

$$\begin{aligned}
&\langle \phi_1 | \mathcal{F}_3(\phi_3\phi_2) \rangle = \langle \mathcal{F}_3^{-1}(\phi_1) | \phi_3\phi_2 \rangle = \langle \mathcal{F}_3^{-1}(\phi_1)\phi_2^* | \phi_3 \rangle \\
=&\langle \mathcal{F}_3^{-1}\mathcal{F}_3(\mathcal{F}_3^{-1}(\phi_1)\phi_2^*) | \phi_3 \rangle \stackrel{(a)}{=} \langle \mathcal{F}_3^{-1}(\phi_1 * \mathcal{F}_3(\phi_2^*)) | \phi_3 \rangle \\
\stackrel{(b)}{=}&\langle \mathcal{F}_3^{-1}(\phi_1 * \mathcal{F}_3^{-1}(\phi_2)^*) | \phi_3 \rangle,
\end{aligned} \tag{5.106}$$

and

$$\langle \phi_1 | \phi_4 * \mathcal{F}_3(\phi_2) \rangle \overset{(c)}{=} \langle \phi_1 * \mathcal{F}_3^{-1}(\phi_2)^* | \phi_4 \rangle. \tag{5.107}$$

where (a), (b), and (c) follow from Exercises 3.22, 3.12, and 5.13, respectively.

Since the set $\mathscr{S}(\mathbb{R}^3)$ is closed for Fourier transform, product and convolution (see Exercises 1.11, 1.12, 3.14, and 3.21), it is enough to show that

$$\langle \mathcal{F}_3^{-1}(\phi_5) | \phi_3 \rangle = \langle \phi_5 | \phi_4 \rangle \tag{5.108}$$

for any $\phi_5 \in \mathscr{S}(\mathbb{R}^3)$. Now, we choose the function $\phi_\delta(x) := e^{-\delta|x|}$. For any $\phi_6 \in \mathscr{S}(\mathbb{R}^3)$, we have $\langle \phi_6 - \phi_\delta \phi_6 | \phi_3 \rangle \to 0$ as $\delta \to 0$. Since

$$\lim_{\delta \to 0} \mathcal{F}_3(\phi_\delta \phi_3) = \phi_4, \tag{5.109}$$

(see Exercise 5.14), we have

$$\begin{aligned} \langle \mathcal{F}_3^{-1}(\phi_5) | \phi_3 \rangle &= \lim_{\delta \to 0} \langle \mathcal{F}_3^{-1}(\phi_5) \phi_\delta | \phi_3 \rangle = \lim_{\delta \to 0} \langle \mathcal{F}_3^{-1}(\phi_5) | \phi_\delta \phi_3 \rangle \\ &= \lim_{\delta \to 0} \langle \phi_5 | \mathcal{F}_3(\phi_\delta \phi_3) \rangle = \langle \phi_5 | \lim_{\delta \to 0} \mathcal{F}_3(\phi_\delta \phi_3) \rangle = \langle \phi_5 | \phi_4 \rangle. \end{aligned} \tag{5.110}$$

The proof is completed. ■

Exercise 5.13 Show (5.107).

Exercise 5.14 Show (5.109) by following steps below.

a: Choose the spherical coordinate $x = r \sin\theta \cos\phi$, $y = r \sin\theta \sin\phi$, and $z = r \cos\theta$. Show that

$$\int_{\mathbb{R}^3} e^{-\delta|x|} \frac{e^{-i(xp_1 + yp_2 + zp_3)}}{\sqrt{x^2 + y^2 + z^2}} dxdydz = e^{-\delta r} \frac{4\pi}{|p|} \int_0^\infty \sin(|p|r) dr. \tag{5.111}$$

b: Show that

$$\lim_{\delta \to 0} \frac{4\pi}{|p|} \int_0^\infty e^{-\delta r} \sin(|p|r) dr = \frac{4\pi}{|p|^2}. \tag{5.112}$$

c: Show (5.109).

5.4 Spin-Orbit Interaction

In the previous section, we have discussed the freedom of the orbit of an electron. In addition to this freedom, an electron has another internal freedom, **spin**, which is represented by a two-dimensional space $\mathbb{C}^2$ and is spanned by $|\uparrow\rangle$ and $|\downarrow\rangle$ (Fig. 5.7).

In the spin system $\mathbb{C}^2$, we consider three Hermitian matrices

$$S_1 := \frac{\hbar}{2}\begin{pmatrix} 0 & 1 \\ 1 & 0 \end{pmatrix}, \quad S_2 := \frac{\hbar}{2}\begin{pmatrix} 0 & -i \\ i & 0 \end{pmatrix}, \quad S_3 := \frac{\hbar}{2}\begin{pmatrix} 1 & 0 \\ 0 & -1 \end{pmatrix}. \tag{5.113}$$

Then, the total system of an electron is given as $L^2(\mathbb{R}^3)\otimes\mathbb{C}^2 = L^2(\mathbb{R}_+)\otimes L^2(S^2)\otimes\mathbb{C}^2$. When an electron has a homogeneous interaction between the orbital angular momentum and the spin, we discuss the following type of Hamiltonian

$$H := \frac{1}{2\mu}\boldsymbol{P}^2 + V_1(r) + V_2(r)\boldsymbol{L}\cdot\boldsymbol{S}, \tag{5.114}$$

where the additional term $V_2(r)\boldsymbol{L}\cdot\boldsymbol{S}$ expresses **spin-orbit interaction**. Many kinds of Hamiltonian can be written as special cases of the above Hamiltonian, e.g., Larmor interaction ($V_2(r) = \frac{2\mu_B}{\hbar m_e^2 c^2 r}\frac{dV_1(r)}{dr}$) and Thomas interaction [101]. Since matrices $\{-\frac{i}{\hbar}S_k\}_{k=1}^3$ satisfy the commutation relation (4.4), they give a skew-adjoint representation of $\mathfrak{su}(2)$. When we define the **total angular momentum** $J_k := S_k + L_k$, the matrices $\{-\frac{i}{\hbar}J_k\}_{j=1}^3$ give a skew-adjoint representation of $\mathfrak{su}(2)$ on the composite system. Since

$$\boldsymbol{J}^2 = \sum_{k=1}^{3} J_k^2 = \boldsymbol{L}^2 + 2\boldsymbol{L}\cdot\boldsymbol{S} + \boldsymbol{S}^2 = \boldsymbol{L}^2 + 2\boldsymbol{L}\cdot\boldsymbol{S} + \frac{3}{2}, \tag{5.115}$$

we have

$$\boldsymbol{L}\cdot\boldsymbol{S} = \frac{1}{2}(\boldsymbol{J}^2 - \boldsymbol{L}^2) - \frac{3}{4}. \tag{5.116}$$

By using (5.33), the Hamiltonian is simplified to

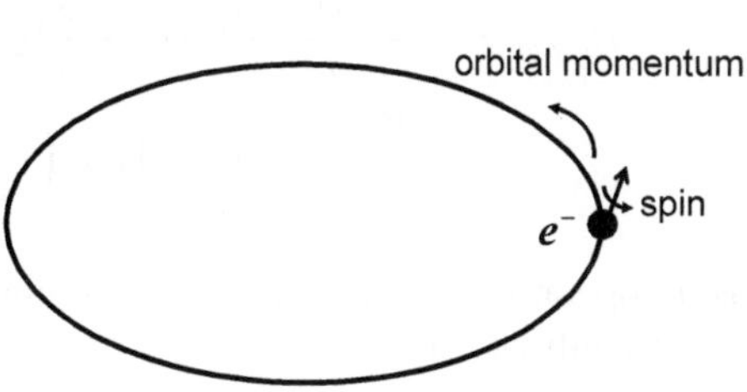

Fig. 5.7 Spin and orbital angular momentum

$$
\begin{aligned}
H = & -\frac{\hbar^2}{2\mu}\Big(\frac{\partial^2}{\partial r^2}+\frac{2}{r}\frac{\partial}{\partial r}\Big)+V_1(r)-\frac{3}{4}V_2(r) \\
& +\frac{V_2(r)}{2}\boldsymbol{J}^2+\Big(\frac{1}{2\mu r^2}-\frac{V_2(r)}{2}\Big)\boldsymbol{L}^2. \qquad (5.117)
\end{aligned}
$$

as an operator on $L^2(\mathbb{R}^3)\otimes\mathbb{C}^2$. However, we need to treat the Hamiltonian on the space $L^2(\mathbb{R}_+)\otimes L^2(S^2)\otimes\mathbb{C}^2$. Then, we have

$$
H=-\frac{\hbar^2}{2\mu}\frac{\partial^2}{\partial r^2}+V_1(r)-\frac{3}{4}V_2(r)+\frac{V_2(r)}{2}\boldsymbol{J}^2+\Big(\frac{1}{2\mu r^2}-\frac{V_2(r)}{2}\Big)\boldsymbol{L}^2. \qquad (5.118)
$$

as an operator on $L^2(\mathbb{R}_+)\otimes L^2(S^2)\otimes\mathbb{C}^2$.

The operators H, $\boldsymbol{J}^2$, $\boldsymbol{L}^2$, and J_3 are commutative with each other and $\boldsymbol{J}^2$, $\boldsymbol{L}^2$, and J_3 are operators on $L^2(S^2)\otimes\mathbb{C}^2$. Based on the discussion in Sect. 4.2.4, in particular, Example 4.1, we focus on the integer ℓ identifying the irreducible representation for the orbital angular momentum $\boldsymbol{L}$ and the integer j identifying the irreducible representation for the total angular momentum $\boldsymbol{J}$. When the irreducible representation for the orbital angular momentum is identified with ℓ, the vector $|j;m\rangle$ by $|\ell;j;m\rangle$. That is, the basis $\{|\ell;j;m\rangle\}_{\ell=1,2,\ldots,j=\ell\pm\frac{1}{2},m=-j,\ldots,j}$ forms a CONS of the space $L^2(S^2)\otimes\mathbb{C}^2$. In fact, $|\ell;\ell\pm\frac{1}{2};m\rangle$ is given as a linear combination of $|\ell,\frac{1}{2};m+\frac{1}{2},-\frac{1}{2}\rangle$ and $|\ell,\frac{1}{2};m-\frac{1}{2},\frac{1}{2}\rangle$, whose coefficients (Clebsch-Gordan coefficients) are given in Example 4.1. Then, we simultaneously diagonalize $\boldsymbol{J}^2$, $\boldsymbol{L}^2$, and J_3 as

$$
\boldsymbol{J}^2=\sum_{\ell,j,m} j(j+1)|\ell;j;m\rangle\langle\ell;j;m| \qquad (5.119)
$$

$$
\boldsymbol{L}^2=\sum_{\ell,j,m} \ell(\ell+1)|\ell;j;m\rangle\langle\ell;j;m| \qquad (5.120)
$$

$$
J_3=\sum_{\ell,j,m} m|\ell;j;m\rangle\langle\ell;j;m|, \qquad (5.121)
$$

which gives the simultaneous measurement of $\boldsymbol{J}^2$, $\boldsymbol{L}^2$, and J_3.

Now, we apply the discussion in Sect. 1.4.2. From (5.118), we have

$$
\begin{aligned}
H_{\ell,j,m} = & -\frac{\hbar^2}{2\mu}\frac{\partial^2}{\partial r^2}+V_1(r)-\frac{3}{4}V_2(r) \\
& +\frac{V_2(r)}{2}\hbar^2 j(j+1)+\Big(\frac{1}{2\mu r^2}-\frac{V_2(r)}{2}\Big)\hbar^2\ell(\ell+1). \qquad (5.122)
\end{aligned}
$$

as an operator on $L^2(\mathbb{R}_+)$. Hence, we obtain the following Schrödinger equation dependently of ℓ and j;

$$H_{\ell,j,m}\chi_{E,\ell,j} = E\chi_{E,\ell,j}. \tag{5.123}$$

That is, when the function $\chi_{E,\ell,j}$ in $L^2(\mathbb{R}_+)$ is a normalized solution of the above equation, the wave function $\chi_{E,\ell,j}|\ell; j; m\rangle$ on $L^2(\mathbb{R}_+) \otimes L^2(S^2) \otimes \mathbb{C}^2$ is an eigenvector of the Hamiltonian H given in (5.114) associated to the eigenvalue E.

5.5 Quark Model

5.5.1 Quark and Anti-quark

In particle physics, the atom is not the most fundamental particle. An atom is composed of protons and neutrons and electrons. While the electrons are elementary particles, the protons and neutrons are composed of more fundamental constituents called **quarks**. These quarks constitute the sub-structure of a proton or a neutron. More generally, composite particles that are made up from these quarks are called "hadrons". Typical hadrons are made up of two or three quarks. Hadrons with two quarks held together are called mesons, whereas those with three quarks are termed baryons. Hence, we introduce quarks as substructures of a proton and a neutron. Firstly, we introduce **up quark** u and **down quark** d, and explain that a proton is composed of two up quarks and a down quark, and a neutron is composed of an up quark and two down quarks. In this explanation, an up quark u has an electric charge of $+\frac{2}{3}e$ and a down quark d has an electric charge of $-\frac{2}{3}e$ so that a proton has an electric charge of e and a neutron has an electric charge of 0 as Fig. 5.8. Besides the proton and the neutron, other exotic hadrons exists. To explain the strangeness of these particles, a **strange quark** is introduced. A strange quark s has an electric charge of $+\frac{2}{3}e$. To discuss the strangeness, we introduce a new quantum number strangeness S so that only a strange quark has strangeness -1. For simplicity, we introduce the quantum number, electric charge Q, whose e times is the electric charge. Since we have three quarks, up u, down d, and strange s, the quantum system of one quark is $\mathbb{C}^3$ and has the fundamental representation of SU(3). The space is spanned by $|u\rangle$, $|d\rangle$, $|s\rangle$, and the Dynkin index is $[1, 0]$ so that the system is $\mathcal{U}_{[1,0]}(3)$. This freedom is called **flavor**. To clarify the group action SU(3) for flavor, we denote the group by $\mathrm{SU}_\mathrm{f}(3)$. The representation space $\mathcal{U}_{[n_1,n_2]}(3)$ is written as $\mathcal{U}_{[n_1,n_2]}(3)_\mathrm{f}$.

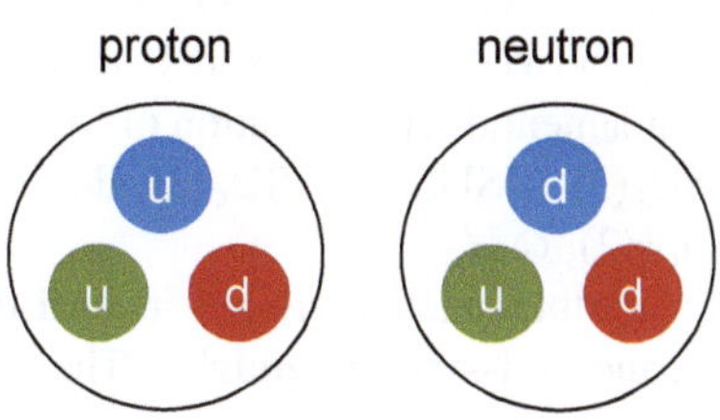

Fig. 5.8 Proton and neutron

Now, the quantum numbers, **electric charge** Q and **strangeness** S are mathematically defined as the following Hermitian matrices;

$$Q := \frac{2}{3}|u\rangle\langle u| - \frac{1}{3}|d\rangle\langle d| + \frac{1}{3}|s\rangle\langle s| \quad (5.124)$$

$$S := -|s\rangle\langle s|. \quad (5.125)$$

We also employ quantum numbers, **isospin** I_3 and **hypercharge** Y, which are defined as

$$I_3 := \frac{1}{2}|u\rangle\langle u| - \frac{1}{2}|d\rangle\langle d| \quad (5.126)$$

$$Y := \frac{1}{3}|u\rangle\langle u| - \frac{1}{3}|d\rangle\langle d| + \frac{2}{3}|s\rangle\langle s|. \quad (5.127)$$

So, $Q = I_3 + \frac{1}{2}Y$. The skew-adjoint representation $\mathfrak{f}_{[1,0]}$of $\mathfrak{su}(3)$ is given as

$$\mathfrak{f}_{[1,0]}(F^x_{j,l}) := \frac{i}{2}(|q_l\rangle\langle q_j| + |q_j\rangle\langle q_l|), \quad \mathfrak{f}_{[1,0]}(F^y_{j,l}) := \frac{1}{2}(|q_j\rangle\langle q_l| - |q_l\rangle\langle q_j|),$$
$$\mathfrak{f}_{[1,0]}(F^z_{1,2}) := -i(|q_1\rangle\langle q_1| - |q_2\rangle\langle q_2|), \quad \mathfrak{f}_{[1,0]}(F^z_{2,3}) := -i(|q_2\rangle\langle q_2| - |q_3\rangle\langle q_3|)$$

for $1 \le j < l \le 3$, where $q_1 = u, q_2 = d$, and $q_3 = s$. In particle physics, to discuss the Lie algebra $\mathfrak{so}(3)$ of diagonal matrices on $\mathbb{C}^3$, we usually use

$$F_3 := \frac{1}{2}|1\rangle\langle 1| - \frac{1}{2}|2\rangle\langle 2|, \quad F_8 := \frac{1}{2\sqrt{3}}|1\rangle\langle 1| + \frac{1}{2\sqrt{3}}|2\rangle\langle 2| - \frac{1}{\sqrt{3}}|2\rangle\langle 2|$$

instead of $F^z_{1,2}$ and $F^z_{2,3}$. So, we have

$$\mathfrak{f}_{[1,0]}(F_3) = I_3, \quad \mathfrak{f}_{[1,0]}(F_8) = \frac{\sqrt{3}}{2}Y, \quad (5.128)$$

Also, these quarks have an internal freedom of spin $\frac{1}{2}$, i.e., the internal freedom for rotation whose quantum system is $\mathbb{C}^2$ with the fundamental representation of SU(2), whose Dynkin index is [1] so that the system is $\mathcal{U}_{[1]}(2)$. Due to relativistic quantum mechanics, when a particle has spin $\frac{1}{2}$, it is needed to be a fermion. So, a quark is a **fermion**. To express this freedom, we add subscript $_{\mathrm{s}}$. Further, quark has another internal freedom for **color**, whose system is $\mathbb{C}^3$ and is spanned by **blue** $|b\rangle$, **green** $|g\rangle$, and **red** $|r\rangle$. To express this freedom, we add subscript $_{\mathrm{c}}$. This space is the fundamental representation of $\mathrm{SU_c}(3)$. Hence, under the representation of the group $\mathrm{SU_f}(3) \times \mathrm{SU_s}(2) \times \mathrm{SU_c}(3)$, the total one-quark system is written as $\mathcal{U}_{[1,0]}(3)_{\mathrm{f}} \otimes \mathcal{U}_{[1]}(2)_{\mathrm{s}} \otimes \mathcal{U}_{[1,0]}(3)_{\mathrm{c}}$.

Further, each quark has its **anti-quark**, i.e., **anti-up quark** $\bar{u}$, **anti-down quark** $\bar{d}$, and **anti-strange quark** $\bar{s}$. The anti-quark has the electric charge and strangeness, i.e., Q and S that are opposite to those of the original quark, and has a spin $\frac{1}{2}$

as well as the original quark. Since their charge and strangeness are opposite, the representation of $SU_f(3)$ in this system is the complex conjugate representation of the the fundamental representation. That is, the Dynkin index is [0, 1] so that the system is $\mathcal{U}_{[0,1]}(3)_f$. The skew-adjoint representation $\mathfrak{f}_{[0,1]}(3)_f$ of $\mathfrak{su}(3)$ is given as

$$\mathfrak{f}_{[0,1]}(F^x_{j,l}) := -\frac{i}{2}(|\bar{q}_l\rangle\langle\bar{q}_j| + |\bar{q}_j\rangle\langle\bar{q}_l|), \quad \mathfrak{f}_{[0,1]}(F^y_{j,l}) := -\frac{1}{2}(|\bar{q}_j\rangle\langle\bar{q}_l| - |\bar{q}_l\rangle\langle\bar{q}_j|),$$
$$\mathfrak{f}_{[0,1]}(F^z_{1,2}) := i(|\bar{q}_1\rangle\langle\bar{q}_1| - |\bar{q}_2\rangle\langle\bar{q}_2|), \quad \mathfrak{f}_{[0,1]}(F^z_{2,3}) := i(|\bar{q}_2\rangle\langle\bar{q}_2| - |\bar{q}_3\rangle\langle\bar{q}_3|)$$

for $1 \le j < l \le 3$, where $\bar{q}_1 = \bar{u}, \bar{q}_2 = \bar{d}$, and $\bar{q}_3 = \bar{s}$. The quantum numbers I_3 and Y are characterized as

$$\mathfrak{f}_{[0,1]}(F_3) = I_3, \quad \mathfrak{f}_{[0,1]}(F_8) = \frac{\sqrt{3}}{2}Y. \tag{5.129}$$

Since the complex conjugate representation of the fundamental representation of $SU_s(2)$ is the fundamental representation of $SU_s(2)$, the anti-quark has the fundamental representation of $SU_s(2)$ so that the system is $\mathcal{U}_{[1]}(2)_s$ because the complex conjugate of $\mathcal{U}_{[1]}(2)_s$ is also $\mathcal{U}_{[1]}(2)_s$. Further, anti-quark has another internal freedom for color, whose system is $\mathbb{C}^3$ and is spanned by **anti-blue** $|\bar{b}\rangle$, **anti-green** $|\bar{g}\rangle$, and **anti-red** $|\bar{r}\rangle$. This space is the complex conjugate of the fundamental representation space $\mathcal{U}_{[0,1]}(3)_c$ with respect to $SU(3)_c$. Hence, under the representation of the group $SU_f(3) \times SU_s(2)$, the total one-anti-quark system is written as $\mathcal{U}_{[0,1]}(3)_f \otimes \mathcal{U}_{[1]}(2)_s \otimes \mathcal{U}_{[0,1]}(3)_c$.

5.5.2 Meson

Hadrons are classified to two types, **mesons** and **baryons**. A meson is composed of a quark and an anti-quark. For example, the meson π^- is composed of d and $\bar{u}$ as Fig. 5.9. First, we omit the freedom of color for simplicity. Then, since a quark is different from an anti-quark, under the representation of the group $SU_f(3) \times SU_s(2)$, the one-meson system is given as $(\mathcal{U}_{[1,0]}(3)_f \otimes \mathcal{U}_{[1]}(2)_s) \otimes (\mathcal{U}_{[0,1]}(3)_f \otimes \mathcal{U}_{[1]}(2)_s)$.

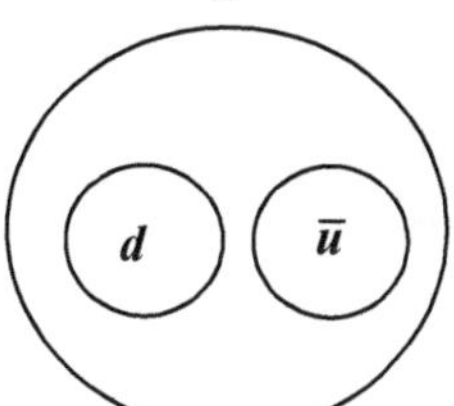

Fig. 5.9 Meson π^-

Due to (4.54), the irreducible decomposition is

$$\begin{aligned}
&(\mathcal{U}_{[1,0]}(3)_{\mathrm{f}} \otimes \mathcal{U}_{[1]}(2)_{\mathrm{s}}) \otimes (\mathcal{U}_{[0,1]}(3)_{\mathrm{f}} \otimes \mathcal{U}_{[1]}(2)_{\mathrm{s}}) \\
=&(\mathcal{U}_{[1,0]}(3)_{\mathrm{f}} \otimes \mathcal{U}_{[0,1]}(3)_{\mathrm{f}}) \otimes \mathcal{U}_{[1]}(2)_{\mathrm{s}}^{\otimes 2} \\
=&(\mathcal{U}_{[0,0]}(3)_{\mathrm{f}} \oplus \mathcal{U}_{[1,1]}(3)_{\mathrm{f}}) \otimes (\mathcal{U}_{[0]}(2)_{\mathrm{s}} \oplus \mathcal{U}_{[2]}(2)_{\mathrm{s}}) \\
=&\mathcal{U}_{[0,0]}(3)_{\mathrm{f}} \otimes \mathcal{U}_{[0]}(2)_{\mathrm{s}} \oplus \mathcal{U}_{[0,0]}(3)_{\mathrm{f}} \otimes \mathcal{U}_{[2]}(2)_{\mathrm{s}} \\
&\oplus \mathcal{U}_{[1,1]}(3)_{\mathrm{f}} \otimes \mathcal{U}_{[0]}(2)_{\mathrm{s}} \oplus \mathcal{U}_{[1,1]}(3)_{\mathrm{f}} \otimes \mathcal{U}_{[2]}(2)_{\mathrm{s}}. \qquad (5.130)
\end{aligned}$$

As $\mathcal{U}_{[1,1]}(3)_{\mathrm{f}}$ is the adjoint representation, the representation space can be identified with the Lie algebra (3). The dimensions are calculated as $\dim \mathcal{U}_{[0,0]}(3)_{\mathrm{f}} = 1$ and $\dim \mathcal{U}_{[1,1]}(3)_{\mathrm{f}} = 8$. The total dimension is $6 \times 6 = 36 = 1 \times 1 + 1 \times 3 + 8 \times 1 + 8 \times 3$.

Since $\dim \mathcal{U}_{[0,0]}(3)_{\mathrm{f}} = 1$ and $\dim \mathcal{U}_{[1,1]}(3)_{\mathrm{f}} = 8$, the systems $\mathcal{U}_{[0,0]}(3)_{\mathrm{f}}$ and $\mathcal{U}_{[1,1]}(3)_{\mathrm{f}}$ are called **meson singlet** and **meson octet**, respectively. A meson particle is an orthogonal base under this irreducible decomposition. When a meson particle belongs to a system with $\mathcal{U}_{[0]}(2)_{\mathrm{s}}$ it has spin 0 and is called a **pseudoscalar meson**, and when a meson particle belongs to a system with $\mathcal{U}_{[2]}(2)_{\mathrm{s}}$ it has spin 1 and is called a **vectorial meson**. Mesons are listed as Table 5.4.

Therefore, the values of the quantum numbers I_3 and Y are summarized as Figs. 5.10 and 5.11.

Now, we consider the freedom of color. As a physical law, the state of color in a meson is restricted to $|\psi'\rangle := \frac{1}{\sqrt{6}}(|b\bar{b}\rangle + |g\bar{g}\rangle + |r\bar{r}\rangle - |\bar{b}b\rangle - |\bar{g}g\rangle - |\bar{r}r\rangle)$, which is called **color confinement**. Since a quark is a fermion, the total system is restricted to the two-fold anti-symmetric space. Even though we take account into the color freedom, Theorem 4.3 with $r = 3$ and $d = 3$ and the color confinement guarantee that the system of meson is reduced to $(\mathcal{U}_{[1,0]}(3)_{\mathrm{f}} \otimes \mathcal{U}_{[1]}(2)_{\mathrm{s}}) \otimes (\mathcal{U}_{[0,1]}(3)_{\mathrm{f}} \otimes \mathcal{U}_{[1]}(2)_{\mathrm{s}})$.

Table 5.4 Lits of mesons: the wave function $|spin\rangle$ is chosen to be $\frac{1}{\sqrt{2}}(|\uparrow\downarrow\rangle + |\downarrow\uparrow\rangle)$ for a pseudoscalar meson, and is chosen to be $|\uparrow\uparrow$ for a vector meson. However, for a vector meson, the wave function $|spin\rangle$ can be chosen to be any element of $\mathcal{U}_{[2]}(2)_{\mathrm{s}}$

Pseudoscalar meson	Vectorial meson	Wave function
K^+	K^{*+}	$\|u\bar{s}\rangle\|spin\rangle$
K^0	K^{*0}	$\|d\bar{s}\rangle\|spin\rangle$
K^-	K^{*-}	$\|s\bar{u}\rangle\|spin\rangle$
$\bar{K}^0$	$\bar{K}^{*0}$	$\|s\bar{d}\rangle\|spin\rangle$
π^+	ρ^+	$\|u\bar{d}\rangle\|spin\rangle$
π^-	ρ^-	$\|d\bar{u}\rangle\|spin\rangle$
π^0	ρ^0	$\frac{1}{\sqrt{2}}(\|u\bar{u}\rangle - \|d\bar{d}\rangle)\|spin\rangle$
η_8	ω_8	$\frac{1}{\sqrt{6}}(\|u\bar{u}\rangle + \|d\bar{d}\rangle - 2\|s\bar{s}\rangle)\|spin\rangle$
η_1	ω_1	$\frac{1}{\sqrt{3}}(\|u\bar{u}\rangle + \|d\bar{d}\rangle + \|s\bar{s}\rangle)\|spin\rangle$

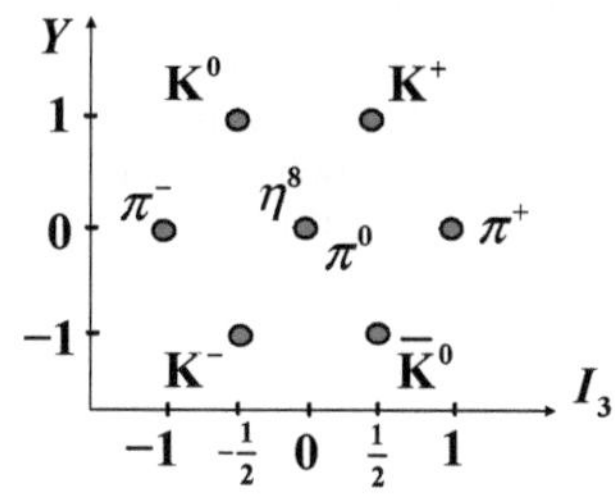

Fig. 5.10 Pseudoscalar meson

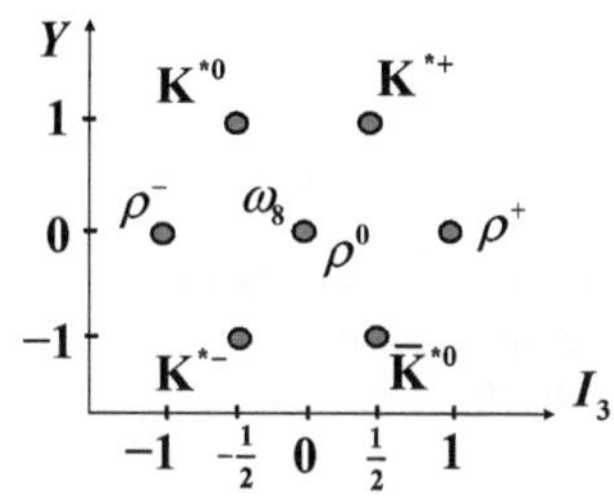

Fig. 5.11 Vectorial meson

5.5.3 *Baryon*

Another type of hadrons, a **baryon** is composed of three quarks, and an anti-baryon is composed of three anti-quarks. To discuss baryon, we need to discuss the freedom of color, first. Due to a physical law, the state of color of a baryon is restricted to $|\psi\rangle := \frac{1}{\sqrt{6}}(|bgr\rangle + |grb\rangle + |rbg\rangle - |gbr\rangle - |rgb\rangle - |brg\rangle)$, which is called **color confinement**. That is, a baryon is composed of three different colors as Fig. 5.8.

We apply Theorem 4.2 with $r = 6$ and $n = 3$ to the system $\mathcal{H} := \mathcal{U}_{[1,0]}(3)_f \otimes \mathcal{U}_{[1]}(2)_s \otimes \mathcal{U}_{[1,0]}(3)_c$. Then, our system of possible state is the three-fold symmetric tensor product system $\mathcal{U}_{[1,0]}(3)_f \otimes \mathcal{U}_{[1]}(2)_s$, which is a 6-dimensional space. That is, due to the color confinement, a quark behaves as a boson only with the freedoms of flavor and spin while it is truly a fermion when including the freedom of color.

Under the group SU(6), we denote the total one-quark system by $\mathcal{U}_{[1,0,0,0,0]}(6)$, Then, the one-baryon system is given as $\mathcal{U}_{[3,0,0,0,0]}(6)$, which is an irreducible space of the representation of SU(6). However, it is not irreducible for the subgroup $SU_f(3) \times SU_s(2) \subset SU(6)$. Due to Theorem 4.1, the system $\mathcal{U}_{[3,0,0,0,0]}(6)$ is irreducibly decomposed as

$$\mathcal{U}_{[3,0,0,0,0]}(6) = \mathcal{U}_{[3,0]}(3)_f \otimes \mathcal{U}_{[3]}(2)_s \oplus \mathcal{U}_{[1,1]}(3)_f \otimes \mathcal{U}_{[1]}(2)_s \tag{5.131}$$

in the sense of the representation of $SU_f(3) \times SU_s(2)$. The dimensions are calculated as $\dim \mathcal{U}_{[3,0]}(3)_f = 10$ and $\dim \mathcal{U}_{[1,1]}(3)_f = 8$. Due to Exercise 4.9, the total dimension is $\frac{8\times7\times6}{3\times2} = 56 = 10 \times 4 + 8 \times 2$.

The space $\mathcal{U}_{[3,0]}(3)_f \otimes \mathcal{U}_{[3]}(2)_s$ is called **baryon decuplet**, and is generated by the vector $|uuu\ \uparrow\uparrow\uparrow\rangle$ with the action of $SU_f(3) \times SU_s(2)$. the other space

Table 5.5 Lits of baryon octet: It is possible to exchange $\uparrow$ and $\downarrow$. This exchange corresponds to the freedom in $\mathcal{U}_{[1]}(2)_s$

Name	Wave function
p	$\|duu\rangle\| \uparrow'\rangle$ + cyclic terms
n	$\|udd\rangle\| \uparrow'\rangle$ + cyclic terms
Λ^0	(5.132)
Σ^+	$\|suu\rangle\| \uparrow'\rangle$ + cyclic terms
Σ^0	(5.133)
Σ^-	$\|sdd\rangle\| \uparrow'\rangle$ + cyclic terms
Ξ^0	$\|uss\rangle\| \uparrow'\rangle$ + cyclic terms
Ξ^-	$\|dss\rangle\| \uparrow'\rangle$ + cyclic terms

Table 5.6 Lits of baryon decuplet: $|spin\rangle$ is any element of $\mathcal{U}_{[3]}(2)_s$

Name	Wave function
Δ^{++}	$\|uuu\rangle\|spin\rangle$
Δ^+	$\|uud\rangle\|spin\rangle$ + cyclic terms
Δ^0	$\|udd\rangle\|spin\rangle$ + cyclic terms
Δ^-	$\|ddd\rangle\|spin\rangle$
Σ^{*+}	$\|uus\rangle\|spin\rangle$ + cyclic terms
Σ^{*0}	$\|uds\rangle\|spin\rangle$ + cyclic terms
Σ^{*-}	$\|dds\rangle\|spin\rangle$ + cyclic terms
Ξ^{*0}	$\|uss\rangle\|spin\rangle$ + cyclic terms
Ξ^{*-}	$\|dss\rangle\|spin\rangle$ + cyclic terms
Ω^-	$\|sss\rangle\|spin\rangle$

$\mathcal{U}_{[1,1]}(3)_f \otimes \mathcal{U}_{[1]}(2)_s$ is called **baryon octet**, and, with the action of $SU_f(3) \times SU_s(2)$, is generated by the symmetrization of the vector $|duu\rangle| \uparrow'\rangle$, where $| \uparrow'\rangle := \sqrt{\frac{2}{3}}| \downarrow\uparrow\uparrow\rangle - \sqrt{\frac{1}{6}}(| \uparrow\downarrow\uparrow\rangle + | \uparrow\uparrow\downarrow\rangle)$. This vector is the proton $|p\rangle$. Baryons are summarized as Tables 5.5 and 5.6. However, Λ^0 and Σ^0 are not so simple. They have the same weight, but they are given as orthogonal vectors in the following way.

$$\begin{aligned}|\Lambda^0\rangle =& \frac{1}{2\sqrt{3}}\Big((|uds\rangle - |dus\rangle)(| \uparrow\downarrow\uparrow\rangle - | \downarrow\uparrow\uparrow\rangle)\\ &+ (|sud\rangle - |sdu\rangle)(| \uparrow\uparrow\downarrow\rangle - | \uparrow\downarrow\uparrow\rangle)\\ &+ (|dsu\rangle - |usd\rangle)(| \downarrow\uparrow\uparrow\rangle - | \uparrow\uparrow\downarrow\rangle)\Big), \end{aligned} \tag{5.132}$$

$$|\Sigma^0\rangle = \frac{1}{6}\Big((|uds\rangle - |dus\rangle)(| \uparrow\downarrow\uparrow\rangle + | \downarrow\uparrow\uparrow\rangle - 2| \uparrow\uparrow\downarrow\rangle)$$

$$+ (|sud\rangle - |sdu\rangle)(|\uparrow\uparrow\downarrow\rangle + |\uparrow\downarrow\uparrow\rangle - 2|\downarrow\uparrow\uparrow\rangle)$$
$$+ (|dsu\rangle - |usd\rangle)(|\downarrow\uparrow\uparrow\rangle + |\uparrow\uparrow\downarrow\rangle - 2|\uparrow\downarrow\uparrow\rangle)\Big). \tag{5.133}$$

Hence, these two vectors give an example for an eigenspace $\mathcal{H}_{\boldsymbol{m}}$ of a non-highest weight $\boldsymbol{m}$ that is not one-dimensional, which is discussed in Sect. 4.3.3.

Therefore, the values of the quantum numbers I_3 and Y are summarized as Figs. 5.12 and 5.13.

Similarly, the anti-one-baryon system is given as $\mathcal{U}_{[0,0,0,0,3]}(6)$. Due to Exercise 4.12, the system $\mathcal{U}_{[0,0,0,0,3]}(6)$ is irreducibly decomposed as

$$\mathcal{U}_{[0,0,0,0,3]} = \mathcal{U}_{[0,3]} \otimes \mathcal{U}_{[3]} \oplus \mathcal{U}_{[1,1]} \otimes \mathcal{U}_{[1]}. \tag{5.134}$$

Exercise 5.15 Show that $|n\rangle$ and $|\Delta^0\rangle$ are orthogonal to each other even though they have the same spin.

Exercise 5.16 Show that $|\Lambda^0\rangle$, $|\Sigma^0\rangle$, and $|\Sigma^{*0}\rangle$ are orthogonal to each other even though they have the same spin.

5.5.4 Charm Quark

The existence of a fourth quark had been discussed by several authors [18, 35], and is named to be a **charm quark** c. The first charmed particle (a particle containing a charm quark) was discovered in 1974 [11, 10]. A charm quark has the quantum

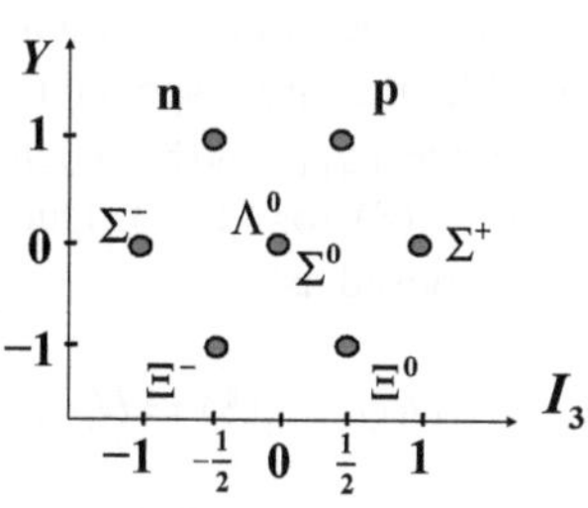

Fig. 5.12 Baryon octet

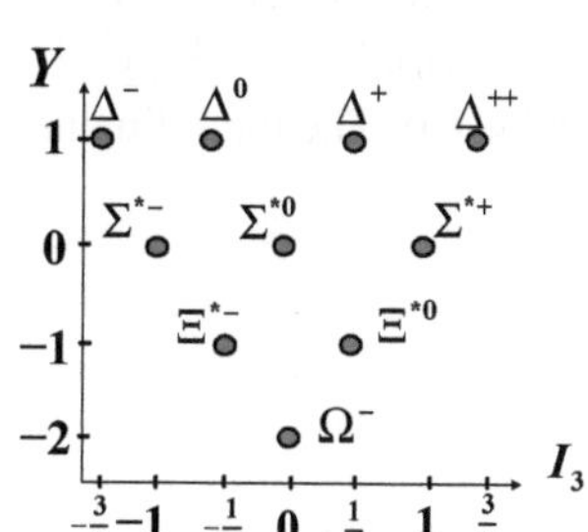

Fig. 5.13 Baryon decuplet

numbers $Q = \frac{2}{3}$, $I_3 = 0$, $S = 0$, and $C = 1$, where C is the new quantum number and is taken to be 0 in other quarks. So, charm quark has the quantum number $Y = -\frac{2}{3}$. When we include a charm quark, the quantum system of one quark is $\mathbb{C}^4$ and has the fundamental representation of SU(4). The flavor space is spanned by $|u\rangle$, $|d\rangle$, $|s\rangle$, $|c\rangle$, and the Dynkin index is $[1, 0, 0]$ so that the system is $\mathcal{U}_{[1,0,0]}$. When we include the freedoms of spin and color, the total one-quark system is written as $\mathcal{U}_{[1,0,0]}(4)_{\mathrm{f}} \otimes \mathcal{U}_{[1]}(2)_{\mathrm{s}} \otimes \mathcal{U}_{[1,0]}(3)_{\mathrm{c}}$. under the representation of the group $\mathrm{SU}_{\mathrm{f}}(4) \times \mathrm{SU}_{\mathrm{s}}(2) \times \mathrm{SU}_{\mathrm{c}}(3)$. The flavor space of one-anti quark is spanned by $|\bar{u}\rangle$, $|\bar{d}\rangle$, $|\bar{s}\rangle$, $|\bar{c}\rangle$, and the total one-anti-quark system is written as $\mathcal{U}_{[0,0,1]}(4)_{\mathrm{f}} \otimes \mathcal{U}_{[1]}(2)_{\mathrm{s}} \otimes \mathcal{U}_{[0,1]}(3)_{\mathrm{c}}$.

A meson is composed of a quark and an anti-quark. Due to the color confinement and Theorem 4.3, under the representation of the group $\mathrm{SU}_{\mathrm{f}}(4) \times \mathrm{SU}_{\mathrm{s}}(2)$, the one-meson system is essentially given as $(\mathcal{U}_{[1,0,0]}(4)_{\mathrm{f}} \otimes \mathcal{U}_{[1]}(2)_{\mathrm{s}}) \otimes (\mathcal{U}_{[0,0,1]}(4)_{\mathrm{f}} \otimes \mathcal{U}_{[1]}(2)_{\mathrm{s}})$. Due to (4.54), the irreducible decomposition is

$$\begin{aligned}
&(\mathcal{U}_{[1,0,0]}(4)_{\mathrm{f}} \otimes \mathcal{U}_{[1]}(2)_{\mathrm{s}}) \otimes (\mathcal{U}_{[0,0,1]}(4)_{\mathrm{f}} \otimes \mathcal{U}_{[1]}(2)_{\mathrm{s}}) \\
=& (\mathcal{U}_{[1,0,0]}(4)_{\mathrm{f}} \otimes \mathcal{U}_{[0,0,1]}(4)_{\mathrm{f}}) \otimes \mathcal{U}_{[1]}(2)_{\mathrm{s}}^{\otimes 2} \\
=& (\mathcal{U}_{[0,0,0]}(4)_{\mathrm{f}} \oplus \mathcal{U}_{[1,0,1]}(4)_{\mathrm{f}}) \otimes (\mathcal{U}_{[0]}(2)_{\mathrm{s}} \oplus \mathcal{U}_{[2]}(2)_{\mathrm{s}}) \\
=& \mathcal{U}_{[0,0,0]}(4)_{\mathrm{f}} \otimes \mathcal{U}_{[0]}(2)_{\mathrm{s}} \oplus \mathcal{U}_{[0,0,0]}(4)_{\mathrm{f}} \otimes \mathcal{U}_{[2]}(2)_{\mathrm{s}} \\
&\oplus \mathcal{U}_{[1,0,1]}(4)_{\mathrm{f}} \otimes \mathcal{U}_{[0]}(2)_{\mathrm{s}} \oplus \mathcal{U}_{[1,0,1]}(4)_{\mathrm{f}} \otimes \mathcal{U}_{[2]}(2)_{\mathrm{s}}.
\end{aligned} \tag{5.135}$$

As $\mathcal{U}_{[1,0,1]}(4)_{\mathrm{f}}$ is the adjoint representation, the representation space can be identified with the Lie algebra (4). The dimensions are calculated as $\dim \mathcal{U}_{[0,0,0]}(4)_{\mathrm{f}} = 1$ and $\dim \mathcal{U}_{[1,0,1]}(4)_{\mathrm{f}} = 15$. The total dimension is $8 \times 8 = 64 = 1 \times 1 + 1 \times 3 + 15 \times 1 + 15 \times 3$.

Since a baryon is composed of three quarks, the color confinement and Theorem 4.2 guarantee that our analysis is reduced to the analysis on the three-fold symmetric space of the 8-dimensional system $\mathcal{U}_{[1,0,0]}(4)_{\mathrm{f}} \otimes \mathcal{U}_{[1]}(2)_{\mathrm{s}}$. Hence, we need to consider the larger special unitary group SU(8) Then, we denote the total one-quark system by $\mathcal{U}_{[1,0,\ldots,0]}(8)$. Then, the one-baryon system is given as the irreducible space $\mathcal{U}_{[3,0,\ldots,0]}(8)$ for SU(8). Due to Theorem 4.1, the system $\mathcal{U}_{[3,0,\ldots,0]}(8)$ is irreducibly decomposed as

$$\mathcal{U}_{[3,0,\ldots,0]}(8) = \mathcal{U}_{[3,0,0]}(4)_{\mathrm{f}} \otimes \mathcal{U}_{[3]}(2)_{\mathrm{s}} \oplus \mathcal{U}_{[1,1,0]}(4)_{\mathrm{f}} \otimes \mathcal{U}_{[1]}(2)_{\mathrm{s}} \tag{5.136}$$

in the sense of $\mathrm{SU}_{\mathrm{f}}(4) \times \mathrm{SU}_{\mathrm{s}}(2)$. Since the dimensions are calculated as $\dim \mathcal{U}_{[3,0,0]}(4)_{\mathrm{f}} = \frac{5\times4\times3}{3\times2} = 20$ and $\dim \mathcal{U}_{[1,1,0]}(4)_{\mathrm{f}} = \frac{5\times4\times3\times2}{3\times2} = 20$ (Exercises 4.9 and 4.10). The total dimension is $\frac{10\times9\times8}{3\times2} = 120 = 20 \times 4 + 20 \times 2$.

5.5.5 *Six Quarks*

In the current standard model, there are 6 quarks, i.e., we have **top quark** t and **bottom quark** b in addition to up quark u, down quark d, charm quark c, and strange quark s so that the flavor system is 6-dimensional. So, the total one-quark system and the total one-anti-quark system are written as $\mathcal{U}_{[1,0,0,0,0]}(6)_{\mathrm{f}} \otimes \mathcal{U}_{[1]}(2)_{\mathrm{s}} \otimes \mathcal{U}_{[1,0,0]}(3)_{\mathrm{c}}$ and $\mathcal{U}_{[0,0,0,0,1]}(6)_{\mathrm{f}} \otimes \mathcal{U}_{[1]}(2)_{\mathrm{s}} \otimes \mathcal{U}_{[0,0,1]}(3)_{\mathrm{c}}$, respectively, under the representation of the group $\mathrm{SU}_{\mathrm{f}}(6) \times \mathrm{SU}_{\mathrm{s}}(2) \times \mathrm{SU}_{\mathrm{c}}(3)$.

Since a meson is composed of a quark and an anti-quark, due to the color confinement and Theorem 4.3, under the representation of the group $\mathrm{SU}_{\mathrm{f}}(6) \times \mathrm{SU}(2)_{\mathrm{s}}$, the one-meson system is essentially given as $(\mathcal{U}_{[1,0,0,0,0]}(6)_{\mathrm{f}} \otimes \mathcal{U}_{[1]}(2)_{\mathrm{s}}) \otimes (\mathcal{U}_{[0,0,0,0,1]}(6)_{\mathrm{f}} \otimes \mathcal{U}_{[1]}(2)_{\mathrm{s}})$. Due to (4.54), the irreducible decomposition is

$$\begin{aligned}
&(\mathcal{U}_{[1,0,0,0,0]}(6)_{\mathrm{f}} \otimes \mathcal{U}_{[1]}(2)_{\mathrm{s}}) \otimes (\mathcal{U}_{[0,0,0,0,1]}(6)_{\mathrm{f}} \otimes \mathcal{U}_{[1]}(2)_{\mathrm{s}}) \\
=&(\mathcal{U}_{[1,0,0,0,0]}(6)_{\mathrm{f}} \otimes \mathcal{U}_{[0,0,0,0,1]}(6)_{\mathrm{f}}) \otimes \mathcal{U}_{[1]}(2)_{\mathrm{s}}^{\otimes 2} \\
=&(\mathcal{U}_{[0,0,0,0,0]}(6)_{\mathrm{f}} \oplus \mathcal{U}_{[1,0,0,0,1]}(6)_{\mathrm{f}}) \otimes (\mathcal{U}_{[0]}(2)_{\mathrm{s}} \oplus \mathcal{U}_{[2]}(2)_{\mathrm{s}}) \\
=&\mathcal{U}_{[0,0,0,0,0]}(6)_{\mathrm{f}} \otimes \mathcal{U}_{[0]}(2)_{\mathrm{s}} \oplus \mathcal{U}_{[0,0,0,0,0]}(6)_{\mathrm{f}} \otimes \mathcal{U}_{[2]}(2)_{\mathrm{s}} \\
&\oplus \mathcal{U}_{[1,0,0,0,1]}(6)_{\mathrm{f}} \otimes \mathcal{U}_{[0]}(2)_{\mathrm{s}} \oplus \mathcal{U}_{[1,0,0,0,1]}(6)_{\mathrm{f}} \otimes \mathcal{U}_{[2]}(2)_{\mathrm{s}}.
\end{aligned} \tag{5.137}$$

The dimensions are calculated as $\dim \mathcal{U}_{[0,0,0,0,0]}(6)_{\mathrm{f}} = 1$ and $\dim \mathcal{U}_{[1,0,0,0,1]}(6)_{\mathrm{f}} = 35$. The total dimension is $12 \times 12 = 144 = 1 \times 1 + 1 \times 3 + 35 \times 1 + 35 \times 3$.

Since a baryon is composed of three quarks, due to the color confinement and Theorem 4.2, we need to consider the three-fold symmetric space of $\mathcal{U}_{[1,0,0,0,0]}(6)_{\mathrm{f}} \otimes \mathcal{U}_{[1]}(2)_{\mathrm{s}}$, whose dimension is 12. That is, the one-baryon system is given as the irreducible space $\mathcal{U}_{[3,0,\ldots,0]}(12)$ for $\mathrm{SU}(12)$. Due to Theorem 4.1, the system $\mathcal{U}_{[3,0,\ldots,0]}(12)$ is irreducibly decomposed as

$$\mathcal{U}_{[3,0,\ldots,0]}(12) = \mathcal{U}_{[3,0,0,0,0]}(6)_{\mathrm{f}} \otimes \mathcal{U}_{[3]}(2)_{\mathrm{s}} \oplus \mathcal{U}_{[1,1,0,0,0]}(6)_{\mathrm{f}} \otimes \mathcal{U}_{[1]}(2)_{\mathrm{s}} \tag{5.138}$$

in the sense of $\mathrm{SU}_{\mathrm{f}}(6) \times \mathrm{SU}_{\mathrm{s}}(2)$. Since the dimensions are calculated as $\dim \mathcal{U}_{[3,0,0,0,0]}(6)_{\mathrm{f}} = \frac{8\times 7\times 6}{3\times 2} = 56$ and $\dim \mathcal{U}_{[1,1,0,0,0]}(6)_{\mathrm{f}} = \frac{7\times 6\times 5}{3} = 70$ (Exercises 4.9 and 4.10). The total dimension is $\frac{14\times 13\times 12}{3\times 2} = 364 = 56 \times 4 + 70 \times 2$. That is, we have 364 baryons and 364 anti-baryons.

5.6 Uncertainty Relations

In this section, we discuss the uncertainty relation for various types of wave functions. Firstly, we discuss the uncertainty relation for wave functions on the real line $\mathbb{R}$. Then, to address wave functions on the one-dimensional and three-dimensional spheres S^1 and S^3, we prepare several knowledges for periodic function space and Mathieu function. After this preparation, we derive the uncertainty relation for wave functions on the one-dimensional and three-dimensional spheres S^1 and S^3. The uncertainty relation on the one-dimensional sphere S^1 is directly derived from Mathieu function. However, the uncertainty relation on the three-dimensional sphere S^3 is more complicated because Mathieu function is defined only for the one-dimensional system. To fill the gap, we employ the representation theory of SU(2), which is mainly discussed in Lemma 5.8.

5.6.1 Uncertainty Relation on $\mathbb{R}$

In the quantum system $L^2(\mathbb{R})$, it is impossible to determine simultaneously the values of the position Q and the operator P, which equals the momentum when $\hbar = 1$. To clarify this issue, we discuss the trade-off relation between $\Delta_\rho \mathsf{P}$ and $\Delta_\rho \mathsf{Q}$ defined in (1.8). For this purpose, we introduce two operators $\mathsf{a} := \frac{1}{\sqrt{2}}(\mathsf{Q} + i\mathsf{P})$ and $\mathsf{a}^\dagger := \frac{1}{\sqrt{2}}(\mathsf{Q} - i\mathsf{P})$, which implies the relations $\mathsf{Q} = \frac{1}{\sqrt{2}}(\mathsf{a} + \mathsf{a}^\dagger)$ and $\mathsf{P} = \frac{-i}{\sqrt{2}}(\mathsf{a} - \mathsf{a}^\dagger)$. Now, we focus on Hermitian inner product $(X, Y) := \frac{1}{2}\operatorname{Tr}(XY^\dagger + YX^\dagger)\rho$ defined for two operators X and Y. We apply Schwarz inequality for the inner product to the case with $X = \mathsf{Q} - (\operatorname{Tr}\mathsf{Q}\rho)I$ and $Y = i(\mathsf{P} - (\operatorname{Tr}\mathsf{P}\rho)I)$.

Then, we obtain the Kennard–Robertson type **uncertainty relation** [63, 91]

$$\Delta_\rho \mathsf{P}\, \Delta_\rho \mathsf{Q} \ge \left| \frac{1}{2} i \operatorname{Tr}[\mathsf{P} - (\operatorname{Tr}\mathsf{P}\rho)I, \mathsf{Q} - (\operatorname{Tr}\mathsf{Q}\rho)I]\rho \right| = \frac{1}{2}. \tag{5.139}$$

Since the inequality $x + y \ge 2xy$ holds for $x, y \ge 0$, we have

$$\Delta_\rho^2 \mathsf{P} + \Delta_\rho^2 \mathsf{Q} \ge 2\Delta_\rho \mathsf{P}\, \Delta_\rho \mathsf{Q} \ge 1. \tag{5.140}$$

The equality condition will be discussed in Sect. 7.1.2.

More precisely, the determinant of the matrix $\Delta_\rho[\mathsf{Q} : \mathsf{P}] := \begin{pmatrix} \Delta_\rho \mathsf{Q} & \Delta_\rho \mathsf{Q} \circ \mathsf{P} \\ \Delta_\rho \mathsf{Q} \circ \mathsf{P} & \Delta_\rho \mathsf{P} \end{pmatrix}$ defined by the notations given in (1.10) satisfies

$$\sqrt{\det \Delta_\rho[\mathsf{Q} : \mathsf{P}]} \ge \frac{1}{2}. \tag{5.141}$$

To show (5.141), we consider the Hermitian matrix

$$\begin{aligned} J_\rho := \begin{pmatrix} \operatorname{Tr}\mathsf{Q}_0^2\rho & \operatorname{Tr}\mathsf{Q}_0\mathsf{P}_0\rho \\ \operatorname{Tr}\mathsf{P}_0\mathsf{Q}_0\rho & \operatorname{Tr}\mathsf{P}_0^2\rho \end{pmatrix} &= \Delta_\rho[\mathsf{Q} : \mathsf{P}] + \frac{1}{2}\begin{pmatrix} 0 & \operatorname{Tr}[\mathsf{Q}_0, \mathsf{P}_0]\rho \\ \operatorname{Tr}[\mathsf{P}_0, \mathsf{Q}_0]\rho & 0 \end{pmatrix} \\ &= \Delta_\rho[\mathsf{Q} : \mathsf{P}] + \frac{i}{2}\begin{pmatrix} 0 & 1 \\ -1 & 0 \end{pmatrix}, \end{aligned}$$

where $\mathsf{Q}_0 := \mathsf{Q} - \operatorname{Tr} \mathsf{Q}\rho$ and $\mathsf{P}_0 := \mathsf{P} - \operatorname{Tr} \mathsf{P}\rho$. Since $J_\rho \geq 0$ (Exercise 5.17), we have $\det \Delta_\rho[\mathsf{Q} : \mathsf{P}] - \frac{1}{4} = \det J_\rho \geq 0$.

Exercise 5.17 Show the matrix inequality $J_\rho \geq 0$.

*5.6.2 Periodic Function Space and Mathieu Equation**

To discuss the uncertainty relation to the wave functions on the one-dimensional and three-dimensional and spheres S^1 and S^3, we prepare several mathematical properties on the space of periodic function. For a positive real number L, we introduce the notations as follows.

$$L_p^2((-L, L]) := \left\{ f | f(x+2L) = f(x), \int_{-L}^{L} |f(x)|^2 \frac{dx}{2L} < \infty \right\}, \tag{5.142}$$

where the subscript p expresses the periodic. As a generalization, we define the space of anti-periodic functions

$$L_a^2((-L, L]) := \left\{ f | f(x+2L) = -f(x), \int_{-L}^{L} |f(x)|^2 \frac{dx}{2L} < \infty \right\}, \tag{5.143}$$

which is a subspace of $L_p^2((-2L, 2L])$. Further, we denote the spaces of even functions and odd functions in $L_p^2((-L, L])$ and $L_a^2((-L, L])$ by $L_{p,\text{even}}^2((-L, L])$, $L_{p,\text{odd}}^2((-L, L])$, $L_{a,\text{even}}^2((-L, L])$, $L_{a,\text{odd}}^2((-L, L])$, respectively. For any $f, g \in L_p^2((-L, L])$, we define the inner product as

$$\langle f | g \rangle := \int_{-L}^{L} \overline{f(x)} g(x) \frac{dx}{2L}. \tag{5.144}$$

The subspaces $L_{p,\text{even}}^2((-L, L])$ and $L_{p,\text{odd}}^2((-L, L])$ (the subspaces $L_{a,\text{even}}^2((-L, L])$ and $L_{a,\text{odd}}^2((-L, L])$) are orthogonal to each other. Also, the two subspaces $L_p^2((-L, L])$ and $L_a^2((-L, L])$ are orthogonal to each other. Therefore, the space $L_p^2((-2L, 2L])$ can be written as $L_{p,\text{even}}^2((-L, L]) \oplus L_{p,\text{odd}}^2((-L, L]) \oplus L_{a,\text{even}}^2((-L, L]) \oplus L_{a,\text{odd}}^2((-L, L])$.

Now, we consider Mathieu equation:

$$\frac{d^2}{d\theta^2}\varphi(\theta) + (a - 2q\cos(2\theta))\varphi(\theta) = 0. \tag{5.145}$$

A function φ satisfies the above equation if and only if the function φ is the eigenfunction of the differential operator $\mathsf{P}^2 + 2q\cos(2\mathsf{Q})$. The operator $X(q) := \mathsf{P}^2 + 2q\cos(2\mathsf{Q})$ preserves the subspaces $L_{p,\text{even}}^2((-\frac{\pi}{2}, \frac{\pi}{2}])$, $L_{p,\text{odd}}^2((-\frac{\pi}{2}, \frac{\pi}{2}])$,

$L^2_{a,\text{even}}((-\frac{\pi}{2}, \frac{\pi}{2}])$, and $L^2_{a,\text{odd}}((-\frac{\pi}{2}, \frac{\pi}{2}])$. Then, we denote the minimum eigenvalues in $L^2_{p,\text{even}}((-\frac{\pi}{2}, \frac{\pi}{2}])$, $L^2_{p,\text{odd}}((-\frac{\pi}{2}, \frac{\pi}{2}])$, $L^2_{a,\text{even}}((-\frac{\pi}{2}, \frac{\pi}{2}])$, and $L^2_{a,\text{odd}}((-\frac{\pi}{2}, \frac{\pi}{2}])$ by $a_0(q)$, $b_2(q)$, $a_1(q)$, and $b_1(q)$, respectively [109, Sect. 28.2]. We call their eigenfunctions Mathieu functions $\text{ce}_0(\theta, q)$, $\text{se}_2(\theta, q)$, $\text{ce}_1(\theta, q)$, and $\text{se}_1(\theta, q)$. The eigenvalues $a_0(q)$, $b_2(q)$, $a_1(q)$, and $b_1(q)$ satisfy the conditions [109, Sect. 28.2(v)]

$$a_0(q) = a_0(-q), \quad b_2(q) = b_2(-q), \quad a_1(q) = b_1(-q). \tag{5.146}$$

When $q < 0$, the ordering relation $a_0(q) < a_1(q) < b_1(q) < b_2(q)$ holds [109, Sect. 28.2(v)].

According to the reference [109, Sect. 28.2(v)], let $a_0(q)$ be the minimum value of a having the solution in $L^2((-\pi/2, \pi/2])$ of the above differential equation, and $b_2(q)$ be the minimum a having the odd solution in $L^2((-\pi/2, \pi/2])$ of the above differential equation. The solution with $a_0(q)$ is Mathieu function $\text{ce}_0(\theta, q)$ and the solution with $b_2(q)$ is Mathieu function $\text{se}_2(\theta, q)$ [109, Sect. 28.2(vi)]. Further, for a large q, the functions a_0 and b_2 have the following asymptotic expansions for a large h as

$$a_0(h^2) \cong -2h^2 + 2h - \frac{1}{4} - \frac{1}{2^5 h} - \frac{3}{2^8 h^2} \tag{5.147}$$

$$b_2(h^2) \cong -2h^2 + 6h - \frac{5}{4} - \frac{9}{2^5 h} - \frac{45}{2^8 h^2}. \tag{5.148}$$

Further expansion is available in [109, Sect. 28.8].

For a small q, the functions $a_0(q)$ and $b_2(q)$ have the following asymptotic expansions as

$$a_0(q) \cong -\frac{1}{2}q^2 + \frac{7}{128}q^4 \tag{5.149}$$

$$b_2(q) \cong 4 - \frac{1}{12}q^2 + \frac{5}{13824}q^4. \tag{5.150}$$

Further expansion is available in [109, Sect. 28.6].

5.6.3 Uncertainty Relation on S^1*

Now, we extend the uncertainty relation to the wave functions on the one-dimensional sphere S^1. A wave function on the one-dimensional sphere S^1 can be identified with a periodic function of the period 2π on $\mathbb{R}$. In this space, we can consider the pair of operators $(\cos \mathsf{Q}, \sin \mathsf{Q})$. For a function $\varphi \in L^2_p((-\pi, \pi])$, instead of the uncertainty $\Delta^2_\varphi \mathsf{Q}$, we focus on the uncertainty

$$\begin{aligned}
&\Delta_\varphi^2(\cos \mathsf{Q}, \sin \mathsf{Q}) := \Delta_\varphi^2 \cos \mathsf{Q} + \Delta_\varphi^2 \sin \mathsf{Q} \\
= &\langle \varphi | \cos^2 \mathsf{Q} | \varphi \rangle + \langle \varphi | \sin^2 \mathsf{Q} | \varphi \rangle - \langle \varphi | \cos \mathsf{Q} | \varphi \rangle^2 - \langle \varphi | \sin \mathsf{Q} | \varphi \rangle^2 \\
= &1 - \langle \varphi | \cos \mathsf{Q} | \varphi \rangle^2 - \langle \varphi | \sin \mathsf{Q} | \varphi \rangle^2.
\end{aligned}$$

Then, as the uncertainty relation, we consider the trade-off relation between $\Delta_\varphi^2(\cos \mathsf{Q}, \sin \mathsf{Q})$ and $\Delta_\varphi^2 \mathsf{P}$. For this purpose, we consider the minimum

$$\begin{aligned}
\gamma_{\mathrm{U}(1)}(s) := &\min_{\varphi \in L^2((-\pi,\pi]):\|\varphi\|=1} \langle \varphi | (I - \cos(\mathsf{Q})) + s\mathsf{P}^2 | \varphi \rangle \\
= &\min_{\varphi \in L^2((-\pi/2,\pi/2]):\|\varphi\|=1} \langle \varphi | (I - \cos(2\mathsf{Q})) + \frac{s\mathsf{P}^2}{4} | \varphi \rangle.
\end{aligned}$$

So, $\gamma_{\mathrm{U}(1)}(s)$ can be characterized as the minimum value of $\gamma_{\mathrm{U}(1)}$ having the solution in $L^2((-\pi/2, \pi/2])$ of the following differential equation.

$$\frac{s}{4}\frac{d^2}{d\theta^2}\varphi(\theta) + (\gamma_{\mathrm{U}(1)} - 1 + \cos(2\theta))\varphi(\theta) = 0, \tag{5.151}$$

which is equivalent to

$$\frac{d^2}{d\theta^2}\varphi(\theta) + \left(\frac{4(\gamma_{\mathrm{U}(1)} - 1)}{s} + \frac{4}{s}\cos(2\theta)\right)\varphi(\theta) = 0, \tag{5.152}$$

In order to find the minimum $\gamma_{\mathrm{U}(1)}(s)$, we employ Mathieu equation (5.145), whose detail is summarized in Sect. 5.6.2. Hence, using the function a_0 given in Sect. 5.6.2, we have $\gamma_{\mathrm{U}(1)}(s) = \frac{s a_0(-\frac{2}{s})}{4} + 1 = \frac{s a_0(\frac{2}{s})}{4} + 1$. Now, we prepare the following lemmas.

Lemma 5.5 *Given two self-adjoint operators X and Y on a Hilbert space $\mathcal{H}$, we consider the functions*

$$\kappa(E) := \inf_{\phi \in \mathcal{H}:\|\phi\|=1} \{\langle \phi | X \| \phi \rangle | \langle \phi | Y \| \phi \rangle \le E\} \tag{5.153}$$

$$\gamma(s) := \inf_{\phi \in \mathcal{H}:\|\phi\|=1} \langle \phi | X \| \phi \rangle + s \langle \phi | Y \| \phi \rangle. \tag{5.154}$$

Assume that $X \ge 0$ Then, $\kappa(E)$ is convex and monotone increasing, and $\gamma(s)$ is concave and monotone increasing. When $\kappa(E) > \lim_{E \to \infty} \kappa(E)$ and $E > 0$, there exists s_E such that $\gamma'(s_E) = E$, where γ' is the derivative of γ. Then, we have

$$\kappa(E) = \gamma(s_E) - s_E E = \max_{s>0} \gamma(s) - sE. \tag{5.155}$$

Further, s_E is positive and monotone decreasing with respect to E.

Lemma 5.5 will be shown in the end of this subsection. Applying Lemma 5.5 to the case when $X = I - \cos(\mathsf{Q})$ and $Y = \mathsf{P}^2$, we obtain the following lemma.

Lemma 5.6 ([49]) *We have the following relation:*

$$\kappa_{\mathrm{U}(1)}(E) := \min_{\varphi \in L^2_p((-\pi,\pi]):\|\varphi\|=1} \{\langle \varphi | I - \cos(\mathsf{Q}) | \varphi \rangle | \langle \varphi | \mathsf{P}^2 | \varphi \rangle \le E\} \tag{5.156}$$

$$= \max_{s>0} \frac{s a_0(\frac{2}{s})}{4} + 1 - sE. \tag{5.157}$$

The minimum (5.156) is attained if and only if $\varphi(\theta) = \mathrm{ce}_0(\frac{\theta}{2}, -\frac{2}{s_E})$, *where* s_E *is* $\mathrm{argmax}_{s>0} \frac{s a_0(\frac{2}{s})}{4} + 1 - sE$ *and the function* ce_0 *is given in Sect. 5.6.2.*

Using Lemma 5.6, we derive the trade-off relation between both uncertainties $\Delta^2_\varphi(\cos \mathsf{Q}, \sin \mathsf{Q})$ and $\Delta^2_\varphi \mathsf{P}$ as follows.

Theorem 5.4 ([49]) *The minimum of the uncertainty* $\Delta^2_\varphi(\cos \mathsf{Q}, \sin \mathsf{Q})$ *under the constraint for* $\Delta^2_\varphi \mathsf{P}$ *is calculated as*

$$\min_{\varphi \in L^2_p([-\pi,\pi]):\|\varphi\|=1} \{\Delta^2_\varphi(\cos \mathsf{Q}, \sin \mathsf{Q}) | \Delta^2_\varphi \mathsf{P} \le E\}$$
$$= 1 - \Big(\max_{s>0} \frac{s a_0(\frac{2}{s})}{4} - sE\Big)^2 = \max_{s>0} 1 - \Big(sE - \frac{s a_0(\frac{2}{s})}{4}\Big)^2. \tag{5.158}$$

The minimum in (5.158) is realized by φ *if and only if* φ *is given as a shift of the Mathieu function* $\mathrm{ce}_0(\frac{\theta}{2}, -\frac{2}{s_E})$.

Proof Due to the symmetry, without loss of generality, we can assume that $\langle \varphi | \sin \mathsf{Q} | \varphi \rangle = 0$ and $\langle \varphi | \mathsf{P} | \varphi \rangle = 0$. Hence, we obtain

$$\min_{\varphi \in L^2_p([-\pi,\pi]):\|\varphi\|=1} \{\Delta^2_\varphi(\cos \mathsf{Q}, \sin \mathsf{Q}) | \Delta^2_\varphi \mathsf{P} \le E\}$$
$$= \min_{\varphi \in L^2_p([-\pi,\pi]):\|\varphi\|=1} \{1 - \langle \varphi | \cos \mathsf{Q} | \varphi \rangle^2 | \langle \varphi | \mathsf{P}^2 | \varphi \rangle \le E\}$$
$$= 1 - \Big(\max_{\varphi \in L^2_p([-\pi,\pi]):\|\varphi\|=1} \{\langle \varphi | \cos \mathsf{Q} | \varphi \rangle | \langle \varphi | \mathsf{P}^2 | \varphi \rangle \le E\}\Big)^2.$$

Thanks to Lemma 5.6, we have

$$1 - \Big(\max_{\varphi \in L^2_p([-\pi,\pi]):\|\varphi\|=1} \{\langle \varphi | \cos \mathsf{Q} | \varphi \rangle | \langle \varphi | \mathsf{P}^2 | \varphi \rangle \le E\}\Big)^2$$
$$= 1 - \Big(1 - \max_{s>0} \frac{s a_0(\frac{2}{s})}{4} + 1 - sE\Big)^2 = 1 - \Big(\max_{s>0} \frac{s a_0(\frac{2}{s})}{4} - sE\Big)^2.$$

Since $\max_{s>0} \frac{s a_0(\frac{2}{s})}{4} + 1 - sE \le 1$, $sE - \frac{s a_0(\frac{2}{s})}{4} \ge 0$. Hence,

$$1 - \left(\max_{s>0} \frac{s a_0(\frac{2}{s})}{4} - sE \right)^2 = 1 - \left(\min_{s>0} sE - \frac{s a_0(\frac{2}{s})}{4} \right)^2$$
$$= 1 - \min_{s>0} \left(sE - \frac{s a_0(\frac{2}{s})}{4} \right)^2 = \max_{s>0} 1 - \left(sE - \frac{s a_0(\frac{2}{s})}{4} \right)^2,$$

which implies (5.158). Then, the minimum uncertainty in (5.158) is realized by φ if and only if φ is given as a shift of the Mathieu function $\mathrm{ce}_0(\frac{\theta}{2}, -\frac{2}{s_E})$. ∎

Now, for the asymptotic expansion of $1 - (\max_{s>0} \frac{s a_0(\frac{2}{s})}{4} - sE)^2$, we discuss the behavior of $\kappa_{\mathrm{U}(1)}(E)$ when E approaches infinity. By using the expansion (5.147) for a_0, as $s \to 0$, $\gamma_{\mathrm{U}(1)}(s)$ can be expanded to

$$\gamma_{\mathrm{U}(1)}(s) \cong \frac{s\left(-2\frac{2}{s} + 2\sqrt{\frac{2}{s}} - \frac{1}{4} \right)}{4} + 1 = \sqrt{\frac{s}{2}} - \frac{s}{16}.$$

As is shown in Lemma 5.5, s_E is decreasing as a function of E. Hence, when E is large, solving the equation $\gamma'_{\mathrm{U}(1)}(s_E) = E$, we approximately obtain $s_E \cong \frac{1}{8(E+1/16)^2}$. Hence,

$$\kappa_{\mathrm{U}(1)}(E) = \gamma_{\mathrm{U}(1)}(s_E) - s_E E \cong \sqrt{\frac{s_E}{2}} - \frac{s_E}{16} - s_E E = \sqrt{\frac{s_E}{2}} - s_E \left(E + \frac{1}{16} \right)$$
$$\cong \frac{1}{8(E+1/16)} \cong \frac{1}{8E} - \frac{1}{128E^2}. \tag{5.159}$$

So, we have

$$1 - \left(\max_{s>0} \frac{s a_0(\frac{2}{s})}{4} - sE \right)^2 = 2\kappa_{\mathrm{U}(1)}(E) - \kappa_{\mathrm{U}(1)}(E)^2 \cong \frac{1}{4E} - \frac{1}{128E^2}. \tag{5.160}$$

Next, we consider the case when E is small. When s is large, by using the expansion (5.149) for a_0, $\gamma_{\mathrm{U}(1)}(s)$ can be expanded to

$$\gamma_{\mathrm{U}(1)}(s) \cong \frac{s(-\frac{1}{2}(\frac{2}{s})^2 + \frac{7}{128}(\frac{2}{s})^4)}{4} + 1 = 1 - \frac{1}{2s} + \frac{7}{32s^3}.$$

When E is small, solving the equation $\gamma'_{\mathrm{U}(1)}(s_E) = E$, we approximately obtain $s_E \cong \sqrt{\frac{1}{2E} + \frac{21}{128}} \cong \frac{1}{\sqrt{2E}}(1 + \frac{21}{32}E)$. Hence,

$$
\begin{aligned}
\kappa_{\mathrm{U}(1)}(E) &= \gamma_{\mathrm{U}(1)}(s_E) - s_E E \cong 1 - \frac{1}{2s_E} - s_E E + \frac{7}{32 s_E^3} \\
&\cong 1 - \frac{\sqrt{E}}{\sqrt{2}}\Big(1 - \frac{21}{32}E\Big) - \frac{\sqrt{E}}{\sqrt{2}}\Big(1 + \frac{21}{32}E\Big) + \frac{7}{32}\sqrt{2E}^3 \cong 1 - \sqrt{2E} + \frac{7\sqrt{2}}{16}E^{\frac{3}{2}}.
\end{aligned} \tag{5.161}
$$

Thus,

$$
1 - \Big(\max_{s>0} \frac{s a_0(\frac{2}{s})}{4} - sE\Big)^2 = 1 - \Big(1 - \kappa_{\mathrm{U}(1)}(E)\Big)^2 = 1 - 2E + \frac{7}{4}E^2. \tag{5.162}
$$

Proof of Lemma 5.5 To show the convexity of κ, we prepare the following lemma.

Lemma 5.7 *Le $\mathcal{V}$ be a two-dimensional system, and A and B be two Hermitian matrices on $\mathcal{V}$. For any density matrix $\rho \in \mathcal{S}(\mathcal{V})$, there exists a normalized vector $\phi \in \mathcal{V}$ such that*

$$
\operatorname{Tr} \rho A = \langle \phi | A | \phi \rangle, \quad \operatorname{Tr} \rho B = \langle \phi | B | \phi \rangle. \tag{5.163}
$$

Proof of Lemma 5.7 We discuss only the case when A and B are not constant. Otherwise, this statement is trivial. In this case, it is enough to show the case when the eigenvalues of A and B are 1 and -1. In this case, we can choose a suitable basis such that A and B can be written as sums of σ_1 and σ_3, σ_j are Pauli matrices defined in (5.175) in the next subsection. When ρ is $\frac{1}{2}(I + a\sigma_1 + b\sigma_2 + c\sigma_3)$, we choose the pure state $|\phi\rangle\langle\phi| := \frac{1}{2}(I + a\sigma_1 + \hat{b}\sigma_2 + c\sigma_3)$ such that $a^2 + \hat{b}^2 + c^2 = 1$. Then, we have (5.163). ∎

Lemma 5.7 guarantees that

$$
\begin{aligned}
\kappa(E) &= \inf_{\phi \in \mathcal{H}: = \|\phi\| = 1} \{ \langle \phi | X | \phi \rangle | \langle \phi | Y | \phi \rangle \le E \} \\
&= \inf_{\rho} \{ \operatorname{Tr} \rho X | \operatorname{Tr} \rho Y \le E \}
\end{aligned} \tag{5.164}
$$

Since $\kappa(E)$ is decreasing and the limit $\lim_{E\to\infty} \kappa(E) \ge 0$ exists, when $\kappa(E) > \lim_{E\to\infty} \kappa(E)$, we have

$$
\kappa(E) = \inf_{\phi \in \mathcal{H}: = \|\phi\| = 1} \{ \langle \phi | X | \phi \rangle | \langle \phi | Y | \phi \rangle = E \}. \tag{5.165}
$$

Now, for simplicity, we show the convexity for κ only for the case when the minimum exists in (5.153). For two real numbers E_1 and E_2, and a real number $p \in [0, 1]$, we assume that $\kappa(E_i) > \lim_{E\to\infty} \kappa(E)$ for $i = 1, 2$. Then, we choose $\phi_1, \phi_2 \in \mathcal{H}$ such that

$$
\kappa(E_i) = \langle \phi_i | X | \phi_i \rangle, \quad \langle \phi_i | Y | \phi_i \rangle = E_i \tag{5.166}
$$

So, we have

$$\mathrm{Tr}(p|\phi_1\rangle\langle\phi_1| + (1-p)|\phi_2\rangle\langle\phi_2|)Y = pE_1 + (1-p)E_2 \tag{5.167}$$

$$\mathrm{Tr}(p|\phi_1\rangle\langle\phi_1| + (1-p)|\phi_2\rangle\langle\phi_2|)X = p\kappa(E_1) + (1-p)\kappa(E_2), \tag{5.168}$$

which implies

$$\inf_{\rho}\{\mathrm{Tr}\,\rho X | \mathrm{Tr}\,\rho Y \le E\} \le p\kappa(E_1) + (1-p)\kappa(E_2). \tag{5.169}$$

So, κ is convex. Thus, κ' monotone increasing. We choose s_E such that $-s_E = \kappa'(E)$.

Now, we fix $E_0 := \sup_E\{E|\kappa(E) > \lim_{E\to\infty}\kappa(E)\}$. So, the minimization in (5.154) is restricted to the case when $\langle\phi|Y|\phi\rangle \le E_0$. Thus, for $s > 0$, we have

$$\gamma(s) = \inf_{E\le E_0} \kappa(E) + sE = \inf_{E} \kappa(E) + sE, \tag{5.170}$$

where the second equation follows from the relation $\kappa(E) = \kappa(E_0)$ for $E > E_0$. Thus, γ is also concave. Generally, as the duality of Legendre transformation [92, Theorem 26.6], any convex function f satisfies

$$f(x) = \sup_y(\inf_{x'} f(x') + x'y) - yx. \tag{5.171}$$

So, we obtain

$$\kappa(E) = \sup_{s>0}\gamma(s) - sE. \tag{5.172}$$

Due to the concavity of $\gamma(s)$, there exists s_E such that $\gamma'(s_E) = E$. So, we have

$$\kappa(E) = \gamma(s_E) - s_E E = \max_{s>0}\gamma(s) - sE. \tag{5.173}$$

Since γ' is monotone decreasing, s_E is positive and monotone decreasing with respect to E. ■

5.6.4 Uncertainty Relation on S^3*

Next, we consider the relation between the above results and the uncertainty relation on wave functions on the three-dimensional sphere $L^2(S^3)$. For this issue, we focus on the fact that the group SU(2) is isomorphic to the 3-dimensional sphere S^3 by the correspondence $g \mapsto (x_0(g), x_1(g), x_2(g), x_3(g))$, where $x_j(g)$ $(j = 0, 1, 2, 3)$ is given as $g = \begin{pmatrix} x_0 + ix_1 & x_2 + ix_3 \\ x_2 - ix_3 & x_0 - ix_1 \end{pmatrix}$ for $g \in \mathrm{SU}(2)$. That is, we have $L^2(\mathrm{SU}(2)) =$

$L^2(S^3)$. Then, we define the operator Q_j as the multiplication of $x_j(g)$. For the set of operators $\mathbf{Q} := (\mathsf{Q}_0, \mathsf{Q}_1, \mathsf{Q}_2, \mathsf{Q}_3)$, we focus on the uncertainty

$$\begin{aligned}\Delta^2_{\varphi^*}\mathbf{Q} :=& \sum_{j=0}^{3} \Delta^2_{\varphi^*}\mathsf{Q}_j = \sum_{j=0}^{3} \langle \varphi^* | \mathsf{Q}_j^2 | \varphi^* \rangle - \sum_{j=0}^{3} \langle \varphi^* | \mathsf{Q}_j | \varphi^* \rangle^2 \\ =& 1 - \sum_{i=0}^{3} \langle \varphi^* | \mathsf{Q}_j | \varphi^* \rangle^2\end{aligned}$$

for $\varphi^* \in L^2(\mathrm{SU}(2))$. Next, we define the momentum operator P_j on $L^2(\mathrm{SU}(2))$ as

$$\mathsf{P}_j \varphi^*(g) := i \frac{d\varphi^*(e^{-it\frac{\sigma_j}{2}} g)}{dt}|_{t=0}, \tag{5.174}$$

where Pauli matrices σ_j is defined as

$$\sigma_1 := \begin{pmatrix} 0 & 1 \\ 1 & 0 \end{pmatrix}, \quad \sigma_2 := \begin{pmatrix} 0 & -i \\ i & 0 \end{pmatrix}, \quad \sigma_3 := \begin{pmatrix} 1 & 0 \\ 0 & -1 \end{pmatrix}. \tag{5.175}$$

For the set of momentum operators $\mathbf{P} := (\mathsf{P}_1, \mathsf{P}_2, \mathsf{P}_3)$, we also consider the uncertainty

$$\Delta^2_{\varphi^*}\mathbf{P} := \sum_{j=1}^{3} \Delta^2_{\varphi^*}\mathsf{P}_j = \sum_{j=1}^{3} \left(\frac{\sigma_j}{2}\right)^2 = \sum_{j=1}^{3} \left(F_1^j\right)^2 = \sum_{k=0}^{\infty} \frac{k}{2}\left(\frac{k}{2}+1\right) I_{\frac{k}{2}},$$

where $I_{\frac{k}{2}}$ is the projection to the irreducible space with the highest weight $\frac{k}{2}$. Then, as the uncertainty relation, we consider the trade-off between $\Delta^2_{\varphi^*}\mathbf{Q}$ and $\Delta^2_{\varphi^*}\mathbf{P}$.

Theorem 5.5 ([49]) *The minimum of the uncertainty $\Delta^2_{\varphi^*}\mathbf{Q}$ under the constraint for $\Delta^2_{\varphi^*}\mathbf{P}$ is calculated as*

$$\min_{\varphi^* \in L^2(\mathrm{SU}(2)): \|\varphi^*\|=1} \{\Delta^2_{\varphi^*}\mathbf{Q} | \Delta^2_{\varphi^*}\mathbf{P} \le E\} = 1 - \left(\min_{s>0} s\left(E + \frac{1}{4}\right) - \frac{s b_2(\frac{8}{s})}{16} \right)^2. \tag{5.176}$$

The minimum in (5.158) is realized by φ^* if and only if there exists $g \in \mathrm{SU}(2)$ such that $\varphi^*(g\tilde{g}_{\theta,\eta_1,\eta_2}) = \frac{\mathrm{se}_2(\frac{\theta}{4}, -\frac{8}{s_E})}{\sin\frac{\theta}{2}}$, where s_E is given in Lemma 5.9.

Further, the RHS of (5.176) has the asymptotic expansion $\frac{9}{16E} - \frac{5\cdot 3^3}{2^9 E^2}$ as $E \to \infty$ and $1 - \frac{4}{3}E + \frac{8}{5}E^2$ as $E \to 0$.

Proof of Theorem 5.5 To prove Theorem 5.5, we prepare several steps. For a given normalized function $\varphi^* \in L^2(\mathrm{SU}(2))$, we can choose an element $g_0 \in \mathrm{SU}(2)$ such that $\langle\varphi^* \circ g_0|\mathsf{Q}_j|\varphi^* \circ g_0\rangle = 0$ for $j = 1, 2, 3$, where $\varphi^* \circ g_0(g) := \varphi^*(g_0 g)$. Since $\Delta^2_{\varphi^*}\mathsf{P} = \Delta^2_{\varphi^* \circ g}\mathsf{P}$, without loss of generality, we can assume that $\langle\varphi^*|\mathsf{Q}_j|\varphi^*\rangle = 0$ for $j = 1, 2, 3$. In this case, $\Delta^2_{\varphi^*}\mathsf{Q} = 1 - \langle\varphi^*|\mathsf{Q}_0|\varphi^*\rangle^2$. Thus, we obtain

$$
\begin{aligned}
&\min_{\varphi^* \in L^2(\mathrm{SU}(2)):\|\varphi^*\|=1} \{\Delta^2_{\varphi^*}\mathsf{Q} | \Delta^2_{\varphi^*}\mathsf{P} \le E\} \\
=& \min_{\varphi^* \in L^2(\mathrm{SU}(2)):\|\varphi^*\|=1} \{\Delta^2_{\varphi^*}\mathsf{Q} | \Delta^2_{\varphi^*}\mathsf{P} \le E, \langle\varphi^*|\mathsf{Q}_j|\varphi^*\rangle = 0, j = 1, 2, 3\} \\
=& \min_{\varphi^* \in L^2(\mathrm{SU}(2)):\|\varphi^*\|=1} \left\{ 1 - \langle\varphi^*|\mathsf{Q}_0|\varphi^*\rangle^2 \middle| \begin{array}{l} \Delta^2_{\varphi^*}\mathsf{P} \le E, \\ \langle\varphi^*|\mathsf{Q}_j|\varphi^*\rangle = 0, j = 1, 2, 3 \end{array} \right\} \\
=& 1 - \left(\max_{\varphi^* \in L^2(\mathrm{SU}(2)):\|\varphi^*\|=1} \left\{ \langle\varphi^*|\mathsf{Q}_0|\varphi^*\rangle \middle| \begin{array}{l} \Delta^2_{\varphi^*}\mathsf{P} \le E, \\ \langle\varphi^*|\mathsf{Q}_j|\varphi^*\rangle = 0, j = 1, 2, 3 \end{array} \right\} \right)^2 .
\end{aligned}
\tag{5.177}
$$

Now, we consider another maximum

$$
\max_{\varphi^* \in L^2(\mathrm{SU}(2)):\|\varphi^*\|=1} \{\langle\varphi^*|\mathsf{Q}_0|\varphi^*\rangle | \Delta^2_{\varphi^*}\mathsf{P} \le E\}. \tag{5.178}
$$

This maximum can be attained when $\langle\varphi^*|\mathsf{Q}_j|\varphi^*\rangle = 0$ for $j = 1, 2, 3$ due to the following reason. If not, we can choose g_0 such that $\langle\varphi^* \circ g_0|\mathsf{Q}_0|\varphi^* \circ g_0\rangle > \langle\varphi^*|\mathsf{Q}_0|\varphi^*\rangle$ and $\langle\varphi^* \circ g_0|\mathsf{Q}_j|\varphi^* \circ g_0\rangle = 0$ for $j = 1, 2, 3$. Thus, we have

$$
\begin{aligned}
&\max_{\varphi^* \in L^2(\mathrm{SU}(2)):\|\varphi^*\|=1} \{\langle\varphi^*|\mathsf{Q}_0|\varphi^*\rangle | \Delta^2_{\varphi^*}\mathsf{P} \le E, \langle\varphi^*|\mathsf{Q}_j|\varphi^*\rangle = 0, j = 1, 2, 3\} \\
=& \max_{\varphi^* \in L^2(\mathrm{SU}(2)):\|\varphi^*\|=1} \{\langle\varphi^*|\mathsf{Q}_0|\varphi^*\rangle | \Delta^2_{\varphi^*}\mathsf{P} \le E\}.
\end{aligned}
\tag{5.179}
$$

To discuss this maximum, we employ a parametrization of SU(2) as

$$
\tilde{g}_{\theta,\eta_1,\eta_2} := \begin{pmatrix} x_0 + i x_1 & x_2 + i x_3 \\ x_2 - i x_3 & x_0 - i x_1 \end{pmatrix}, \tag{5.180}
$$

where $x_0 = \cos\frac{\theta}{2}$, $x_1 = \sin\frac{\theta}{2}\cos\eta_1$, $x_2 = \sin\frac{\theta}{2}\sin\eta_1\cos\eta_2$, $x_3 = \sin\frac{\theta}{2}\sin\eta_1\sin\eta_2$ with the range $\theta \in (-2\pi, 2\pi]$, $\eta_1 \in (-\frac{\pi}{2}, \frac{\pi}{2}]$, $\eta_2 \in (-\frac{\pi}{2}, \frac{\pi}{2}]$.

Then, we employ the following lemma.

Lemma 5.8 *We consider the Fourier transform of a normalized wave function $\varphi^* \in L^2(\mathrm{SU}(2))$ as $\mathcal{F}[\varphi^*] = \oplus_{k=0}^{\infty}\beta_{\frac{k}{2}}\Phi_{\frac{k}{2}}$, where $\beta_{\frac{k}{2}} \ge 0$ and $\|\Phi_{\frac{k}{2}}\| = 1$. Then,*

$$
\begin{aligned}
\langle\varphi^*|\mathsf{Q}_0|\varphi^*\rangle =& \int_{\mathrm{SU}(2)} \frac{1}{2}\chi_{\frac{1}{2}}(g)|\varphi^*(g)|^2 \mu_{\mathrm{SU}(2)}(dg) \\
\le& \frac{1}{2}\sum_{k,k'=0}^{\infty} |\beta_{\frac{k}{2}}\beta_{\frac{k'}{2}}| = \int_{-2\pi}^{2\pi} \cos\frac{\theta}{2}|\varphi(\theta)|^2 \frac{d\theta}{2\pi},
\end{aligned}
\tag{5.181}
$$

where $\varphi(\theta)$ *is defined as the odd function* $\sum_{k=0}^{\infty}\beta_{\frac{k}{2}}\sin\frac{k+1}{2}\theta$.

When $\varphi^*(\tilde{g}_{\theta,\eta_1,\eta_2})\sin\frac{\theta}{2}$ *is written as* $\sum_{k=0}^{\infty}\beta_{\frac{k}{2}}\sin\frac{k+1}{2}\theta$ *with* $\beta_{\frac{k}{2}}\ge 0$, *the equality holds. This condition is equivalent to the condition that* $\Phi_{\frac{k}{2}}$ *is* $|\frac{1}{\sqrt{k+1}}I_{\frac{k}{2}}\rangle\rangle$.

This lemma will be shown in the end of this subsection. When φ^* is written as above way,

$$\Delta^2_{\varphi^*}\mathsf{P}=\sum_{k=0}^{\infty}\frac{k}{2}\Big(\frac{k}{2}+1\Big)\beta^2_{\frac{k}{2}}=\langle\varphi|\mathsf{P}^2|\varphi\rangle-\frac{1}{4}. \tag{5.182}$$

Since the function φ with the above form is an odd periodic function with period 4π, we have

$$\begin{aligned}&\max_{\varphi^*\in L^2(\mathrm{SU}(2)):\|\varphi^*\|=1}\{\langle\varphi^*|\mathsf{Q}_0|\varphi^*\rangle\,|\,\Delta^2_{\varphi^*}\mathsf{P}\le E\}\\ =&\max_{\varphi\in L^2_{p,\mathrm{odd}}([-2\pi,2\pi]):\|\varphi\|=1}\Big\{\langle\varphi|\cos\frac{\mathsf{Q}}{2}|\varphi\rangle\Big|\langle\varphi|\mathsf{P}^2|\varphi\rangle\le E+\frac{1}{4}\Big\}.\end{aligned} \tag{5.183}$$

To calculate the RHS of (5.183), we consider the minimum

$$\begin{aligned}\gamma_{\mathrm{SU}(2)}(s):=&\min_{\varphi\in L^2_{p,\mathrm{odd}}((-2\pi,2\pi]):\|\varphi\|=1}\Big\langle\varphi\Big|(I-\cos\frac{\mathsf{Q}}{2})+s\mathsf{P}^2\Big|\varphi\Big\rangle\\ =&\min_{\varphi\in L^2_{p,\mathrm{odd}}((-\pi/2,\pi/2]):\|\varphi\|=1}\Big\langle\varphi\Big|(I-\cos(2\mathsf{Q}))+\frac{s\mathsf{P}^2}{16}\Big|\varphi\Big\rangle.\end{aligned}$$

So, $\gamma_{\mathrm{SU}(2)}(s)$ can be characterized as the minimum value of $\gamma_{\mathrm{SU}(2)}$ having the solution in $L^2_{p,\mathrm{odd}}((-\pi/2,\pi/2])$ of the following differential equation.

$$\frac{s}{16}\frac{d^2}{d\theta^2}\varphi(\theta)+(\gamma_{\mathrm{SU}(2)}-1+\cos(2\theta))\varphi(\theta)=0, \tag{5.184}$$

which is equivalent to

$$\frac{d^2}{d\theta^2}\varphi(\theta)+\Big(\frac{16(\gamma_{\mathrm{SU}(2)}-1)}{s}+\frac{16}{s}\cos(2\theta)\Big)\varphi(\theta)=0. \tag{5.185}$$

In order to find the minimum $\gamma_{\mathrm{SU}(2)}(s)$, we employ Mathieu equation (5.145), whose detail is summarized in Sect. 5.6.2. Hence, using the function b_2 given in Sect. 5.6.2, we have $\gamma_{\mathrm{SU}(2)}(s)=\frac{sb_2(-\frac{8}{s})}{16}+1=\frac{sb_2(\frac{8}{s})}{16}+1$. So, applying Lemma 5.5 to the case when $X=I-\cos(\mathsf{Q})$ and $Y=\mathsf{P}^2$, we obtain the following lemma.

Lemma 5.9

$$\max_{\varphi \in L^2_{p,\mathrm{odd}}((-2\pi,2\pi]):\|\varphi\|=1} \left\{ \left\langle \varphi \left| I - \cos\frac{\mathsf{Q}}{2} \right| \varphi \right\rangle \middle| \langle \varphi | \mathsf{P}^2 | \varphi \rangle \le E + \frac{1}{4} \right\} \tag{5.186}$$

$$= \max_{s>0} \frac{s b_2\left(\frac{8}{s}\right)}{16} + 1 - s\left(E + \frac{1}{4}\right). \tag{5.187}$$

The minimum (5.186) is attained and $\langle \varphi | \mathsf{Q}_j | \varphi \rangle = 0$ *for* $j = 1, 2, 3$ *if and only if* $\varphi(\tilde{\rho}_{\theta,\eta_1,\eta_2}) = \frac{\mathrm{se}_2(\frac{\theta}{4}, -\frac{8}{s_E})}{\sin\frac{\theta}{2}}$*, where* s_E *is* $\operatorname{argmax}_{s>0} \frac{s b_2(\frac{8}{s})}{16} + 1 - s(E + \frac{1}{4})$ *and Mathieu function* se_2 *is given in Sect. 5.6.2.*

This lemma shows that

$$\min_{\varphi \in L^2_{p,\mathrm{odd}}((-2\pi,2\pi]):\|\varphi\|=1} \left\{ \langle \varphi | \cos(\frac{\mathsf{Q}}{2}) | \varphi \rangle | \langle \varphi | \mathsf{P}^2 | \varphi \rangle \le E + \frac{1}{4} \right\}$$

$$= \min_{s>0} s\big(E + \frac{1}{4}\big) - \frac{s b_2(\frac{8}{s})}{16} \ge 1. \tag{5.188}$$

The equality condition is the same as Lemma 5.9. Combining (5.177), (5.179), (5.183) and (5.188), we obtain (5.176).

Next, we give the asymptotic expansion. By using the expansion (5.148) for b_2, as $s \to 0$, $\gamma_{\mathrm{SU}(2)}(s)$ can be expanded to

$$\gamma_{\mathrm{SU}(2)}(s) \cong \frac{s(-2\frac{8}{s} + 6\sqrt{\frac{8}{s}} - \frac{5}{4})}{16} + 1 = \frac{3}{2}\sqrt{\frac{s}{2}} - \frac{5s}{64}.$$

As is shown in Lemma 5.5, s_E is decreasing as a function of E. Hence, when E is large, solving the equation $\gamma'_{\mathrm{SU}(2)}(s_E) = E + \frac{1}{4}$, we approximately obtain $s_E \cong 2 \cdot (\frac{3}{8E+21/8})^2$. Hence,

$$\kappa_{\mathrm{SU}(2)}(E) = \gamma_{\mathrm{SU}(2)}(s_E) - s_E\left(E + \frac{1}{4}\right) \cong \frac{3}{2\sqrt{2}}\sqrt{s_E} - \frac{5 s_E}{64} - s_E\left(E + \frac{1}{4}\right)$$

$$= \frac{3}{2\sqrt{2}}\sqrt{s_E} - s_E\left(\frac{21}{64} + E\right) \cong \frac{9}{2(8E + 21/8)} - 2 \cdot \left(\frac{3}{8E + 21/8}\right)^2 \frac{(\frac{21}{8} + 8E)}{8}$$

$$= \frac{9}{2(8E + 21/8)} - \frac{9}{4(8E + 21/8)} = \frac{9}{4(8E + 21/8)} \cong \frac{9}{32E} - \frac{7 \cdot 3^3}{2^{11} E^2}. \tag{5.189}$$

Thus, we have

$$1-\left(\min_{s>0} s\left(E+\frac{1}{4}\right)-\frac{sb_2(\frac{8}{s})}{16}\right)^2 \cong 1-\left(1-\left(\frac{9}{32E}-\frac{7\cdot 3^3}{2^{11}E^2}\right)\right)^2$$
$$\cong \frac{9}{16E}-\frac{5\cdot 3^3}{2^9E^2}, \tag{5.190}$$

which is the asymptotic expansion for the RHS of (5.176) when E approaches to infinity.

Next, we consider the case when E is small. By using the expansion (5.150) for b_2, when s is large, $\gamma_{\mathrm{SU}(2)}(s)$ can be expanded to

$$\gamma_{\mathrm{SU}(2)}(s) \cong \frac{s(4-\frac{1}{12}(\frac{8}{s})^2+\frac{5}{13824}(\frac{8}{s})^4)}{16}+1=\frac{s}{4}+1-\frac{1}{3s}+\frac{5}{54s^3}.$$

When E is small, since $\gamma'_{\mathrm{SU}(2)}(s) \cong \frac{1}{4}+\frac{1}{3s^2}-\frac{5}{18s^4}$ solving the equation $\gamma'_{\mathrm{SU}(2)}(s_E)=E+\frac{1}{4}$, we approximately obtain $s_E \cong \sqrt{\frac{1}{3E}+\frac{5}{6}} \cong \sqrt{\frac{1}{3E}}(1+\frac{5}{4}E)$. Hence,

$$\kappa_{\mathrm{SU}(2)}(E)=\gamma_{\mathrm{SU}(2)}(s_E)-s_E\left(E+\frac{1}{4}\right)$$
$$\cong \frac{s_E}{4}+1-\frac{1}{3s_E}+\frac{5}{54s_E^3}-s_E\left(E+\frac{1}{4}\right)=1+\frac{1}{3s_E}\left(\frac{5-18s_E^2}{18s_E^2}\right)-s_EE$$
$$=1+\frac{1}{3\left(\sqrt{\frac{1}{3E}}\left(1+\frac{5}{4}E\right)\right)}\left(\frac{5-18(\frac{1}{3E}+\frac{5}{6})}{18(\frac{1}{3E}+\frac{5}{6})}\right)-\left(\sqrt{\frac{1}{3E}}\left(1+\frac{5}{4}E\right)\right)E$$
$$=1-\frac{\sqrt{E}}{\sqrt{3}\left(1+\frac{5}{4}E\right)}\left(\frac{1+\frac{5E}{3}}{1+\frac{5E}{2}}\right)-\left(\sqrt{\frac{E}{3}}\left(1+\frac{5}{4}E\right)\right)$$
$$\cong 1-\frac{\sqrt{E}}{\sqrt{3}}\left(1-\frac{5}{4}E+\frac{5E}{3}-\frac{5E}{2}\right)-\left(\sqrt{\frac{E}{3}}\left(1+\frac{5}{4}E\right)\right)$$
$$=1-\frac{\sqrt{E}}{\sqrt{3}}\left(1-\frac{25}{12}E\right)-\sqrt{\frac{E}{3}}\left(1+\frac{5}{4}E\right)=1-\frac{2}{\sqrt{3}}\sqrt{E}+\frac{5}{6\sqrt{3}}E^{\frac{3}{2}} \tag{5.191}$$

That is,

$$1-\kappa_{\mathrm{SU}(2)}(E) \cong \frac{2}{\sqrt{3}}\sqrt{E}-\frac{5}{6\sqrt{3}}E^{\frac{3}{2}}. \tag{5.192}$$

Thus, we obtain the asymptotic expansion for the RHS of (5.176) with $E \to 0$ as

$$1-(1-\kappa_{\mathrm{SU}(2)}(E))^2 \cong 1-\frac{4}{3}E+\frac{8}{5}E^2. \tag{5.193}$$

Now, our Proof of Theorem 5.5 is completed. ■

Proof of Lemma 5.8 Since $x_0(g) = \frac{1}{2}\chi_{\frac{1}{2}}(g)$, we have

$$
\begin{aligned}
\langle \varphi^* | \mathsf{Q}_0 | \varphi^* \rangle &= \int_{\mathrm{SU}(2)} \mu_{\mathrm{SU}(2)}(dg) \frac{\chi_{\frac{1}{2}}(g)}{2} |\varphi^*(g)|^2 \\
&= \int_{\mathrm{SU}(2)} \mu_{\mathrm{SU}(2)}(dg) \frac{\chi_{\frac{1}{2}}(g)}{2} \Big| \sum_{\lambda \in \hat{G}[\mathcal{L}]} \operatorname{Tr} \mathsf{f}_{\frac{k}{2}}(g) \beta_{\frac{k}{2}} d_{\frac{k}{2}} \Phi_{\frac{k}{2}} \Big|^2 \\
&= \int_{\mathrm{SU}(2)} \mu_{\mathrm{SU}(2)}(dg) \frac{\chi_{\frac{1}{2}}(g)}{2} \\
&\quad \cdot \sum_{\lambda, \lambda' \in \hat{G}[\mathcal{L}]} \beta_{\frac{k}{2}} \beta_{\frac{k'}{2}} \operatorname{Tr} \mathsf{f}_{\lambda}(g) \otimes \mathsf{f}_{\lambda'^*}(g) (k+1) \Phi_{\frac{k}{2}} \otimes (k'+1) \Phi_{\frac{k'}{2}}^{\dagger} \\
&= \sum_{k,k'} \beta_{\frac{k}{2}} \beta_{\frac{k'}{2}} \sqrt{(k+1)(k'+1)} \\
&\quad \cdot \operatorname{Tr} \Big[\int_{\mathrm{SU}(2)} \mu_{\mathrm{SU}(2)}(dg) \frac{\chi_{\frac{1}{2}}(g)}{2} \mathsf{f}_{\lambda}(g) \otimes \mathsf{f}_{\lambda'^*}(g) \Big] \sqrt{k+1} \Phi_{\frac{k}{2}} \otimes \sqrt{k'+1} \Phi_{\frac{k'}{2}}^{\dagger} \\
&= \sum_{k,k'} \beta_{\frac{k}{2}} \beta_{\frac{k'}{2}} \sqrt{(k+1)(k'+1)} \operatorname{Tr} \Xi_{k,k'} \sqrt{k+1} \Phi_{\frac{k}{2}} \otimes \sqrt{k'+1} \Phi_{\frac{k'}{2}}^{\dagger}
\end{aligned}
\tag{5.194}
$$

hold, where we use the matrix

$$
\Xi_{k,k'} := \int_G -\frac{1}{2} \chi_{\frac{1}{2}}(g) \mathsf{f}_{\lambda}(g) \otimes \mathsf{f}_{\lambda'^*}(g) \mu_G(dg) = \begin{cases} \frac{1}{4} I_{\frac{1}{2}} & \text{when} |k - k'| = \frac{1}{2} \\ 0 & \text{otherwise.} \end{cases}
\tag{5.195}
$$

The last equation follows from (4.22) and the formula $\int_{\mathrm{SU}(2)} \chi_{\frac{k}{2}}(g) \mathsf{f}_{\frac{k'}{2}}(g) \mu_{\mathrm{SU}(2)}(dg) = \int_{\mathrm{SU}(2)} \overline{\chi_{\frac{k}{2}}(g)} \mathsf{f}_{\frac{k'}{2}}(g^{-1}) \mu_{\mathrm{SU}(2)}(dg) = \frac{\delta_{\frac{k}{2}, \frac{k'}{2}}}{k+1} I_{\frac{k}{2}}$, which is a special case of (3.77).

Hence, applying Schwarz inequality with respect to the inner product $\langle A, B \rangle := \operatorname{Tr} \Xi_{k,k'} A^{\dagger} B$, to the case $A := I_{\frac{k}{2}} \otimes \sqrt{k'+1} \Phi_{\frac{k'}{2}}$, $B := \sqrt{k+1} \Phi_{\frac{k}{2}} \otimes I_{\frac{k'}{2}}$, we obtain

$$
\begin{aligned}
&\operatorname{Tr} \Xi_{k,k'} \sqrt{k+1} \Phi_{\frac{k}{2}} \otimes \sqrt{k'+1} \Phi_{\frac{k'}{2}}^{\dagger} \\
\le &\sqrt{\operatorname{Tr} \Xi_{k,k'} (k+1) \Phi_{\frac{k}{2}}^{\dagger} \Phi_{\frac{k}{2}} \otimes I_{k'}} \sqrt{\operatorname{Tr} \Xi_{k,k'} I_{\lambda} \otimes (k'+1) \Phi_{\frac{k'}{2}}^{\dagger} \Phi_{\frac{k'}{2}}}.
\end{aligned}
\tag{5.196}
$$

Since $\Xi_{k,k'}$ is invariant with respect to the action of G,

$$
\begin{aligned}
&\mathrm{Tr}\, \Xi_{k,k'}(k+1)\Phi_{\frac{k}{2}}^{\dagger}\Phi_{\frac{k}{2}} \otimes I_{\frac{k'}{2}} \\
&= \mathrm{Tr}\, \Xi_{k,k'} \int_{\mathrm{SU}(2)} (\mathsf{f}_{\lambda}(g) \otimes \mathsf{f}_{\frac{k'}{2}}(g))(k+1)(\Phi_{\frac{k}{2}}^{\dagger}\Phi_{\frac{k}{2}} \otimes I_{\frac{k'}{2}}) \\
&\qquad \cdot (\mathsf{f}_{\frac{k}{2}}(g) \otimes \mathsf{f}_{\frac{k'}{2}}(g))^{\dagger} \mu_{\mathrm{SU}(2)}(dg) \\
&= \mathrm{Tr}\, \Xi_{k,k'} I_{\frac{k}{2}} \otimes I_{\frac{k'}{2}} = \begin{cases} \frac{1}{2} \text{ when} |k-k'| = \frac{1}{2} \\ 0 \text{ otherwise,} \end{cases}
\end{aligned} \tag{5.197}
$$

where we used the condition $\mathrm{Tr}\, \Phi_{\frac{k}{2}}^{\dagger}\Phi_{\frac{k}{2}} = 1$.

Similarly, we have

$$
\mathrm{Tr}\, \Xi_{k,k'} I_{\frac{k}{2}} \otimes d_{\frac{k'}{2}} \Phi_{\frac{k'}{2}}^{\dagger}\Phi_{\frac{k'}{2}} = \begin{cases} \frac{1}{2} \text{ when} |k-k'| = \frac{1}{2} \\ 0 \text{ otherwise.} \end{cases} \tag{5.198}
$$

Combining (5.196)–(5.198), we have

$$
\mathrm{Tr}\, \Xi_{k,k'} \sqrt{k+1}\Phi_{\frac{k}{2}} \otimes \sqrt{k'+1}\Phi_{\frac{k'}{2}}^{\dagger} \le \begin{cases} \frac{1}{2} \text{ when } |k-k'| = \frac{1}{2} \\ 0 \text{ otherwise.} \end{cases}
$$

Hence, combining the above relation with (5.194), we obtain (5.181). Due to the equality condition for Schwarz inequality, the equality in (5.181) holds when $\mathcal{F}[\varphi^*] = \oplus_{k=0}^{\infty} \beta_{\frac{k}{2}} | \frac{1}{\sqrt{k+1}} I_{\frac{k}{2}} \rangle\rangle$.

When $\varphi^*(\tilde{g}_{\theta,\eta_1,\eta_2}) \sin\frac{\theta}{2}$ is written as $\sum_{k=0}^{\infty} \beta_{\frac{k}{2}} \sin\frac{k+1}{2}\theta$ with $\beta_{\frac{k}{2}} \ge 0$, we have

$$
\begin{aligned}
&\int_{\mathrm{SU}(2)} x_0(g) |\varphi^*(g)|^2 \mu_{\mathrm{SU}(2)}(dg) = \int_{-2\pi}^{2\pi} \cos\frac{\theta}{2} |\varphi^*(\tilde{g}_{\theta,\eta_1,\eta_2})|^2 \sin^2\frac{\theta}{2}\frac{d\theta}{2\pi} \\
&= \int_{-2\pi}^{2\pi} \cos\frac{\theta}{2} \Big| \sum_{k=0}^{\infty} \beta_{\frac{k}{2}} \sin\frac{k+1}{2}\theta \Big|^2 \frac{d\theta}{2\pi} = \int_{-2\pi}^{2\pi} \cos\frac{\theta}{2} |\varphi(\theta)|^2 \frac{d\theta}{2\pi}.
\end{aligned} \tag{5.199}
$$

■

Chapter 6
Representation of General Lie Groups and General Lie Algebras

Abstract Based on root systems, this chapter deals with representations of a Lie group via those of the corresponding Lie algebra. This chapter discusses a non-compact Lie group as well as a compact Lie group. To discuss both, we focus on the relation between a (real) Lie algebra and a complex Lie algebra. Since representations of a non-compact group are very complicated, this chapter treats only a part of its representations that are related to quantum theory. As such special representation has analogies with representations of a compact Lie group, they can be more easily understood than the general case. Since this chapter is composed of very advanced topics and such sections are labeled with *, the reader can omit this chapter in the first time.

6.1 Root System and Its Classification*

The previous section discusses specific types of Lie groups and Lie algebras. This section deals with Lie groups and Lie algebras in more general forms. In Sect. 4.3, we have discussed an irreducible skew-Hermitian representation of $\mathfrak{su}(r)$ based on a root system. In the following, based on a root system, we discuss compact semi simple Lie algebras in the general setting. For this purpose, we give the definition of a root system in the general form, and discuss its properties.

Definition 6.1 A subset Φ of a real linear space V with a positive definite inner product $(\cdot, \cdot)$ is called a **root system** when the following conditions hold.

(R1) $|\Phi| < \infty$ and the set Φ does not contain 0 and spans V.
(R2) For an element $\alpha \in \Phi$, the element $-\alpha$ is contained in $\in \Phi$, but $\pm 2\alpha$, $\pm 3\alpha, \ldots$ are not contained in Φ.
(R3) We define the bilinear form $\langle \beta, \gamma \rangle := \frac{2(\beta,\gamma)}{(\gamma,\gamma)}$ for two elements β and γ of $\mathfrak{h}_0^*$. Then, $\langle \alpha, \alpha' \rangle$ is an integer for any elements $\alpha, \alpha' \in \Phi$.
(R4) The map $W_\alpha(\beta) := \beta - \langle \beta, \alpha \rangle \alpha$ is closed with respect to Φ.

Here, we should remark that $\langle \beta, \gamma \rangle$ is not necessarily symmetric. For a root system Φ, the set $\mathcal{W} := \{W_\alpha\}_{\alpha \in \Phi}$ of maps forms a group, and is called the **Weyl group**.

The following theorem is known for a root system (Exercise 6.4).

M. Hayashi, *Group Representation for Quantum Theory*,
DOI 10.1007/978-3-319-44906-7_6

Theorem 6.1 *For a root system Φ, there exists a subset $\Delta = \{\alpha_1, \dots, \alpha_r\}$ satisfying the following conditions. The subset is called a* **fundamental system** *or a* **root basis***.*

(B1) *Δ forms a basis of the linear space V.*
(B2) *For any element $\alpha \in \Phi$, we have $\alpha = \sum_{j=1}^{r} m_j \alpha_j$, where all of m_j are positive integers or negative integers.*

Definition 6.2 For a root system $\Phi \subset V$ and its fundamental system $\Delta = \{\alpha_1, \dots, \alpha_r\}$, an element of Δ is called a **simple root**. An element of Φ is called a **positive root(negative root)** when it is a linear sum of Δ with positive (negative) coefficients. Their set is denoted by $\Phi^+(\Delta)(\Phi^-(\Delta))$. When there is no possibility for confusion, we omit (Δ).

The vector expression $[\langle\beta, \alpha_1\rangle, \dots, \langle\beta, \alpha_r\rangle]$ of $\beta \in V$ is called **Dynkin index** (under the fundamental system Δ). For two vectors β and γ, we write $\beta \geq \gamma$ when the relation $\beta - \gamma = \sum_{j=1}^{r} c_j \alpha_j$ holds with $c_1, \dots, c_r \geq 0$. The relation $\geq$ satisfies the condition for **partial order**.

Then, we have the following proposition. (Exercise 6.4)

Proposition 6.1 *For a given root system $\Phi \subset V$, we choose an element $\beta \in V$ satisfying $(\alpha, \beta) \neq 0$ for $\alpha \in \Phi$. Then, we can choose a fundamental system Δ satisfying*

$$\Phi^+ = \{\alpha \in \Phi | (\alpha, \beta) > 0\}. \tag{6.1}$$

Conversely, given a fundamental system Δ, there exists an element $\beta \in V$ satisfying (6.1).

For a fundamental system $\Delta = \{\alpha_j\}_{j=1}^{r}$ of a root system Φ, the matrix $M := (\langle\alpha_l, \alpha_{l'}\rangle)$ is called **Cartan matrix**. The Dynkin index of the j-th element α_j of the fundamental system is given as j-th row vector in Cartan matrix. Then, Conditions **(R1-4)** guarantees that the real number $\langle\alpha, \alpha'\rangle$ takes one of $0, \pm 1, \pm 2$, and ± 3 for any two elements $\alpha, \alpha' \in \Phi$ (Exercise 6.1). In particular, when $\alpha \neq \alpha'$ are elements of the fundamental system, since the relation $(\alpha, \alpha') \leq 0$ is known (Exercise 6.5), the real number $\langle\alpha, \alpha'\rangle$ takes one of $0, -1, -2, -3$. Hence, any diagonal entry of Cartan matrix is 2, and its non-diagonal entry takes one of $0, -1, -2, -3$.

In the following, we study Cartan matrix. When the real number $\langle\alpha_l, \alpha_{l'}\rangle$ is -2 or -3 for $l \neq l'$, it is known that $\langle\alpha_{l'}, \alpha_l\rangle$ is -1 (Exercise 6.1). For the classification of Cartan matrix, we connect each point α_l based on the following rule. When $\langle\alpha_l, \alpha_{l'}\rangle = \langle\alpha_{l'}, \alpha_l\rangle = -1$, we connect α_l and $\alpha_{l'}$ with one line. When $\langle\alpha_l, \alpha_{l'}\rangle = -2$ and $\langle\alpha_{l'}, \alpha_l\rangle = -1$, we connect α_l and $\alpha_{l'}$ with two lines, and write the inequality $>$ on the two lines to clarify that l is larger. This condition implies that $(\alpha_l, \alpha_l) > (\alpha_{l'}, \alpha_{l'})$. When $\langle\alpha_l, \alpha_{l'}\rangle = -3$ and $\langle\alpha_{l'}, \alpha_l\rangle = -1$, we connect α_l and $\alpha_{l'}$ with three lines, and write the inequality $>$ on the three lines to clarify that l is larger. This diagram is called a **Dynkin diagram**.

In particular, when the all points are connected via other points, the Dynkin diagram is called **connected**. When the Dynkin diagram corresponding to a root system Φ is connected, the root system Φ is called connected.

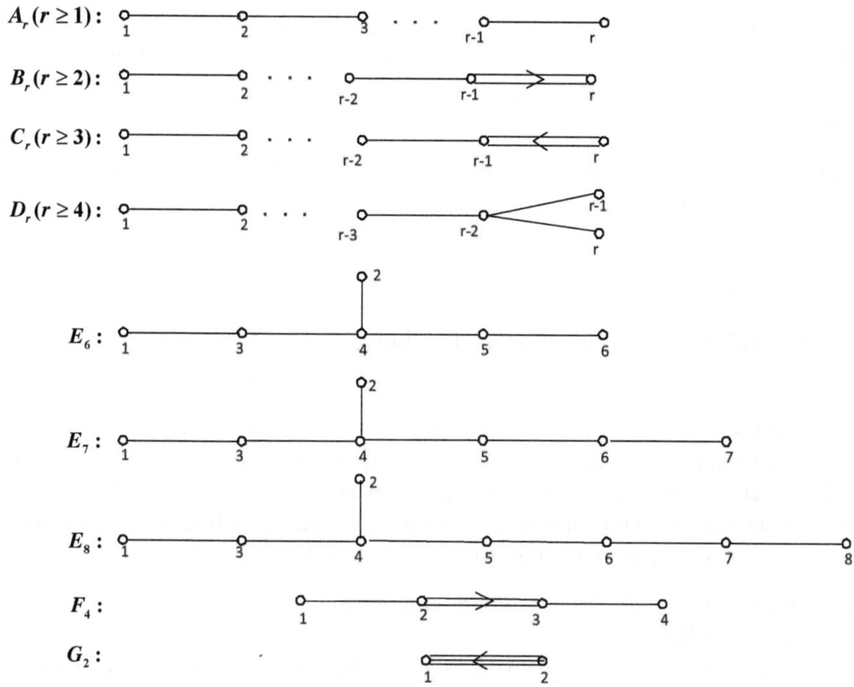

Fig. 6.1 Dynkin diagram: One line between l and l' means $\langle \alpha_l, \alpha_{l'} \rangle = \langle \alpha_{l'}, \alpha_l \rangle = -1$. Two lines between l and l' mean $\langle \alpha_l, \alpha_{l'} \rangle = -2$ and $\langle \alpha_{l'}, \alpha_l \rangle = -1$, in which, we write the inequality $>$ on the two lines because l is larger. Three lines between l and l' mean $\langle \alpha_l, \alpha_{l'} \rangle = -3$ and $\langle \alpha_{l'}, \alpha_l \rangle = -1$, in which, we write the inequality $>$ on the three lines because l is larger

Proposition 6.2 *There exists a root $\boldsymbol{\lambda}(\Phi)$ in a connected root system Φ such that $\boldsymbol{\lambda}(\Phi) \geq \boldsymbol{\alpha}$ for any root $\boldsymbol{\alpha} \in \Phi$. Such a root $\boldsymbol{\lambda}(\Phi) \in \Phi$ is called the* ***maximum root*** *of Φ.*

For its proof, see Exercise 6.6. A connected Dynkin diagram is limited to one of A_r, B_r, C_r, D_r, E_6, E_7, E_8, F_4, and G_2, which are given in Fig. 6.1 [60, Theorem in Sect. 11.4]. Especially, as shown in Fig. 6.1, we have $A_1 = B_1 = C_1$, $B_2 = C_2$, and $A_3 = D_3$.

Cartan matrix $M = (\langle \alpha_l, \alpha_{l'} \rangle)$ of A_r is the matrix given in (4.41). When $\boldsymbol{\lambda}_l := M^{-1}\boldsymbol{\alpha}_l$, we have $\langle \boldsymbol{\lambda}_l, \boldsymbol{\alpha}_{l'} \rangle = \delta_{l,l'}$, which implies the relations $\boldsymbol{\beta} = \sum_{l=1}^{r} \langle \boldsymbol{\lambda}_l, \boldsymbol{\beta} \rangle \boldsymbol{\alpha}_l$ and $\boldsymbol{\beta} = \sum_{l=1}^{r} \langle \boldsymbol{\beta}, \boldsymbol{\alpha}_l \rangle \boldsymbol{\lambda}_l$. Especially, given two elements $\boldsymbol{\beta}$ and $\boldsymbol{\gamma}$ in $\mathfrak{h}$, we write $\boldsymbol{\beta} \geq \boldsymbol{\gamma}$ when the relation $\langle \boldsymbol{\lambda}_l, \boldsymbol{\beta} \rangle \geq \langle \boldsymbol{\lambda}_l, \boldsymbol{\gamma} \rangle$ holds for $l = 1, \ldots, r$. Then, we define two sets $\Lambda(\Delta)$ and $\Lambda_+(\Delta)$ as follows.

$$\Lambda(\Delta) := \{\boldsymbol{\beta} \in \mathfrak{h}^* \,|\langle \boldsymbol{\beta}, \boldsymbol{\alpha}_l \rangle \in \mathbb{Z}, \forall \boldsymbol{\alpha}_l \in \Delta\}$$
$$\Lambda_+(\Delta) := \{\boldsymbol{\beta} \in \Lambda(\Delta) | \langle \boldsymbol{\beta}, \boldsymbol{\alpha}_l \rangle \geq 0, \forall \boldsymbol{\alpha}_l \in \Delta\}. \tag{6.2}$$

Exercise 6.1 ([60, Table 1 of Sect. 9.4]) Show that the pair $(\langle\alpha, \beta\rangle, \langle\beta, \alpha\rangle)$ is limited to $(0, 0)$, $(\pm 1, \pm 1)$, $(\pm 1, \pm 2)$, $(\pm 1, \pm 3)$, and $(\pm 2, \pm 2)$ for $\alpha, \beta \in \Phi$ with double sign in same order when $\|\alpha\| \leq \|\beta\|$. Show that the pair $(\pm 2, \pm 2)$ is limited to the case when $\alpha = \pm\beta$.

Exercise 6.2 Assume that $\alpha, \beta \in \Phi$ satisfy $\alpha \neq \pm\beta$. Show the following items. (Use Exercise 6.1.)

a: When $(\alpha, \beta) > 0$, $\alpha - \beta \in \Phi$.
b: When $(\alpha, \beta) < 0$, $\alpha + \beta \in \Phi$.

Exercise 6.3 Let r be the dimension of V. Let $V^+(\gamma) := \{\gamma' \in V | (\gamma, \gamma') > 0\}$. For any r vectors $\gamma_1, \ldots, \gamma_r$, show that $\cap_{j=1}^r V^+(\gamma_j)$ is not empty.

Exercise 6.4 ([60, Sect. 10.1]) For a vector $\gamma \in V$, we define $P_\gamma := \{\gamma' \in V | (\gamma, \gamma') = 0\}$ and $\Phi^+(\gamma) := \{\alpha \in \Phi | (\gamma, \alpha) > 0\}$. For a vector $\gamma \in V \setminus \cup_{\alpha\in\Phi} P_\alpha$, we call $\alpha \in \Phi^+(\gamma)$ decomposable when there exist $\beta_1, \beta_2 \in \Phi^+(\gamma)$ such that $\alpha = \beta_1 + \beta_2$, indecomposable otherwise, and we denote the set of indecomposable roots by $\Delta(\gamma)$. Show the following items for a vector $\gamma \in V \setminus \cup_{\alpha\in\Phi} P_\alpha$.

a: Any element $\alpha \in \Phi^+(\gamma)$ is written as non-negative $\mathbb{Z}$-linear combination of $\Delta(\gamma)$. (Use contradiction.)
b: When $\alpha, \beta \in \Delta(\gamma)$, $(\alpha, \beta) \leq 0$ or $\alpha = \beta$. (Use contradiction and Exercise 6.2.)
c: $\Delta(\gamma)$ is a liner independent set. (Use **b** and contradiction.)
d: $\Delta(\gamma)$ is a base of Φ.
e: When Δ is a base of Φ, there exists a vector $\gamma' \in V \setminus \cup_{\alpha\in\Phi} P_\alpha$ such that $\Delta = \Delta(\gamma')$ and $\Phi^+ = \Phi^+(\gamma')$. (Use Exercise 6.3.)

Exercise 6.5 ([60, Sect. 10.1]) Assume that Δ is a fundamental system of Φ and $\alpha \neq \beta \in \Delta$. Show that $(\alpha, \beta) \leq 0$ and $\alpha - \beta$ is not a root. (Use contradiction and Exercise 6.2.)

Exercise 6.6 ([60, Lemma A in Sect. 10.4]) Assume that Φ is a connected root system. Show Proposition 6.1 by showing the following items for $\beta, \beta' \in \Phi$.

a: When there is no element $\alpha \in \Phi$ such that $\alpha \geq \beta$, we have $\beta \in \Phi^+$.
b: When there is no element $\alpha \in \Phi$ such that $\alpha \geq \beta$, we have $(\alpha, \beta) \geq 0$ for $\alpha \in \Delta$ and there exists an element $\alpha \in \Delta$ such that $(\alpha, \beta) > 0$. (Use contradiction and Exercise 6.2.)
c: When there is no element $\alpha \in \Phi$ such that $\alpha \geq \beta$, the root β is written as $\beta = \sum_{\alpha\in\Delta} r_\alpha \alpha$ with $r_\alpha > 0$. (Use contradiction, **a**, **b**, **(B2)**, and Exercise 6.5.)
d: When there is no element $\alpha \in \Phi$ such that $\alpha \geq \beta$ or $\alpha \geq \beta'$, we have $\beta = \beta'$. (Use contradiction, **b**, **c**, and Exercise 6.2.)

6.2 Structure of Compact Lie Algebra*

In this section, we focus on a compact Lie group G and its compact Lie algebra G without any condition for G and $\mathfrak{g}$. An irreducible unitary representation of the compact Lie group G can be given from a skew-Hermitian representation of its compact Lie algebra G. For analysis of a real Lie algebra $\mathfrak{g}$, we often focus on its Lie subalgebra $\mathfrak{h}$ satisfying $[X, Y] = 0$ for any two elements $X, Y \in \mathfrak{h}$. Such a Lie subalgebra $\mathfrak{h}$ is called a **Cartan subalgebra** of $\mathfrak{g}$. Especially, a Cartan subalgebra $\mathfrak{h}$ is called a **maximal Cartan subalgebra** when a Cartan subalgebra including $\mathfrak{h}$ is limited only to $\mathfrak{h}$ itself. There exist more than one maximal Cartan subalgebras, however, its dimension does not depend on the choice of the maximal Cartan subalgebra, and is called the **rank** of the Lie algebra $\mathfrak{g}$.

The structure of the Lie algebra mainly is decided by the maximal Cartan subalgebra $\mathfrak{h}$ and its relation to the Lie algebra. Especially, when the Lie algebra $\mathfrak{g}$ is compact, all of maximal Cartan subalgebras are isomorphic to each other. As mentioned in (3.47), the compact Lie algebra $\mathfrak{g}$ can be decomposed to the direct sum of the semi simple part $\mathfrak{g}^{(1)} := [\mathfrak{g}, \mathfrak{g}]$ and the commutative part $\mathfrak{c}(\mathfrak{g}) = \{X \in \mathfrak{g} \,|\, \mathsf{ad}(X) = 0\}$. Letting $\mathfrak{h}$ be a maximal Cartan subalgebra of the semi simple part $\mathfrak{g}^{(1)}$, we find that a maximal Cartan subalgebra of $\mathfrak{g}$ is given as $\mathfrak{h} \oplus \mathfrak{c}(\mathfrak{g})$. Then, the structure of the compact Lie algebra $\mathfrak{g}$ can be characterized by that of the compact semi simple Lie algebra $\mathfrak{g}^{(1)}$ as follows.

In the following, based on the root system, we investigate the structure of the compact semi simple Lie algebra. Let $\mathfrak{g}$ be a compact real semi simple Lie algebra and $\mathfrak{h}$ be its maximal Cartan subalgebra. Then, for an element β of the dual space $\mathfrak{h}^*$ of $\mathfrak{h}$, we choose an element $Z_\beta \in \mathfrak{h}$ such that

$$\beta(X) = (X, Z_\beta)_{\mathfrak{g}}, \quad \forall X \in \mathfrak{h}. \tag{6.3}$$

Using the element Z_β, we define a positive definite inner product of $\mathfrak{h}^*$ as

$$(\beta, \gamma) := -(Z_\beta, Z_\gamma)_{\mathfrak{g}} = -\beta(Z_\gamma), \quad \forall \beta, \gamma \in \mathfrak{h}^*. \tag{6.4}$$

Since all elements of $\mathfrak{h}$ are commutative with each other, we can choose a basis of $\mathfrak{g}$ such that all elements $X \in \mathfrak{h}$ simultaneously take the form given in (3.46) of Lemma 3.9. Hence, there exist elements $\{\alpha_t\} \in \mathfrak{h}^*$ and the basis $\{F_t^x, F_t^y\}_t$ of the orthogonal complement of $\mathfrak{h}$ such that

$$\mathsf{ad}(X)F_t^x = \alpha_t(X)F_t^y, \quad \mathsf{ad}(X)F_t^y = -\alpha_t(X)F_t^x, \quad \forall X \in \mathfrak{h}, \tag{6.5}$$

which is equivalent to

$$\mathsf{ad}(X)(F_t^x + iF_t^y) = i\alpha_t(X)(F_t^x + iF_t^y), \quad \forall X \in \mathfrak{h}. \tag{6.6}$$

Here, we remark that when F_t^y is replaced by $-F_t^y$, instead of the element α_t, the element $-\alpha_t$ satisfies the above condition. In the following, we denote the set of α_t and $-\alpha_t$ by $\Phi_{\mathfrak{g}}$.

Theorem 6.2 *When $\mathfrak{g}$ is a compact semi simple Lie algebra, the relation $\alpha_t \neq \alpha_{t'}$ holds for $t \neq t'$.*

Theorem 6.3 *The set $\Phi_{\mathfrak{g}}$ is a root system when $\mathfrak{g}$ is a compact semi simple Lie algebra. Conversely, for any root system Φ, there exists a compact semi simple Lie algebra $\mathfrak{g}$ such that $\Phi = \Phi_{\mathfrak{g}}$.*

Theorems 6.2 and 6.3 follow from Theorems 6.6 and 6.7, which will be given later.

Since $\Phi_{\mathfrak{g}}$ is a root system due to the above theorems, we choose a fundamental system Δ of the root system. When $-\alpha_t \in \Phi_{\mathfrak{g}}^+(\Delta)$, we replace F_t^y by $-F_t^y$. Then, $\Phi_{\mathfrak{g}}^+(\Delta)$ equals the set $\{\alpha_t\}_t$. Hence, without loss of generality, we can assume that $\Phi_{\mathfrak{g}}^+(\Delta) = \{\alpha_t\}_t$ by choosing the proper sign of F_t^y. In the following, using the element α_t of Φ, we denote $\{F_t^x, F_t^y\}$ by $\{F_{\alpha_t}^x, F_{\alpha_t}^y\}$. Then, the relations

$$[F_\alpha^x, F_\alpha^y] = -(F_\alpha^x, F_\alpha^x)_{\mathfrak{g}} Z_\alpha = -(F_\alpha^y, F_\alpha^y)_{\mathfrak{g}} Z_\alpha \tag{6.7}$$

hold, and the subspace generated by $\{F_\alpha^x, F_\alpha^y, Z_\alpha\}$ is a Lie subalgebra isomorphic to $\mathfrak{su}(2)$. Since the relation (6.7) implies that $(F_\alpha^x, F_\alpha^x)_{\mathfrak{g}} = (F_\alpha^y, F_\alpha^y)_{\mathfrak{g}}$, we choose F_α^x, F_α^y such that

$$(F_\alpha^x, F_\alpha^x)_{\mathfrak{g}} = (F_\alpha^y, F_\alpha^y)_{\mathfrak{g}} = -\frac{1}{(\alpha, \alpha)} = \frac{1}{(Z_\alpha, Z_\alpha)_{\mathfrak{g}}}. \tag{6.8}$$

Hence, we have

$$[F_\alpha^x, F_\alpha^y] = \frac{1}{(\alpha, \alpha)} Z_\alpha. \tag{6.9}$$

Thus, letting $\mathfrak{g}[\alpha]$ be the linear space generated by $\{F_\alpha^x, F_\alpha^y\}$, we have $\mathfrak{g} = \mathfrak{h} \oplus (\bigoplus_{\alpha \in \Phi^+} \mathfrak{g}[\alpha])$. Since the set $\{Z_{\alpha_1}, \ldots, Z_{\alpha_r}\} \cup \bigcup_{\alpha \in \Phi^+} \{F_\alpha^x, F_\alpha^y\}$ is a basis of the linear space $\mathfrak{g}$, it is called a **basis** of the compact semi simple Lie algebra $\mathfrak{g}$. These equations express the structure of the compact semi simple Lie algebra $\mathfrak{g}$. Further, the relation (6.6) and Jacobi law imply

$$\mathrm{ad}(X)[F_\alpha^x + iF_\alpha^y, F_{\alpha'}^x + iF_{\alpha'}^y] = i(\alpha + \alpha')(X)[F_\alpha^x + iF_\alpha^y, F_{\alpha'}^x + iF_{\alpha'}^y]$$

for any element $X \in \mathfrak{h}$. Describing the real coefficient parts and the pure imaginary coefficient parts, separately, we obtain the following theorem.

Theorem 6.4 *When $\alpha + \alpha' \in \Phi^+$, the elements $[F_\alpha^x, F_{\alpha'}^x] - [F_\alpha^y, F_{\alpha'}^y]$ and $[F_\alpha^x, F_{\alpha'}^y] + [F_\alpha^y, F_{\alpha'}^x]$ are constant times of $F_{\alpha+\alpha'}^x$ and $F_{\alpha+\alpha'}^y$ under the above basis. When $\alpha + \alpha' \notin \Phi^+$, both of the above elements are 0.*

On the other hand, for a compact Lie algebra $\mathfrak{g}$, we choose a maximal Cartan subalgebra $\mathfrak{h}$ and Φ^+ for a compact semi simple Lie algebra $[\mathfrak{g}, \mathfrak{g}]$. Then, we obtain $\mathfrak{g} = \mathfrak{h} \oplus \mathfrak{c}(\mathfrak{g}) \oplus (\bigoplus_{\alpha\in\Phi^+} \mathfrak{g}[\alpha])$. Hence, the analysis of a compact semi simple Lie algebra can be reduced to the analysis of the root system. Given a fundamental system Δ, Weyl group $\mathcal{W}$ is generated by the subset $\{W_\alpha\}_{\alpha\in\Delta}$, and the image of Δ with respect to $\mathcal{W}$ is Φ. Hence, all informations for the root system are contained in the fundamental system. For the analysis of the fundamental system, it is sufficient to analyze Cartan matrix. So, the analysis of the compact semi simple Lie algebra is reduced to that of Cartan matrix. Since the classification of Cartan matrices can be reduced to that of Dynkin diagrams, the classification of compact semi simple Lie algebras can be reduced to that of Dynkin diagrams. Then, the following theorem holds.

Theorem 6.5 *A compact semi simple Lie algebra $\mathfrak{g}$ is simple if and only if its Dynkin diagram is connected.*

The proof of this theorem will be given in the next section because it is easier to show the same argument with its complexified Lie algebra. The connected Dynkin diagrams A_r, B_r, C_r, and D_r given in the previous section are called **classical type**, they are the Dynkin diagrams of the root systems of $\mathfrak{su}(r+1)$, $\mathfrak{so}(2r+1, \mathbb{R})$, $\mathfrak{sp}(r)$, and $\mathfrak{so}(2r, \mathbb{R})$, respectively.

The remaining connected Dynkin diagrams E_6, E_7, E_8, F_4, and G_2 are called **exceptional type**, the corresponding Lie algebras are given as Lie subalgebras of $\mathfrak{u}(r)$. The structure of the classical type of root systems are summarized in Table 6.1 [66, pp. 509–510].

Table 6.1 Root systems of A_r, B_r, C_r, D_r: indices j, j', and l of positive roots are chosen as $j = 1, \ldots, r-1$, $j' = 1, \ldots, r$, $l = j+1, \ldots, r$. The number j' of simple root $\alpha_{j'}$ ($j' = 1, \ldots, r$) corresponds to the number in Fig. 6.1

	Lie algebra	Simple root $\alpha_j, j = 1, \ldots, r-1$ α_r	Positive root	Norm of e_j	Number of roots	Dimension
A_r	$\mathfrak{su}(r+1)$	$e_j - e_{j+1}$, $e_r - e_{r+1}$	$e_j - e_l$	$\frac{1}{\sqrt{2(r+1)}}$	$r^2 + r$	$r^2 + 2r$
B_r	$\mathfrak{so}(2r+1, \mathbb{R})$	$e_j - e_{j+1}$, e_r	$e_j - e_l$, $e_{j'}$, $e_j + e_l$	$\frac{1}{\sqrt{2(2r-1)}}$	$2r^2$	$2r^2 + r$
C_r	$\mathfrak{sp}(r)$	$e_j - e_{j+1}$, $2e_{2r}$	$e_j - e_l$, $e_j + e_l$, $2e_{2j'}$	$\frac{1}{2\sqrt{r+1}}$	$2r^2$	$2r^2 + r$
D_r	$\mathfrak{so}(2r, \mathbb{R})$	$e_j - e_{j+1}$, $e_{r-1} + e_r$	$e_j - e_l$, $e_j + e_l$	$\frac{1}{\sqrt{2(2r-1)}}$	$2r^2$	$2r^2 + r$

In the following, we discuss the relation between the elements of the root system A_r and the elements of the Lie algebra $\mathfrak{su}(r+1)$. Since the root system Φ is determined by the real numbers $\langle \beta, \gamma \rangle = \frac{2(\beta,\gamma)}{(\gamma,\gamma)}$, the set $\{c\alpha\}_{\alpha \in \Phi}$ also satisfies the conditions for a root system. However, the root system generates a compact simple Lie algebra, it needs to satisfy the condition (6.5) under the inner product given in (6.4), in which, we adopt the Killing form as the inner product of the space $\mathfrak{h}$. Taking account into this condition, we can uniquely determine the inner product of $\mathfrak{h}^*$ and the root system corresponding the compact simple Lie algebra. By using $\alpha_{j,l} \in \mathfrak{h}^*$ defined in (4.38), the root system Φ is given as $\{\alpha_{j,l}, -\alpha_{j,l}\}_{1 \le j < l \le r+1}$. Since $F^x_{j,l}$, $F^y_{j,l}$, and $F^z_{j,l}$ given in Sect. 4.3 satisfy

$$(F^x_{j,l}, F^x_{j,l})_{\mathfrak{su}(r+1)} = (F^y_{j,l}, F^y_{j,l})_{\mathfrak{su}(r+1)} = \frac{1}{4}(F^z_{j,l}, F^z_{j,l})_{\mathfrak{su}(r+1)} = -(r+1),$$

$Z_{\alpha_{j,l}} := \frac{1}{2(r+1)} F^z_{j,l}$ satisfies (6.3). Since $(\alpha_{j,l}, \alpha_{j,l}) = \frac{1}{r+1}$, $F^x_{\alpha_{j,l}} := F^x_{j,l}$ and $F^y_{\alpha_{j,l}} := F^y_{j,l}$ satisfy (6.8). Hence, when $\mathfrak{h}^*$ is regarded as a subspace of the $r+1$-dimensional real linear space V_r with the CONS $\{e_j\}_{1 \le j \le r+1}$, we obtain Table 6.1.

Exercise 6.7 Show (6.7).

Exercise 6.8 Assume that a fundamental system Δ has more than two connected components $\Delta_1, \Delta_2, \ldots$. Define $\Phi_i^+ := \{\sum_{\alpha \in \Delta_i} m_\alpha \alpha\} \cap \Phi^+$. Show that $\alpha_i + \alpha_j$ $(i \ne j)$ is not contained in the root system Φ for any element $\alpha_i \in \Phi_i^+$.

6.3 Structure of Complex Semi Simple Lie Algebra*

6.3.1 Complex Semi Simple Lie Algebra

The analysis of a complex semi simple Lie algebra $\mathfrak{g}$ is easier than that of a compact semi simple Lie algebra. In the case of complex Lie algebra, a Cartan subalgebra is defined as a complex Lie subalgebra whose elements are commutative with each other and are diagonalizable under the adjoint representation. Similarly, we define its maximal Cartan subalgebra $\mathfrak{h}$ and the **rank** of the complex semi simple Lie algebra $\mathfrak{g}$. Given any element β of the dual space $\mathfrak{h}^*$ of $\mathfrak{h}$ and any element $X \in \mathfrak{h}$, we define $E_\beta \in \mathfrak{h}$ such that $\beta(X) = (X, E_\beta)$, and define the inner product $(\beta, \gamma) := (E_\beta, E_\gamma) = \beta(E_\gamma)$ for two elements β and γ of $\mathfrak{h}^*$.

Since $\{\mathrm{ad}(X)\}_{X \in \mathfrak{h}}$ is simultaneously diagonalizable, we choose a basis $\{K_t\}$ of the orthogonal complementary space to $\mathfrak{h}$ among eigenvectors of $\{\mathrm{ad}(X)\}_{X \in \mathfrak{h}}$ and choose an element α_t of the dual space $\mathfrak{h}^*$ such that $\mathrm{ad}(X)K_t = \alpha_t(X)K_t$ for any t. Such an element α_t is called a **root**, and we denote the set by $\Phi_{\mathfrak{g}}$. The following theorem holds.

Theorem 6.6 ([60, Proposition (a) of Sect. 8.4]) *When $\mathfrak{g}$ is a complex semi simple Lie algebra, we have $\alpha_t \ne \alpha_{t'}$ for $t \ne t'$.*

The following theorem also holds for the set $\Phi_{\mathfrak{g}}$.

Theorem 6.7 *$\Phi_{\mathfrak{g}}$ is a root system for a complex semi simple Lie algebra $\mathfrak{g}$. Conversely, for an arbitrary root system Φ, there exists a complex semi simple Lie algebra $\mathfrak{g}$ such that $\Phi = \Phi_{\mathfrak{g}}$.*[1]

Due to Theorem 6.6, for an element α of $\Phi_{\mathfrak{g}}$, there uniquely exists K_α such that

$$\mathsf{ad}(X)K_\alpha = \alpha(X)K_\alpha, \quad \forall X \in \mathfrak{h}. \tag{6.10}$$

So, we can uniquely choose E_α such that

$$(K_\alpha, K_{-\alpha})_{\mathfrak{g}} = \frac{2}{(E_\alpha, E_\alpha)_{\mathfrak{g}}}. \tag{6.11}$$

Then,

$$[K_\alpha, K_{-\alpha}] = \frac{2}{(E_\alpha, E_\alpha)_{\mathfrak{g}}} E_\alpha. \tag{6.12}$$

Then, $\mathfrak{n}_+ :=< \{K_\alpha\}_{\alpha\in\Phi^+} >$ and $\mathfrak{n}_- :=< \{K_{-\alpha}\}_{\alpha\in\Phi^+} >$ are nilpotent Lie subalgebras. We obtain the **Gauss decomposition** $\mathfrak{g} = \mathfrak{n}_+ \oplus \mathfrak{h} \oplus \mathfrak{n}_-$ of the Lie algebra $\mathfrak{g}$. It is known that a similar fact holds for the corresponding Lie group G. Since the set $\{E_\alpha\}_{\alpha\in\Delta} \cup \{K_\alpha\}_{\alpha\in\Phi}$ is a basis of the the complex semi simple Lie algebra $\mathfrak{g}$, it is called a **basis** of the complex semi simple Lie algebra $\mathfrak{g}$. Using complex vectors $\{\zeta_\alpha\}_{\alpha\in\Phi^+}$ and $\{\zeta'_\alpha\}_{\alpha\in\Phi^+}$, we can define the Lie groups $N_+ := \{\exp(\sum_{\alpha\in\Phi^+} \zeta'_\alpha K_\alpha)\}_{\zeta'}$ and $N_- := \{\exp(\sum_{\alpha\in\Phi^-} \zeta_\alpha K_\alpha)\}_\zeta$ that correspond to the nilpotent Lie subalgebras, where ζ_α is a complex number for any $\alpha \in \Phi^+$ and $\{\zeta_\alpha\}_{\alpha\in\Phi^+}$ is a complex vector with $|\Phi^+|$ entries. Now, we consider the right hand side action of N_+ and the left hand side action of N_-. Since both actions are commutative with each other, we can define the quotient space for the combined action. Then, $\mathfrak{h}$ gives a set of representatives of the tangent space of the quotient space. So, $\{\exp(\sum_{j=1}^r \eta_\alpha E_{\alpha_j})\}_\eta$ gives a set of representatives of the quotient space. Hence, as a generalization of (4.2), for an element $g \in G$ except for measure 0 points, we can choose complex vectors $\{\zeta_\alpha\}_{\alpha\in\Phi^+}$, $\{\zeta'_\alpha\}_{\alpha\in\Phi^+}$, and $\{\eta_j\}_{j=1}^r$ such that [87, p. 474]

$$g = \exp\left(\sum_{\alpha\in\Phi^+} \zeta_\alpha K_{-\alpha}\right) \exp\left(\sum_{j=1}^r \eta_\alpha E_{\alpha_j}\right) \exp\left(\sum_{\alpha\in\Phi^+} \zeta'_\alpha K_\alpha\right), \tag{6.13}$$

which is called the **Gauss decomposition** of the Lie group G.

[1] The former part is shown in [60, Theorem in Sect. 8.5], and the latter part can be checked by considering the fact that the corresponding simple Lie algebras are given for all of simple root systems in [60].

6.3.2 Relation to Compact Semi Simple Lie Algebra

From now, to show Theorem 6.5, we denote a compact semi simple Lie algebra by $\mathfrak{g}$, and its complexification by $\mathfrak{g}_{\mathbb{C}}$, which is a complex semi simple Lie algebra. For an element $\alpha \in \Phi_{\mathfrak{g}}^+$, the elements K_α and $E_\alpha \in \mathfrak{g}_{\mathbb{C}}$ are given from the elements F_α^x, F_α^y, and Z_α of $\mathfrak{g}$ as follows

$$E_\alpha = -iZ_\alpha, \quad K_\alpha = -iF_\alpha^x - F_\alpha^y, \quad K_{-\alpha} = -iF_\alpha^x + F_\alpha^y. \tag{6.14}$$

The relations (6.5), (6.7) and (6.9) guarantee that the above given elements K_α and E_α satisfy (6.10), (6.11) and (6.12).

Now, define the one-dimensional linear subspace $\mathfrak{g}_\alpha$ generated by K_α. (Remark that $\mathfrak{g}_\alpha$ is different from $\mathfrak{g}[\alpha]$.) Then, we have $\mathfrak{g} = \mathfrak{h}_{\mathbb{C}} \oplus \bigoplus_{\alpha\in\Phi} \mathfrak{g}_\alpha$. Theorem 6.4 yields

$$[\mathfrak{g}_\alpha, \mathfrak{g}_{\alpha'}] = \begin{cases} \mathfrak{g}_{\alpha+\alpha'} & \alpha + \alpha' \in \Phi \\ \{0\} & \alpha + \alpha' \notin \Phi \end{cases} \tag{6.15}$$

for $\alpha' \neq -\alpha$. When $\alpha' = -\alpha$, we have

$$[\mathfrak{g}_\alpha, \mathfrak{g}_{\alpha'}] \subset \mathfrak{h}_{\mathbb{C}}. \tag{6.16}$$

Conversely, based on the basis $\{E_\alpha\}_{\alpha\in\Delta} \cup \{K_\alpha\}_{\alpha\in\Phi}$ of the complex semi simple Lie algebra, we define

$$Z_\alpha := iE_\alpha, \quad F_\alpha^x := \frac{i}{2}(K_\alpha + K_{-\alpha}), \quad F_\alpha^y := \frac{1}{2}(-K_\alpha + K_{-\alpha}),$$

the real linear space generated by F_α^x, F_α^y, Z_α forms a compact semi simple Lie algebra. That is, there exists a one-to-one correspondence between complex semi simple Lie algebras and compact semi simple Lie algebras. In the next two sections, using this relation, we investigate the properties of a compact semi simple Lie algebra.

Here, we give the proof of Theorem 6.5. Firstly, we remark that given a compact semi simple Lie algebra $\mathfrak{g}$, the compact semi simple Lie algebra $\mathfrak{g}$ is simple when its complexification $\mathfrak{g}_{\mathbb{C}}$ is simple. Then, K_α is an eigenvector of the maximal Cartan subalgebra $\mathfrak{h}$ under the adjoint representation of the semi simple complex Lie algebra $\mathfrak{g}_{\mathbb{C}}$. The action of $K_{-\alpha}$ on the semi simple complex Lie algebra $\mathfrak{g}_{\mathbb{C}}$ exchanges E_α and K_α under the identification of the scalar times. Hence, for $\alpha \in \Phi^+$, the subspace spanned by $\{E_\alpha, K_\alpha, K_{-\alpha}\}$ is included in the invariant subspace of the semi simple complex Lie algebra $\mathfrak{g}_{\mathbb{C}}$ with respect to the adjoint representation. Further, when elements $\alpha, \alpha' \in \Phi^+$ satisfy $\alpha'' = \alpha + \alpha' \in \Phi^+$, K_α and $K_{\alpha''}$ are exchanged to each other under the adjoint representation. Since, similarly, $K_{\alpha'}$ and $K_{\alpha''}$ are exchanged to each other, these three elements are exchanged to each other.

We choose two elements $\alpha, \alpha' \in \Delta$ of the fundamental system such that $\langle \alpha, \alpha' \rangle \neq 0$. Since $\langle \alpha, \alpha' \rangle$ is a non-negative integer, we have $\alpha - \langle \alpha, \alpha' \rangle \alpha' \in \Phi^+$, which implies that K_α and $K_{\alpha'}$ are exchanged to each other. Hence, the complex Lie algebra corresponding to a connected Dynkin diagram does not have a non-trivial invariant subspace with respect to the adjoint representation. That is, the corresponding complex Lie algebra is simple. Hence, the corresponding compact Lie algebra is simple.

On the other hand, we assume that the corresponding fundamental system has more than two components that are not connected to each other. Then, due to Exercise 6.8, the element $\boldsymbol{\alpha}_i + \boldsymbol{\alpha}_j$ is not contained in Φ for any element $\boldsymbol{\alpha}_i \in \Phi_i^+$. Now, let $\tilde{\mathfrak{g}}_{\mathbb{C}}$ be a complex Lie subalgebra generated by one connected component of the corresponding fundamental system. $K_{\pm\alpha'}$ maps any elements of $\tilde{\mathfrak{g}}_{\mathbb{C}}$ to 0. Similarly, E_α maps them to 0. Hence, the complex Lie subalgebra $\tilde{\mathfrak{g}}_{\mathbb{C}}$ is an invariant subspace of the complex Lie algebra $\mathfrak{g}_{\mathbb{C}}$ with respect to the adjoint representation. Thus, the corresponding compact Lie subalgebra $\tilde{\mathfrak{g}}$ is also an invariant subspace of the compact Lie algebra $\mathfrak{g}$ with respect to the adjoint representation.

These facts indicate the equivalence relation among the following three arguments; (1) The semi simple complex Lie algebra $\mathfrak{g}_{\mathbb{C}}$ is simple. (2) The semi simple compact Lie algebra $\mathfrak{g}_{\mathbb{C}}$ is simple. (3) The corresponding Dynkin diagram is connected. Hence, we obtain Theorem 6.5.

6.4 Skew-Hermitian Representation of Compact Semi Simple Lie Algebra*

Next, to study a unitary representation of an arbitrary compact Lie group, we investigate a skew-Hermitian representation of the compact simple Lie algebra $\mathfrak{g}$. Similar to representations of $\mathfrak{su}(2)$ and $\mathfrak{su}(d)$, for an irreducible skew-Hermitian representation f of $\mathfrak{g}$ on $\mathcal{H}$, we define

$$\mathsf{E}_j := -i\mathsf{f}(Z_{\alpha_j}), \quad \boldsymbol{\alpha}_j \in \Delta_{\mathfrak{g}}, \quad \mathsf{K}_{\pm\alpha} := -i\mathsf{f}(F_\alpha^x) \mp \mathsf{f}(F_\alpha^y), \quad \boldsymbol{\alpha} \in \Phi_{\mathfrak{g}}.$$

Considering the complexification $\mathfrak{g}_{\mathbb{C}}$, we have $\mathsf{E}_j = \mathsf{f}(E_{\alpha_j})$ for $\boldsymbol{\alpha}_j \in \Delta_{\mathfrak{g}}$ and $\mathsf{K}_{\pm\alpha} = \mathsf{f}(F_{\pm\alpha})$ for $\boldsymbol{\alpha} \in \Phi_{\mathfrak{g}}$. Then, the relations (6.10) and (6.12) imply

$$[\mathsf{E}_{j'}, \mathsf{E}_j] = 0 \tag{6.17}$$

$$[\mathsf{E}_{j'}, \mathsf{K}_\alpha] = \boldsymbol{\alpha}(Z_j)\mathsf{K}_\alpha \tag{6.18}$$

$$[\mathsf{K}_{\alpha_j}, \mathsf{K}_{-\alpha_j}] = \frac{2}{(\boldsymbol{\alpha}_j, \boldsymbol{\alpha}_j)}\mathsf{E}_j.$$

Then, since the representation is skew-Hermitian, the relation $(\mathsf{K}_\alpha)^\dagger = \mathsf{K}_{-\alpha}$ holds and E_j is Hermitian. Hence, due to (6.17), $\mathcal{H}$ is decomposed to the direct sum space of the simultaneous eigenspaces with respect to elements of $\mathfrak{h}$. The eigenvalues can be written by using an element $\boldsymbol{m}$ of the dual space $\mathfrak{h}^*$ of $\mathfrak{h}$. The simultaneous eigenspace corresponding to the eigenvalues given by $\boldsymbol{m}$ is given as $\mathcal{U}_{\boldsymbol{m}}^w := \{v \in \mathcal{H} | \mathsf{f}(X)v =$

$i\boldsymbol{m}(X)v, \quad \forall X \in \mathfrak{h}\}$. The condition given in the set $\mathcal{U}^w_{\boldsymbol{m}}$ is equivalent to the condition $\mathsf{E}_j v = \boldsymbol{m}(Z_{\alpha_j})v$. An element $\boldsymbol{m} \in \mathfrak{h}^*$ is called a **weight** for a representation f on a representation space $\mathcal{H}$ when $\mathcal{U}^w_{\boldsymbol{m}} \neq \{0\}$. Then, due to (6.18), $\mathsf{K}_\alpha v$ is an element of $\mathcal{U}^w_{\boldsymbol{m}+\alpha}$ for $v \in \mathcal{U}^w_{\boldsymbol{m}}$. The action of an element $e^X \in G$ $(X \in \mathfrak{h})$ satisfies

$$\mathsf{f}(e^X)v = e^{i\boldsymbol{m}(X)}v, \quad \forall v \in \mathcal{U}^w_{\boldsymbol{m}}. \tag{6.19}$$

When $\mathcal{H}$ is an irreducible skew-Hermitian representation space, we can uniquely choose a weight $\boldsymbol{\lambda}$ such that

(1) $\mathcal{U}^w_{\boldsymbol{\lambda}}$ is a one-dimensional space.
(2) An element $|\boldsymbol{\lambda}; \boldsymbol{\lambda}\rangle$ of $\mathcal{U}^w_{\boldsymbol{\lambda}}$ satisfies $\mathsf{K}_\alpha|\boldsymbol{\lambda}; \boldsymbol{\lambda}\rangle = 0$ for $\boldsymbol{\alpha} \in \Phi^+$.

Such a weight $\boldsymbol{\lambda}$ is called a **highest weight**. Especially, a vector $|\boldsymbol{\lambda}; \boldsymbol{\lambda}\rangle$ satisfying Condition **(2)** is called the **highest weight vector**. The representation space with the highest weight $\boldsymbol{\lambda}$ is generated by $\{\prod_{\alpha\in\Phi^+}(\mathsf{K}_{-\alpha})^{l_\alpha}v_{\boldsymbol{\lambda}}\}$ and is uniquely determined by the highest weight $\boldsymbol{\lambda}$. Hence, the highest weight $\boldsymbol{\lambda}$ is regarded as the label (i.e., an element of $\hat{\mathfrak{g}}$) to identify the irreducible skew-Hermitian representation. So, we denote the representation space by $\mathcal{U}_{\boldsymbol{\lambda}}$, and the representation on $\mathcal{U}_{\boldsymbol{\lambda}}$ by $\mathsf{f}_{\boldsymbol{\lambda}}$.

There exists a skew-Hermitian representation with the highest weight $\boldsymbol{\lambda}$ if and only if all of entries of Dynkin index of $\boldsymbol{\lambda}$ are non-negative integers $\langle\boldsymbol{\lambda}, \boldsymbol{\alpha}_j\rangle$, i.e., $\boldsymbol{\lambda} \in \Lambda_+$. Since the fundamental system Δ is given as $\{\boldsymbol{\alpha}_j\}$, the definition of $\Lambda_+(\Delta)$ given in (6.2) guarantees that the latter condition is equivalent to the condition $\boldsymbol{\lambda} \in \Lambda_+(\Delta)$. That is, $\Lambda_+(\Delta)$ can be identified with the set $\hat{\mathfrak{g}}$ of irreducible skew-Hermitian representations. In particular, $\mathfrak{g}\,\bar{\otimes}\mathbb{C}$ is also an irreducible representation space under the adjoint representation, and its highest weight is the maximal root $\boldsymbol{\lambda}(\Phi)$ given in Proposition 6.2.

To characterize the set $\Lambda(\boldsymbol{\lambda}) := \{\boldsymbol{\beta} \in \mathfrak{h}^* \,|\mathcal{U}^w_{\boldsymbol{\beta}} \neq \{0\}\}$, we define the following map as a generalization of the map $W_{\boldsymbol{\alpha}}$.

$$W_{\boldsymbol{\alpha},k}(\boldsymbol{\beta}) := \begin{cases} \boldsymbol{\beta} - k\boldsymbol{\alpha} & \text{when } k \text{ is an integer between 0 and } \langle\boldsymbol{\beta}, \boldsymbol{\alpha}\rangle, \\ \boldsymbol{\beta} & \text{otherwise.} \end{cases}$$

Note that $\langle\boldsymbol{\beta}, \boldsymbol{\alpha}\rangle = \frac{2(\boldsymbol{\beta},\boldsymbol{\alpha})}{(\boldsymbol{\alpha},\boldsymbol{\alpha})} = -2\boldsymbol{\beta}([F^x_{\boldsymbol{\alpha}}, F^y_{\boldsymbol{\alpha}}])$ and the Lie subalgebra generated by $\{F^x_{\boldsymbol{\alpha}}, F^y_{\boldsymbol{\alpha}}, [F^x_{\boldsymbol{\alpha}}, F^y_{\boldsymbol{\alpha}}]\}$ is isomorphic to $\mathfrak{su}(2)$. Since the action of the Lie subalgebra characterizes the action of $W_{\boldsymbol{\alpha},k}$, we find that the set $\Lambda(\boldsymbol{\lambda})$ is closed with respect to $W_{\boldsymbol{\alpha},k}$. Hence, the set $\Lambda(\boldsymbol{\lambda})$ is $\{W_{\boldsymbol{\beta}_1,k_1} \circ W_{\boldsymbol{\beta}_2,k_2} \circ \cdots \circ W_{\boldsymbol{\beta}_n,k_n}(\boldsymbol{\lambda})\}_{\boldsymbol{\beta}_j\in\Phi,k_j\in\mathbb{Z}}$. Further, the dimension of $\mathcal{U}_{\boldsymbol{\lambda}}$ is characterized by the following **Weyl's dimension formula** [60, Corollary in Sect. 24.3];

$$\dim \mathcal{U}_{\boldsymbol{\lambda}} = \frac{\prod_{\alpha\in\Phi^+}(\boldsymbol{\lambda}+\boldsymbol{\delta}, \boldsymbol{\alpha})}{\prod_{\alpha\in\Phi^+}(\boldsymbol{\delta}, \boldsymbol{\alpha})} = \frac{\prod_{\alpha\in\Phi^+}\langle\boldsymbol{\lambda}+\boldsymbol{\delta}, \boldsymbol{\alpha}\rangle}{\prod_{\alpha\in\Phi^+}\langle\boldsymbol{\delta}, \boldsymbol{\alpha}\rangle}, \tag{6.20}$$

where $\boldsymbol{\delta} := \frac{1}{2}\sum_{\alpha\in\Phi^+}\boldsymbol{\alpha}$. By using $\boldsymbol{\lambda}_j$ defined in Sect. 6.1, $\boldsymbol{\delta}$ can be written as $\boldsymbol{\delta} = \sum_{j=1}^r \boldsymbol{\lambda}_j$ [60, Lemma A in Sect. 13.3].

When $\mathfrak{g}$ is a compact real semi simple Lie algebra, the root system can be decomposed to connected components $\Phi_1, \ldots, \Phi_l$ and each connected component Φ_j is a root system. Let Δ_j be the fundamental system of Φ_j. Then, the set of weights of $\mathfrak{g}$ is given as $\Lambda_+(\Delta_1) \times \cdots \times \Lambda_+(\Delta_l)$. Further, due to Lemma 3.3, the set $\hat{\mathfrak{g}}$ of irreducible skew-Hermitian representations is given as $\Lambda_+(\Delta_1) \times \cdots \times \Lambda_+(\Delta_l)$. When $\mathfrak{g}$ is a compact Lie algebra, $\mathfrak{g}$ can be decomposed to the direct sum of the compact real semi simple Lie algebra $[\mathfrak{g}, \mathfrak{g}]$ and the center $\mathfrak{c}(\mathfrak{g})$. Since any irreducible representation space of the center $\mathfrak{c}(\mathfrak{g})$ is one-dimensional, an irreducible representation space of $\mathfrak{g}$ is also an irreducible representation space of the compact real semi simple Lie algebra $[\mathfrak{g}, \mathfrak{g}]$. Thus, an irreducible skew-Hermitian representation of $\mathfrak{g}$ can be determined by the action of $[\mathfrak{g}, \mathfrak{g}]$ on the irreducible representation space $\mathcal{U}_\lambda$ and the scalar action of the center $\mathfrak{c}(\mathfrak{g})$ on the space $\mathcal{U}_\lambda$.

Exercise 6.9 Show that $\delta = \frac{1}{2} \sum_{j=1}^{r-1} j(r-j)\alpha_j$ in the case of $\mathrm{SU}(r)$.

Exercise 6.10 Show that (6.20) equals (4.47) in the case of $\mathrm{SU}(r)$.

6.5 Subalgebra and Root Subsystem*

When a homogeneous space has a symmetry with respect to the action of the Lie group G, the homogeneous space is written as the quotient space G/H_1 by using the stabilizer H_1, which is a subgroup of the group G. Hence, to investigate the structure of the homogeneous space, we need to study the relation between the group G and the subgroup H_1. This discussion plays an important role for the generalization of the concept of coherent states. In the following, we focus on the case when H_1 is a Lie subgroup. However, since it might be difficult to study the structure of the Lie group directly, we study that of the corresponding Lie algebra. Hence, this section deals with the compact real semi simple Lie algebra $\mathfrak{g}$ and its Lie subalgebra $\mathfrak{h}_1$.

This section discusses only a compact real semi simple Lie algebra, however, this discussion works as a preparation of our discussion of a non-compact Lie algebra. Hence, the discussion in this section is essential for that for a non-compact Lie algebra.

When a maximal Cartan subalgebra $\mathfrak{h}_2$ of $\mathfrak{h}_1$ is a Cartan subalgebra of $\mathfrak{g}$, $\mathfrak{h}_2$ is a subspace of a maximal Cartan subalgebra $\mathfrak{h}$ of $\mathfrak{g}$. Hence, $\mathfrak{h}_2^*$ can be regarded as a subspace of $\mathfrak{h}^*$ under the inner product of $\mathfrak{h}^*$. When a root system $\Phi' := \Phi(\mathfrak{h}_1)$ of $\mathfrak{h}_1$ is a subset of $\Phi := \Phi_{\mathfrak{g}}$, $\mathfrak{h}_1$ is called a **regular subalgebra** of $\mathfrak{g}$. When a Lie subalgebra $\mathfrak{h}_1$ contains a maximal Cartan subalgebra $\mathfrak{h}$ of $\mathfrak{g}$, $\mathfrak{h}$ is a maximal Cartan subalgebra of $\mathfrak{h}_1$. Hence, for the common maximal Cartan subalgebra $\mathfrak{h}$, we can define their root systems $\Phi := \Phi_{\mathfrak{g}}$ and $\Phi' := \Phi(\mathfrak{h}_1)$. Then, Φ' is a subset of Φ. In this case, $\mathfrak{h}_1$ is called a **maximal regular subalgebra** of $\mathfrak{g}$. In the following, we discuss only the case when $\mathfrak{h}_1$ is a maximal regular subalgebra of $\mathfrak{g}$.

Then, since $\mathfrak{h}_1$ is a subalgebra, i.e.,

$$[\mathfrak{h}_1, \mathfrak{h}_1] \subset \mathfrak{h}_1, \tag{6.21}$$

we have

$$(\Phi' + \Phi') \cap \Phi \subset \Phi', \tag{6.22}$$

where we define $S + S' := \{\alpha + \alpha' | \alpha \in S, \alpha' \in S'\}$ for two subsets S and S' of a linear space V. In this case, Φ' is called a **root subsystem** of Φ.

Letting Φ^+ be the set of positive roots of the root system Φ, we give the set of positive roots of the root subsystem Φ' by $\Phi'^+(\Delta) := \Phi^+(\Delta) \cap \Phi'$. Then, since the set $\mathfrak{h}_1 := \mathfrak{h} \oplus (\bigoplus_{\alpha \in \Phi'^+} \mathfrak{g}[\alpha])$ satisfies the condition (6.21) due to the condition (6.22), $\mathfrak{h}_1$ is a Lie subalgebra including the maximal Cartan subalgebra $\mathfrak{h}$ of the compact semi simple Lie algebra $\mathfrak{g} := \mathfrak{h} \oplus (\bigoplus_{\alpha \in \Phi^+} \mathfrak{g}[\alpha])$. Hence, the analysis of the Lie subalgebra $\mathfrak{h}_1$ including the maximal Cartan subalgebra can be reduced to that of the root subsystem.

Now, let $\mathfrak{p}$ be the orthogonal complement of $\mathfrak{h}_1$ with respect to Killing form. Given $X, Y \in \mathfrak{h}_1$ and $Z \in \mathfrak{p}$, as $[X, Y] \in \mathfrak{h}_1$, we have $(X, [Y, Z])_{\mathfrak{g}} = ([X, Y], Z)_{\mathfrak{g}} = 0$, which implies that

$$[\mathfrak{h}_1, \mathfrak{p}] \subset \mathfrak{p}\,. \tag{6.23}$$

Since the set of roots contained in $\mathfrak{p}$ is $\Phi \backslash \Phi'$, we obtain

$$(\Phi' + (\Phi \backslash \Phi')) \cap \Phi \subset (\Phi \backslash \Phi'). \tag{6.24}$$

That is, (6.24) is the property derived from (6.22).

The following equivalence relation holds for the root subsystem.

Theorem 6.8 *Let Φ' be a root subsystem of a root system Φ whose fundamental system is a subset Δ of a real linear space V. The following relations are equivalent to each other.*

(1) *$\Phi'^+(\Delta)$ is the set of positive roots of Φ', and its fundamental system Δ' is included in the fundamental system $\Delta = \{\alpha_1, \dots, \alpha_r\}$ of the root system Φ. The relation $\Delta' = \Delta \cap \Phi'^+(\Delta)$ also holds. Further, there exists an element $\boldsymbol{\lambda} \in \Lambda_+(\Delta)$ such that $\Phi' = \{\alpha \in \Phi | \langle \boldsymbol{\lambda}, \alpha \rangle = 0\}$.*

(2) *There exists an element $X \in \mathfrak{h}$ such that*

$$\begin{aligned} \Phi^+(\Delta) \backslash \Phi' &= \{\alpha \in \Phi | \alpha(X) > 0\} \\ \Phi' &= \{\alpha \in \Phi | \alpha(X) = 0\}. \end{aligned} \tag{6.25}$$

(3) *The set $\Phi^+(\Delta)$ satisfies $((\Phi^+(\Delta) \backslash \Phi') + \Phi') \cap \Phi \subset (\Phi^+(\Delta) \backslash \Phi')$.*

(4) *The set $\Phi^+(\Delta) \backslash \Phi'$ is invariant for Weyl group $\mathcal{W}(\Phi')$ (the group generated by $\{W_\alpha\}_{\alpha \in \Phi'}$) with respect to Φ'.*

Proof **(2)⇒(3)** is trivial.

(3)⇒(4): For $\beta \in \Phi^+(\Delta) \backslash \Phi'$ and $\alpha \in \Phi'$, $\langle \beta, \alpha \rangle$ is an integer. Hence, **(3)** guarantees that $W_\alpha(\beta) = \beta - \langle \beta, \alpha \rangle \alpha$ is an element of $\Phi^+(\Delta) \backslash \Phi'$.

(4)⇒(2): The set $cone(\Phi^+(\Delta)\backslash\Phi')^* := \{X \in \mathfrak{h} \mid \alpha(X) > 0, \quad \forall\alpha \in \Phi^+(\Delta)\backslash\Phi'\}$ is a convex cone. Especially, since $cone(\Phi^+(\Delta)\backslash\Phi') \cap cone(\Phi^+(\Delta)\backslash\Phi') = \emptyset$, the set $cone(\Phi^+(\Delta)\backslash\Phi')^*$ is not the empty set. Next, we define the action of an element W of Weyl group $\mathcal{W}(\Phi')$ for $X \in \mathfrak{h}$ as $\alpha(W(X)) := W(\alpha)(X)$. Then, we have $W_\alpha(X) := X - \alpha(X)Z_\alpha$. Since the set $\Phi^+(\Delta)\backslash\Phi'$ is invariant with respect to the action of Weyl group $\mathcal{W}(\Phi')$, $cone(\Phi^+(\Delta)\backslash\Phi')^*$ is also invariant with respect to the action of Weyl group $\mathcal{W}(\Phi')$. Hence, $cone(\Phi^+(\Delta)\backslash\Phi')^*$ contains a non-zero fixed point X with respect to the action of Weyl group $\mathcal{W}(\Phi')$. Thus, for $\alpha \in \Phi'$ we have $\alpha(X) = 0$, which implies **(2)**.
(2)⇒(1): For $\alpha \in \Delta'$, we choose elements $\alpha_{i_1}, \ldots, \alpha_{i_m}$ of Δ and positive integers $m_{i_1}, \ldots, m_{i_m}$ such that $\alpha = m_{i_1}\alpha_{i_1} + \ldots m_{i_m}\alpha_{i_m}$. Condition **(2)** yields that $\alpha_{i_1}, \ldots, \alpha_{i_m}$ are elements of Φ'. The definition of the fundamental system guarantees that $l = 1$ and $\alpha \in \Delta$. That is, we have $\Delta' = \Delta \cap \Phi'^+(\Delta)$. Since the set $\Delta = \{\alpha_j\}$ is a basis of the linear space spanned by Δ, we can choose $\boldsymbol{\lambda}_j \in \mathfrak{h}^*$ such that $\langle \boldsymbol{\lambda}_j, \alpha_i \rangle = \delta_{i,j}$.

Now, we assume that $\Delta\backslash\Delta' = \{\alpha_{j_1}, \ldots, \alpha_{j_l}\}$ and we define $\boldsymbol{\lambda} := \boldsymbol{\lambda}_{j_1} + \ldots + \boldsymbol{\lambda}_{j_l}$. Then, we have $\boldsymbol{\lambda} \in \Lambda_+(\Delta)$. So, we obtain the relation $\Phi' = \{\alpha \in \Phi \mid \langle \boldsymbol{\lambda}, \alpha \rangle = 0\}$.
(1)⇒(2) is trivial. ■

When a root subsystem Φ' of a root system Φ with fundamental system Δ satisfies the above condition, $(\Phi^+(\Delta), \Phi' \subset \Phi)$ is called a **hyperbolic system**[79, Definition VII.2.6.]. When $(\Phi^+(\Delta), \Phi' \subset \Phi)$ is a hyperbolic system, Condition **(2)** implies that

$$((\Phi^+(\Delta)\backslash\Phi') + (\Phi^+(\Delta)\backslash\Phi')) \cap \Phi \subset (\Phi^+(\Delta)\backslash\Phi'). \tag{6.26}$$

Hence, the following theorem holds even though a Lie subalgebra $\mathfrak{h}_1$ of $\mathfrak{g}$ does not include the maximal Cartan subalgebra.

Theorem 6.9 *The following conditions are equivalent to each other for a compact real semi simple Lie algebra $\mathfrak{g}$ and its Lie subalgebra $\mathfrak{h}_1$.*

(1) $\mathfrak{c}_\mathfrak{g}(\mathfrak{c}(\mathfrak{h}_1)) = \mathfrak{h}_1$.
(2) *The Lie subalgebra $\mathfrak{h}_1$ contains a maximal Cartan subalgebra $\mathfrak{h}$ of $\mathfrak{g}$.*

Further, when the above condition holds and Φ' is the root system of $\mathfrak{h}_1$, there exists a fundamental system Δ of Φ such that $(\Phi^+(\Delta), \Phi' \subset \Phi)$ is a hyperbolic system.

Proof **(1)⇒(2)**: Since an element of a maximal Cartan subalgebra $\mathfrak{h}'$ of the semi simple Lie algebra $[\mathfrak{h}_1, \mathfrak{h}_1]$ is commutative with any element of $\mathfrak{c}(\mathfrak{h}_1)$, $\mathfrak{h}' + \mathfrak{c}(\mathfrak{h}_1)$ is a maximal Cartan subalgebra of $\mathfrak{h}_1$. The discussion in the beginning part of Sect. 6.2 guarantees that $\mathfrak{h}_1$ is a maximal Cartan subalgebra. Let $\mathfrak{h}$ be a maximal Cartan subalgebra of $\mathfrak{g}$ including $\mathfrak{h}' + \mathfrak{c}(\mathfrak{h}_1)$. Since $\mathfrak{c}(\mathfrak{h}_1) \subset \mathfrak{h}$, we have $\mathfrak{h} = \mathfrak{c}_\mathfrak{g}(\mathfrak{h}) \subset \mathfrak{c}_\mathfrak{g}(\mathfrak{c}(\mathfrak{h}_1)) = \mathfrak{h}_1$, which implies that $\mathfrak{h}$ is a Cartan subalgebra of $\mathfrak{h}_1$. Since $\mathfrak{h}' + \mathfrak{c}(\mathfrak{h}_1)$ is a maximal Cartan subalgebra of $\mathfrak{g}$, $\mathfrak{h}' + \mathfrak{c}(\mathfrak{h}_1)$ is $\mathfrak{h}$. That is, $\mathfrak{h}_1$ contains a maximal Cartan subalgebra $\mathfrak{h}$ of $\mathfrak{g}$.
(2)⇒(1): Since the definition of the center implies that $\mathfrak{h}_1 \subset \mathfrak{c}_\mathfrak{g}(\mathfrak{c}(\mathfrak{h}_1))$, it is sufficient to show that $\mathfrak{p} \cap \mathfrak{c}_\mathfrak{g}(\mathfrak{c}(\mathfrak{h}_1)) = \{0\}$. Because $\mathfrak{c}(\mathfrak{h}_1)$ is included in a maximal Cartan

subalgebra, we have $\mathfrak{c}(\mathfrak{h}_1) = \{X \in \mathfrak{h} \,|\alpha(X) = 0, \forall \alpha \in \Phi'\}$. We choose an element $X \in \mathfrak{h}$ satisfying Condition **(2)** of Theorem 6.8. So, we have $X \in \mathfrak{c}(\mathfrak{h}_1)$. Also, (6.25) guarantees that $Y \in \mathfrak{p}$ satisfies $[X, Y] \neq 0$. Hence, we obtain $Y \notin \mathfrak{c}_{\mathfrak{g}}(\mathfrak{c}(\mathfrak{h}_1))$, which yields Condition **(1)**.

Now, we assume Conditions **(1)** and **(2)**, and let Φ' be the root system of $\mathfrak{h}_1$. Based on the discussion of the proof of **(1)⇒(2)**, we choose two kinds of elements F_α^x and F_α^y of $\mathfrak{h}_1$ as elements of a basis of the orthogonal complement of $\mathfrak{h}$. Since F_α^x and F_α^y belong to $\mathfrak{g}$, a root of $\mathfrak{h}_1$ is also a root of $\mathfrak{g}$. That is, we have $\Phi_{\mathfrak{h}_1} \subset \Phi_{\mathfrak{g}}$. Hence, $X \in \mathfrak{c}(\mathfrak{h}_1)$ satisfies

$$\alpha(X) = 0, \forall \alpha \in \Phi(\mathfrak{h}_1). \tag{6.27}$$

On the other hand, since $\mathfrak{c}_{\mathfrak{g}}(\mathfrak{c}(\mathfrak{h}_1)) = \mathfrak{h}_1$, for a root $\alpha \in \Phi_{\mathfrak{g}} \setminus \Phi_{\mathfrak{h}_1}$, there exists an element $X_\alpha \in \mathfrak{c}(\mathfrak{h}_1)$ such that $\alpha(X_\alpha) \neq 0$. Thus, there exists an element $X \in \mathfrak{c}(\mathfrak{h}_1)$ such that the relation $\alpha(X) \neq 0$ holds for any root $\alpha \in \Phi_{\mathfrak{g}} \setminus \Phi_{\mathfrak{h}_1}$. Hence, Proposition 6.1 guarantees that there exists a fundamental system Δ of $\Phi_{\mathfrak{g}}$ such that the set of positive roots $\Phi_{\mathfrak{g}}^+$ satisfies

$$\alpha(X) > 0, \quad \alpha \in \Phi_{\mathfrak{g}}^+ \setminus \Phi_{\mathfrak{h}_1}. \tag{6.28}$$

Due to (6.27) and (6.28), any element of $\Phi_{\mathfrak{h}_1}^+ := \Phi_{\mathfrak{h}_1} \cap \Phi_{\mathfrak{g}}^+$ is given as a sum of elements of $\Delta(\mathfrak{h}_1) := \Phi_{\mathfrak{h}_1} \cap \Delta$, which is a fundamental system of $\Phi_{\mathfrak{h}_1}$. Since $\Delta = \{\alpha_1, \ldots, \alpha_r\}$ forms a basis of a linear space, we can choose elements $\{\boldsymbol{\lambda}_1, \ldots, \boldsymbol{\lambda}_r\}$ such that $\langle \boldsymbol{\lambda}_i, \alpha_j \rangle = \delta_{i,j}$. Therefore, describing all of elements of $\Delta \setminus \Delta(\mathfrak{h}_1)$ as $\{\alpha_{i_1}, \ldots, \alpha_{i_l}\}$, we have $\Phi_{\mathfrak{h}_1} = \{\alpha \in \Phi_{\mathfrak{g}} | \langle \boldsymbol{\lambda}_{i_1} + \cdots + \boldsymbol{\lambda}_{i_l}, \alpha \rangle = 0\}$. Thus, we obtain Condition **(1)** of Theorem 6.8, which yields **(2)**. ■

Next, we consider the complexifications $\mathfrak{h}_{1,\mathbb{C}}$ and $\mathfrak{p}_{\mathbb{C}}$ of $\mathfrak{h}_1$ and $\mathfrak{p}$. Then, the conditions (6.21) and (6.23) yield

$$[\mathfrak{h}_{1,\mathbb{C}}, \mathfrak{h}_{1,\mathbb{C}}] \subset \mathfrak{h}_{1,\mathbb{C}}, \quad [\mathfrak{h}_{1,\mathbb{C}}, \mathfrak{p}_{\mathbb{C}}] \subset \mathfrak{p}_{\mathbb{C}}. \tag{6.29}$$

Now, we focus on the complex Lie algebras $\mathfrak{p}_{+,\mathbb{C}} := \oplus_{\alpha \in \Phi^+(\Delta) \setminus \Phi'} \mathfrak{g}_\alpha$, $\mathfrak{p}_{-,\mathbb{C}} := \oplus_{\alpha \in \Phi^+(\Delta) \setminus \Phi'} \mathfrak{g}_{-\alpha}$, $\mathfrak{h}_{1,+,\mathbb{C}} := \oplus_{\alpha \in \Phi'^+(\Delta)} \mathfrak{g}_\alpha$, and $\mathfrak{h}_{1,-,\mathbb{C}} := \oplus_{\alpha \in \Phi'^+(\Delta)} \mathfrak{g}_{-\alpha}$ generated by $\{K_\alpha\}_{\alpha \in \Phi^+(\Delta) \setminus \Phi'}$, $\{K_{-\alpha}\}_{\alpha \in \Phi^+(\Delta) \setminus \Phi'}$, $\{K_\alpha\}_{\alpha \in \Phi'^+(\Delta)}$, and $\{K_{-\alpha}\}_{\alpha \in \Phi'^+(\Delta)}$, respectively. The relations (6.15) and (6.16) and the property of $\Phi'^+(\Delta)$ imply that

$$[\mathfrak{h}_{1,+,\mathbb{C}}, \mathfrak{h}_{1,+,\mathbb{C}}] \subset \mathfrak{h}_{1,+,\mathbb{C}}, \quad [\mathfrak{h}_{1,-,\mathbb{C}}, \mathfrak{h}_{1,-,\mathbb{C}}] \subset \mathfrak{h}_{1,-,\mathbb{C}}. \tag{6.30}$$

Similarly, the relations (6.15), (6.16) and (6.26) yield that

$$[\mathfrak{p}_{+,\mathbb{C}}, \mathfrak{p}_{+,\mathbb{C}}] \subset \mathfrak{p}_{+,\mathbb{C}}, \quad [\mathfrak{p}_{-,\mathbb{C}}, \mathfrak{p}_{-,\mathbb{C}}] \subset \mathfrak{p}_{-,\mathbb{C}}. \tag{6.31}$$

These facts show that $\mathfrak{p}_{+,\mathbb{C}}$, $\mathfrak{p}_{-,\mathbb{C}}$, $\mathfrak{h}_{1,+,\mathbb{C}}$, and $\mathfrak{h}_{1,-,\mathbb{C}}$ are complex Lie subalgebras of $\mathfrak{g}_{\mathbb{C}}$.

Here, let us consider the adjoint representation of the Lie subalgebra $\mathfrak{h}_1$ including a maximal Cartan subalgebra of $\mathfrak{g}$ on $\mathfrak{g}$. Then, $\mathfrak{h}_1 \subset \mathfrak{g}$ is an invariant subspace. Hence, we can define a real representation of $\mathfrak{h}_1$ on $\mathfrak{g} / \mathfrak{h}_1$. Since $\mathfrak{g} / \mathfrak{h}_1$ is isomorphic to $\mathfrak{p}$, this representation can be regarded as a real representation of $\mathfrak{h}_1$ on $\mathfrak{p}$.

On the other hand, the root subsystem Φ' of the root system Φ satisfies

$$((\Phi \backslash \Phi') + (\Phi \backslash \Phi')) \cap \Phi \subset \Phi' \tag{6.32}$$

if and only if the Lie subalgebra $\mathfrak{h}_1$ including the maximal Cartan subalgebra corresponding to Φ' and the orthogonal complement $\mathfrak{p}$ satisfy $[\mathfrak{p}, \mathfrak{p}] \subset \mathfrak{h}_1$. In the following, the root subsystem Φ' of Φ is called **irreducible** when the condition holds. The condition is equivalent to the condition $[\mathfrak{p}_{\mathbb{C}}, \mathfrak{p}_{\mathbb{C}}] \subset \mathfrak{h}_{1,\mathbb{C}}$ for the complexifications $\mathfrak{p}_{\mathbb{C}}$ and $\mathfrak{h}_{1,\mathbb{C}}$.

When $(\Phi^+(\Delta), \Phi' \subset \Phi)$ is a parabolic system for an irreducible root subsystem Φ' of Φ, $(\Phi^+(\Delta), \Phi' \subset \Phi)$ is called an **irreducible parabolic system**.

Theorem 6.10 *The following conditions are equivalent to each other for a root subsystem Φ' of a root system Φ with a fundamental system Δ in a real linear space V.*

(1) $(\Phi^+(\Delta), \Phi' \subset \Phi)$ *is an irreducible parabolic system.*

(2)

$$((\Phi^+(\Delta) \backslash \Phi') + (\Phi^+(\Delta) \backslash \Phi')) \cap \Phi = \emptyset \tag{6.33}$$

$$((\Phi^+(\Delta) \backslash \Phi') + \Phi') \cap \Phi \subset (\Phi^+(\Delta) \backslash \Phi'). \tag{6.34}$$

Proof **(1)⇒(2)**: The relations (6.26) and (6.32) yield the relation (6.33). The relation (6.34) is none other than the Condition **(3)** for a parabolic system.
(2)⇒(1): Due to the relation (6.34), $(\Phi^+(\Delta), \Phi' \subset \Phi)$ is a parabolic system. The relation (6.33) yields the relation (6.32). So, Φ' is irreducible. ∎

When a Lie algebra $\mathfrak{g}$ is simple, its root system Φ is connected. An irreducible parabolic system $(\Phi^+(\Delta), \Phi' \subset \Phi)$ is called **connected irreducible parabolic system** when Φ is connected. Then, Condition **(2)** guarantees that there exists an element of $\Delta \backslash \Phi'$. If there exist two elements β_1 and β_2 of $\Delta \backslash \Phi'$, we can choose elements $\alpha_1, \ldots, \alpha_l \in \Delta'$ that connect the two elements β_1 and β_2 because the fundamental system is connected, where the inner product between two distinct roots α_i and α_j is assumed to be zero unless $|i - j| = 1$. Further, we assume that $\langle \beta_1, \alpha_1 \rangle = \langle \alpha_l, \beta_2 \rangle = -1$. If these relations do not hold, they hold by exchanging β_1 and β_2 because of the classification given in Table 6.1. Hence, due to the condition (6.24), $\beta_1' := W_{\alpha_l} \circ \cdots \circ W_{\alpha_1}(\beta_1)$ is contained in $\Phi^+ \backslash \Phi'$. Thus, we have $W_{\beta_2}(\beta_1') = \beta_1' + \beta_2$, which contradicts (6.33). So, the set $\Delta \backslash \Phi'$ has the unique element β. Thus, the set $\Phi^+(\Delta) \backslash \Phi'$ is given as $(\beta + span_{\mathbb{Z}_{\geq 0}}(\Delta \backslash \{\beta\})) \cap \Phi$. Due to this fact, all of connected irreducible parabolic systems can be classified by use of the Dynkin diagram describing Δ and the unique simple root β [79, Theorem A.V.1 (c)] [66, Appendix C]. The pair of the Dynkin diagram and the simple root β is

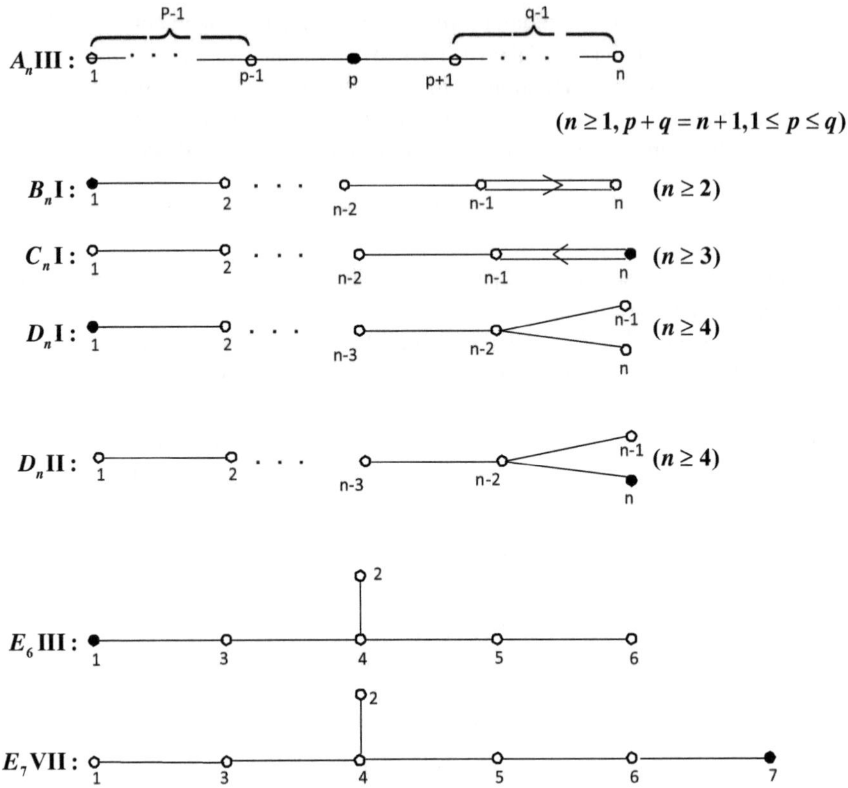

Fig. 6.2 Hermitian Dynkin diagram

classified as Fig. 6.2, in which, the black point corresponds to the simple root β. The figures given in Fig. 6.2 is called a **Hermitian Dynkin diagram**.

When $(\Phi^+(\Delta), \Phi' \subset \Phi)$ is a connected irreducible parabolic system, there exists the maximum root $\lambda(\Phi) \in \Phi$ because the Lie algebra $\mathfrak{g}$ corresponding to Φ is simple, but $\lambda(\Phi)$ does not belong to Φ'. This fact is useful for latter discussions. These discussions yield the following Theorem.

Theorem 6.11 *When a root subsystem Φ' of Φ is irreducible and $\mathfrak{g}$ is simple, we obtain the following items.* $[\mathfrak{p}, \mathfrak{h}_1] = \mathfrak{p}$, $|\Delta \backslash \Delta'| = 1$, *and the real representation of* $\mathfrak{h}_1$ *on* $\mathfrak{p}$ *is irreducible.*

Proof Since $\mathfrak{p} + \mathfrak{h}_1 = [\mathfrak{p} + \mathfrak{h}_1, \mathfrak{p} + \mathfrak{h}_1] = [\mathfrak{p}, \mathfrak{h}_1] + [\mathfrak{p}, \mathfrak{p}] + [\mathfrak{h}_1, \mathfrak{h}_1]$ and $[\mathfrak{p}, \mathfrak{p}] + [\mathfrak{h}_1, \mathfrak{h}_1] = \mathfrak{h}_1$, we have $[\mathfrak{p}, \mathfrak{h}_1] = \mathfrak{p}$. Further, the above discussion yields the relation $|\Delta \backslash \Delta'| = 1$, any element of $\Phi^+ \backslash \Phi'$ can be written as $\beta + m_1\alpha_1 + \ldots + m_l\alpha_l$. Hence, the orbit of K_β with respect to $\mathfrak{h}_{1,\mathbb{C}}$ contains K_α for any $\alpha \in \Phi^+ \backslash \Phi'$. Thus, the representation of $\mathfrak{h}_{1,\mathbb{C}}$ on $\mathfrak{p}_\mathbb{C}$ is irreducible.

When the representation of $\mathfrak{h}_1$ on $\mathfrak{p}$ has a non-trivial invariant subspace, the complexification of the invariant subspace is also an invariant subspace for complexification of the representation, which contradicts the above fact. So, the representation of $\mathfrak{h}_1$ on $\mathfrak{p}$ is also irreducible. ∎

6.6 Structure of $\mathfrak{sp}(r)$*

Section 3.1.2 mentioned that the real symplectic group $\mathrm{Sp}(2r,\mathbb{R})$ inevitably appears in the symmetry related to analytical mechanics. However, since $\mathrm{Sp}(2r,\mathbb{R})$ is a non-compact Lie group, there are several difficulties to treat it. Since it is impossible to deal with it in this section,[2] we discuss the Lie algebra $\mathfrak{sp}(r)$ of the compact Lie group $\mathrm{Sp}(r)$. As both complexifications of the Lie groups $\mathrm{Sp}(r)$ and $\mathrm{Sp}(2r,\mathbb{R})$ are the complex Lie group $\mathrm{Sp}(2r,\mathbb{C})$, both Lie groups have common structures. Since the Lie algebra $\mathfrak{sp}(r)$ is an example of a compact semi simple Lie group belonging to the Case C_r in Sect. 6.5, it is a good example other than $\mathfrak{su}(r)$ to understand the general discussion in Sect. 6.5.

Any matrix on $\mathbb{C}^r$ can be converted to a matrix on $\mathbb{R}^{2r}$ by replacing the multiplication of the pure imaginary number i by that of the matrix $I \otimes J$. An element $X = X_1 + iX_2$ (X_1 and X_2 are real-valued matrices) of $\mathfrak{u}(r)$ can be regarded as an element of $\mathfrak{sp}(r)$ as follows. Since the matrix $\tilde{\kappa}(X) := X_1 \otimes I_2 + X_2 \otimes J$ satisfies

$$\begin{aligned}
&(X_2 \otimes I_2 + X_2 \otimes J)^T J_n = (X_1^T \otimes I_2 - X_2^T \otimes J) I_n \otimes J \\
&= (-X_1 \otimes I_2 - X_2 \otimes J) I_n \otimes J = (-X_1 \otimes I_2 - X_2 \otimes J) I_n \otimes J \\
&= -I_n \otimes J (X_1 \otimes I_2 + X_2 \otimes J),
\end{aligned}$$

it is an element of $\mathfrak{sp}(r)$. As an extension of the map κ defined in Sect. 4.1, using W defined in (4.9), we define

$$\kappa(Y) := (I_r \otimes W) Y (I_r \otimes W)^{-1} \tag{6.35}$$

for a square matrix Y of order $2r$. The map $\tilde{\kappa}$ satisfies $\tilde{\kappa}(X) = \kappa(X \oplus \overline{X})$. Then, the relations $\tilde{\kappa}(\mathfrak{u}(r)) = \mathfrak{sp}(r) \cap \mathfrak{sp}(2r,\mathbb{R})$, $\tilde{\kappa}(\mathrm{U}(r)) = \mathrm{Sp}(r) \cap \mathrm{Sp}(2r,\mathbb{R})$ hold.

Hence, we can regard $\mathfrak{u}(r)$ as a subalgebra of $\mathfrak{sp}(r)$. In the following, we discuss $\mathfrak{sp}(2r,\mathbb{R})$ by using $F^x_{j,l}$, $F^y_{j,l}$, $F^z_{j,l}$, and $F^{z,u}_j$ defined in (4.35), (4.36), (4.37) and (4.44). Then, a maximal Cartan subalgebra $\mathfrak{h}$ of $\mathfrak{sp}(r)$ is given by $\mathfrak{h} := \{X_2 \otimes J | X_2 \text{ is a diagonal matrix of size } r\}$, and is a subalgebra of $\mathfrak{u}(r)$. That is, since $\mathfrak{u}(r)$ is a maximal regular subalgebra of $\mathfrak{sp}(r)$, a root of $\mathfrak{u}(r)$ is a root of $\mathfrak{sp}(r)$. Hence, $\alpha_{j,l} \in \Phi_{\mathfrak{su}(r)}$ can be regarded as an element of $\Phi_{\mathfrak{sp}(r)}$ as follows.

$$\alpha_{j,l}(X_2 \otimes J) = -\langle j|X_2|j\rangle + \langle l|X_2|l\rangle, \quad \forall X_2 \otimes J \in \mathfrak{h}.$$

[2] $\mathrm{Sp}(2r,\mathbb{R})$ will be discussed in Sect. 7.8.

Then, the weight β is defined so that the relation $\beta(X_2 \otimes J) = -2\langle r|X_2|r\rangle$ holds for an element $X_2 \otimes J \in \mathfrak{h}$.

We have

$$(\tilde{\kappa}(X), \tilde{\kappa}(Y))_{\mathfrak{sp}(r)} = -4(r+1)\,\mathrm{Tr}\, XY^\dagger, \quad \forall X, Y \in \mathfrak{u}(r)$$
$$(|r\rangle\langle r| \otimes F_1^z, |r\rangle\langle r| \otimes F_1^z)_{\mathfrak{sp}(r)} = (|r\rangle\langle r| \otimes F_1^y, |r\rangle\langle r| \otimes F_1^y)_{\mathfrak{sp}(r)} = -(r+1).$$

The set $\Phi^+_{\mathfrak{u}(r)}$ of positive roots and the fundamental system Δ of $\Phi_{\mathfrak{sp}(r)}$ are given by $\Phi^+_{\mathfrak{u}(r)} = \{\alpha_{j,l}\}_{1\le j<lr}$ and $\Delta = \{\alpha_{j,j+1}\}_{j=1,\dots,r-1} \cup \{\beta\}$, respectively. Hence,

$$Z_{\alpha_{j,l}} := \frac{-i}{4(r+1)} F^z_{j,l} \otimes J = \frac{1}{4(r+1)} \tilde{\kappa}(F^z_{j,l})$$

$$Z_\beta := \frac{1}{2(r+1)} |r\rangle\langle r| \otimes J = \frac{-1}{4(r+1)} \tilde{\kappa}(2F_r^{z,u})$$

satisfy the condition (6.3). Since $(\alpha_{j,l}, \alpha_{j,l}) = \frac{1}{2(r+1)}$ and $(\beta, \beta) = \frac{1}{r+1}$, we have $\langle \beta, \alpha_{r-1,r}\rangle = -2$. Thus, positive roots except for $\Phi_{\mathfrak{u}(r)}$ are written as $e_j + e_l$ in Table 6.1, and is given as

$$\Phi^+_{\mathfrak{sp}(r)} \backslash \Phi^+_{\mathfrak{u}(r)} = \{\beta_{j,l} := \alpha_{j,l} + 2\alpha_{l,r} + \beta\}_{1\le j\le l<r}, \tag{6.36}$$

where we assume that $\alpha_{j,l} = 0$ for $j = l$. On the other hand, the roots $\{\alpha_{j,j+1}\}_{j=1,\dots,r-1}$ are given in the same way as $\mathfrak{su}(r)$. Here, we remark that the inner product of $\mathfrak{h}$ is different from that of $\mathfrak{su}(r)$ because the inner product is given as the Killing form defined in the Lie algebra $\mathfrak{sp}(r)$, which is larger than $\mathfrak{su}(r)$. Due to (6.36), $(\Phi^+_{\mathfrak{sp}(r)}, \Phi_{\mathfrak{u}(r)} \subset \Phi_{\mathfrak{sp}(r)})$ is a connected irreducible parabolic system. Further, the matrix

$$Z_{\beta_{j,l}} := \frac{-1}{4(r+1)} (|j\rangle\langle j| + |l\rangle\langle l|) \otimes J = \frac{1}{4(r+1)} \tilde{\kappa}(F_j^{z,u} + F_l^{z,u})$$

satisfies (6.3), and

$$F^x_{\alpha_{j,l}} := -iF^x_{j,l} \otimes J = \tilde{\kappa}(F^x_{j,l}), \quad F^y_{\alpha_{j,l}} := F^y_{j,l} \otimes I = \tilde{\kappa}(F^y_{j,l})$$

$$F^x_{\beta_{j,l}} := \frac{1}{\sqrt{2}} (|j\rangle\langle l| + |l\rangle\langle j|) \otimes F_1^z, \quad F^y_{\beta_{j,l}} := \frac{1}{\sqrt{2}} (|j\rangle\langle l| + |l\rangle\langle j|) \otimes F_1^x$$

satisfy (6.7) and (6.8).

6.7 Complex Structure and Coherent State*

Section 6.5 dealt with a Lie subalgebra of $\mathfrak{g}$ including its maximal Cartan subalgebra $\mathfrak{h}$ and its root subsystem. Using this discussion, we extend coherent states to a more general setting because coherent states play an important role in a quantum system. Remember that coherent states are given as an orbit of the ground state with respect to a representation of a group of our interest. The ground state is a stable state and invariant for a subgroup of our group. Hence, the discussions in Sect. 6.5 play an important role. When the root systems of a compact semi simple Lie algebra $\mathfrak{g}$ and its Lie subalgebra $\mathfrak{h}_1$ including its maximal Catran subalgebra, are characterized as a parabolic system($\Phi^+(\Delta)$, $\Phi' \subset \Phi$), as explained later, the homogeneous space G/H_1 has the structure of a complex manifold, where G and H_1 are the Lie groups corresponding to $\mathfrak{g}$ and $\mathfrak{h}_1$.

As an example, we consider a representation space $\mathcal{U}_{\boldsymbol{\lambda}}$ with an arbitrary highest weight $\boldsymbol{\lambda} \in \Lambda_+(\Delta)$. Assume that the stabilizer H_1 is the invariant Lie subgroup $\{g \in G | \mathsf{f}_{\boldsymbol{\lambda}}(g)|\boldsymbol{\lambda}; \boldsymbol{\lambda}\rangle\langle\boldsymbol{\lambda}; \boldsymbol{\lambda}|\mathsf{f}_{\boldsymbol{\lambda}}(g)^\dagger = |\boldsymbol{\lambda}; \boldsymbol{\lambda}\rangle\langle\boldsymbol{\lambda}; \boldsymbol{\lambda}|\}$ of the highest weight vector $|\boldsymbol{\lambda}; \boldsymbol{\lambda}\rangle$. Then, the corresponding Lie subalgebra $\mathfrak{h}_1$ is the invariant Lie subalgebra $\{X \in \mathfrak{g} | \mathsf{f}_{\boldsymbol{\lambda}}(X)|\boldsymbol{\lambda}; \boldsymbol{\lambda}\rangle\langle\boldsymbol{\lambda}; \boldsymbol{\lambda}| - |\boldsymbol{\lambda}; \boldsymbol{\lambda}\rangle\langle\boldsymbol{\lambda}; \boldsymbol{\lambda}|\mathsf{f}_{\boldsymbol{\lambda}}(X) = 0\} \subset \mathfrak{g}$ with respect to the highest weight vector $|\boldsymbol{\lambda}; \boldsymbol{\lambda}\rangle$. Letting Φ' be the root system of the Lie subalgebra $\mathfrak{h}_1$, we see that Φ' is $\{\alpha \in \Phi | \langle\boldsymbol{\lambda}, \boldsymbol{\alpha}\rangle = 0\}$. Condition **(1)** of Theorem 6.8 guarantees that $(\Phi^+(\Delta), \Phi' \subset \Phi)$ is a parabolic system. Hence, coherent states are given as the orbit of the highest weight state $|\boldsymbol{\lambda}; \boldsymbol{\lambda}\rangle\langle\boldsymbol{\lambda}; \boldsymbol{\lambda}|$ with respect to the representation of the Lie group G. Then, the set of coherent states is characterized by the homogeneous space G/H_1, where H_1 is the Lie group corresponding to $\mathfrak{h}_1$. The set $\{F^x_\alpha, F^y_\alpha\}_{\alpha\in\Phi^+_{\boldsymbol{\lambda}}}$ generates the orthogonal complement $\mathfrak{p}$ to $\mathfrak{h}_1$.

For an arbitrary complex vector $\xi = \{\xi_\alpha = \xi^x_\alpha + \xi^y_\alpha i\}_{\alpha\in\Phi^+}$, we define $g(\xi) := \exp(\sum_{\alpha\in\Phi^+} 2i(\xi^x_\alpha F^y_\alpha - \xi^y_\alpha F^x_\alpha))$ and apply Gauss decomposition (6.13) to the complex Lie group corresponding to the complex Lie algebra $\mathfrak{g}_\mathbb{C}$. Then, there exist complex vectors $\tilde{\zeta} = \{\tilde{\zeta}_\alpha\}_{\alpha\in\Phi^+}$, $\zeta' = \{\zeta'_\alpha\}_{\alpha\in\Phi^+}$, and $\eta = \{\eta_j\}_{j=1}^r$ such that

$$\mathsf{f}_{\boldsymbol{\lambda}}(g(\xi)) = \exp\left(\sum_{\alpha\in\Phi^+} \tilde{\zeta}_\alpha \mathsf{K}_{-\alpha}\right) \exp\left(\sum_{j=1}^r \eta_j \mathsf{E}_j\right) \exp\left(\sum_{\alpha\in\Phi^+} \zeta'_\alpha \mathsf{K}_\alpha\right). \tag{6.37}$$

Since, as mentioned in Sect. 6.5, $\mathfrak{h}_{1,-,\mathbb{C}}$ and $\mathfrak{p}_{-,\mathbb{C}}$ are complex Lie subalgebras of $\mathfrak{n}_- = \mathfrak{h}_{1,-,\mathbb{C}} \oplus \mathfrak{p}_{-,\mathbb{C}}$, there exists a complex vector $\{\zeta_\alpha\}_{\alpha\in\Phi^+}$ such that

$$\exp\left(\sum_{\alpha\in\Phi^+} \tilde{\zeta}_\alpha \mathsf{K}_{-\alpha}\right) = \exp\left(\sum_{\alpha\in\Phi^+\setminus\Phi'} \zeta_\alpha \mathsf{K}_{-\alpha}\right) \exp\left(\sum_{\alpha\in\Phi'^+} \zeta_\alpha \mathsf{K}_{-\alpha}\right). \tag{6.38}$$

As a generalization of $U_{\zeta,s}$ given in (4.31), we define U_ζ as the substitute of the right hand side of (6.38) into the right hand side of (6.37). Note that this decomposition holds independently of the choice of the representation $\mathsf{f}_{\boldsymbol{\lambda}}$. Applying them to the vector $|\boldsymbol{\lambda}', \boldsymbol{\lambda}'\rangle$, we have

$$\mathsf{f}_{\boldsymbol{\lambda}}(g(\xi))|\boldsymbol{\lambda}'; \boldsymbol{\lambda}'\rangle \propto \exp\left(\sum_{\alpha\in\Phi^+\setminus\Phi'} \zeta_\alpha \mathsf{K}_{-\alpha}\right)|\boldsymbol{\lambda}'; \boldsymbol{\lambda}'\rangle. \tag{6.39}$$

The complex vector ζ is given as a continuous function of the complex vector ξ. G and H_1 are Lie groups corresponding to $\mathfrak{g}$ and $\mathfrak{h}_1$, respectively, and H_1 is the stabilizer of the state $|\boldsymbol{\lambda}'; \boldsymbol{\lambda}'\rangle\langle\boldsymbol{\lambda}'; \boldsymbol{\lambda}'|$ with respect to the representation of the Lie group G. Hence, due to (6.39), the homogeneous space G/H_1 can be identified with set of ζ except for points with measure 0. Thus, the homogeneous space G/H_1 is isomorphic to the complex Lie group $P_{-,\mathbb{C}}$ generated by the complex Lie algebra $\mathfrak{p}_{-,\mathbb{C}}$. Hence, the homogeneous space G/H_1 has the structure of a complex manifold. This discussion holds independently of the choice of the highest weight $\boldsymbol{\lambda} \in \mathfrak{h}^*$ as long as H_1 is the stabilizer of $|\boldsymbol{\lambda}'; \boldsymbol{\lambda}'\rangle$.

Then, given a highest weight $\boldsymbol{\lambda} \in \mathfrak{h}^*$ and a complex vector $\zeta = \{\zeta_\alpha\}_{\alpha\in\Phi^+\setminus\Phi'}$, we define the coherent vector $|\boldsymbol{\lambda} : \zeta\rangle := C_\zeta \exp(\sum_{\alpha\in\Phi^+\setminus\Phi'} \zeta_\alpha \mathsf{K}_{-\alpha})|\boldsymbol{\lambda}; \boldsymbol{\lambda}\rangle$, where C_ζ is the normalizing constant. By choosing the complex vector ξ satisfying (6.37) and (6.38), the **coherent state** $\rho_{\boldsymbol{\lambda},\zeta} := |\boldsymbol{\lambda} : \zeta\rangle\langle\boldsymbol{\lambda} : \zeta|$ satisfies $\rho_{\boldsymbol{\lambda},\zeta}=\mathsf{f}_{\boldsymbol{\lambda}}(g(\xi))|\boldsymbol{\lambda}; \boldsymbol{\lambda}\rangle\langle\boldsymbol{\lambda}; \boldsymbol{\lambda}|\mathsf{f}_{\boldsymbol{\lambda}}(g(\xi))^\dagger$. The closure of the set $\{\rho_{\boldsymbol{\lambda},\zeta}\}$ has a one-to-one correspondence to the orbit of $|\boldsymbol{\lambda}; \boldsymbol{\lambda}\rangle\langle\boldsymbol{\lambda}; \boldsymbol{\lambda}|$ with respect to the representation of G, i.e., the homogeneous space G/H_1. Due to (3.56), as a generalization of (4.34) to a compact semi simple Lie group, the invariant measure μ_Θ on $\Theta := G/H_1$ defined in the end of Sect. 3.7 satisfies

$$d_{\boldsymbol{\lambda}} \int_\Theta |\boldsymbol{\lambda} : \zeta\rangle\langle\boldsymbol{\lambda} : \zeta|\mu_\Theta(d\zeta) = I, \tag{6.40}$$

which gives a POVM.

Conversely, when the root systems of a compact semi simple Lie algebra $\mathfrak{g}$ and a Lie subalgebra $\mathfrak{h}_1$ including its maximal Catran subalgebra are characterized by a parabolic system $(\Phi^+(\Delta), \Phi' \subset \Phi)$, due to Condition **(1)** of Theorem 6.8, there exists a positive root $\boldsymbol{\lambda} \in \Lambda_+(\Delta)$ such that $\Phi' = \{\alpha \in \Phi|\langle\boldsymbol{\lambda}, \alpha\rangle = 0\}$. In this general setting, the homogeneous space G/H_1 is characterized as the orbit of the state $|\boldsymbol{\lambda}; \boldsymbol{\lambda}\rangle\langle\boldsymbol{\lambda}; \boldsymbol{\lambda}|$ on the representation space of the highest weight $\boldsymbol{\lambda}$. Hence, applying the above discussion, we obtain the following theorem.

Theorem 6.12 *Let $\mathfrak{g}$ and $\mathfrak{h}_1$ be the Lie algebras corresponding to a compact semi simple Lie group G and its Lie subgroup H_1. When the Lie algebra $\mathfrak{h}_1$ includes a maximal Catran subalgebra of $\mathfrak{g}$ and its root system is characterized by an irreducible parabolic system $(\Phi^+(\Delta), \Phi' \subset \Phi)$, the homogeneous space G/H_1 has the structure of a complex manifold.*

Next, let $(\Phi^+(\Delta), \Phi' \subset \Phi)$ be a connected irreducible parabolic system with $\Delta \setminus \Phi' = \{\beta\}$. Also, let $\mathfrak{g}$ and $\mathfrak{h}_1$ be the compact semi simple Lie algebra and its Lie subalgebra that correspond to the connected irreducible parabolic system. When the irreducible representation space $\mathcal{U}_{n\beta}$ of $\mathfrak{g}$ has the irreducible decomposition $\bigoplus_{j=0}^n \mathcal{V}[\mathsf{K}_{\beta,-}^j|n\beta, n\beta\rangle]$ with respect to the representation of $\mathfrak{h}_1$, $\mathcal{U}_{n\beta}$ is called

$\mathfrak{h}_1$-**irreducible**, where $\mathcal{V}[\mathsf{K}^j_{\beta,-}|n\beta, n\beta\rangle]$ is the irreducible representation space of the Lie algebra $\mathfrak{h}_1$ that contains $\mathsf{K}^j_{\beta,-}|n\beta, n\beta\rangle$. For example, in any Hermitian Dynkin diagram, $\mathcal{U}_\beta$ is $\mathfrak{h}_1$-**irreducible** because $\mathfrak{g} / \mathfrak{h}_1$ is irreducible with respect to the representation of $\mathfrak{h}_1$. Under A_nIII Type of Hermitian Dynkin diagram, when $p = 1$, $\mathcal{U}_{m\beta}$ is $\mathfrak{u}(n-1)$-**irreducible** for any integer m.

6.8 Structure of Non-compact Semi Simple Lie Algebra*

All of groups discussed in relation to analytical mechanics in Sect. 3.1.2 are non-compact groups. Hence, for an application to quantum system, we need to treat a non-compact group. So, we investigate the structure of a non-compact semi simple Lie group G. For this study, we discuss the corresponding Lie algebra $\mathfrak{g}$. Since it is semi simple, the Killing form is non-degenerate. However, it is not negative definite because the Lie algebra $\mathfrak{g}$ is not compact. Thus, we focus on the complex Lie algebra $\mathfrak{g}_{\mathbb{C}} := \mathfrak{g} \otimes \mathbb{C}$, which is a complexification of $\mathfrak{g}$. Especially, the complexification of a simple Lie algebra is a complex simple Lie algebra. Non-compact semi simple Lie algebras whose complexification is a classical type of simple Lie algebra are limited to $\mathfrak{sl}(n, \mathbb{R})$, $\mathfrak{su}(p, q)$, $\mathfrak{su}^*(2m)$, $\mathfrak{so}(p, q)$, $\mathfrak{so}^*(2m)$, $\mathfrak{sp}(2n, \mathbb{R})$, and $\mathfrak{sp}(p, q)$ given in Table 3.6 [67, p. 6]. In this section, we avoid to treat all of non-compact semi simple Lie algebras, and discuss only a special class of non-compact semi simple Lie algebras, whose analysis is not so difficult.

For this purpose, for a compact subalgebra $\mathfrak{h}_0$ of $\mathfrak{g}$, we choose a real linear subspace $\mathfrak{u}$ of $\mathfrak{g}_{\mathbb{C}}$ such that its dimension is $\mathfrak{g}$, its Killing form is negative definite, and $\mathfrak{h}_0 \subset \mathfrak{u}$. Then, $\mathfrak{u}$ is a compact semi simple Lie algebra. Letting $\mathfrak{k} := \mathfrak{g} \cap \mathfrak{u}$ and $\mathfrak{p} := \mathfrak{g} \cap i\,\mathfrak{u}$, we find that $\mathfrak{k}$ is a compact Lie algebra and is called a **maximal compact subalgebra**. Then, we obtain the direct sum decomposition $\mathfrak{g} = \mathfrak{k} + \mathfrak{p}$, which is called the **Cartan decomposition** of $\mathfrak{g}$ [67, p. 3]. Then, the Killing form is negative definite on $\mathfrak{k}$ and is positive definite on $\mathfrak{p}$. Since $[\mathfrak{u}, \mathfrak{u}] \subset \mathfrak{u}$, $[\mathfrak{u}, i\,\mathfrak{u}] \subset i\,\mathfrak{u}$, and $[i\,\mathfrak{u}, i\,\mathfrak{u}] \subset \mathfrak{u}$, we have

$$[\mathfrak{k}, \mathfrak{k}] \subset \mathfrak{k}, \quad [\mathfrak{k}, \mathfrak{p}] \subset \mathfrak{p}, \quad [\mathfrak{p}, \mathfrak{p}] \subset \mathfrak{k}. \tag{6.41}$$

Since $[\mathfrak{g}, \mathfrak{g}] = \mathfrak{g}$, we also have $[\mathfrak{k}, \mathfrak{k}] + [\mathfrak{p}, \mathfrak{p}] = \mathfrak{k}$ and $[\mathfrak{k}, \mathfrak{p}] = \mathfrak{p}$. Using the map $\tau : X_1 + X_2 \mapsto X_1 - X_2$ for $X_1 \in \mathfrak{k}$ and $X_2 \in \mathfrak{p}$, we define the inner product $B(X, Y) := -(X, \tau(Y))_{\mathfrak{g}}$, which is positive definite. For $X_1 \in \mathfrak{k}$ and $X_2 \in \mathfrak{p}$, $\mathsf{ad}(X_1)$ and $\mathsf{ad}(X_2)$ are alternative and symmetric matrices under the CONS with respect to the inner product, respectively. Define $K := \{\exp(X)\}_{X \in \mathfrak{k}}$ and $P := \{\exp(X)\}_{X \in \mathfrak{p}}$. Then, K is a maximal compact subgroup of G, and we obtain the **Cartan decomposition** $G = KP$ of the Lie group G.

However, since a non-compact simple Lie group G and its Lie algebra $\mathfrak{g}$ have too complicated structure, it is too difficult to discuss the general theory. To avoid such a difficulty, we assume that there exists a compact maximal Cartan subalgebra $\mathfrak{h}$. Since many physically important examples satisfy this condition, our analysis is restricted to this case.

Similar to the case when $\mathfrak{g}$ is compact, which is discussed in Sect. 6.2, we define the inner product of $\mathfrak{h}^*$ and define $Z_\beta \in \mathfrak{h}$ for $\beta \in \mathfrak{h}^*$. Then, there exists a basis of $\mathfrak{g}$ such that an element of $\mathfrak{h}$ is written as an alternative matrix given in (3.46) of Lemma 3.9 under the basis. Hence, similar to the compact case, as subsets of $\mathfrak{h}^*$, we can define the root system $\Phi_\mathfrak{g}$ of $\mathfrak{g}$ and the root system $\Phi_\mathfrak{k}$ of $\mathfrak{k}$, and choose $\{F_\alpha^x, F_\alpha^y\}$ for $\alpha \in \Phi_\mathfrak{g}$ such that

$$[X, F_\alpha^x] = \alpha(X) F_\alpha^y, \quad [X, F_\alpha^y] = -\alpha(X) F_\alpha^x, \quad \forall X \in \mathfrak{h}.$$

Since $\mathfrak{g}$ is semi simple, Theorem 6.6 guarantees that a different pair of basis $F_\alpha^{x\prime}$ and $F_\alpha^{y\prime}$ does not correspond to the same root α. Similar to (6.7), we can show that $[F_\alpha^x, F_\alpha^y] = -(F_\alpha^x, F_\alpha^x)_\mathfrak{g} Z_\alpha = -(F_\alpha^y, F_\alpha^y)_\mathfrak{g} Z_\alpha$, which implies that $(F_\alpha^x, F_\alpha^x)_\mathfrak{g} = (F_\alpha^y, F_\alpha^y)_\mathfrak{g}$.

Since $\mathfrak{k}$ is a compact Lie algebra, the real numbers $(F_\alpha^x, F_\alpha^x)_\mathfrak{g}$ and $(F_\alpha^y, F_\alpha^y)_\mathfrak{g}$ are negative values for $\alpha \in \Phi_\mathfrak{k}$. So, the subalgebra generated by $\{F_\alpha^x, F_\alpha^y, Z_\alpha\}$ is isomorphic to $\mathfrak{su}(2)$. Similar to (6.8), we choose the constant factors of F_α^x, F_α^y such that $(F_\alpha^x, F_\alpha^x)_\mathfrak{g} = (F_\alpha^y, F_\alpha^y)_\mathfrak{g} = -\frac{1}{(\alpha,\alpha)}$. On the other hand, since F_α^x and F_α^x belong to $\mathfrak{p}$ for $\alpha \in \Phi_\mathfrak{p} := \Phi_\mathfrak{g} \setminus \Phi_\mathfrak{k}$, the real number $(F_\alpha^x, F_\alpha^x)_\mathfrak{g}$ is positive. So, the subalgebra generated by $\{F_\alpha^x, F_\alpha^y, Z_\alpha\}$ is isomorphic to $\mathfrak{sl}(2, \mathbb{R})$. Then, we choose the constant factors of F_α^x and F_α^y such that $(F_\alpha^x, F_\alpha^x)_\mathfrak{g} = (F_\alpha^y, F_\alpha^y)_\mathfrak{g} = \frac{1}{(\alpha,\alpha)}$. So, the basis of $\mathfrak{g}$ is uniquely determined from $\mathfrak{h}$.

Also, due to the final property $[\mathfrak{p}, \mathfrak{p}] \subset \mathfrak{k}$ of (6.41), the root system $\Phi_\mathfrak{k}$ of $\mathfrak{k}$ is irreducible as a root subsystem of the root system Φ of $\mathfrak{g}$. In the discussions up to the previous section, we have assumed that $\mathfrak{g}$ is compact. However, the most crucial assumption is the non-degeneracy of killing form, i.e., the semi-simplicity of $\mathfrak{g}$. In this case, since $\mathfrak{g}$ is not compact, but semi simple, a similar discussion is possible. Then, by the same discussion as the previous section, we define the root system Φ of $\mathfrak{g}$, the fundamental system $\Delta = \{\alpha, \ldots, \alpha_r\}$, the set $\Phi^+(\Delta)$ of positive roots, and the root system $\Phi_\mathfrak{k}$ of $\mathfrak{k}$.

Then, since the Lie algebra $\mathfrak{k}$ is assumed to contain the maximal Catran subalgebra $\mathfrak{h}$ of $\mathfrak{g}$, due to Theorem 6.9, $(\Phi^+(\Delta), \Phi_\mathfrak{k} \subset \Phi)$ is a parabolic system. Further, the third relation (6.41) guarantees that $(\Phi^+(\Delta), \Phi_\mathfrak{k} \subset \Phi)$ is an irreducible parabolic system.

Such a semi simple Lie algebra $\mathfrak{g}$ is called **Hermitian type** and is classified in Table 6.2 [79, Theorem A.V.1(c)] [66, Appendix C]. That is, a Hermitian type Lie algebra $\mathfrak{g}$ is described by an irreducible parabolic system. Conversely, for an irreducible parabolic system $(\Phi^+(\Delta), \Phi_\mathfrak{k} \subset \Phi)$, we can construct a semi simple Lie algebra $\mathfrak{g}$ and its maximal compact Lie subalgebra $\mathfrak{k}$ whose root systems are Φ and $\Phi_\mathfrak{k}$, by tracing the above discussion. Hence, the classification of Hermitian type simple Lie algebras can be reduced to the classification of connected irreducible parabolic systems.

Table 6.2 Hermitian type Lie algebra

Type	$\mathfrak{g}$	$\mathfrak{g}_{\mathbb{C}}$	$\mathfrak{k}$
A_nIII $p+q=n$	$\mathfrak{su}(p, q)$	$\mathfrak{sl}(n, \mathbb{C})$	$\mathfrak{su}(p) \oplus \mathfrak{su}(q) \oplus \mathfrak{u}(1)$
B_nI	$\mathfrak{so}(2n-1, 2)$	$\mathfrak{so}(2n+1, \mathbb{C})$	$\mathfrak{so}(2n-1)$
C_nI	$\mathfrak{sp}(2n, \mathbb{R})$	$\mathfrak{sp}(2n, \mathbb{C})$	$\mathfrak{u}(n)$
D_nI	$\mathfrak{so}(2n-2, 2)$	$\mathfrak{so}(2n, \mathbb{C})$	$\mathfrak{so}(2n-2)$
D_nIII	$\mathfrak{so}^*(2n)$	$\mathfrak{so}(2n, \mathbb{C})$	$\mathfrak{u}(n)$
E_6III	$\mathfrak{e}_6^{-14}$	$\mathfrak{e}_6$	$\mathfrak{so}(10) \oplus \mathbb{R}$
E_7VII	$\mathfrak{e}_7^{-25}$	$\mathfrak{e}_7$	$\mathfrak{e}_6 \oplus \mathbb{R}$

6.9 Skew-Hermitian Representation of Non-compact Simple Lie Algebra*

Next, we address a skew-Hermitian representation of a non-compact simple Lie algebra $\mathfrak{g}$. A skew-Hermitian representation of non-compact simple Lie algebra $\mathfrak{g}$ does not necessarily have the highest weight in general. However, a real quantum system is usually characterized by a representation that has the highest weight. Such a representation is called **highest weight representation**. It is often called a **discrete series representation** [67].

When a non-compact simple Lie algebra $\mathfrak{g}$ has a compact maximal Cartan subalgebra, it has a highest weight representation [67, Theorem 9.20]. In the following, for simplicity, we address a Hermitian type simple Lie algebra $\mathfrak{g}$ as a special case of simple Lie algebras. Also, we deal with a highest weight skew-Hermitian representation of $\mathfrak{g}$.

Under a highest weight skew-Hermitian representation f of $\mathfrak{g}$, for a simple root $\alpha_j \in \Delta$ and a positive root $\alpha \in \Phi^+(\Delta)$, we define

$$\mathsf{E}_j := -i\mathsf{f}(Z_{\alpha_j}), \quad \mathsf{K}_{\pm\alpha} := -i\mathsf{f}(F_\alpha^x) \mp \mathsf{f}(F_\alpha^y). \tag{6.42}$$

Then, for an arbitrary root $\alpha \in \Phi$, we have

$$[\mathsf{E}_{j'}, \mathsf{E}_j] = 0 \tag{6.43}$$

$$[\mathsf{E}_{j'}, \mathsf{K}_\alpha] = \alpha(Z_{\alpha_j})\mathsf{K}_\alpha. \tag{6.44}$$

Since the representation f is skew-Hermitian, we find that $(\mathsf{K}_\alpha)^\dagger = \mathsf{K}_{-\alpha}$ and E_j is self-adjoint. In the following, we assume that the skew-Hermitian representation f is irreducible, the relation $\mathsf{K}_\alpha|\boldsymbol{\lambda}; \boldsymbol{\lambda}\rangle = 0$ holds for any positive root $\alpha \in \Phi^+$, and there exists a normalized simultaneous eigenvector $|\boldsymbol{\lambda}; \boldsymbol{\lambda}\rangle$ of $\mathsf{E}_1, \ldots, \mathsf{E}_r$ (**highest weight vector**). Then, the representation f is a **highest weight representation**.

In this case, an element $m \in \mathfrak{h}^*$ of $\mathfrak{h}^*$ is called a **weight** when

$$\mathcal{U}_m^w := \{v \in \mathcal{U} | \mathsf{f}(X)v = im(X)v, \quad \forall X \in \mathfrak{h}\} \neq \{0\}.$$

In particular, a weight $\boldsymbol{\lambda} \in \mathfrak{h}^*$ is called the **highest weight** when the relation $\mathsf{f}(X)|\boldsymbol{\lambda}; \boldsymbol{\lambda}\rangle = i\boldsymbol{\lambda}(X)|\boldsymbol{\lambda}; \boldsymbol{\lambda}\rangle$ holds for any element $X \in \mathfrak{h}$. The highest weight uniquely determines the irreducible highest weight skew-Hermitian representation, and the relation $\mathcal{U}_{\boldsymbol{\lambda}}^w = \mathbb{R}|\boldsymbol{\lambda}; \boldsymbol{\lambda}\rangle$ holds. Then, we denote the representation space of the representation by $\mathcal{U}_{\boldsymbol{\lambda}}$, and the representation by $\mathsf{f}_{\boldsymbol{\lambda}}$. Then, the set of weights is written as $\Lambda(\boldsymbol{\lambda})$.

Similar to the discussion in the previous sections, the representation space $\mathcal{U}_{\boldsymbol{\lambda}}$ is generated by $\{\prod_{\alpha \in \Phi^-}(\mathsf{K}_\alpha)^m|\boldsymbol{\lambda}; \boldsymbol{\lambda}\rangle\}$. Hence, due to (6.43), $\mathcal{U}_{\boldsymbol{\lambda}}$ is given as $\mathcal{U}_{\boldsymbol{\lambda}} = \oplus_m \mathcal{U}_m^w$ as the decomposition to simultaneous eigenspace of elements of $\mathfrak{h}$ by using weights m.

For $\boldsymbol{\alpha} \in \Phi_{\mathfrak{k}}^+(\Delta)$, the subspace generated by $\{F_{\boldsymbol{\alpha}}^x, F_{\boldsymbol{\alpha}}^y, [F_{\boldsymbol{\alpha}}^x, F_{\boldsymbol{\alpha}}^y]\}$ is isomorphic to $\mathfrak{su}(2)$. Hence, since the action of these elements are isomorphic to the skew-Hermitian representation of $\mathfrak{su}(2)$, the highest weight $\boldsymbol{\lambda} \in \mathfrak{h}^*$ satisfies

$$\langle \boldsymbol{\lambda}, \boldsymbol{\alpha} \rangle = -2\boldsymbol{\lambda}([F_{\boldsymbol{\alpha}}^x, F_{\boldsymbol{\alpha}}^y]) \in \mathbb{Z}_{\geq 0}. \tag{6.45}$$

On the other hand, for any element $\boldsymbol{\alpha} \in \Phi^+(\Delta) \setminus \Phi_{\mathfrak{k}}^+(\Delta)$, the subspace generated by $\{F_{\boldsymbol{\alpha}}^x, F_{\boldsymbol{\alpha}}^y, [F_{\boldsymbol{\alpha}}^x, F_{\boldsymbol{\alpha}}^y]\}$ is isomorphic to $\mathfrak{sl}(2, \mathbb{R})$. Hence, since the action of these elements is isomorphic to the skew-Hermitian representation of $\mathfrak{sl}(2, \mathbb{R})$, the relation $\langle m, \boldsymbol{\alpha} \rangle \leq 0$ holds for an arbitrary element $m \in \Lambda(\boldsymbol{\lambda})$. Thus, unless $\boldsymbol{\lambda} = 0$, the relation

$$\langle \boldsymbol{\lambda}, \boldsymbol{\alpha} \rangle < 0 \tag{6.46}$$

holds.

Conversely, when $\boldsymbol{\lambda} \in \mathfrak{h}^*$ satisfies the **Harish-Chandra condition**

$$\langle \boldsymbol{\lambda} + \delta, \boldsymbol{\lambda}(\Phi) \rangle < 0, \tag{6.47}$$

there exists an irreducible skew-Hermitian representation with the highest weight $\boldsymbol{\lambda} \in \mathfrak{h}^*$ [79, p. 547] [67, p. 158] [44, Lemma 27]. Further, the formal dimension $d_{\boldsymbol{\lambda}}$ of the representation space $\mathcal{U}_{\boldsymbol{\lambda}}$ satisfies the formula by Harish-Chandra

$$d_{\boldsymbol{\lambda}} = \frac{\prod_{\alpha \in \Phi^+(\Delta)} |\langle \boldsymbol{\lambda} + \delta, \boldsymbol{\alpha} \rangle|}{\prod_{\alpha \in \Phi^+(\Delta)} \langle \delta, \boldsymbol{\alpha} \rangle},$$

which can be regarded as a non-compact version of Weyl's dimension formula [44, Theorem 4] [79, p. 547].

Assume that the highest weight $\boldsymbol{\lambda}$ of the irreducible skew-Hermitian representation satisfies

$$\langle \boldsymbol{\lambda}, \boldsymbol{\alpha} \rangle = 0, \forall \boldsymbol{\alpha} \in \Delta \cap \Phi_{\mathfrak{k}}. \tag{6.48}$$

Then, similar to the compact case, the representation space $\mathcal{U}_{\boldsymbol{\lambda}}$ is called **$\mathfrak{k}$-irreducible** when the irreducible decomposition of the representation of $\mathfrak{k}$ is given as $\bigoplus_{j=0}^{\infty} \mathcal{V}[\mathsf{K}_{-\beta}^{j}|\boldsymbol{\lambda};\boldsymbol{\lambda}\rangle]$. For example, when $\mathfrak{g}$ is $\mathfrak{su}(1, n-1)$, the highest weight representation satisfying (6.48) is $\mathfrak{u}(n-1)$-irreducible.

Next, similar to Sect. 6.7, we define the coherent states for irreducible skew-Hermitian representation of $\mathfrak{g}$ with non-trivial (non-zero) highest weight $\boldsymbol{\lambda} \in \mathfrak{h}^*$. Let $\mathfrak{h}_1$ be the subalgebra that does not change the highest weight state $|\boldsymbol{\lambda};\boldsymbol{\lambda}\rangle\langle\boldsymbol{\lambda};\boldsymbol{\lambda}|$. Then, $\mathfrak{h}_1$ contains a compact maximal Cartan subalgebra $\mathfrak{h}$ of $\mathfrak{g}$, and its root system $\Phi_{\mathfrak{h}_1}$ is given as $\{\alpha \in \Phi | (\alpha, \boldsymbol{\lambda}) = 0\}$.

We define $\Phi_{\boldsymbol{\lambda}}^{+} := \{\alpha \in \Phi^{+} | (\alpha, \boldsymbol{\lambda}) \neq 0\}$. For a complex vector $\zeta = \{\zeta_\alpha\}_{\alpha \in \Phi_{\boldsymbol{\lambda}}^{+}}$, we define a **coherent vector** $|\boldsymbol{\lambda} : \zeta\rangle := C_\zeta \exp(\sum_{\alpha \in \Phi_{\boldsymbol{\lambda}}^{+}} \zeta_\alpha \mathsf{K}_{-\alpha})|\boldsymbol{\lambda};\boldsymbol{\lambda}\rangle$, where the normalizing constant C_ζ is chosen such that the norm of $|\boldsymbol{\lambda} : \zeta\rangle$ is 1. Then, similar to the discussion in Sect. 6.7, applying Gauss decomposition (6.13) to the complex Lie group corresponding to the complex Lie algebra $\mathfrak{g}_{\mathbb{C}}$, we can show that there exist complex vectors ξ, ζ', η satisfying (6.37) for a highest weight $\boldsymbol{\lambda} \in \mathfrak{h}^*$ and a complex vector ζ. Hence, the **coherent state** $\rho_{\boldsymbol{\lambda},\zeta} := |\boldsymbol{\lambda} : \zeta\rangle\langle\boldsymbol{\lambda} : \zeta|$ satisfies $\rho_{\boldsymbol{\lambda},\zeta} = \mathsf{f}_{\boldsymbol{\lambda}}(g(\xi))|\boldsymbol{\lambda};\boldsymbol{\lambda}\rangle\langle\boldsymbol{\lambda};\boldsymbol{\lambda}|\mathsf{f}_{\boldsymbol{\lambda}}(g(\xi))^{\dagger}$. When Harish-Chandra condition (6.47) holds, due to (3.56), as a generalization of (6.40) to a non-compact semi simple Lie group, we have

$$d_{\boldsymbol{\lambda}} \int_{\Theta} |\boldsymbol{\lambda} : \zeta\rangle\langle\boldsymbol{\lambda} : \zeta| \mu_{\Theta}(d\zeta) = I, \tag{6.49}$$

where μ_Θ is the invariant measure of $\Theta := G/H_1$ defined in Sect. 3.7, and $d_{\boldsymbol{\lambda}}$ is the formal dimension of the representation $\mathsf{f}_{\boldsymbol{\lambda}}$ [44, Theorem 4]. This relation gives a POVM by coherent states.

Further, we choose a highest weight representation such that the stabilizer of the highest weight state is a maximal compact Lie subgroup K. Then, similar to Theorem 6.12, we can show the following theorem.

Theorem 6.13 *Let G be a Hermitian type Lie group, and K be its maximal compact Lie subgroup K. Then, the homogeneous space G/K has the structure of a complex manifold.*

6.10 General Non-compact Lie Algebra and Its Skew-Hermitian Representation*

Up to the previous section, we imposed the semi simple condition for our Lie algebra $\mathfrak{g}$. In the following, we deal with a real Lie algebra that is not necessarily semi simple. However, since it is quite difficult to discuss a real Lie algebra $\mathfrak{g}$ without any additional condition, we assume that the Lie algebra $\mathfrak{g}$ has a compact maximal Cartan subalgebra $\mathfrak{h}$. Then, there exists a maximal compact subalgebra $\mathfrak{k}$ including $\mathfrak{h}$. Any element of the center $\mathfrak{c}(\mathfrak{g}) = \{X \in \mathfrak{h} | (X, X)_{\mathfrak{g}} = 0\}$ of $\mathfrak{h}$ is commutative with any element of $\mathfrak{g}$. We choose a subspace $\mathfrak{g}_0$ of $\mathfrak{g}$ that is isomorphic to the quotient space $\mathfrak{g}/\mathfrak{c}(\mathfrak{g})$, and define $\mathfrak{h}_0 := \mathfrak{h} \cap \mathfrak{g}_0$. Similar to the semi simple case, we define the element $Z_\beta \in \mathfrak{h}_0$ for $\beta \in \mathfrak{h}_0^*$ and the positive definite inner product (,) on $\mathfrak{h}_0^*$. Since all of elements of $\mathfrak{h}_0$ are commutative with each other, we can choose the basis $\{F_j^x, F_j^y\}$ of the orthogonal complement of $\mathfrak{h}_0$ in the space $\mathfrak{g}_0$ such that all of elements of $\mathfrak{h}_0$ are written as the form (3.46) of Lemma 3.9. Then, the relation

$$\mathsf{ad}(X)F_j^x = \alpha_j(X)F_j^y, \quad \mathsf{ad}(X)F_j^y = -\alpha_j(X)F_j^x \tag{6.50}$$

holds for any element $X \in \mathfrak{h}_0$.

Let $\Phi_\mathfrak{g}^+$ be the set of elements α_j satisfying the above condition, and $\Phi_\mathfrak{g}$ be the set of elements α_j and $-\alpha_j$. Then, we choose the signs of F_j^y such that there exists an element of $X_1 \in \mathfrak{h}_0$ satisfying that $\Phi_\mathfrak{g}^+ = \{\alpha \in \Phi | \alpha(X_1) > 0\}$. Here, Φ generates $\mathfrak{h}_0^*$, but $\mathfrak{g}$ is not necessarily semi simple. Hence, Φ does not necessarily satisfy the condition for a root system. So, it is called a **root set**. The root set $\Phi_\mathfrak{k}$ of a maximally compact subalgebra $\mathfrak{k}$ is a subset of Φ. We define $\Phi_\mathfrak{p} := \Phi \backslash \Phi_\mathfrak{k}$.

Then, we consider the decomposition $\mathfrak{g}_0 = \mathfrak{h}_0 \oplus \bigoplus_{\alpha\in\Phi_\mathfrak{k}^+} \mathfrak{g}[\alpha] \oplus \bigoplus_{\alpha\in\Phi_\mathfrak{p}^+} \mathfrak{g}[\alpha]$ of $\mathfrak{g}_0$, where $\Phi_\mathfrak{k}^+ := \Phi_\mathfrak{k} \cap \Phi^+$ and $\Phi_\mathfrak{p}^+ := \Phi_\mathfrak{p} \cap \Phi^+$. For the pair $\Phi_\mathfrak{k} \subset \Phi$, we can show the equivalence among Conditions **(2)**, **(3)** and **(4)** of Theorem 6.8. The pair $(\Phi^+, \Phi_\mathfrak{k} \subset \Phi)$ is called a **parabolic system** when one of the above conditions holds. Further, the adjoint representation of $\mathfrak{g}$ induces a representation of $\mathfrak{k}$ on $\mathfrak{g}/\mathfrak{k} \cong \mathfrak{p} := \bigoplus_{\alpha\in\Phi_\mathfrak{p}^+} \mathfrak{g}[\alpha]$. Additionally, when the induced representation of $\mathfrak{k}$ is irreducible, $\mathfrak{g}$ is Hermitian-type. Then, the highest weight γ with respect to the representation of $\mathfrak{k}$ on $\mathfrak{g}/\mathfrak{k}$ is the maximum element of $\Phi_\mathfrak{p}^+$. So, we have $\Phi_\mathfrak{p}^+ = (\gamma - span_{\mathbb{Z}_{\geq 0}} \Delta_\mathfrak{k}) \cap \Phi^+$, where $\Delta_\mathfrak{k}$ is a fundamental system of $\Phi_\mathfrak{k}$.

When we define $\mathsf{K}_{\pm\alpha}$ for a representation f of $\mathfrak{g}$ in the same way as (6.42), the relation $[-i\mathsf{f}(X), \mathsf{K}_\alpha] = \alpha(X)\mathsf{K}_\alpha$ holds for an arbitrary element $X \in \mathfrak{h}$ in the same way. In the following, we consider the highest weight representation characterized by the highest weight vector. However, it is not so easy to write down the necessary and sufficient condition of the highest weight $\boldsymbol{\lambda}$ for the representation to be skew-Hermitian in the general case. Especially, the representation space $\mathcal{U}_{\boldsymbol{\lambda}}$ is called $\mathfrak{k}$-irreducible when the following three conditions hold; (1) The Lie algebra $\mathfrak{g}$ is Hermitian-type. (2) The highest weight $\boldsymbol{\lambda}$ is orthogonal to $\Delta_\mathfrak{k}$. (3) The irreducible decomposition of the representation space $\mathcal{U}_{\boldsymbol{\lambda}}$ is given as $\bigoplus_{j=0}^\infty \mathcal{V}[\mathsf{K}_{-\gamma}^j|\boldsymbol{\lambda}; \boldsymbol{\lambda}\rangle]$ by using the highest weight vector $|\boldsymbol{\lambda}; \boldsymbol{\lambda}\rangle$.

Next, we define the coherent state for the irreducible skew-Hermitian representation of $\mathfrak{g}$ with the highest weight $\boldsymbol{\lambda} \in \mathfrak{h}_0^*$ in the same way as Sect. 6.7. Let $\mathfrak{h}_1$ be the subalgebra of elements of $\mathfrak{g}$ that does not change the highest weight state $|\boldsymbol{\lambda}; \boldsymbol{\lambda}\rangle\langle\boldsymbol{\lambda}; \boldsymbol{\lambda}|$. We assume that $\mathfrak{h}_1$ is included in $\mathfrak{k}$. Then, $\mathfrak{h}_1$ includes a compact maximal Cartan subalgebra $\mathfrak{h}$ of $\mathfrak{g}$, and the root set $\Phi_{\mathfrak{h}_1}$ is given as $\{\alpha \in \Phi | \boldsymbol{\lambda}(Z_\alpha) = 0\}$. Using $\Phi_{\boldsymbol{\lambda}}^+ := \{\alpha \in \Phi_\mathfrak{k}^+ | \boldsymbol{\lambda}(Z_\alpha) \neq 0\}$, we define the coherent vector $|\boldsymbol{\lambda} : \zeta\rangle := C_\zeta \exp(\sum_{\alpha\in\Phi_{\boldsymbol{\lambda}}^+} \zeta_\alpha \mathsf{K}_{-\alpha})|\boldsymbol{\lambda}; \boldsymbol{\lambda}\rangle$ for a complex vector $\zeta = \{\zeta_\alpha\}_{\alpha\in\Phi_{\boldsymbol{\lambda}}^+}$, where C_ζ is chosen such that the norm of $|\boldsymbol{\lambda} : \zeta\rangle$ is 1.

Let H_1 be the Lie group corresponding to the Lie algebra $\mathfrak{h}_1$. Then, the homogeneous space G/H_1 is given as the closure of the set $\{|\boldsymbol{\lambda} : \zeta\rangle\langle\boldsymbol{\lambda} : \zeta|\}_\zeta$. Hence, when the formal dimension $d_{\boldsymbol{\lambda}}$ can be defined by (3.54), we regard ζ as an element of the homogeneous space G/H_1 and focus on the invariant measure μ_Θ on the $\Theta := G/H_1$ defined in the end of Sect. 3.7. As a generalization of (6.49) to the case when $\mathfrak{g}$ is not necessarily semi simple, we obtain the relation $d_{\boldsymbol{\lambda}} \int_\Theta |\boldsymbol{\lambda} : \zeta\rangle\langle\boldsymbol{\lambda} : \zeta|\mu_\Theta(d\zeta) = I$. This relation shows that the resolution of the identity I by the Coherent state coherent states $\rho_{\boldsymbol{\lambda},\zeta} := |\boldsymbol{\lambda} : \zeta\rangle\langle\boldsymbol{\lambda} : \zeta|$ forms a POVM.

In this way, the discussion given in the previous section can be generalized to the case when $\mathfrak{g}$ is not necessarily semi simple. The next chapter discusses an example of such a Lie algebra $\mathfrak{g}$.

Chapter 7
Bosonic System and Quantum Optics

Abstract When more than one particles can simultaneously share the same state of the quantum system, the particle is called a boson. A photon is a typical example of a boson. In the bosonic system, a unitary representation of Heisenberg group is given. In the photonic case, the coherent state of the representation is called coherent light and is physically implementable. The operation corresponding to the representation of SU(1, 1) is called squeezing. When we apply the squeezing to the vacuum state, which is the state with no photon, the resultant state is called a squeezed state and plays an important role in the quantum optical system. In fact, the Heisenberg group corresponds to the transformation of the position and the momentum in analytical mechanics. The canonical transformation is a typical transformation preserving the product of the Heisenberg group. This chapter discusses the representation that describes the transformation for the position and the momentum by Heisenberg group as well as the canonical transformation.

7.1 One-Mode Bosonic System

7.1.1 One-Variable System

This section addresses the one-mode bosonic system. As the first step, we focus on the quantum system of a particle with one-dimensional freedom, i.e., the one-variable system, which is the quantum system with respect to the position of a particle on the one-dimensional space $\mathbb{R}$ and is given as the linear space $L^2(\mathbb{R})$ of square integrable functions from $\mathbb{R}$ to $\mathbb{C}$ with the norm $\|f\|^2 := \frac{1}{\sqrt{2\pi}} \int_{-\infty}^{\infty} |f(x)|^2 dx$. In this system, the physical quantity describing the position is given as the multiplication operator Q defined as[1]

$$(\mathsf{Q}f)(x) := xf(x). \tag{7.1}$$

[1]For the precise definition of the operators in this chapter, see Sect. 1.6.

M. Hayashi, *Group Representation for Quantum Theory*,
DOI 10.1007/978-3-319-44906-7_7

On the other hand, the momentum is described by the operator $\hat{\mathsf{P}} := \hbar \mathsf{P}$ by using the **Plank constant** $\hbar$ and the operator P defined by

$$(\mathsf{P} f)(x) := -i\frac{d}{dx} f(x). \tag{7.2}$$

Since the operators Q and P satisfy the conditions $\mathsf{Q}^\dagger = \mathsf{Q}$ and $\mathsf{P}^\dagger = \mathsf{P}$, they are self-adjoint operators[84].[2] Then, these operators satisfy the following commutation relation

$$[\mathsf{Q}, \mathsf{P}] = iI. \tag{7.3}$$

That is, two physical quantities satisfy the commutation relation

$$[\mathsf{Q}, \hat{\mathsf{P}}] = i\hbar I. \tag{7.4}$$

In the following, we write Plank constant $\hbar$ as λ for the sake of the consistency with the representation-theoretical notation. Hence, we define the skew-Hermitian representation f of $\mathfrak{h}(2, \mathbb{R})$ as

$$\mathsf{f}(P) = i\lambda \mathsf{P},\ \mathsf{f}(Q) = i\mathsf{Q},\ \mathsf{f}(Z) = i\lambda I. \tag{7.5}$$

This representation is isomorphic to the following representation via the unitary transformation $f(x) \mapsto \lambda^{1/4} f(\lambda^{1/2} x)$ on $L^2(\mathbb{R})$.

$$\mathsf{f}(P) = i\sqrt{\lambda}\mathsf{P},\ \mathsf{f}(Q) = i\sqrt{\lambda}\mathsf{Q},\ \mathsf{f}(Z) = i\lambda I. \tag{7.6}$$

When $\lambda \neq 0$, this representation is an irreducible representation. An arbitrary irreducible skew-Hermitian representation of $\mathfrak{h}(2, \mathbb{R})$ is limited to the representation isomorphic to the above representation, which is known as Stone-von Neumann Theorem [65], [31, Sect. 1.5].

When the relation $[X, [X, Y]] = [Y, [X, Y]] = 0$ holds, we have the **Hausdorff formula**

$$e^X e^Y = e^{\frac{1}{2}[X,Y]} e^{X+Y}. \tag{7.7}$$

When $\zeta = \frac{1}{\sqrt{2}}(q + ip)$ and $\zeta' = \frac{1}{\sqrt{2}}(q' + ip')$, due to the above relation (7.7) and the commutation relation (7.11), the above representation f of Heisenberg group $\mathrm{H}(2, \mathbb{R})$ satisfies the relation

[2] The precise definitions of Q and P are given in Sect. 1.6.

$$\mathsf{W}(\zeta) := \mathsf{f}(\exp(pQ - qP)) = \exp(i(p\mathsf{Q} - q\mathsf{P})) = \exp(\zeta \mathsf{a}^\dagger - \overline{\zeta}\mathsf{a})$$
$$= e^{-\frac{|\zeta|^2}{2}} \exp(\zeta \mathsf{a}^\dagger)\exp(-\overline{\zeta}\mathsf{a}) \tag{7.8}$$
$$\mathsf{W}(\zeta)\mathsf{W}(\zeta') = e^{\frac{1}{2}(\zeta\overline{\zeta'} - \zeta'\overline{\zeta})}\mathsf{W}(\zeta + \zeta') = e^{\frac{i}{2}(q'p - qp')}\mathsf{W}(\zeta + \zeta'). \tag{7.9}$$

Now, we rewrite the operator $\mathsf{W}(\zeta)$ as $\mathsf{W}(\vec{z})$ by using the real vector $\vec{z} = (q, p)^T$. Then, $\mathsf{W}(\sqrt{\lambda}\vec{z})e^{i\lambda\theta}$ is a unitary representation of Heisenberg group $\mathrm{H}(2, \mathbb{R}) = \{g_h(\vec{z}, \theta)\}$ and is called **Heisenberg representation**.

On the other hand, the map $\mathbb{R}^2 \ni (q, p) \mapsto \mathsf{W}(\sqrt{\lambda}\zeta)$ is can be regarded as a projective unitary representation of group $\mathbb{R}^2$ with factor system $\{e^{\frac{i\lambda}{2}(q'p - qp')}\}_{(q,p),(q',p')}$. The representation is called **Heisenberg representation**. When we need to distinguish the representation from the Heisenberg representation of Heisenberg group $\mathrm{H}(2, \mathbb{R}) = \{g_h(\vec{z}, \theta)\}$, it is called Heisenberg representation of $\mathbb{R}^2$. Then, Stone-von Neumann Theorem [65], [31, Sect. 1.5] guarantees that a projective irreducible unitary representation of $\mathbb{R}^2$ with factor system $\{e^{\frac{i\lambda}{2}(q'p - qp')}\}_{(q,p),(q',p')}$ is limited to Heisenberg representation $\mathsf{W}(\sqrt{\lambda}\vec{z})$.

Since this representation satisfies

$$\mathsf{W}(\sqrt{\lambda}\zeta)^{-1}\mathsf{Q}\mathsf{W}(\zeta) = \mathsf{Q} + \sqrt{\lambda}qI, \quad \mathsf{W}(\sqrt{\lambda}\zeta)^{-1}\mathsf{P}\mathsf{W}(\zeta) = \mathsf{P} + \sqrt{\lambda}pI, \tag{7.10}$$

it can be regarded as the representation corresponding to the displacement of position and momentum. Hence, the unitary operator $\mathsf{W}(\sqrt{\lambda}\zeta)$ is called **displacement transformation**.

Exercise 7.1 Show that $\mathsf{W}(\sqrt{\lambda}\vec{z})e^{i\lambda\theta}$ is a unitary representation of Heisenberg group $\mathrm{H}(2, \mathbb{R}) = \{g_h(\vec{z}, \theta)\}$.

7.1.2 One-Mode Bosonic System and One-Variable System

Next, we discuss the bosonic system, in which, the number of possible states to be taken by a particle is only one. Such a quantum system is called the **one-mode bosonic system**. For example, when several photons proceed to the same direction with the same frequency and are localized in the same place, we say that these photons take the same mode. In the bosonic case, many particles can take the same mode simultaneously. Hence, even though only one mode is available for each particle, the quantum system is composed of more than one states because the states are assigned according to the number n of the particles taking the unique mode. We denote the state vector (**number vector**) describing n particles by $|n\rangle$. Here, we assume that $|n\rangle$ is a normalized vector and $|n\rangle$ is orthogonal to $|m\rangle$ for distinct integers n and m. In particular, the density operator $|0\rangle\langle 0|$ expresses the **vacuum state** corresponding situation with no photon. Then, the one-mode bosonic system is given as the CONS $\ell^2 := \left\{\{x_n\}_{n=0}^\infty \middle| \sum_{n=0}^\infty |x_n|^2 < \infty\right\}$ generated by $\{|n\rangle\}_{n=0}^\infty$, where x_n is a complex number.

The **annihilation operator** a on ℓ^2 is defined by $\mathsf{a}|n\rangle = \sqrt{n}|n-1\rangle$. The adjoint operator $\mathsf{a}^\dagger$ is called the **creation operator** and satisfies $\mathsf{a}^\dagger|n\rangle = \sqrt{n+1}|n+1\rangle$. Then, based on the definition, we can show the following relation

$$[\mathsf{a}, \mathsf{a}^\dagger] = I, \tag{7.11}$$

where I is the identity operator. Since the operator $\mathsf{N} := \mathsf{a}^\dagger\mathsf{a}$ satisfies

$$\mathsf{N}|n\rangle = \mathsf{a}^\dagger\mathsf{a}|n\rangle = \mathsf{a}^\dagger\sqrt{n}|n-1\rangle = n|n\rangle, \tag{7.12}$$

it is called the **number operator**.

However, when we treat the one-mode bosonic system only as the sequence space ℓ^2, we have some inconveniences. Identifying the sequence space ℓ^2 with the functional space $L^2(\mathbb{R})$, we can employ several properties and concepts defined in $L^2(\mathbb{R})$ so that we have much wider application. Now, we identify $|n\rangle$ with the n-th Hermitian function $\psi_n(x) := \frac{1}{\sqrt{n!2^n\sqrt{\pi}}} e^{\frac{x^2}{2}} \frac{d^n}{dx^n} e^{-x^2}$. Then, we have naturally the isomorphism from the sequence space ℓ^2 to the space $L^2(\mathbb{R})$ of square integrable functions on $\mathbb{R}$. In the following, we identify the one-mode bosonic system ℓ^2 and the one-variable system $L^2(\mathbb{R})$ via the isomorphism. Under the isomorphism, we have the relations $\mathsf{Q} = \frac{1}{\sqrt{2}}(\mathsf{a}+\mathsf{a}^\dagger)$ and $\mathsf{P} = \frac{-i}{\sqrt{2}}(\mathsf{a}-\mathsf{a}^\dagger)$. Similarly, the relations $\mathsf{a} = \frac{1}{\sqrt{2}}(\mathsf{Q}+i\mathsf{P})$, $\mathsf{a}^\dagger = \frac{1}{\sqrt{2}}(\mathsf{Q}-i\mathsf{P})$, and $\mathsf{N} = \frac{1}{2}(\mathsf{P}^2+\mathsf{Q}^2-I)$ hold.

As a typical example, we focus on the quantum system of photons with frequency $f = \frac{c}{\lambda}$(angular frequency $\omega = 2\pi f$, c is velocity of light), i.e., the electromagnetic field with wave length λ as Fig. 7.1. Here, the operator Q corresponds to the vector potential A of electromagnetic field and the operator P does to the electric field E of electromagnetic field. That is, the physical quantities A and E are given by using the dielectric constant ϵ_0 and the Planck constant h as [105, Sect. 2.1]

$$A = \sqrt{\frac{h}{2\pi\omega\epsilon_0}}\mathsf{Q}, \ E = -\sqrt{\frac{h\omega}{2\pi\epsilon_0}}\mathsf{P}. \tag{7.13}$$

By using a, the relations (7.10) can be unified to the following relation

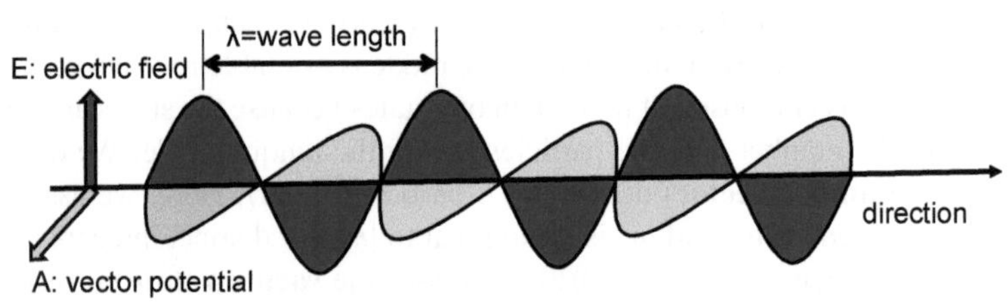

Fig. 7.1 Electromagnetic field with wave length λ

$$W(\zeta)^{-1}aW(\zeta) = a + \zeta I. \tag{7.14}$$

In general, the Hamiltonian in the one-mode bosonic system is often given as the constant times of N. Hence, the state $\rho_N := \sum_{n=0}^{\infty} \frac{N^n}{(N+1)^n} |n\rangle\langle n|$ is called the **thermal equilibrium state** and plays an important role in the quantum optics [29]. Since N satisfies the relation $N = \mathrm{Tr}\, \rho_N \mathsf{N}$, it is called the **average number**. We choose the parameter t dependently of the average number N such that $\rho_N = \frac{e^{t\mathsf{N}}}{\mathrm{Tr}\, e^{t\mathsf{N}}}$. Since $\zeta = \frac{q+ip}{\sqrt{2}}$, we have $\rho_N = \int_{\mathbb{C}} |\zeta\rangle\langle\zeta| e^{-\frac{|\zeta|^2}{N}} \frac{dqdp}{2\pi N}$. So, the thermal equilibrium state is the average state of the coherent states under the Gaussian integral.

As fundamental measurements on the quantum system $L^2(\mathbb{R})$, we can list the measurements E_{Q} and E_{P} based on the spectral measures of the self-adjoint operators P and Q and the measurement E_{N} based on the spectral measure of N. Due to (7.12), the spectral decomposition of N is $\sum_{n=0}^{\infty} n|n\rangle\langle n|$, and its eigenvalues are limited to non-negative integers. Then, using the above relation, we define the coherent vector under the Heisenberg representation of the Heisenberg group $\mathrm{H}(2, \mathbb{R})$ as

$$|\zeta\rangle := e^{-\frac{|\zeta|^2}{2}} \exp(\zeta a^\dagger)|0\rangle = e^{-\frac{|\zeta|^2}{2}} \sum_{n=0}^{\infty} \frac{\zeta^n}{\sqrt{n!}} |n\rangle. \tag{7.15}$$

Since (7.8) guarantees the relation $|\zeta\rangle = W(\zeta)|0\rangle$, the coherent vector can be written by using the unitary representation. As the vector $|\zeta\rangle$ satisfies $a|\zeta\rangle = \zeta|\zeta\rangle$, we have $\langle\zeta|a|\zeta\rangle = \zeta$, i.e., $\langle\zeta|Q|\zeta\rangle = x$ and $\langle\zeta|P|\zeta\rangle = y$ because $\zeta = \frac{q+ip}{\sqrt{2}}$.

Now, we discuss the equality condition of the uncertainty relation (5.140). The relation (7.14) yields that

$$\begin{aligned} 0 = \langle 0|a^\dagger a|0\rangle &= \langle\zeta|W(\zeta)a^\dagger W(\zeta)^\dagger W(\zeta)aW(\zeta)^\dagger|\zeta\rangle \\ &= \langle\zeta|\frac{1}{2}((Q-q)^2 + (P-p)^2 - i[Q-qI, P-pI])|\zeta\rangle \\ &= \frac{1}{2}(\Delta^2_{|\zeta\rangle\langle\zeta|}P + \Delta^2_{|\zeta\rangle\langle\zeta|}Q - 1), \end{aligned} \tag{7.16}$$

which implies the equality of (5.140) when $\rho = |\zeta\rangle\langle\zeta|$. Especially, the equality condition of the first and second inequalities in (5.140) yield that

$$\Delta^2_{|\zeta\rangle\langle\zeta|}P = \Delta^2_{|\zeta\rangle\langle\zeta|}Q = \frac{1}{2}. \tag{7.17}$$

That is, the dispersion for P and Q of the coherent state $|\zeta\rangle$ does not depend on α.

Here, we compare the coherent state defined here with the coherent states of $\mathfrak{su}(2)$, $\mathfrak{su}(1,1)$ defined in Sect. 4.2. Using (7.15), we have

$$\int_{\mathbb{C}} W(\zeta)|0\rangle\langle 0|W(\zeta)^\dagger \frac{dqdp}{2\pi} = \int_{\mathbb{C}} |\zeta\rangle\langle\zeta| \frac{dqdp}{2\pi} = I, \tag{7.18}$$

which gives a measurement based on the coherent state $|\zeta\rangle\langle\zeta|$. This measurement is called the **heterodyne measurement**. The equation (7.18) corresponds to (4.34) for the coherent states of $\mathfrak{su}(2), \mathfrak{su}(1,1)$. Since $\frac{dqdp}{2\pi}$ is the invariant measure on $\mathbb{C}$, (7.18) implies that the formal dimension is 1, as discussed in the relation (3.56). The invariant measure $\frac{dqdp}{2\pi}$ is given as $\frac{dqdp}{2\pi(1+s|\zeta|^2)^2}$ with substituting 0 into s. So, we denote $\mathbb{C}$ by Θ_0.

Now, we introduce two types of functions by using the coherent states $|\zeta\rangle$. We call $Q_X(\zeta) := \langle\zeta|X|\zeta\rangle$ the **Q-function** (**Husimi function**) of the operator X [61]. When X is a density operator, i.e., a state, the Q-function gives the probability density function of the outcome ζ of the heterodyne measurement. On the other hand, for an operator X, there uniquely exists a function $P_X(\zeta)$ such that $X = \int_{\mathbb{C}} P_X(\zeta)|\zeta\rangle\langle\zeta|\frac{dqdp}{2\pi}$, and such a function $P_X(\zeta)$ is called the **P-function**. When we perform the heterodyne measurement to the system whose state is a coherent state $|\alpha\rangle$, the outcome $\zeta = \frac{q+ip}{\sqrt{2}}$ obeys the Gaussian distribution $e^{|\zeta-\alpha|^2}\frac{dadb}{2\pi}$ (Exercise 7.2), where $\zeta = \frac{a+ib}{\sqrt{2}}$. So, we have

$$
\begin{aligned}
Q_X(\zeta) =& \langle\zeta|\Big(\int_{\mathbb{C}} P_X(\alpha)|\alpha\rangle\langle\alpha|\frac{dadb}{2\pi}\Big)|\zeta\rangle = \int_{\mathbb{C}} P_X(\alpha)|\langle\zeta|\alpha\rangle|^2\frac{dadb}{2\pi} \\
=& \int_{\mathbb{C}} P_X(\alpha)e^{|\zeta-\alpha|^2}\frac{dadb}{2\pi}.
\end{aligned}
\tag{7.19}
$$

That is, the Q-function is given as the output distribution when the information is generated by the P-function with the additive white Gaussian noisy operation.

For the operator X, **Wigner function** $W_X(\zeta)$ is defined to be $\mathrm{Tr}\, X\mathsf{W}(\zeta)$. When $\alpha := \frac{1}{\sqrt{2}}(a+ib)$, Wigner function $W_{|\alpha\rangle\langle\alpha|}(\zeta)$ of the coherent state $|\alpha\rangle\langle\alpha|$ is calculated as

$$
W_{|\alpha\rangle\langle\alpha|}(\zeta) = e^{\zeta\bar{\alpha}-\alpha\bar{\zeta}}e^{-\frac{|\zeta|^2}{2}} = e^{-\frac{|\zeta|^2}{2}+i(ap-qb)}.
\tag{7.20}
$$

That is, the absolute square of Wigner function $|W_{|\alpha\rangle\langle\alpha|}(\zeta)|^2$ is the standard Gaussian distribution for p and q. The reason why we consider the absolute square of Wigner function is given in Sect. 7.7. Generally, Wigner function $W_X(\zeta)$ can be characterized by P-function as follows.

$$
\begin{aligned}
& W_X(\zeta) = \mathrm{Tr}\Big(\int_{\mathbb{C}} P_X(\alpha)|\alpha\rangle\langle\alpha|\frac{dadb}{2\pi}\Big)\mathsf{W}(\zeta) \\
=& \int_{\mathbb{C}} P_X(\alpha)\langle\alpha|\mathsf{W}(\zeta)|\alpha\rangle\frac{dadb}{2\pi} = e^{\zeta\bar{\alpha}-\alpha\bar{\zeta}}\int_{\mathbb{C}} P_X(\alpha)e^{-\frac{|\zeta|^2}{2}}\frac{dadb}{2\pi}.
\end{aligned}
\tag{7.21}
$$

Q-function is also characterized by Wigner function as (Exercise 7.11)

$$
Q_X(\zeta) = \int_{\mathbb{C}} W_X(\alpha)e^{-\zeta\bar{\alpha}+\alpha\bar{\zeta}}e^{-\frac{|\zeta|^2}{2}}\frac{dadb}{2\pi}.
\tag{7.22}
$$

When $\zeta = \frac{q+ip}{\sqrt{2}}$, Wigner function is written as

$$
\begin{aligned}
& W_X\left(\frac{q+ip}{\sqrt{2}}\right) = \operatorname{Tr} X \mathsf{W}\left(\frac{q+ip}{\sqrt{2}}\right) = \operatorname{Tr} X e^{-i\frac{qp}{2}} \mathsf{W}\left(\frac{ip}{\sqrt{2}}\right) \mathsf{W}\left(\frac{q}{\sqrt{2}}\right) \\
= & \operatorname{Tr} \mathsf{W}\left(\frac{q}{\sqrt{2}}\right) X e^{-i\frac{qp}{2}} \mathsf{W}\left(\frac{ip}{\sqrt{2}}\right) \stackrel{(a)}{=} e^{-i\frac{qp}{2}} \int_{-\infty}^{\infty} X(x-q, x) e^{ipx} dx \\
= & \int_{-\infty}^{\infty} X(x-q, x) e^{ip(x-\frac{q}{2})} dx = \int_{-\infty}^{\infty} X\left(t - \frac{q}{2}, t + \frac{q}{2}\right) e^{ipt} dt,
\end{aligned} \tag{7.23}
$$

which is the definition of Wigner function in many literatures [108, 55]. Here, (a) follows from (3.65). Overall, the functions, Q-function, P-function, and Wigner function are used for visualization of the state in the phase space, which describes the position space and the momentum space equally.

Exercise 7.2 Show that the distribution of the outcome $\zeta = \frac{q+ip}{\sqrt{2}}$ of the heterodyne measurement is $e^{|\zeta-\alpha|^2} \frac{dadb}{2\pi}$ when the state is the coherent state $|\alpha\rangle$.

Exercise 7.3 Show (7.20).

Exercise 7.4 Calculate the distribution of the outcome $\zeta = \frac{q+ip}{\sqrt{2}}$ of the heterodyne measurement when the state is the number state $|n\rangle$.

Exercise 7.5 Calculate the distribution of the outcome n when we measure the number operator N and the state is the coherent state $|\alpha\rangle$.

7.1.3 Extended Heisenberg Representation

In the above Heisenberg representation, the coherent vector cannot be characterized by using a compact maximal Cartan subalgebra of the the Lie algebra corresponding to Heisenberg group $\mathrm{H}(2, \mathbb{R})$. Hence, the characterization of coherent vectors given in Chap. 3 cannot be applied. To realize such a characterization, we consider a larger group than $\mathrm{H}(2, \mathbb{R})$.

Firstly, we define the action T of $g \in \mathrm{U}(1)$ on $\mathrm{H}(2, \mathbb{R})$ as

$$
T(e^{i\eta}) g_h(q, p, \theta) = g_h(\cos \eta q - \sin \eta p, \sin \eta q + \cos \eta p, \theta). \tag{7.24}
$$

The semi direct product Lie group $\mathrm{H}(2, \mathbb{R}) \rtimes \mathrm{U}(1)$ based on the action T is called the **extended Heisenberg group with degree 1**. By letting F be the generator of $\mathfrak{u}(1)$, the Lie algebra $\mathfrak{h}(2, \mathbb{R}) \rtimes \mathfrak{u}(1)$ of the extended Heisenberg group of degree 1 is generated by the elements F, Q, P, and Z satisfying the following commutation relation.

$$
[F, Q] = -P, \quad [F, P] = Q, \quad [F, Z] = 0. \tag{7.25}
$$

Hence, due to the discussion in Sect. 7.1.1, the irreducible skew-Hermitian representation f of the Lie algebra $\mathfrak{h}(2,\mathbb{R}) \rtimes \mathfrak{u}(1)$ is isomorphic to the following irreducible skew-Hermitian representation given by $\lambda_Z, \lambda_F \in \mathbb{R}$.

$$-i\mathsf{f}(Q) = \sqrt{\lambda_Z}\mathsf{Q}, \quad -i\mathsf{f}(P) = \sqrt{\lambda_Z}\mathsf{P}, \quad -i\mathsf{f}(Z) = \lambda_Z I. \tag{7.26}$$

The operator $\mathsf{E} := -i\mathsf{f}(F)$ satisfies the relation $[-i\mathsf{E}, \mathsf{a}] = -i\mathsf{a}$ due to the commutation relation (7.25). So, we obtain

$$\mathsf{E} = -i\mathsf{f}(F) = -\mathsf{a}^\dagger\mathsf{a} + \lambda_F. \tag{7.27}$$

The subspace $\langle F, Z\rangle$ spanned by $\{F, Z\}$ is a Cartan subalgebra of the Lie algebra $\mathfrak{h}(2,\mathbb{R}) \rtimes \mathfrak{u}(1)$. Hence, in the representation (7.26) and (7.27), the vectors $\{|n\rangle\}_{n=0}^{\infty}$ are the simultaneous eigenvectors (weight vector) of $-i\mathsf{f}(Z)$ and E. The eigenvalues (weight) of $|n\rangle$ with respect to $-i\mathsf{f}(Z)$ and E are λ_Z and $\lambda_F - n$, respectively. In this representation, the highest weight is (λ_Z, λ_F) and the highest weight vector is the vacuum vector $|0\rangle$. Since any representation can be characterized by (7.26) and (7.27), a skew-Hermitian representation of Lie algebra $\mathfrak{h}(2,\mathbb{R}) \rtimes \mathfrak{u}(1)$ is determined by the highest weight (λ_Z, λ_F) and the representation space is denoted by $\mathcal{U}_{\lambda_Z,\lambda_F}$. When we need to identify the Lie algebra, the representation space is denoted by $\mathcal{U}_{\lambda_Z,\lambda_F}(\mathfrak{h}(2,\mathbb{R}) \rtimes \mathfrak{u}(1))$. Since the case of $\lambda_Z \neq 1$ can be reduced to the case of $\lambda_Z = 1$ via a suitable conversion with respect to (q, p), in the following, we discuss only the case of $\lambda_Z = 1$, in which, $\mathcal{U}_{\lambda_Z,\lambda_F}$ is simplified to $\mathcal{U}_{\lambda_F}$.

Next, we investigate what representation of the Lie algebra $\mathfrak{h}(2,\mathbb{R}) \rtimes \mathfrak{u}(1)$ gives a representation of the extended Heisenberg group of degree 1 $\mathrm{H}(2,\mathbb{R}) \rtimes \mathrm{U}(1)$. For this purpose, we need to define the map $\mathsf{U} : e^{i\theta} \mapsto e^{i\theta\mathsf{E}}$. This map can be defined only when λ_F is an integer. Then, due to (7.27), U satisfies

$$\mathsf{U}(e^{i\theta})\mathsf{W}(\sqrt{\lambda_Z}\zeta)\mathsf{U}(e^{i\theta})^{-1} = \mathsf{W}(e^{i\theta}\sqrt{\lambda_Z}\zeta), \tag{7.28}$$

which is the condition (2.37) given in Sect. 2.3. Hence, based on the discussion in Sect. 2.3, W is a unitary representation of the extended Heisenberg group with degree 1 $\mathrm{H}(2,\mathbb{R}) \rtimes \mathrm{U}(1)$, which is a semi direct product Lie group. This unitary representation is called the **extended Heisenberg representation**. By letting $\zeta = \frac{q+ip}{\sqrt{2}}$, the relation (7.28) is equivalent to the following relation

$$\begin{aligned}&\mathsf{U}(e^{i\theta})\mathsf{W}(\sqrt{\lambda_Z}x, \sqrt{\lambda_Z}y)\mathsf{U}(e^{i\theta})^{-1} \\ =&\mathsf{W}(\cos\theta\sqrt{\lambda_Z}x - \sin\theta\sqrt{\lambda_Z}y, \sin\theta\sqrt{\lambda_Z}x + \cos\theta\sqrt{\lambda_Z}y).\end{aligned}$$

Hence, when $\lambda_F \in \mathbb{R}$ gives the irreducible skew-Hermitian representation $\mathsf{f}_{\lambda_Z,\lambda_F}$ of the Lie algebra $\mathfrak{h}(2,\mathbb{R}) \rtimes \mathfrak{u}(1)$, the representation $\mathsf{f}_{\lambda_Z,\lambda_F}$ can be extended to an irreducible unitary representation of the Lie group $\mathrm{H}(2,\mathbb{R}) \rtimes \mathrm{U}(1)$ if and only if λ_F is an integer.

Further, by letting $\check{U(1)}$ be the double covering of $U(1)$, (which is isomorphic to $U(1)$), the double covering group of the Lie group $H(2,\mathbb{R}) \rtimes U(1)$ is the Lie group $H(2,\mathbb{R}) \rtimes \check{U(1)}$. An irreducible skew-Hermitian representation $\mathfrak{f}_{\lambda_F}$ of the Lie algebra $\mathfrak{h}(2,\mathbb{R}) \rtimes \mathfrak{u}(1)$ cannot be extended to an irreducible unitary representation of the Lie group $H(2,\mathbb{R}) \rtimes U(1)$, but can be extended to an irreducible unitary representation of the double covering group if and only if λ_F is a half integer, i.e., $2\lambda_F$ is an integer and λ_F is not an integer.

Further, the coherent vector given in Sect. 6.10 for the representation $\mathfrak{f}_{\lambda_Z,\lambda_F}$ is $|\sqrt{\lambda_Z}\zeta\rangle$ defined in (7.15). Hence, when $\lambda_Z \frac{dxdy}{2\pi}$ is the normalized measure, the formal dimension of the representation space $\mathcal{U}_{\lambda_Z,\lambda_F}$ is λ_Z, which can be checked by (7.18) with a suitable change of variables from p and q to x and y.

Taking the derivative in (7.28) for the complex number ζ, we have $\mathsf{U}(e^{i\eta})\mathsf{a}\,\mathsf{U}(e^{i\eta})^{-1} = e^{-i\eta}\mathsf{a}$, which is equivalent to

$$\mathsf{U}(e^{i\eta})\mathsf{Q}\mathsf{U}(e^{i\eta})^{-1} = \cos\eta\mathsf{Q} + \sin\eta\mathsf{P},\ \mathsf{U}(e^{i\eta})\mathsf{P}\mathsf{U}(e^{i\eta})^{-1} = -\sin\eta\mathsf{Q} + \cos\eta\mathsf{P}.$$

For example, since the Fourier transform $\mathcal{F}$ on $\mathbb{R}$ satisfies $\mathcal{F}\mathsf{Q}\mathcal{F}^{-1} = \mathsf{P}$ and $\mathcal{F}\mathsf{P}\mathcal{F}^{-1} = -\mathsf{Q}$, it equals $\mathsf{U}(e^{i\pi/2})$. The tensor product space $\mathcal{U}_{\lambda_Z,\lambda_F} \otimes \mathcal{U}_{\lambda'_Z,\lambda'_F}$ of two representation spaces $\mathcal{U}_{\lambda_Z,\lambda_F}$ and $\mathcal{U}_{\lambda'_Z,\lambda'_F}$ has the following irreducible decomposition

$$\mathcal{U}_{\lambda_Z,\lambda_F} \otimes \mathcal{U}_{\lambda'_Z,\lambda'_F} = \bigoplus_{n=0}^{\infty} \mathcal{U}_{\lambda_Z+\lambda'_Z,\lambda_F+\lambda'_F-n}. \tag{7.29}$$

This fact can be easily understood because the weight $(\lambda_Z + \lambda'_Z, \lambda_F + \lambda'_F - n)$ has n orthogonal eigenvectors. In general, in the representation space $\mathcal{U}_{\lambda_Z+\lambda'_Z,\lambda_F+\lambda'_F-n}$, possible weights are limited to $(\lambda_Z + \lambda'_Z, \lambda_F + \lambda'_F - m)$, where m is an integer satisfying $m \geq n$. Each weight has only one eigenvector. These facts show the decomposition (7.29).

7.2 Multi-mode Bosonic System

7.2.1 Multi-variable System

This section addresses the multi-mode bosonic system, whose physical explanation in Sect. 7.2.2. For this purpose, firstly we focus on the multi-variable system, which is the same mathematical structure as the multi-mode bosonic system. The multi-variable system is the quantum system describing the position of a particle in the r-dimensional space $\mathbb{R}^r$, and is mathematically given as the Hilbert space $L^2(\mathbb{R}^r)$ of $\mathbb{C}$-valued square integrable functions on $\mathbb{R}^r$, which is isomorphic to the space $L^2(\mathbb{R})^{\otimes r}$. In this system, the physical quantity describing the position of the j-th coordinate is given as the multiplication operator Q_j

$$(\mathsf{Q}_j f)(x) := x_j f(x). \tag{7.30}$$

The physical quantity describing the momentum of the j-th coordinate is given as the operator P_j

$$(\mathsf{P}_j f)(x) := -i\frac{\partial}{\partial x_j} f(x). \tag{7.31}$$

Both Q_j and P_j are self-adjoint operators because they satisfy the conditions $\mathsf{Q}_j^\dagger = \mathsf{Q}_j$ and $\mathsf{P}_j^\dagger = \mathsf{P}_j$ [84]. They also satisfy the following commutation relations

$$[\mathsf{Q}_j, \mathsf{P}_{j'}] = i\delta_{j,j'} I, \ [\mathsf{P}_j, \mathsf{P}_{j'}] = [\mathsf{Q}_j, \mathsf{Q}_{j'}] = 0. \tag{7.32}$$

Hence, a skew-Hermitian representation f of $\mathfrak{h}(2r, \mathbb{R})$ can be defined as

$$\mathsf{f}(P_j) = i\sqrt{\lambda}\mathsf{P}_j, \ \mathsf{f}(Q_j) = i\sqrt{\lambda}\mathsf{Q}_j, \ \mathsf{f}(Z) = i\lambda I. \tag{7.33}$$

The representation is an irreducible representation, and it is known as Stone-von Neumann Theorem [65] [31, Sect. 1.5] that any irreducible skew-Hermitian representation of $\mathfrak{h}(2r, \mathbb{R})$ is isomorphic to the above representation.

Exercise 7.6 Show that $[Q \otimes I + I \otimes Q, P \otimes I - I \otimes P] = 0$.

7.2.2 Relation Between Multi-mode Bosonic System and Multi-variable System

Next, we address the bosonic system when the number of possible states of particles is more than one. Such a quantum system is called the **multi-mode bosonic system**. For example, in the quantum optical system, such a system is constructed by the collection of several optical paths. That is, when we focus only on photons with one frequency and we discuss several paths for photon, the quantum system is the multi-mode bosonic system. When we have r paths, the quantum system is the r-mode bosonic system and the representation space is $(\ell^2)^{\otimes r}$ as Fig. 7.2. Then, we denote the number of particles in the j-th path (mode) by n_j. The state vectors $|n_1, \dots, n_r\rangle$ form a CONS of $(\ell^2)^{\otimes r}$. The annihilation operator a_j of the j-th mode is defined as

$$\mathsf{a}_j |n_1, \dots, n_j, \dots, n_r\rangle := \sqrt{n_j}|n_1, \dots, n_j - 1, \dots, n_r\rangle,$$

and its adjoint operator, the creation operator $\mathsf{a}_j^\dagger$ of the j-th mode satisfies

$$\mathsf{a}_j^\dagger |n_1, \dots, n_j, \dots, n_r\rangle := \sqrt{n_j + 1}|n_1, \dots, n_j + 1, \dots, n_r\rangle.$$

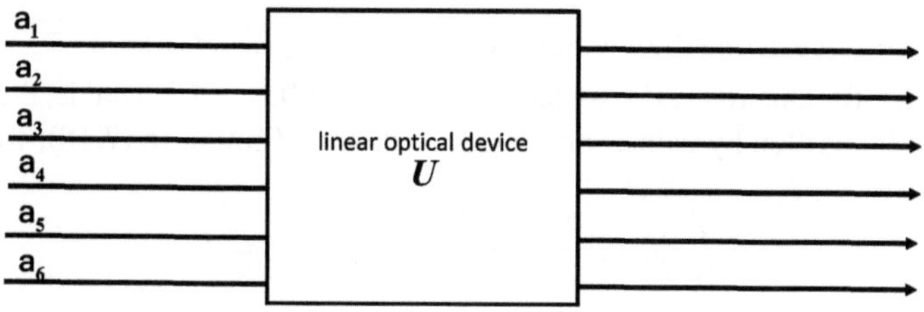

Fig. 7.2 Quantum system of 6 optical paths

Thus, the commutation relation

$$[\mathsf{a}_j, \mathsf{a}_{j'}^\dagger] = \delta_{j,j'} I \tag{7.34}$$

holds. Then, the sum $\mathsf{N} = \mathsf{N}_1 + \cdots + \mathsf{N}_r$ of operators $\mathsf{N}_j := \mathsf{a}_j^\dagger \mathsf{a}_j$ gives the number of particles of all modes.

In the following, similar to the case with one-mode, we identify the sequence space $(\ell^2)^{\otimes r}$ with $L^2(\mathbb{R})^{\otimes r}$ and $L^2(\mathbb{R}^r)$ via Hermitian functions. Then, we obtain the relations $\mathsf{Q}_j = \frac{1}{\sqrt{2}}(\mathsf{a}_j + \mathsf{a}_j^\dagger)$, $\mathsf{P}_j = -\frac{i}{\sqrt{2}}(\mathsf{a}_j - \mathsf{a}_j^\dagger)$, $\mathsf{a}_j = \frac{1}{\sqrt{2}}(\mathsf{Q}_j + i\mathsf{P}_j)$, and $\mathsf{a}_j^\dagger = \frac{1}{\sqrt{2}}(\mathsf{Q}_j - i\mathsf{P}_j)$. Next, we define an irreducible unitary representation of the Lie group $\mathrm{H}(2r, \mathbb{R})$ from the irreducible skew-Hermitian representation of the Lie algebra $\mathfrak{h}(2r, \mathbb{R})$ defined in (7.33). Using a $2r$-dimensional real-valued vector $\vec{z} := (q_1, \ldots, q_r, p_1, \ldots, p_r)^T$ and a $2r$-dimensional operator-valued vector $\vec{\mathsf{R}} := (\mathsf{Q}_1, \ldots, \mathsf{Q}_r, \mathsf{P}_1, \ldots, \mathsf{P}_r)^T$, we obtain the relation $\sum_{j=1}^r p_j \mathsf{Q}_j - q_j \mathsf{P}_j = (\vec{z}, J_r \vec{\mathsf{R}})$. Letting $\zeta_j = \frac{1}{\sqrt{2}}(q_j + ip_j)$ and using the Hausdorff formula (7.7) and the commutation relation (7.34), we find that the operator $\mathsf{W}^r(\zeta) := \mathsf{f}(\exp(\sum_{j=1}^r p_j Q_j - q_j P_j))$ satisfies the relations

$$\begin{aligned} \mathsf{W}^r(\zeta) &= \exp\left(\sum_{j=1}^r \zeta_j \mathsf{a}_j^\dagger - \overline{\zeta}_j \mathsf{a}_j\right) \\ &= e^{-\frac{\|\zeta\|^2}{2}} \exp\left(\sum_{j=1}^r \zeta_j \mathsf{a}_j^\dagger\right) \exp\left(-\sum_{j=1}^r \overline{\zeta}_j \mathsf{a}_j\right) \end{aligned} \tag{7.35}$$

$$\mathsf{W}^r(\sqrt{\lambda}\zeta)\mathsf{W}^r(\sqrt{\lambda}\alpha) = e^{\frac{-1}{2}(\lambda \sum_{j=1}^r \zeta_j \overline{\alpha}_j + \zeta_j' \overline{\zeta}_j)} \mathsf{W}^r(\sqrt{\lambda}(\zeta + \alpha)).$$

When the input variable of $\mathsf{W}^r(\zeta)$ is $\vec{z} = (x_1, \ldots, x_r, y_1, \ldots, y_r)$ and $\vec{\mathsf{R}}$ is the $2r$-dimensional vector $(\mathsf{Q}_1, \ldots, \mathsf{Q}_r, \mathsf{P}_1, \ldots, \mathsf{P}_r)^T$ whose entries are the elements $\mathsf{Q}_1, \ldots, \mathsf{Q}_r, \mathsf{P}_1, \ldots, \mathsf{P}_r$ of the Lie algebra, the operator $\mathsf{W}^r(\zeta)$ equals the operator $\exp(i(\vec{z}, J_r\vec{\mathsf{R}}))$ and is denoted by $\mathsf{W}^r(\vec{z})$. Denoting $\exp(i(\vec{z}, J_r\vec{\mathsf{R}}) + \theta)$ by $\mathsf{W}^r(\vec{z}, \theta)$, $\mathsf{W}^r(\sqrt{\lambda}\vec{z}, \lambda\theta)$ gives a unitary representation of $\mathrm{H}(2r, \mathbb{R})$.

When $\lambda=1$, we define the coherent vector $|\zeta\rangle := |\zeta_1,\ldots,\zeta_r\rangle := e^{-\frac{\|\zeta\|^2}{2}}\exp(\sum_{j=1}^r \zeta_j \mathsf{a}_j^\dagger)|0,\ldots,0\rangle$. Due to (7.35), we have $|\zeta\rangle = \mathsf{W}^r(\zeta)|0,\ldots,0\rangle$. That is, the coherent vector can be written by using unitary representation. Similar to (7.18), we have

$$\int_{\mathbb{C}^r} |\zeta\rangle\langle\zeta| \frac{dx_1\cdots dx_r dy_1\cdots dy_r}{(2\pi)^r} = I. \tag{7.36}$$

7.3 Transformation by Unitary Group U(r)

7.3.1 Correlation by Beam Splitter

In quantum optics, a beam splitter is known as an important optical device in the multimode bosonic system of multiple optical paths. A beam splitter can be constructed by glass plates and makes an interaction between two optical paths. As Fig. 7.2, the unitary operator describing the time evolution by the beam splitter is given as the representation $\mathsf{U}(g)$ of $g \in \mathrm{U}(r)$ satisfying the condition

$$\mathsf{U}(g)\mathsf{a}_j\mathsf{U}(g)^{-1} = \sum_{j'} (g^\dagger)_{j,j'}\mathsf{a}_{j'}. \tag{7.37}$$

When the matrix $g \in \mathrm{U}(r)$ is decomposed into the real part and the imaginary part as $g = g_1 + ig_2$, where g_1 and g_2 are real-valued matrices, we can define $\tilde{\kappa}(g) := g_1 \otimes I_2 + g_2 \otimes J \in \mathrm{Sp}(2r,\mathbb{R})$. By using the map $\tilde{\kappa}$, the action of g on a $2r$-dimensional vector $\vec{\mathsf{R}}$ with operator entries is given as

$$\mathsf{U}(g)\vec{\mathsf{R}}\mathsf{U}(g)^{-1} = \tilde{\kappa}(g)^{-1}\vec{\mathsf{R}}. \tag{7.38}$$

So, the unitary matrix g acts on a unitary operator $\mathsf{W}^r(\zeta)$ as

$$\mathsf{U}(g)\mathsf{W}^r(\zeta)\mathsf{U}(g)^{-1} = \mathsf{W}^r(g\zeta). \tag{7.39}$$

When $\zeta_j = \frac{x_j+iy_j}{\sqrt{2}}$, the relation (7.39) is equivalent to the relation $\mathsf{U}(g)\mathsf{W}^r(\vec{z})\mathsf{U}(g)^{-1} = \mathsf{W}^r(\tilde{\kappa}(g)\vec{z})$.

In the following, we concretely give a skew-Hermitian representation of $\mathfrak{u}(r)$ satisfying (7.37) as follows. Taking the derivative at (7.37), for $X \in \mathfrak{u}(r)$, we have

$$[\mathsf{U}(X), \mathsf{a}_j] = \sum_{l=1}^r -X_{j,l}\mathsf{a}_l. \tag{7.40}$$

Taking the derivative at (7.38), we have $[\mathsf{U}(X), \vec{\mathsf{R}}] = -\tilde{\kappa}(X)\vec{\mathsf{R}}$. Hence, we define $\mathsf{E}_j := i\mathsf{U}(-i|j\rangle\langle j|)$ and $\mathsf{K}_{j,l;\pm} := -i\mathsf{U}(F^x_{j,l}) \mp \mathfrak{su}(F^y_{j,l})$ according to the discussion in Sect. 4.3. So, we have $[\mathsf{E}_j, \mathsf{a}_l] = \delta_{j,l}\mathsf{a}_l$, $[\mathsf{K}_{j,l;+}, \mathsf{a}_{j'}] = -\delta_{j,j'}\mathsf{a}_l$, and $[\mathsf{K}_{j,l;-}, \mathsf{a}_{j'}] = -\delta_{j',l}\mathsf{a}_j$. Further, the properties of $\mathfrak{u}(r)$ yield that

$$[\mathsf{E}_{j'}, \mathsf{E}_j] = 0,\ [-\mathsf{E}_{j'} + \mathsf{E}_{j'+1}, \mathsf{K}_{j,l;\pm}] = \pm\alpha_{j,l}(F^z_{j',j'+1})\mathsf{K}_{j,l;\pm},$$
$$[\mathsf{K}_{j,l;+}, \mathsf{K}_{j,l;-}] = -\mathsf{E}_j\mathsf{E}_l.$$

Thus, we have $\mathsf{E}_j = -\mathsf{a}_j^\dagger\mathsf{a}_j + \lambda_F I$, $\mathsf{K}_{j,l;+} = \mathsf{a}_j^\dagger\mathsf{a}_l$, and $\mathsf{K}_{j,l;-} = \mathsf{a}_l^\dagger\mathsf{a}_j$. Therefore, we obtain a skew-Hermitian representation of $\mathfrak{u}(r)$.

Here, the condition (7.40) for the semi direct product Lie algebra $\mathfrak{g} = \mathfrak{h}(2r, \mathbb{R}) \rtimes \mathfrak{u}(r)$ corresponds to the condition (3.42) of the general case. Hence, combining the above skew-Hermitian representation of $\mathfrak{u}(r)$ and the preceding skew-Hermitian representation of $\mathfrak{h}(2r, \mathbb{R})$, we can define the semi direct product representation f of the Lie algebra $\mathfrak{g} = \mathfrak{h}(2r, \mathbb{R}) \rtimes \mathfrak{u}(r)$.

Then, discussing the representation of the subgroup U(r), we obtain the irreducible decomposition of the representation space $L^2(\mathbb{R})^{\otimes r}$ as

$$L^2(\mathbb{R})^{\otimes r} = \bigoplus_{n=0}^{\infty} \mathcal{U}_{s,n}, \tag{7.41}$$

where $\mathcal{U}_{s,n}$ is the space generated by $\{|n_1, \ldots, n_r\rangle\}_{n_1+\cdots+n_r=n}$, and isomorphic to n-fold symmetric tensor product space of $\mathbb{C}^r$ as representation spaces with respect to SU(r). Hence, the relation $\mathsf{U}(g)\mathsf{N}\mathsf{U}(g)^{-1} = \mathsf{N}$ holds for an arbitrary element $g \in \mathrm{U}(r)$.

When the Hamiltonian for the interaction is given as the sum of the product $\mathsf{a}_j\mathsf{a}_{j'}^\dagger$ of the annihilation and creation operators, the time evolution is written as the action of $\mathfrak{u}(r)$. When the operators $\mathsf{U}(g)$ and $\mathsf{U}(g)^{-1}$ are applied to the operators $I, \mathsf{Q}_1, \ldots, \mathsf{Q}_r, \mathsf{P}_1, \ldots, \mathsf{P}_r, \mathsf{N}$ from both sides, the resultant operators are given as the linear sum of these operators. Since $\mathsf{U}(g)$ transforms these operators linearly, $\mathsf{U}(g)$ is called a **linear operation**. Also, when the interaction is written as the products $\mathsf{a}_j\mathsf{a}_{j'}^\dagger$ of the annihilation and creation operators, it is called a **linear interaction**.

When $g \in \mathrm{O}(r) \subset \mathrm{U}(r)$ and we regard $L^2(\mathbb{R})^{\otimes r}$ as the Hilbert space $L^2(\mathbb{R}^r)$ of n-variable square integrable functions, we find that $(\mathsf{U}(g)f)(x) = f(gx)$ by considering the coordinate changes of $Q_1, \ldots, Q_r$. For example, when we apply a beam splitter to two optical paths and the interaction between two paths is given as $\begin{pmatrix} a & b \\ c & d \end{pmatrix} \in \mathrm{SU}(2)$, the coherent vectors are changed as Fig. 7.3.

Especially, the beam splitter described by the unitary matrix $b_\eta := \begin{pmatrix} \sqrt{1-\eta} & \sqrt{\eta} \\ -\sqrt{\eta} & \sqrt{1-\eta} \end{pmatrix} \in \mathrm{SU}(2)$ is called the beam splitter of **transmittance** η from system 1 to system 2. In particular, $\mathsf{U}(b_{1/2})$ is simplified to $\mathsf{U}_{1,2}$. Similarly, we can define $\mathsf{U}_{k,j}$. Using the beam splitter of transmittance η, we define the quantum communication channel Att_η as

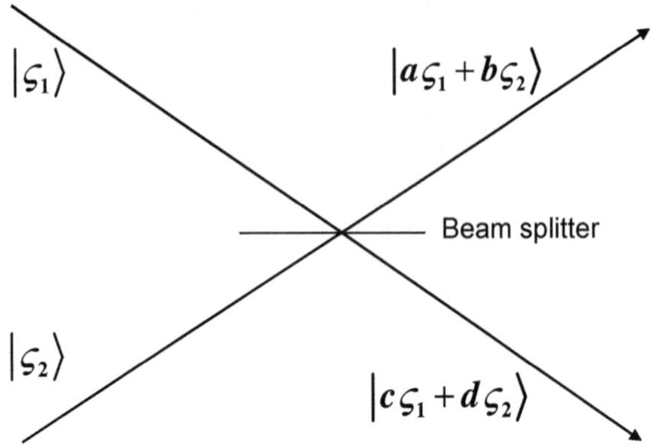

Fig. 7.3 Beam splitter

$$\mathrm{Att}_\eta(\rho) := \mathrm{Tr}_2\, \mathsf{U}(b_\eta)\rho \otimes |0\rangle\langle 0|\mathsf{U}(b_\eta)^\dagger,$$

which is called the **attenuation channel** of transmittance $1-\eta$. The noisy process caused by thermal noise is given by the following trace-preserving completely positive map (TP-CP map) Λ_N [29]

$$\Lambda_N(|\zeta\rangle\langle\zeta|) = \int_{\mathbb{C}} |\alpha\rangle\langle\alpha| e^{-\frac{|\alpha-\zeta|^2}{N}} \frac{dxdy}{2\pi N}. \tag{7.42}$$

Since any state is determined by the P-function, (7.42) uniquely determines the TP-CP map. Since Λ_N satisfies

$$\Lambda_N(\rho) = \lim_{\eta\to+0} \mathrm{Tr}_2\, \mathsf{U}(b_\eta)\rho \otimes \rho_{N/\eta}\mathsf{U}(b_\eta)^\dagger, \tag{7.43}$$

it can be regarded as a small interaction with a large system 2. In general, the large system 2 is called the **environment system**.

Now, to implement the heterodyne measurement $\{|\alpha\rangle\langle\alpha|\}$, we set the states of the systems 1 and 2 to be ρ and the vacuum state $|0\rangle\langle 0|$. Then, we insert them into the beam splitter of transmittance $\frac{1}{2}$. Finally, we apply the measurements F_{Q} and F_{P} of Q and P to the output systems of the systems 1 and 2, respectively. Using these outcomes x and y, we define the final outcome $\alpha := x + iy$. So, the measurement with the final outcome is the heterodyne measurement $\{|\alpha\rangle\langle\alpha|\}$ for the system 1. This fact can be shown by checking that both measurements have the same expectation of $\alpha^n\overline{\alpha}^m = (x+iy)^n(x-iy)^m$ for any integers n and m as follows.

In the case of the heterodyne measurement $\{|\alpha\rangle\langle\alpha|\}$, we have

$$\begin{aligned}
&\frac{1}{2\pi}\int_{\mathbb{C}} \alpha^n\overline{\alpha}^m \langle\alpha|\rho|\alpha\rangle dqdp = \mathrm{Tr}\,\rho\frac{1}{2\pi}\int_{\mathbb{C}} \mathsf{a}^n|\alpha\rangle\langle\alpha|(\mathsf{a}^\dagger)^m dqdp \\
=&\mathrm{Tr}\,\rho\mathsf{a}^n\frac{1}{2\pi}\int_{\mathbb{C}} |\alpha\rangle\langle\alpha| dqdp(\mathsf{a}^\dagger)^m = \mathrm{Tr}\,\rho\mathsf{a}^n(\mathsf{a}^\dagger)^m.
\end{aligned}$$

When the system 1 is input to the beam splitter of transmittance $\frac{1}{2}$, we have

$$
\begin{aligned}
&\operatorname{Tr} \mathsf{U}(b_{\frac{1}{2}})\rho \otimes |0\rangle\langle 0|\mathsf{U}(b_{\frac{1}{2}})^{\dagger}(\mathsf{Q}_1 + i\mathsf{P}_2)^n(\mathsf{Q}_1 - i\mathsf{P}_2)^m \\
=&\operatorname{Tr} \rho \otimes |0\rangle\langle 0|(\mathsf{U}_{1,2}^{\dagger}(\mathsf{Q}_1 + i\mathsf{P}_2)\mathsf{U}_{1,2})^n(\mathsf{U}_{1,2}^{\dagger}(\mathsf{Q}_1 - i\mathsf{P}_2)\mathsf{U}_{1,2})^m \\
=&\operatorname{Tr} \rho \otimes |0\rangle\langle 0|\left(\frac{1}{\sqrt{2}}(\mathsf{Q}_1 + \mathsf{Q}_2)\right) + \frac{i}{\sqrt{2}}(\mathsf{P}_1 - \mathsf{P}_2))^n \\
&\cdot\left(\frac{1}{\sqrt{2}}(\mathsf{Q}_1 + \mathsf{Q}_2) - \frac{i}{\sqrt{2}}(\mathsf{P}_1 - \mathsf{P}_2)\right)^m \\
=&\operatorname{Tr} \rho \otimes |0\rangle\langle 0|(\mathsf{a}_1 + \mathsf{a}_2^{\dagger})^n(\mathsf{a}_1^{\dagger} + \mathsf{a}_2)^m = \operatorname{Tr} \rho \mathsf{a}^n(\mathsf{a}^{\dagger})^m.
\end{aligned}
\tag{7.44}
$$

Therefore, we can conclude that the heterodyne measurement can be realized by inputting the state to be measured and the vacuum state $|0\rangle\langle 0|$ into the beam splitter of transmittance $\frac{1}{2}$.

Here, we need to an additional discussion to show the final equation of (7.44). Since any density operator can be written as the sum of pure states, it is sufficient to show the final equation in the pure states case. When $\rho = |\phi\rangle\langle\phi|$, it can be shown as follows

$$
(\mathsf{a}_1^{\dagger} + \mathsf{a}_2)^m|\phi\rangle \otimes |0\rangle = (\mathsf{a}_1^{\dagger} + \mathsf{a}_2)^{m-1}(\mathsf{a}_1^{\dagger}|\phi\rangle) \otimes |0\rangle = ((\mathsf{a}_1^{\dagger})^m|\phi\rangle) \otimes |0\rangle.
$$

Similarly, we have $(\langle\phi| \otimes \langle 0|)(\mathsf{a}_1 + \mathsf{a}_2^{\dagger})^n = (\langle\phi|\mathsf{a}_1^n) \otimes \langle 0|$, which implies the final equation of (7.44).

Exercise 7.7 Show (7.43).

7.3.2 Transformation Corresponding to the Irreducible Decomposition

Using the above beam splitter, we can realize the transformation corresponding to irreducible decomposition given in (7.29) of the Lie group $\mathrm{H}(2,\mathbb{R}) \rtimes \mathrm{U}(1)$ of the one-mode bosonic system. Firstly, we focus on two representation spaces $\mathcal{U}_{\lambda_Z,\lambda_F}$ and $\mathcal{U}_{0,0}$ of the Lie group $\mathrm{H}(2,\mathbb{R}) \rtimes \mathrm{U}(1)$ as one-mode bosonic systems. These two spaces are isomorphic to $L^2(\mathbb{R})$. We denote their annihilation operators by a_1 and a_2. The Lie group $\mathrm{H}(2,\mathbb{R})$ does not act on the second space, and only the Lie algebra $\mathfrak{u}(1)$ acts on it so that the representation of the generator F is given as $-i\mathfrak{f}(F) = -\mathsf{a}_2^{\dagger}\mathsf{a}_2$. Then, the tensor product space $\mathcal{U}_{\lambda_Z,\lambda_F} \otimes \mathcal{U}_{0,0}$ has the following irreducible decomposition

$$
\mathcal{U}_{\lambda_Z,\lambda_F} \otimes \mathcal{U}_{0,0} = \oplus_{n=0}^{\infty} L^2(\mathbb{R}) \otimes |n\rangle = \oplus_{n=0}^{\infty} \mathcal{U}_{\lambda_Z,\lambda_F - n}. \tag{7.45}
$$

When two representation spaces $\mathcal{U}_{\lambda_{Z,1},\lambda_{F,1}}$ and $\mathcal{U}_{\lambda_{Z,2},\lambda_{F,2}}$ of the Lie group $\mathrm{H}(2,\mathbb{R}) \rtimes \mathrm{U}(1)$ are given, the representation space can be reduced to a simple case by applying a suitable beam splitter, as explained later. Then, the tensor product representation on the two-mode bosonic system, i.e., the composite system $\mathcal{U}_{\lambda_{Z,1},\lambda_{F,1}} \otimes \mathcal{U}_{\lambda_{Z,2},\lambda_{F,2}}$ is given as

$$
\begin{aligned}
-i\mathsf{f}(F) &= -(\mathsf{a}_1^\dagger \mathsf{a}_1 + \mathsf{a}_2^\dagger \mathsf{a}_2) + \lambda_{F,1} + \lambda_{F,2}, \\
\frac{1}{\sqrt{2}}(-i\mathsf{f}(Q) + \mathsf{f}(P)) &= \sqrt{\lambda_{Z,1}}\mathsf{a}_1 + \sqrt{\lambda_{Z,2}}\mathsf{a}_2, \\
-i\mathsf{f}(Z) &= (\lambda_{Z,1} + \lambda_{Z,2})I.
\end{aligned}
$$

When we choose $g_0 = \begin{pmatrix} \sqrt{\frac{\lambda_{Z,1}}{\lambda_{Z,1}+\lambda_{Z,2}}} & -\sqrt{\frac{\lambda_{Z,2}}{\lambda_{Z,1}+\lambda_{Z,2}}} \\ \sqrt{\frac{\lambda_{Z,2}}{\lambda_{Z,1}+\lambda_{Z,2}}} & \sqrt{\frac{\lambda_{Z,1}}{\lambda_{Z,1}+\lambda_{Z,2}}} \end{pmatrix}$, the representation $\mathsf{U}(g_0)$ satisfies $\mathsf{U}(g_0)\mathsf{a}_1\mathsf{U}(g_0)^\dagger = \sqrt{\frac{\lambda_{Z,1}}{\lambda_{Z,1}+\lambda_{Z,2}}}\mathsf{a}_1 + \sqrt{\frac{\lambda_{Z,2}}{\lambda_{Z,1}+\lambda_{Z,2}}}\mathsf{a}_2$ and $\mathsf{U}(g_0)\mathsf{a}_2\mathsf{U}(g_0)^\dagger = \sqrt{\frac{\lambda_{Z,2}}{\lambda_{Z,1}+\lambda_{Z,2}}}\mathsf{a}_1 - \sqrt{\frac{\lambda_{Z,1}}{\lambda_{Z,1}+\lambda_{Z,2}}}\mathsf{a}_2$. So, we have

$$
\begin{aligned}
\mathsf{U}(g_0)(-i\mathsf{f}(F))\mathsf{U}(g_0)^\dagger &= -\mathsf{a}_1^\dagger \mathsf{a}_1 - \mathsf{a}_2^\dagger \mathsf{a}_2 + \lambda_{F,1} + \lambda_{F,2}, \\
\mathsf{U}(g_0)\frac{1}{\sqrt{2}}(-i\mathsf{f}(Q) + \mathsf{f}(P))\mathsf{U}(g_0)^\dagger &= \sqrt{\lambda_{Z,1} + \lambda_{Z,2}}\mathsf{a}_1 \\
\mathsf{U}(g_0)(-i\mathsf{f}(Z))\mathsf{U}(g_0)^\dagger &= (\lambda_{Z,1} + \lambda_{Z,2})I.
\end{aligned}
$$

That is, the subspace $L^2(\mathbb{R}) \otimes |n\rangle\langle n|$ of $\mathsf{U}(g_0)\mathcal{U}_{\lambda_{Z,1},\lambda_{F,1}} \otimes \mathcal{U}_{\lambda_{Z,2},\lambda_{F,2}}\mathsf{U}(g_0)^\dagger$ is an irreducible representation space with the highest weight $\lambda_{F,1} + \lambda_{F,2} - n$. Hence, we have

$$
\begin{aligned}
\mathsf{U}(g_0)\mathcal{U}_{\lambda_{Z,1},\lambda_{F,1}} \otimes \mathcal{U}_{\lambda_{Z,2},\lambda_{F,2}}\mathsf{U}(g_0)^\dagger &= \bigoplus_{n=0}^{\infty} \mathcal{U}_{\lambda_{Z,1}+\lambda_{Z,2},\lambda_{F,1}+\lambda_{F,2}-n} \\
&= \mathcal{U}_{\lambda_{Z,1}+\lambda_{Z,2},\lambda_{F,1}+\lambda_{F,2}} \otimes \mathcal{U}_{0,0}.
\end{aligned}
$$

7.3.3 Representation of Semi Direct Product Lie Group $\mathbf{H}(2r, \mathbb{R}) \rtimes \mathbf{U}(r)$ and Its Coherent State*

In the discussion of Sect. 7.2, the Lie algebra of the stabilizer of the Heisenberg group $\mathrm{H}(2r, \mathbb{R})$ did not include any compact maximal Cartan subalgebra of the Lie algebra of $\mathrm{H}(2r, \mathbb{R})$. Hence, we cannot characterize the coherent states by using a compact maximal Cartan subalgebra as in the previous chapters. To characterize the coherent states in this way, we need to consider a representation of a larger Lie group. For this purpose, we focus on a unitary representation of the semi direct product Lie group $\mathrm{H}(2r, \mathbb{R}) \rtimes \mathrm{U}(r)$. To address the unitary representation, we discuss skew-Hermitian representations of the corresponding Lie algebra $\mathfrak{h}(2r, \mathbb{R}) \rtimes \mathfrak{u}(r)$. Then, we need to fix the root system of the Lie algebra $\mathfrak{h}(2r, \mathbb{R}) \rtimes \mathfrak{u}(r)$. Since the center of the Lie algebra is the one-dimensional subspace $\mathfrak{c} := \mathbb{R}Z$ generated by Z, we choose the root system for $\mathfrak{h}(2r, \mathbb{R}) \rtimes \mathfrak{u}(r)/\mathfrak{c}$. Then, we fix the root system for the Lie subalgebra $\mathfrak{g}_0 :=< Q_1, \ldots, Q_r, P_1, \ldots, P_r > \rtimes \mathfrak{u}(r)$ of $\mathfrak{h}(2r, \mathbb{R}) \rtimes \mathfrak{u}(r)$ that is isomorphic to $\mathfrak{h}(2r, \mathbb{R})/\mathfrak{c}$. A maximal compact Cartan subalgebra $\mathfrak{h}$ of $\mathfrak{h}(2r, \mathbb{R}) \rtimes \mathfrak{u}(r)$ is given as $\mathfrak{c} \oplus \mathfrak{d}(r)$. Then, $\mathfrak{h}_0 := \mathfrak{g}_0 \cap (\mathfrak{c} \oplus \mathfrak{d}(r))$ equals $\mathfrak{d}(r)$. So, we define $\boldsymbol{\alpha}_{j,l}, \boldsymbol{\alpha}_j \in \mathfrak{d}(r)^*$ as elements of the dual space $\mathfrak{d}(r)^*$ of $\mathfrak{d}(r)$ as follows.

$$\boldsymbol{\alpha}_{j,l}(X) := i(\langle j|X|j\rangle - \langle l|X|l\rangle), \ \boldsymbol{\alpha}_j(X) := i\langle j|X|j\rangle, \ \forall X \in \mathfrak{d}(r). \tag{7.46}$$

When $F^x_{\boldsymbol{\alpha}_{j,l}}, F^y_{\boldsymbol{\alpha}_{j,l}}, F^x_{\boldsymbol{\alpha}_j}, F^y_{\boldsymbol{\alpha}_j}$ are given as $F^x_{j,l}, F^y_{j,l}, -\frac{Q_j}{\sqrt{2}}, -\frac{P_j}{\sqrt{2}}$, we can see that the condition (6.50) holds.

When a maximal compact Lie subalgebra $\mathfrak{k}$ of $\mathfrak{g}$ includes $\mathfrak{h}$, it is the Lie subalgebra $\mathfrak{c} \oplus \mathfrak{u}(r)$. Hence, $\mathfrak{h}_0$ is $\mathfrak{d}(r)$. The set Φ^+_k of positive roots of $\mathfrak{k}$ is $\{\boldsymbol{\alpha}_{j,l}\}_{1\le j<\le r}$, and $\Phi^+_{\mathfrak{p}} = \Phi^+ \setminus \Phi^+_{\mathfrak{k}}$ is $\{\boldsymbol{\alpha}_j\}_{j=1}^r$.

For the semi direct product representation $\mathsf{W} \rtimes \mathsf{U}$ given in Sect. 7.3.1, we have $\mathsf{K}_{\boldsymbol{\alpha}_j} = i(\mathsf{W} \rtimes \mathsf{U})(F^x_{\boldsymbol{\alpha}_j}) - (\mathsf{W} \rtimes \mathsf{U})(F^y_{\boldsymbol{\alpha}_j}) = \mathsf{a}_j, \mathsf{K}_{-\boldsymbol{\alpha}_j} = i(\mathsf{W} \rtimes \mathsf{U})(F^x_{\boldsymbol{\alpha}_j}) + (\mathsf{W} \rtimes \mathsf{U})(F^y_{\boldsymbol{\alpha}_j}) = \mathsf{a}_j^\dagger, \mathsf{K}_{\boldsymbol{\alpha}_{j,l}} = \mathsf{K}_{j,l;+} = \mathsf{a}_j^\dagger \mathsf{a}_l$, and $\mathsf{K}_{-\boldsymbol{\alpha}_{j,l}} = \mathsf{K}_{j,l;-} = \mathsf{a}_l^\dagger \mathsf{a}_j$. Letting $\boldsymbol{v} := \sum_{j=1}^r \boldsymbol{\alpha}_j \in \mathfrak{h}_0^*$, we obtain

$$\boldsymbol{v}\left(\sum_{j=0}^r c_j F_j^{z,u}\right) = \sum_{j=0}^r c_j. \tag{7.47}$$

Since the relation $\mathsf{K}_{\boldsymbol{\alpha}}|0, \ldots, 0\rangle = 0$ holds for an arbitrary element $\boldsymbol{\alpha} \in \Phi^+$, $|0, \ldots, 0\rangle$ is the highest weight vector. Further, the relation (7.47) implies that

$$\begin{aligned} -i\mathsf{f}\left(\sum_{j=0}^r c_j F_j^{z,u}\right)|0, \ldots, 0\rangle &= \sum_{j=0}^r c_j(-\mathsf{a}_j^\dagger \mathsf{a}_j + \lambda_F)|0, \ldots, 0\rangle \\ =&\lambda_F \sum_{j=0}^r c_j|0, \ldots, 0\rangle = \lambda_F \boldsymbol{v}\left(\sum_{j=0}^r c_j F_j^{z,u}\right)|0, \ldots, 0\rangle. \end{aligned}$$

Then, the highest weight is $(1, \lambda_F \boldsymbol{v}) \in (\mathfrak{c} \oplus \mathfrak{d}(r))^*$. We denote the representation space by $\mathcal{U}_{1,\lambda_F \boldsymbol{v}}$.

Since SU(r) is a universal covering group, we have the unitary representation of the subgroup $G = \mathrm{H}(2r, \mathbb{R}) \rtimes \mathrm{SU}(r)$ for any real number λ_F. Although we have the decomposition $\mathrm{U}(r) = \mathrm{SU}(r) \times \mathrm{U}(1)$, in order to define the action of U(1), λ_F needs to be an integer. Hence, the unitary representation on the representation space $\mathcal{U}_{1,\lambda_F \boldsymbol{v}}$ can be extended to a unitary representation of the Lie group $G = \mathrm{H}(2r, \mathbb{R}) \rtimes \mathrm{U}(r)$ if and only if λ_F is an integer. When λ_F is a half integer, using the double covering $\check{\mathrm{U}}(r)$ of U(r), we can extend the representation to a unitary representation of the Lie group $\mathrm{H}(2r, \mathbb{R}) \rtimes \tilde{\mathrm{U}}(r)$.

The Lie algebra of the stabilizer H of the highest weight state $|0,\ldots,0\rangle\langle 0,\ldots,0|$ is $\mathfrak{c}\oplus\mathfrak{u}(r)$. According to the discussion in Sect. 6.10, we construct the coherent vector. We see that the coherent vector is $|\zeta\rangle$ and that the parameter space G/H is $\mathbb{C}^r$. The normalized invariant measure on $\mathbb{C}^r$ is $\mu_{\mathbb{C}}(d\zeta) = \frac{dq_1\cdots dq_r dp_1\cdots dp_r}{(2\pi)^r}$. Hence, the formal dimension of the representation is 1. Especially, since the relation (6.19) holds even for a representation of a non-compact group, denoting the representation of $g = e^X \in \mathrm{U}(r)$ by $\mathsf{U}(g)$, we have

$$\begin{aligned}\mathsf{U}(g)|0,\ldots,0\rangle &= e^{i\lambda_F v(X)}|0,\ldots,0\rangle\\ &= e^{i\lambda_F\cdot i\,\mathrm{Tr}\,X}|0,\ldots,0\rangle = (\det g)^{-\lambda_F}|0,\ldots,0\rangle.\end{aligned}$$

Hence, the element $g\in\mathrm{U}(r)$ acts on the coherent vector $|\zeta\rangle$ as follows

$$\begin{aligned}\mathsf{U}(g)|\zeta\rangle &= \mathsf{U}(g)\mathsf{W}^r(\zeta)|0\rangle = \mathsf{W}^r(g\zeta)\mathsf{U}(g)|0\rangle\\ &= \mathsf{W}^r(g\zeta)(\det g)^{-\lambda_F}|0\rangle = (\det g)^{-\lambda_F}|g\zeta\rangle.\end{aligned}$$

7.4 Second Quantization and Boson

Based on the contents given until now, we explain the structure of the quantum system more deeply. To discuss the quantum system, we need the two-step procedure for the quantization. Firstly, we are given a quantum system $\mathcal{H}$ that describes the possible states for one particle. For example, when we fix a specific polarization with a specific frequency and we prepare r paths for a photon, the quantum system of the possible states of one photon forms a r-dimensional quantum system $\mathcal{H}=\mathbb{C}^r$. Then, the state corresponding each path is called a mode.

As mentioned in Sect. 4.4.2, when more than one particles can share the same mode, the type of the particles is called **boson**, and when more than one particles cannot share the same mode, the type of the particles is called **fermion**. Hence, when the type of the particle is boson, it is possible that two particles take the same mode. In this case, it is impossible to distinguish two particles. We can distinguish only the number of particles in each mode. In the following, we focus only on bosons. As mentioned in Sect. 4.4.2, when the total number of bosons is fixed to n, the quantum system is given as the n-fold symmetric tensor product space $\mathcal{U}_{s,n}(\mathcal{H})$ of $\mathcal{H}$. Hence, when the total number n of bosons is not fixed and may take an arbitrary non-negative integer, the total quantum system is described as the space $L^2(\mathcal{H}) := \oplus_{n=0}^{\infty}\mathcal{U}_{s,n}(\mathcal{H})$. As explained in (7.41), the space $L^2(\mathcal{H})$ is equivalent to the space $L^2(\mathbb{R})^{\otimes r}$. Whereas deriving the quantum system $\mathcal{H}$ of states of one particle is called the **first quantization**, deriving the total quantum system $L^2(\mathcal{H})$ is called the **second quantization**. In this case, when we apply the coordinate transformation $g\in\mathrm{U}(r)$ to the original system $\mathcal{H}$, the system $L^2(\mathcal{H})$ is naturally transformed by the representation $\mathsf{U}(g)$. In the case of fermion, when the total number is fixed to n, the quantum system is given as the n-fold alternative tensor product space $\mathcal{U}_{a,n}(\mathcal{H})$ of $\mathcal{H}$. The total quantum system is described as the space $\oplus_{n=0}^{r}\mathcal{U}_{s,n}(\mathcal{H})$.

Assume that one boson may take a state in the system $\mathcal{H}_A = \{|u_j\rangle\}$ or $\mathcal{H}_B = \{|v_j\rangle\}$. In this case, the system of possible modes of one boson is the product space $\mathcal{H}_A \oplus \mathcal{H}_B$. When more than two particles are possible, the total quantum system is given as $L^2(\mathcal{H}_A \oplus \mathcal{H}_B) = \oplus_{n=0}^{\infty}\mathcal{U}_{s,n}(\mathcal{H}_A \oplus \mathcal{H}_B)$. Since the space $\mathcal{U}_{s,2}(\mathcal{H}_A \oplus \mathcal{H}_B)$ is the symmetric tensor product space, the space is composed of symmetrized vectors in $(\mathcal{H}_A \oplus \mathcal{H}_B)^{\otimes 2}$. Here, we consider the case when the respective systems $\mathcal{H}_A$ and $\mathcal{H}_B$ have one particle. In this case, the total number of particles is 2, and the state is included in the symmetric tensor product space $\mathcal{U}_{s,2}(\mathcal{H}_A \oplus \mathcal{H}_B)$. For example, when the state of the particle in the system $\mathcal{H}_A$ is $|u_{j_1}\rangle$ and the state of the particle in the system $\mathcal{H}_B$ is $|v_{j_2}\rangle$, the state of the total system is given as $\frac{1}{\sqrt{2}}(|u_{j_1}, v_{j_2}\rangle + |v_{j_2}, u_{j_1}\rangle) \in \mathcal{U}_{s,2}(\mathcal{H}_A \oplus \mathcal{H}_B)$. The subspace composed of such elements is isomorphic to the tensor product space $\mathcal{H}_A \otimes \mathcal{H}_B$. That is, the space $\mathcal{H}_A \otimes \mathcal{H}_B$ can be regarded as a subspace of $\mathcal{U}_{s,2}(\mathcal{H}_A \oplus \mathcal{H}_B)$. In other words, the tensor product space $\mathcal{H}_A \otimes \mathcal{H}_B$ can be realized by projection from $L^2(\mathcal{H}_A \oplus \mathcal{H}_B)$ to the subspace $\mathcal{H}_A \otimes \mathcal{H}_B$.

In fact, to realize the composite system $\mathcal{H}_A \otimes \mathcal{H}_B$ of the systems $\mathcal{H}_A$ and $\mathcal{H}_B$ physically, we often employ this method. Even if the particle is a boson whose mode is given as an element of the system $\mathcal{H}_A \oplus \mathcal{H}_B$, the realized system can be regarded as not the space $\mathcal{U}_{s,2}(\mathcal{H}_A \oplus \mathcal{H}_B)$ but the subspace $\mathcal{H}_A \otimes \mathcal{H}_B$ by making several physical constraints artificially, i.e., by assuming that the respective systems $\mathcal{H}_A$ and $\mathcal{H}_B$ have only one particle. Indeed, when both systems $\mathcal{H}_A$ and $\mathcal{H}_B$ are isomorphic to $\mathcal{H}$, we denote $\mathcal{H}_A \otimes \mathcal{H}_B$ by $\mathcal{H}^{\otimes 2}$. However, it does not mean that both systems are the physically same system. When the two particles are in the same system $\mathcal{H}$, in the bosonic case, the real quantum system is described by the space $\mathcal{U}_{s,2}(\mathcal{H})$, which is strictly smaller than the system $\mathcal{H}^{\otimes 2}$. Hence, in this case, the system $\mathcal{H}^{\otimes 2}$ cannot be realized physically.

Similarly, when there exists only one bosonic particle in n respective systems $\mathcal{H}_1, \ldots, \mathcal{H}_n$ so that there exist totally n bonoic particles, the system $\mathcal{H}_1 \otimes \cdots \otimes \mathcal{H}_n$ is realized as a subspace of the n-fold symmetric tensor product space $\mathcal{U}_{s,n}(\mathcal{H}_1 \oplus \cdots \oplus \mathcal{H}_n)$. That is, even though n systems $\mathcal{H}_1, \ldots, \mathcal{H}_n$ are mathematically isomorphic to each other, when they are physically distinguishable as in Sect. 4.4.1, the system $\mathcal{H}_1 \otimes \cdots \otimes \mathcal{H}_n$ can be realized. The same discussion holds in the fermionic case by replacing the symmetric subspace by the alternative subspace.

7.5 Squeezed State on One-Mode Bosonic System

In Sect. 3.1.2, the symplectic group $\mathrm{Sp}(2, \mathbb{R})$ is introduced as a Lie group preserving the product of Heisenberg group $\mathrm{H}(2, \mathbb{R})$. Since the symplectic group $\mathrm{Sp}(2, \mathbb{R})$ corresponds to canonical transformation in analytical mechanics, we can expect that there exists a projective unitary representation of the symplectic group $\mathrm{Sp}(2, \mathbb{R})$ that corresponds to the canonical transformation in the quantum system. Such a projective unitary representation f of $\mathrm{Sp}(2,\mathbb{R})$ is required to satisfy the condition

$$\mathsf{f}(g)\mathsf{W}(q,p,\theta)\mathsf{f}(g)^{-1}=\mathsf{W}(g(q,p,\theta)). \tag{7.48}$$

Indeed, a construction of a projective unitary representation f satisfying the above condition is equivalent to a construction of a projective unitary representation of the semi direct product Lie group $\mathrm{H}(2,\mathbb{R})\rtimes \mathrm{SU}(1,1)$ as an extension of W.

Taking the derivative with respect to p and q, we obtain the following condition

$$\begin{pmatrix}\mathsf{f}(g)\mathsf{Q}\mathsf{f}(g)^{-1}\\ \mathsf{f}(g)\mathsf{P}\mathsf{f}(g)^{-1}\end{pmatrix}=g^{-1}\begin{pmatrix}\mathsf{Q}\\ \mathsf{P}\end{pmatrix}. \tag{7.49}$$

We should remark that the Lie group $\mathrm{Sp}(2,\mathbb{R})$ is the Lie group $\mathrm{SL}(2,\mathbb{R})$, and is isomorphic to $\mathrm{SU}(1,1)$ via the map κ given in (4.8). Then, defining a projective unitary representation S of $\mathrm{SU}(1,1)$ as $\mathsf{S}(g):=\mathsf{f}(\kappa(g))$ for $g\in\mathrm{SU}(1,1)$, we see that the condition (7.49) is equivalent to the following condition

$$\begin{pmatrix}\mathsf{S}(g)\mathsf{Q}\mathsf{S}(g)^{-1}\\ \mathsf{S}(g)\mathsf{P}\mathsf{S}(g)^{-1}\end{pmatrix}=\kappa(g)^{-1}\begin{pmatrix}\mathsf{Q}\\ \mathsf{P}\end{pmatrix}. \tag{7.50}$$

This condition can be expressed as the action on the annihilation and creation operators a and $\mathsf{a}^\dagger$ in the following way

$$\begin{pmatrix}\mathsf{S}(g)\mathsf{a}\mathsf{S}(g)^{-1}\\ \mathsf{S}(g)\mathsf{a}^\dagger\mathsf{S}(g)^{-1}\end{pmatrix}=g^{-1}\begin{pmatrix}\mathsf{a}\\ \mathsf{a}^\dagger\end{pmatrix}. \tag{7.51}$$

When we define the action $g\zeta$ of $g\in\mathrm{SU}(1,1)$ on a complex number ζ by (4.10), the condition (7.48) is equivalent to the following condition

$$\mathsf{S}(g)\mathsf{W}(\zeta)\mathsf{S}(g)^{-1}=\mathsf{W}(g\zeta). \tag{7.52}$$

Then, taking the derivative in (7.51), for an element $X\in\mathfrak{su}(1,1)$, we obtain

$$\begin{pmatrix}[\mathsf{S}(X),\mathsf{a}]\\ [\mathsf{S}(X),\mathsf{a}^\dagger]\end{pmatrix}=-X\begin{pmatrix}\mathsf{a}\\ \mathsf{a}^\dagger\end{pmatrix}.$$

Similarly, taking the derivative (7.50), we also obtain

$$\begin{pmatrix}[\mathsf{S}(X),\mathsf{Q}]\\ [\mathsf{S}(X),\mathsf{P}]\end{pmatrix}=-\kappa(X)\begin{pmatrix}\mathsf{Q}\\ \mathsf{P}\end{pmatrix}. \tag{7.53}$$

Hence, under the condition (7.53), the operators $\mathsf{E}_{0,-1}$ and $\mathsf{K}_{\pm,-1}$ defined in (4.12) are limited to the operators satisfying the following conditions

$$[\mathsf{E}_{0,-1}, \mathsf{a}] = \frac{1}{2}\mathsf{a}, \quad [\mathsf{E}_{0,-1}, \mathsf{a}^\dagger] = \frac{1}{2}\mathsf{a}^\dagger$$

$$[\mathsf{K}_{+,-1}, \mathsf{a}] = 0, \quad [\mathsf{K}_{+,-1}, \mathsf{a}^\dagger] = \mathsf{a} \tag{7.54}$$

$$[\mathsf{K}_{-,-1}, \mathsf{a}] = -\mathsf{a}^\dagger, \quad [\mathsf{K}_{-,-1}, \mathsf{a}^\dagger] = 0. \tag{7.55}$$

The conditions (7.54) and (7.55) imply $\mathsf{K}_{+,-1} = \frac{1}{2}\mathsf{a}^2$ and $\mathsf{K}_{-,-1} = \frac{1}{2}(\mathsf{a}^\dagger)^2$. Further, the second equation in (4.13) yields $\mathsf{E}_{0,-1} = [\mathsf{K}_{+,-1}, \mathsf{K}_{-,-1}] = -\frac{1}{2}\mathsf{a}^\dagger\mathsf{a} - \frac{1}{4}$. Hence, we obtain a skew-Hermitian representation of $\mathfrak{g} = \mathfrak{h}(2, \mathbb{R}) \rtimes \mathfrak{sl}(2, \mathbb{R})$.

The property of the operator $\mathsf{E}_{0,-1}$ guarantees that all of weight vectors are number vectors $|n\rangle$. The highest weight vector is the vacuum state vector $|0\rangle$ and has weight $-\frac{1}{4}$. However, the irreducible space containing the vacuum state vector $|0\rangle$ is the space generated by number vectors $\{|2n\rangle\}$ with even number, and is the even function space $L^2_{\mathrm{even}}(\mathbb{R})$. Hence, there exists an irreducible space containing the highest weight vector $|1\rangle$ with the highest weight $-\frac{3}{4}$. This space is generated by number vectors $\{|2n+1\rangle\}$ with odd numbers, and is the odd function space $L^2_{\mathrm{odd}}(\mathbb{R})$. Since these representations of $\mathfrak{sl}(2, \mathbb{R})$ have the highest weights $-\frac{1}{4}$ and $-\frac{3}{4}$, respectively, there exist the corresponding unitary representations of the double covering group $\mathrm{SL}(\check{2}, \mathbb{R})$ of $\mathrm{SL}(2, \mathbb{R})$, as explained in Sect. 2.4. Hence, there exists a projective representation of $\mathrm{SL}(2, \mathbb{R})$ on $L^2(\mathbb{R}) = L^2_{\mathrm{even}}(\mathbb{R}) \oplus L^2_{\mathrm{odd}}(\mathbb{R})$. Thus, we can construct a projective unitary representation of the Lie group $\mathrm{H}(2, \mathbb{R}) \rtimes \mathrm{SL}(2, \mathbb{R})$ such that the representations of its subgroups satisfy the condition (7.48).

Now, we observe this fact from another viewpoint. This projective representation contains a representation of $\mathfrak{g} = \mathfrak{h}(2, \mathbb{R}) \rtimes \mathfrak{u}(1)$. The operator $\mathsf{S}(F^z_{-1})$ gives the representation of the generator of $\mathfrak{u}(1)$. While $\mathsf{E}_1 = 2i\mathsf{S}(F^z_{-1})$, the highest weight λ_F in the previous section is $\frac{1}{2}$.

Hence, we cannot determine the representation of the subgroup $\mathrm{U}(1)$, and can determine the representation only of the double covering group $\mathrm{U}(\check{1})$. Since this fact appears in the representation of the Lie group $\mathrm{SL}(2, \mathbb{R})$, we can determine the representation only of the double covering group $\mathrm{SL}(\check{2}, \mathbb{R})$. Here, we define a vector by squeezing the coherent vector by the action of $\mathrm{SL}(\check{2}, \mathbb{R})$. For a complex number $\zeta \in D$, we define the **squeezed vector** $|\zeta_2, \zeta_1\rangle := \exp(\frac{\zeta_2}{2}(\mathsf{a}^\dagger)^2 - \frac{\overline{\zeta_2}}{2}\mathsf{a}^2)|\zeta_1\rangle = \mathsf{S}(\hat{g}_{\zeta_2,-1})|\zeta_1\rangle$. Especially, when $\zeta_1 = 0$, the vector is the coherent vector with respect to $\mathrm{SL}(\check{2}, \mathbb{R})$, and is written as $|\zeta_2, 0\rangle = (1 - |\zeta_2|^2)^{\frac{1}{4}} \exp(\frac{\zeta_2}{2}(\mathsf{a}^\dagger)^2)|0\rangle$.

Next, we investigate the dispersions of the physical quantities P and Q for the squeezed vector. To discuss the weighted sum of the components of $\Delta_{|\zeta_2,\zeta_1\rangle}[\mathsf{Q} : \mathsf{P}]$ with an arbitrary weight, we remember the relation

$$\begin{aligned}
&\mathrm{Tr}(\kappa(g)^{-1})^T \kappa(g)^{-1} \Delta_{|\zeta_2,\zeta_1\rangle}[\mathsf{Q} : \mathsf{P}] \\
=& \langle \zeta_2, \zeta_1 | (\mathsf{Q} - qI, \mathsf{P} - pI)(\kappa(g)^{-1})^T \kappa(g)^{-1} \begin{pmatrix} \mathsf{Q} - qI \\ \mathsf{P} - pI \end{pmatrix} |\zeta_2, \zeta_1\rangle
\end{aligned}$$

for $g \in \mathrm{SU}(1,1)$, where $\zeta_1 = \frac{x+iy}{\sqrt{2}}$. Then, we choose $g = \hat{g}_{\zeta_2,-1}$ defined in (4.7). So, we have $\mathsf{S}(\hat{g}_{\zeta_2,-1}) = \exp(\frac{\zeta_2}{2}(\mathsf{a}^\dagger)^2 - \frac{\overline{\zeta_2}}{2}\mathsf{a}^2)$, which implies that

$$
\begin{aligned}
&\kappa(\hat{g}_{\zeta_2,-1})^{-1}\begin{pmatrix}\mathsf{Q}-qI\\ \mathsf{P}-pI\end{pmatrix}|\zeta_2,\zeta_1\rangle\\
=&\kappa(\hat{g}_{\zeta_2,-1})^{-1}\begin{pmatrix}\mathsf{Q}-qI\\ \mathsf{P}-pI\end{pmatrix}\mathsf{W}(\hat{g}_{\zeta_2,-1}\zeta)\mathsf{S}(\hat{g}_{\zeta_2,-1})|0\rangle\\
=&\mathsf{W}(\hat{g}_{\zeta_2,-1}\zeta)\kappa(\hat{g}_{\zeta_2,-1})^{-1}\left[\begin{pmatrix}\mathsf{Q}-qI\\ \mathsf{P}-pI\end{pmatrix}+\kappa(\hat{g}_{\zeta_2,-1})\begin{pmatrix}qI\\ pI\end{pmatrix}\right]\mathsf{S}(\hat{g}_{\zeta_2,-1})|0\rangle\\
=&\mathsf{W}(\hat{g}_{\zeta_2,-1}\zeta)\mathsf{S}(\hat{g}_{\zeta_2,-1})\kappa(\hat{g}_{\zeta_2,-1})^{-1}\\
&\cdot\left[\kappa(\hat{g}_{\zeta_2,-1})\begin{pmatrix}\mathsf{Q}-qI\\ \mathsf{P}-pI\end{pmatrix}+\kappa(\hat{g}_{\zeta_2,-1})\begin{pmatrix}qI\\ pI\end{pmatrix}\right]|0\rangle\\
=&\mathsf{W}(\hat{g}_{\zeta_2,-1}\zeta)\mathsf{S}(\hat{g}_{\zeta_2,-1})\begin{pmatrix}\mathsf{Q}\\ \mathsf{P}\end{pmatrix}|0\rangle.
\end{aligned}
$$

So, the relation (7.16) implies

$$
\begin{aligned}
&\langle\zeta_2,\zeta_1|(\mathsf{Q}-qI,\mathsf{P}-pI)(\kappa(\hat{g}_{\zeta_2,-1})^{-1})^T\kappa(\hat{g}_{\zeta_2,-1})^{-1}\begin{pmatrix}\mathsf{Q}-qI\\ \mathsf{P}-pI\end{pmatrix}|\zeta_2,\zeta_1\rangle\\
=&\langle 0|(\mathsf{Q},\mathsf{P})\mathsf{S}(\hat{g}_{\zeta_2,-1})^{\dagger}\mathsf{W}(\hat{g}_{\zeta_2,-1}\zeta)^{\dagger}\mathsf{W}(\hat{g}_{\zeta_2,-1}\zeta)\mathsf{S}(\hat{g}_{\zeta_2,-1})\begin{pmatrix}\mathsf{Q}\\ \mathsf{P}\end{pmatrix}|0\rangle\\
=&\langle 0|(\mathsf{Q},\mathsf{P})\begin{pmatrix}\mathsf{Q}\\ \mathsf{P}\end{pmatrix}|0\rangle=1.
\end{aligned}
$$

We obtain $\mathrm{Tr}(\kappa(\hat{g}_{\zeta_2,-1})^{-1})^T\kappa(\hat{g}_{\zeta_2,-1})^{-1}\Delta_{|\zeta_2,\zeta_1\rangle}[\mathsf{Q}:\mathsf{P}]=1$. Since $\kappa(\hat{g}_{\zeta_2,-1})\in \mathrm{SL}(2,\mathbb{R})$, the relation (5.141) yields that

$$
\begin{aligned}
&\frac{1}{2}\mathrm{Tr}(\kappa(\hat{g}_{\zeta_2,-1})^{-1})^T\ \kappa(\hat{g}_{\zeta_2,-1})^{-1}\Delta_{\rho}[\mathsf{Q}:\mathsf{P}]\\
\geq&\sqrt{\det(\kappa(\hat{g}_{\zeta_2,-1})^{-1})^T\kappa(\hat{g}_{\zeta_2,-1})^{-1}\Delta_{\rho}[\mathsf{Q}:\mathsf{P}]}\geq\frac{1}{2}
\end{aligned}
$$

for an arbitrary state ρ. When $\rho=|\zeta_2,\zeta_1\rangle\langle\zeta_2,\zeta_1|$, since the equality of the above inequality holds, the equality condition guarantees the relation $\mathrm{Tr}(\kappa(\hat{g}_{\zeta_2,-1})^{-1})^T\kappa(\hat{g}_{\zeta_2,-1})^{-1}\Delta_{\rho}[\mathsf{Q}:\mathsf{P}]=\frac{1}{2}$, i.e., the relation $\Delta_{|\zeta_2,\zeta_1\rangle}[\mathsf{Q}:\mathsf{P}]=\frac{1}{2}\kappa(\hat{g}_{\zeta_2,-1})\kappa(\hat{g}_{\zeta_2,-1})^T$.

In general, a 2×2 strictly positive matrix can be written as a constant times of $(\kappa(g')^{-1})^T\kappa(g')^{-1}$ by using a matrix $g'\in \mathrm{SU}(1,1)$. Hence, we have the element ζ_2 minimizing $\frac{1}{2}\mathrm{Tr}(\kappa(g')^{-1})^T\kappa(g')^{-1}\Delta_{\rho}[\mathsf{Q}:\mathsf{P}]$ dependently of the weight matrix $(\kappa(g')^{-1})^T\kappa(g')^{-1}$.

Finally, we focus on the Wigner function $W_{|\zeta_2,\zeta_1\rangle\langle\zeta_2,\zeta_1|}(\zeta)$ of the squeezed state $|\zeta_2,\zeta_1\rangle\langle\zeta_2,\zeta_1|$. Using (7.52), we can calculate the Wigner function as

$$
\begin{aligned}
&W_{|\zeta_2,\zeta_1\rangle\langle\zeta_2,\zeta_1|}(\zeta)=\mathrm{Tr}\,|\zeta_2,\zeta_1\rangle\langle\zeta_2,\zeta_1|\mathsf{W}(\zeta)\\
=&\mathrm{Tr}\,\mathsf{S}(\hat{g}_{\zeta_2,-1})|\zeta_1\rangle\langle\zeta_1|\mathsf{S}(\hat{g}_{\zeta_2,-1})^{\dagger}\mathsf{W}(\zeta)=\mathrm{Tr}\,|\zeta_1\rangle\langle\zeta_1|\mathsf{S}(\hat{g}_{\zeta_2,-1}^{-1})\mathsf{W}(\zeta)\mathsf{S}(\hat{g}_{\zeta_2,-1}^{-1})^{-1}
\end{aligned}
$$

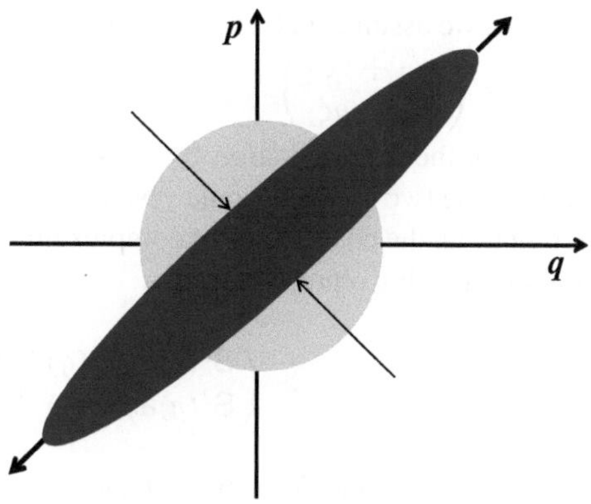

Fig. 7.4 Absolute square of Wigner function: squeezed state and coherent state

$$= \mathrm{Tr}\, |\zeta_1\rangle\langle\zeta_1| \mathsf{W}(\hat{g}_{\zeta_2,-1}^{-1}\zeta) = e^{(\hat{g}_{\zeta_2,-1}^{-1}\zeta)\overline{\zeta_1} - \zeta_1 \overline{(\hat{g}_{\zeta_2,-1}^{-1}\zeta)} - \frac{|(\hat{g}_{\zeta_2,-1}^{-1}\zeta|^2}{2}}. \tag{7.56}$$

Hence, the absolute square of Wigner function is a Gaussian distribution. Since $\kappa(\hat{g}_{\zeta_2,-1})$ is the symmetric matrix $\begin{pmatrix} \cosh\tau + \sinh\tau\cos\phi & \sinh\tau\sin\phi \\ \sinh\tau\sin\phi & \cosh\tau - \sinh\tau\cos\phi \end{pmatrix}$, where $\zeta_2 = e^{i\phi}\tau$ (see (4.8)), the covariance matrix of the Gaussian distribution for p and q ($\zeta = \frac{q+ip}{\sqrt{2}}$) is

$$\kappa((\hat{g}_{\zeta_2,-1})^2 = \begin{pmatrix} \chi_{\tau,1} + 2\chi_{\tau,2}\cos\phi & 2\chi_{\tau,2}\sin\phi \\ 2\chi_{\tau,2}\sin\phi & \chi_{\tau,1} - 2\cos\phi\chi_{\tau,2} \end{pmatrix} \tag{7.57}$$

where $\chi_{\tau,1} := \cosh^2\tau + \sinh^2\tau$ and $\chi_{\tau,2} := \cosh\tau\sinh\tau$. Hence, the squeezed vector is characterized as Fig. 7.4 based on the above covariance matrix.

7.6 Squeezed State on Two-Mode Bosonic System*

In this section, we discuss the squeezing operation on the two-mode bosonic system. For this purpose, we define the action of $g = \begin{pmatrix} \mu & \nu \\ \overline{\nu} & \overline{\mu} \end{pmatrix} \in \mathrm{SU}(1,1)$ for $\zeta = (\zeta_1, \zeta_2)^T \in \mathbb{C}^2$ as follows

$$\begin{pmatrix} \zeta_1 \\ \overline{\zeta_2} \end{pmatrix} \mapsto \begin{pmatrix} \mu & \nu \\ \overline{\nu} & \overline{\mu} \end{pmatrix} \begin{pmatrix} \zeta_1 \\ \overline{\zeta_2} \end{pmatrix}. \tag{7.58}$$

That is, we assume that a vector $\zeta = (\zeta_1, \zeta_2)^T \in \mathbb{C}^2$ is transformed to another vector $g \cdot \zeta := \begin{pmatrix} \mu\zeta_1 + \nu\overline{\zeta}_2 \\ \nu\overline{\zeta}_1 + \mu\zeta_2 \end{pmatrix}$.

Then, the representation S^2 of SU(1, 1) corresponding to the squeezing operation on the two-mode bosonic system is given by $\mathsf{S}^2(g)\mathsf{W}^2(\zeta)\mathsf{S}^2(g)^{-1} = \mathsf{W}^2(g \cdot \zeta)$. Comparing the coefficients of complex numbers $-\overline{\zeta}_1$ and ζ_2, we obtain the following relation for this representation.

$$\begin{pmatrix} \mathsf{S}^2(g)\mathsf{a}_1\mathsf{S}^2(g)^{-1} \\ \mathsf{S}^2(g)\mathsf{a}_2^\dagger\mathsf{S}^2(g)^{-1} \end{pmatrix} = g^{-1}\begin{pmatrix} \mathsf{a}_1 \\ \mathsf{a}_2^\dagger \end{pmatrix}. \tag{7.59}$$

Here, by taking the derivative, an element $X \in \mathfrak{su}(1, 1)$ satisfies

$$\begin{pmatrix} [\mathsf{S}^2(X), \mathsf{a}_1] \\ [\mathsf{S}^2(g), \mathsf{a}_2^\dagger] \end{pmatrix} = -X\begin{pmatrix} \mathsf{a}_1 \\ \mathsf{a}_2^\dagger \end{pmatrix}.$$

Hence, the elements $\mathsf{E}_{0,-1}$ and $\mathsf{K}_{\pm,-1}$ given in (4.12) satisfy

$$[\mathsf{E}_{0,-1}, \mathsf{a}_1] = \frac{-1}{2}\mathsf{a}_1, \quad [\mathsf{E}_{0,-1}, \mathsf{a}_2^\dagger] = \frac{1}{2}\mathsf{a}_2^\dagger$$

$$[\mathsf{K}_{+,-1}, \mathsf{a}_1] = 0, \quad [\mathsf{K}_{+,-1}, \mathsf{a}_2^\dagger] = \mathsf{a}_1 \tag{7.60}$$

$$[\mathsf{K}_{-,-1}, \mathsf{a}_1] = -\mathsf{a}_2^\dagger, \quad [\mathsf{K}_{-,-1}, \mathsf{a}_2^\dagger] = 0. \tag{7.61}$$

Since $\mathsf{K}_{+,-1}^\dagger = \mathsf{K}_{+,-1}$, the relations (7.60) and (7.61) guarantee that $\mathsf{K}_{+,-1}$ does not contain the operators $\mathsf{a}_1^\dagger$ and a_2. So, we obtain $\mathsf{K}_{+,-1} = \mathsf{a}_1\mathsf{a}_2$ and $\mathsf{K}_{-,-1} = \mathsf{a}_1^\dagger\mathsf{a}_2^\dagger$. Further, the second relation of (4.13) yields

$$\mathsf{E}_{0,-1} = -\frac{1}{2}[\mathsf{K}_{+,-1}, \mathsf{K}_{-,-1}] = \frac{-1}{2}(\mathsf{a}_1^\dagger\mathsf{a}_1 + \mathsf{a}_2^\dagger\mathsf{a}_2 + I).$$

Since a number vector $|n, m\rangle$ satisfies

$$\mathsf{K}_{+,-1}|n, m\rangle = \sqrt{n}\sqrt{m}|n-1, m-1\rangle,$$

$$\mathsf{K}_{-,-1}|n, m\rangle = \sqrt{n+1}\sqrt{m+1}|n+1, m+1\rangle,$$

the vectors $|n, 0\rangle$ and $|0, n\rangle$ are highest weight vectors. So, the irreducible space containing $|n, 0\rangle$ or $|0, n\rangle$ has the highest weight $\frac{-n-1}{2}$. In general, the coherent state of the representation is called a **two-mode squeezed state** when the highest weight vector is the vacuum vector $|0, 0\rangle$. Hence, the representation corresponding to a two-mode squeezed state has the highest weight $\frac{-1}{2}$, and can be extended to the representation of the Lie groupSL(2, $\mathbb{R}$).

7.7 Wigner Function and Fourier Transform of Heisenberg Representation*

7.7.1 Fourier Transform of Heisenberg Representation

Next, we consider the Fourier transform based on Heisenberg representation. For this purpose, we remember the discussion in Sect. 7.1.1, which guarantees that the projective unitary representation of the Lie group $\mathbb{R}^2$ with the factor system $\{e^{\frac{i}{2}(q'p-qp')}\}_{(q,p),(q',p')}$ is limited to the unitary projective representation isomorphic to the Heisenberg representation W. Hence, using the notation W appearing the Heisenberg representation W, we denote the Fourier transform of the two-dimensional group $\mathbb{R}^2$ with the factor system by $\mathcal{F}_{\mathsf{W}}$. Remember that in order to define the Fourier transform with respect to projective unitary representations, we need to fix the factor system, as explained in Sects. 2.8 and 3.8. Since the irreducible projective unitary representation with the factor system is unique and has the representation space $L^2(\mathbb{R}^2)$, we consider the space $L^2(L^2(\mathbb{R}))$ of square integrable space on $L^2(\mathbb{R})$ as

$$L^2(L^2(\mathbb{R})) := \{X : \text{ linear operator on } L^2(\mathbb{R}) | \|X\|^2 := \operatorname{Tr} XX^\dagger < \infty\}, \quad (7.62)$$

which is called the space of Hilbert–Schmidt operators. Hence, the Fourier transform is defined as a map $\mathcal{F}_{\mathsf{W}}$ from $L^2(\mathbb{R}^2)$ to $L^2(L^2(\mathbb{R}))$. That is, letting $\zeta := \frac{1}{\sqrt{2}}(q+ip)$, we obtain the Fourier transform of an element $f \in L^2(\mathbb{R}^2)$ by

$$\mathcal{F}_{\mathsf{W}}[f] = \int_{\mathbb{C}} f(\zeta)\mathsf{W}(-\zeta)\frac{dqdp}{2\pi}. \quad (7.63)$$

For the Inverse Fourier transform, is defined as a map $\mathcal{F}_{\mathsf{W}}^{-1}$ from $L^2(L^2(\mathbb{R}))$ to $L^2(\mathbb{R})$. For an operator $X \in L^2(L^2(\mathbb{R}))$, we have

$$\mathcal{F}_{\mathsf{W}}^{-1}[X](\zeta) = \operatorname{Tr} \mathsf{W}(\zeta)X, \quad (7.64)$$

which equals Wigner function $W_X(\zeta)$ of X. The discussion in Sect. 3.8.2 guarantees that $\mathcal{F}_{\mathsf{W}}^{-1} \circ \mathcal{F}_{\mathsf{W}}$ and $\mathcal{F}_{\mathsf{W}} \circ \mathcal{F}_{\mathsf{W}}^{-1}$ are identify maps on $L^2(\mathbb{R}^2)$ and $L^2(L^2(\mathbb{R}))$, respectively. That is, both maps $\mathcal{F}_{\mathsf{W}}$ and $\mathcal{F}_{\mathsf{W}}^{-1}$ are isometric. In this context, the operator $XX^\dagger$ corresponds to a density operator on $L^2(\mathbb{R})$. Thus, the operator $X \in L^2(L^2(\mathbb{R}))$ can be written by use of Wigner function as

$$X = \mathcal{F}_{\mathsf{W}}[W_X] = \int_{\mathbb{C}} W_X(\alpha)\mathsf{W}(-\alpha)\frac{dadb}{2\pi}, \quad (7.65)$$

where $\alpha := \frac{1}{\sqrt{2}}(a+ib)$. So, when $\operatorname{Tr} XX^\dagger = 1$, $|X\rangle\rangle$ is an entangled state with a suitable reference system and the Wigner function W_X contains all information for the entangled state with respect to the symmetry for the Heisenberg Representation. In this section, we employ the notation $\mathcal{F}_{\mathsf{W}}^{-1}[X]$ instead of the notation W_X because it is better to avoid to denote X as the subscript. Further, when the operator $XX^\dagger$ describes a density operator on $L^2(\mathbb{R})$, the vector $|X\rangle\rangle$ on the system $L^2(\mathbb{R})^{\otimes 2}$ describes an entangled state with the reference system $L^2(\mathbb{R})$.

Here, we calculate Wigner functions, i.e., Inverse Fourier transforms of various operators while the Wigner functions for the coherent state $|\alpha\rangle\langle\alpha|$ and the squeezed state $|\zeta_2, \zeta_1\rangle\langle\zeta_2, \zeta_1|$ are calculated in (7.20) and (7.56), respectively. First, we focus on the operators $\mathsf{a}|\alpha\rangle\langle\alpha|$ and $\mathsf{a}^\dagger|\alpha\rangle\langle\alpha|$. Then,

$$\mathcal{F}_{\mathsf{W}}^{-1}[\mathsf{a}|\alpha\rangle\langle\alpha|](\zeta) = \operatorname{Tr}\mathsf{W}(\zeta)\alpha|\alpha\rangle\langle\alpha| = \alpha e^{-\frac{|\zeta|^2}{2}+i(ap-qb)}$$

$$\begin{aligned}\mathcal{F}_{\mathsf{W}}^{-1}[\mathsf{a}^\dagger|\alpha\rangle\langle\alpha|](\zeta) &= \operatorname{Tr}\mathsf{W}(\zeta)\mathsf{a}^\dagger|\alpha\rangle\langle\alpha| = \operatorname{Tr}(\mathsf{a}^\dagger - \overline{\zeta})\mathsf{W}(\zeta)|\alpha\rangle\langle\alpha| \\ &= (\overline{\alpha} - \overline{\zeta})\operatorname{Tr}\mathsf{W}(\zeta)|\alpha\rangle\langle\alpha| = (\overline{\alpha} - \overline{\zeta})e^{-\frac{|\zeta|^2}{2}+i(ap-qb)}.\end{aligned}$$

So, the relations

$$\begin{aligned}\mathcal{F}_{\mathsf{W}}^{-1}\left[\frac{\mathsf{a}+\mathsf{a}^\dagger}{\sqrt{2}}|\alpha\rangle\langle\alpha|\right](\zeta) &= \left(-i\frac{\partial}{\partial p} - \frac{q}{2}\right)e^{-\frac{|\zeta|^2}{2}+i(ap-qb)} \qquad (7.66)\\ &= \left(-i\frac{\partial}{\partial p} - \frac{q}{2}\right)\mathcal{F}_{\mathsf{W}}^{-1}[|\alpha\rangle\langle\alpha|](\zeta)\end{aligned}$$

$$\begin{aligned}\mathcal{F}_{\mathsf{W}}^{-1}\left[-i\frac{\mathsf{a}-\mathsf{a}^\dagger}{\sqrt{2}}|\alpha\rangle\langle\alpha|\right](\zeta) &= \left(i\frac{\partial}{\partial q} - \frac{p}{2}\right)e^{-\frac{|\zeta|^2}{2}+i(ap-qb)} \qquad (7.67)\\ &= \left(i\frac{\partial}{\partial q} - \frac{p}{2}\right)\mathcal{F}_{\mathsf{W}}^{-1}[|\alpha\rangle\langle\alpha|](\zeta)\end{aligned}$$

hold (Exercise 7.9). Since any operator on the space $L^2(\mathbb{R})$ can be written as a linear sum of the coherent states $\{|\alpha\rangle\langle\alpha|\}_\alpha$, we have

$$\mathcal{F}_{\mathsf{W}}^{-1}\left[\frac{\mathsf{a}+\mathsf{a}^\dagger}{\sqrt{2}}X\right](\zeta) = \left(-i\frac{\partial}{\partial p} - \frac{q}{2}\right)\mathcal{F}_{\mathsf{W}}^{-1}[X](\zeta)$$

$$\mathcal{F}_{\mathsf{W}}^{-1}\left[-i\frac{\mathsf{a}-\mathsf{a}^\dagger}{\sqrt{2}}X\right](\zeta) = \left(i\frac{\partial}{\partial q} - \frac{p}{2}\right)\mathcal{F}_{\mathsf{W}}^{-1}[X](\zeta)$$

for any operator X on $L^2(\mathbb{R})$. Since $\zeta = \frac{q+ip}{\sqrt{2}}$, $\mathcal{F}_{\mathsf{W}}^{-1}[X](\zeta)$ can be regarded as an element of the function space $L^2(\mathbb{R}^2)$ with respect to two variables q and p. Then, when q is the variable for the first mode and p is that for the second mode, we denote the operations of multiplying q and p by the operators Q_1 and Q_2, respectively. Similarly, we denote the $-i$ multiplication of the derivatives withe respect to q and p by the operators P_1 and P_2, respectively. Hence, we have

$$\mathcal{F}_{\mathrm{W}}^{-1}[\mathsf{Q}X](\zeta) = \left(\mathsf{P}_2 - \frac{1}{2}\mathsf{Q}_1\right)\mathcal{F}_{\mathrm{W}}^{-1}[X](\zeta)$$
$$\mathcal{F}_{\mathrm{W}}^{-1}[\mathsf{P}X](\zeta) = \left(-\mathsf{P}_1 - \frac{1}{2}\mathsf{Q}_2\right)\mathcal{F}_{\mathrm{W}}^{-1}[X](\zeta).$$

Now, for an operator Y on $L^2(\mathbb{R})$, we define the operator L_Y on the operator space on $L^2(\mathbb{R})$ by $L_Y : X \mapsto YX$. We notice that $\mathcal{F}_{\mathrm{W}}$ is a unitary operator from the function space $L^2(\mathbb{R}^2)$ to the the operator space on $L^2(\mathbb{R})$. Then, we have

$$\mathcal{F}_{\mathrm{W}}^{-1} L_{\mathsf{Q}} \mathcal{F}_{\mathrm{W}} = \mathsf{P}_2 - \frac{1}{2}\mathsf{Q}_1, \quad \mathcal{F}_{\mathrm{W}}^{-1} L_{\mathsf{P}} \mathcal{F}_{\mathrm{W}} = -\mathsf{P}_1 - \frac{1}{2}\mathsf{Q}_2. \tag{7.68}$$

7.7.2 Uncertainty Relation with Wigner Function

In this subsection, we extend the uncertainty relation for wave packets discussed in Sect. 5.6 to the case with Wigner function. For this purpose, we focus on the uncertainty for Wigner function

$$\int_{\mathbb{R}^2} 2|\zeta|^2 |\mathcal{F}_{\mathrm{W}}^{-1}[X](\zeta)|^2 dq dp, \tag{7.69}$$

whose operational meaning is discussed in [50, Sect. 4.3] [48]. To consider a generalization of the uncertainty relation (5.139), we consider the energy constraint

$$\langle\langle X|\mathsf{Q}^2 + \mathsf{P}^2|X\rangle\rangle = \mathrm{Tr}(\mathsf{Q}^2 + \mathsf{P}^2)XX^\dagger \le E \tag{7.70}$$

for $X \in L^2(L^2(\mathbb{R}))$. That is, we consider the following minimization problem;

$$\min_X \left\{ \int_{\mathbb{R}^2} 2|\zeta|^2 |\mathcal{F}_{\mathrm{W}}^{-1}[X](\zeta)|^2 dq dp \middle| \langle\langle X|\mathsf{Q}^2 + \mathsf{P}^2|X\rangle\rangle \le E \right\}. \tag{7.71}$$

In this problem, we address the trade-off between the dispersions of the operators before and after the Inverse Fourier transform.

When we transform the operator X to the element $\phi := \mathcal{F}_{\mathrm{W}}^{-1}[X]$ of $L^2(\mathbb{R}^2)$ via the Inverse Fourier transform, due to the condition (7.68), the condition (7.70) is transformed to the condition $\langle\phi|(\mathsf{P}_2 - \frac{1}{2}\mathsf{Q}_1)^2 + (-\mathsf{P}_1 - \frac{1}{2}\mathsf{Q}_2)^2|\phi\rangle \le E$. Then, the value to be minimized is $\langle\phi|\mathsf{Q}_1^2 + \mathsf{Q}_2^2|\phi\rangle$. To discuss this minimization, we consider the opposite minimization as

$$\min_{\phi:\langle\phi|\mathsf{Q}_1^2+\mathsf{Q}_2^2|\phi\rangle\le C} \langle\phi| \left(\mathsf{P}_2 - \frac{1}{2}\mathsf{Q}_1\right)^2 + \left(-\mathsf{P}_1 - \frac{1}{2}\mathsf{Q}_2\right)^2 |\phi\rangle. \tag{7.72}$$

For this minimization, we employ a method similar to the minimization (5.141). For this purpose, we define the operators $\mathsf{A}_j := \mathsf{Q}_j$ and $A_{2+j} := \mathsf{P}_j$. Then, we have the 4×4 semi-definite matrix $J_\rho := (\mathrm{Tr}\, \mathsf{A}_k \mathsf{A}_j \rho)_{k,j}$. (For its semi-definiteness, see Exercise 7.8.) For a function ϕ, we define $\tilde{\phi}$ as $\tilde{\phi}(q,p) := \phi(p,-q)$. Then, we have $\langle \tilde{\phi} | \mathsf{A}_k \mathsf{A}_j | \tilde{\phi} \rangle = (-1)^{a+b} \langle \phi | \mathsf{A}_{k'} \mathsf{A}_{j'} | \phi \rangle$, where k', j', a, b are given as follows. When $k = 1,2,3,4$, we decide $k' = 2,1,4,3$ and $a = 1,0,1,0$, respectively. j' and b are decided in the same way from j. Since $\langle \tilde{\phi} | \mathsf{Q}_1^2 + \mathsf{Q}_2^2 | \tilde{\phi} \rangle = \langle \phi | \mathsf{Q}_1^2 + \mathsf{Q}_2^2 | \phi \rangle$ and $\langle \tilde{\phi} | (\mathsf{P}_2 - \frac{1}{2}\mathsf{Q}_1)^2 + (-\mathsf{P}_1 - \frac{1}{2}\mathsf{Q}_2)^2 | \tilde{\phi} \rangle = \langle \phi | (\mathsf{P}_2 - \frac{1}{2}\mathsf{Q}_1)^2 + (-\mathsf{P}_1 - \frac{1}{2}\mathsf{Q}_2)^2 | \phi \rangle$, the minimization problem can be converted to

$$\min_{\phi : \mathrm{Tr}(\mathsf{Q}_1^2 + \mathsf{Q}_2^2)\rho_\phi \le C} \mathrm{Tr} \left(\left(\mathsf{P}_2 - \frac{1}{2}\mathsf{Q}_1 \right)^2 + \left(-\mathsf{P}_1 - \frac{1}{2}\mathsf{Q}_2 \right)^2 \right) \rho_\phi, \tag{7.73}$$

where $\rho_\phi := \frac{1}{2}(|\tilde{\phi}\rangle\langle\tilde{\phi}| + |\phi\rangle\langle\phi|)$. Due to the symmetry, by using three real numbers $\alpha \ge 0$, $\beta \ge 0$, and γ, the semi-definite matrix $(\mathrm{Tr}\, \mathsf{A}_k \mathsf{A}_j \rho_\phi)_{k,j}$ can be written as

$$J_{\alpha,\beta,\gamma} := \begin{pmatrix} \beta & 0 & \frac{i}{2} & -\gamma \\ 0 & \beta & \gamma & \frac{i}{2} \\ -\frac{i}{2} & \gamma & \alpha & 0 \\ -\gamma & -\frac{i}{2} & 0 & \alpha \end{pmatrix}. \tag{7.74}$$

Since for any three real numbers $\alpha \ge 0$, $\beta \ge 0$, and γ satisfying $J_{\alpha,\beta,\gamma} \ge 0$, there exists an element $g \in \mathrm{Sp}(4, \mathbb{R})$ such that $g^T J_{\alpha,\beta,\gamma} g = J_{\alpha',\beta',0}$ (Exercise 7.10). As explained in the next section, there exists a unitary operator corresponding to the element g. So, we can choose a state ρ such that $J_\rho = J_{\alpha,\beta,\gamma}$. Thus, the above minimization problem can be converted to

$$\min_{\alpha,\beta,\gamma : 2\beta \le C, J_{\alpha,\beta,\gamma} \ge 0} 2 \left(\alpha + \gamma + \frac{\beta}{4} \right), \tag{7.75}$$

The condition $J_{\rho_\phi} \ge 0$ is equivalent to

$$\begin{pmatrix} \alpha\beta & 0 \\ 0 & \alpha\beta \end{pmatrix} \ge \begin{pmatrix} \frac{i}{2} & -\gamma \\ \gamma & \frac{i}{2} \end{pmatrix}^\dagger \begin{pmatrix} \frac{i}{2} & -\gamma \\ \gamma & \frac{i}{2} \end{pmatrix} = \begin{pmatrix} \frac{1}{4} + \gamma^2 & i\gamma \\ -i\gamma & \frac{1}{4} + \gamma^2 \end{pmatrix}. \tag{7.76}$$

This condition is equivalent to $\alpha\beta \ge |\gamma| + \frac{1}{4} + \gamma^2$. Since it is enough to choose $\alpha = \frac{1}{\beta}(|\gamma| + \frac{1}{4} + \gamma^2)$ and $\beta = \frac{C}{2}$, the minimization is converted to

$$\min_\gamma 2 \left(\frac{2}{C} \left(|\gamma| + \frac{1}{4} + \gamma^2 \right) + \gamma + \frac{C}{8} \right) = \min_\gamma \frac{4}{C} \left(\gamma^2 + |\gamma| + \gamma\frac{C}{2} + \frac{1}{4} + \frac{C^2}{16} \right). \tag{7.77}$$

When $C > 2$, the minimum equals 1 and is realized when $\gamma = -\frac{1}{2}(\frac{C}{2} - 1)$. When $C \le 2$, the minimum equals $\frac{1}{C} + \frac{C}{4}$ and is realized when $\gamma = 0$. Since the inverse function of $C \mapsto \frac{1}{C} + \frac{C}{4}$ is the function $E \mapsto 2(E - \sqrt{E^2 - 1})$ and this minimization still holds by including mixed states, we have

$$
\begin{aligned}
&\min_{\rho:\mathrm{Tr}((\mathsf{P}_2 - \frac{1}{2}\mathsf{Q}_1)^2 + (-\mathsf{P}_1 - \frac{1}{2}\mathsf{Q}_2)^2)\rho \le E} \mathrm{Tr}(\mathsf{Q}_1^2 + \mathsf{Q}_2^2)\rho \\
= &\min_{\phi:\langle\phi|(\mathsf{P}_2 - \frac{1}{2}\mathsf{Q}_1)^2 + (-\mathsf{P}_1 - \frac{1}{2}\mathsf{Q}_2)^2|\phi\rangle \le E} \langle\phi|\mathsf{Q}_1^2 + \mathsf{Q}_2^2|\phi\rangle = 2(E - \sqrt{E^2 - 1})
\end{aligned}
\tag{7.78}
$$

when $E \ge 1$. Since the minimum eigenvalue of $\mathsf{P}^2 + \mathsf{Q}^2$ is 1, we have no minimum for $E < 1$.

So, this discussion is summarized as follows.

Theorem 7.1 ([49])

$$
\begin{aligned}
&\min_X \left\{ \int_{\mathbb{R}^2} 2|\zeta|^2 |\mathcal{F}_{\mathsf{W}}^{-1}[X](\zeta)|^2 dq dp \,\middle|\, \langle\langle X|\mathsf{Q}^2 + \mathsf{P}^2|X\rangle\rangle \le E \right\} \\
= &2(E - \sqrt{E^2 - 1})
\end{aligned}
\tag{7.79}
$$

for $E \ge 1$.

Exercise 7.8 Show that the Hermitian matrix $J_\rho := (\mathrm{Tr}\, \mathsf{A}_k \mathsf{A}_j \rho)_{k,j}$ is positive semi-definite.

Exercise 7.9 Show (7.66) and (7.67).

Exercise 7.10 Show that for any three real numbers $\alpha \ge 0$, $\beta \ge 0$, and γ satisfying $J_{\alpha,\beta,\gamma} \ge 0$, there exists an element $g \in \mathrm{Sp}(4, \mathbb{R})$ such that $g^T J_{\alpha,\beta,\gamma} g = J_{\alpha',\beta',0}$.

Exercise 7.11 Show (7.22) by using (7.65).

7.8 Squeezed State on Multi-mode Bosonic System*

Next, we address squeezing operations on a multi-mode bosonic system. The representation of the Lie group SU(1, 1) is not sufficient for describing a squeezing operation over more than two modes. Alternatively, we employ the representation S^r of the **symplectic group** $\mathrm{Sp}(2r, \mathbb{R})$ when the number of modes is r. In this case, i.e., on the r-mode bosonic system, the representation S^r of an element $g \in \mathrm{Sp}(2r, \mathbb{R})$ is defined as a generalization of (7.51) so that $\mathsf{S}^r(g)\vec{\mathsf{R}}\mathsf{S}^r(g)^{-1} = g^{-1}\vec{\mathsf{R}}$. When we employ the map κ defined in (6.35) with a general form, we have

$$
\begin{pmatrix} \mathsf{S}^r(g)\vec{\mathsf{a}}\mathsf{S}(g)^{-1} \\ \mathsf{S}^r(g)\vec{\mathsf{a}}^\dagger\mathsf{S}(g)^{-1} \end{pmatrix} = \kappa^{-1}(g^{-1}) \begin{pmatrix} \vec{\mathsf{a}} \\ \vec{\mathsf{a}}^\dagger \end{pmatrix}. \tag{7.80}
$$

In the following, to see what representation on the r-mode bosonic system is derived from the above condition, we simply discuss the structure of the Lie algebra $\mathfrak{sp}(2r, \mathbb{R})$ corresponding to $\mathrm{Sp}(2r, \mathbb{R})$. The complexification of $\mathfrak{sp}(2r, \mathbb{R})$ is $\mathfrak{sp}(2r, \mathbb{C})$, which is the complexification of $\mathfrak{sp}(r)$. Hence, the root system $\Phi_{\mathfrak{sp}(2r,\mathbb{R})}$ of $\mathfrak{sp}(2r, \mathbb{R})$ is that of $\mathfrak{sp}(r)$. Further, since a maximal compact subgroup of $\mathrm{Sp}(2r, \mathbb{R})$ is $\mathrm{U}(r)$, the root system $\Phi_{\mathfrak{sp}(2r,\mathbb{R})}$ is divided into the compact part $\Phi_{\mathfrak{u}(r)}$ and the non-compact part $\Phi_{\mathfrak{p}}$. As $\mathfrak{u}(r)$ contains a maximal Cartan subalgebra of $\mathfrak{sp}(r)$, the sets of respective positive roots are given as $\Phi^+_{\mathfrak{u}(r)} = \{\alpha_{j,l}\}_{1\le j<l\le r}$ and $\Phi^+_{\mathfrak{p}} = \{\beta_{j,l}\}_{1\le j\le l\le r}$ by using the notations in Sect. 6.6. Especially, $\mathfrak{sp}(2r, \mathbb{R})$ is Hermitian and $\beta_{r,r}$ is β. Since $Z_{\alpha_{j,l}}$, $Z_{\beta_{j,l}}$, $F^x_{\alpha_{j,l}}$, and $F^y_{\alpha_{j,l}}$ are included in $\mathfrak{u}(r)$, they have the same form as in the case of $\mathfrak{sp}(r)$. However, $F^x_{\beta_{j,l}}$ and $F^y_{\beta_{j,l}}$ have different forms from those in the case of $\mathfrak{sp}(r)$, and are given as

$$F^x_{\beta_{j,l}} = \begin{cases} \frac{1}{\sqrt{2}}(|j\rangle\langle l| + |l\rangle\langle j|) \otimes F^x_{-1} & \text{when } j < l \\ |j\rangle\langle j| \otimes F^x_{-1} & \text{when } j = l \end{cases}$$

$$F^y_{\beta_{j,l}} = \begin{cases} \frac{1}{\sqrt{2}}(|j\rangle\langle l| + |l\rangle\langle j|) \otimes F^y_{-1} & \text{when } j < l \\ |j\rangle\langle j| \otimes F^y_{-1} & \text{when } j = l. \end{cases}$$

Thus, the element of maximal Cartan subalgebra and the representations of $F^x_{\alpha_{j,l}}$ and $F^y_{\alpha_{j,l}}$ on the r-mode bosonic system are given in Sect. 7.3. Although the discussion in Sect. 7.3 gives no constraint for the parameter λ_F corresponding to the highest weight, it is calculated to be $-\frac{1}{2}$ on the r-mode bosonic system due to the following discussion.

We need to fix the representations of $F^x_{\beta_{j,l}}$ and $F^y_{\beta_{j,l}}$ for discussing the representation S^r on the r-mode bosonic system, instead of their representations, we deal with the operators $\mathsf{K}_{\beta_{j,l}}$ and $\mathsf{K}_{-\beta_{j,l}}$. When $j = l$, the commutation relations among $\mathsf{K}_{\beta_{j,l}}$, $\mathsf{K}_{-\beta_{j,l}}$, and $Z_{\beta_{j,l}}$ are reduced to the commutation relations given in the case with $\mathfrak{su}(1, 1)$. Since only the operators a_j and $\mathsf{a}_j^\dagger$ appear in the commutation relations, the analysis of the action of these operators can be reduced to the one-mode bosonic case discussed in Sect. 7.5. Hence, we have $\mathsf{K}_{\beta_{j,l}} = \frac{1}{2}\mathsf{a}_j^2$ and $\mathsf{K}_{-\beta_{j,l}} = \frac{1}{2}(\mathsf{a}_j^\dagger)^2$. Since $-\mathsf{f}(|j\rangle\langle j|) = -\mathsf{a}_j^\dagger\mathsf{a}_j - \frac{1}{2}I$, the highest weight λ_F given in Sect. 7.3 is decided to be $\frac{-1}{2}$.

Similarly, when $j < l$, the commutation relations among $\mathsf{K}_{\beta_{j,l}}$, $\mathsf{K}_{-\beta_{j,l}}$, and $Z_{\beta_{j,l}}$ are reduced to the commutation relations of $\mathfrak{su}(1, 1)$. So, only the operators a_j, a_l, $\mathsf{a}_j^\dagger$, and $\mathsf{a}_l^\dagger$ appear in the commutation relations, the analysis of the action of these operators can be reduced to the two-mode bosonic case discussed in Sect. 7.6. Hence, we obtain $\mathsf{K}_{\beta_{j,l}} = \mathsf{a}_j\mathsf{a}_l$ and $\mathsf{K}_{-\beta_{j,l}} = \mathsf{a}_j^\dagger\mathsf{a}_l^\dagger$.

The actions of $\mathsf{K}_{\alpha_{j,l}}$ and $\mathsf{K}_{-\alpha_{j,l}}$ preserve the number of total photons, that of $\mathsf{K}_{\beta_{j,l}}$ decreases the number by two, and that of $\mathsf{K}_{-\beta_{j,l}}$ increases the number by two. Hence, the representation space $L^2(\mathbb{R})^{\otimes r}$ is divided to the spaces $L^2(\mathbb{R})^{\otimes r:\mathrm{even}}$ and $L^2(\mathbb{R})^{\otimes r:\mathrm{odd}}$, where the space $L^2(\mathbb{R})^{\otimes r:\mathrm{even}}$ ($L^2(\mathbb{R})^{\otimes r:\mathrm{odd}}$) is composed of the states whose total photon number is even (odd). The highest weight vector $L^2(\mathbb{R})^{\otimes r:\mathrm{even}}$ is

$|0,\ldots,0\rangle$ and the highest weight is $-\frac{1}{2}\boldsymbol{v}$. Hence, the representation of the subgroup $\mathrm{U}(r)$ of $\mathrm{Sp}(2r,\mathbb{R})$ cannot be given, and only that of the double covering group $\check{\mathrm{U}(r)}$ can be decided. Similarly, the representation of $\mathrm{Sp}(2r,\mathbb{R})$ cannot be decided on the r-mode bosonic system and that of the double covering group $\mathrm{Sp}(\check{2r},\mathbb{R})$ can be decided on the r-mode bosonic system. The space $\mathcal{V}[\mathsf{K}^j_{-\beta}]|0,\ldots,0\rangle$ is the space of states whose total photon numbers is $2j$, and is irreducible with respect to the representation of $\mathfrak{u}(r)$. Hence, the representation of $L^2(\mathbb{R})^{\otimes r:\mathrm{even}}$ is $\mathfrak{u}(r)$-irreducible. The highest weight vector $|0,\ldots,0\rangle$ of the space $\mathcal{V}[\mathsf{K}^j_{-\beta}]|0,\ldots,0\rangle$ is written as $|-\frac{1}{2}\boldsymbol{v};-\frac{1}{2}\boldsymbol{v}\rangle$ by using the notation given in Sect. 6.9.

Now, we describe the representation of the double covering group $\mathrm{Sp}(\check{2r},\mathbb{R})$ by using Cartan decomposition. Since $\Phi_{\mathfrak{p}}^+ = \{\beta_{j,l}\}_{1\le j\le l\le r}$, any element of $\mathrm{Sp}(\check{2r},\mathbb{R})$ can be written by an element g of $\check{\mathrm{U}(r)}$ and a complex symmetric matrix $\zeta=\{\zeta_{k,l}\}$. So, its representation is given as $\mathsf{U}(g)\exp(\frac{1}{2}\sum_{k,l}\zeta_{k,l}\mathsf{a}_k^\dagger\mathsf{a}_l^\dagger - C.C.)$, where $C.C.$ denotes the adjoint operator. Here, there exists an element $g'\in\mathrm{U}(r)$ such that the matrix $g'\zeta g'^T$ is a diagonal matrix $\begin{pmatrix}\eta_1 & & & \\ & \eta_2 & & \\ & & \vdots & \\ & & & \eta_r\end{pmatrix}$, where η_j is areal number. Hence, letting $\mathsf{S}^r_j(\eta_j) := \exp(\frac{1}{2}\eta_j(\mathsf{a}_j^\dagger)^2 - C.C.)$, we have

$$
\begin{aligned}
&\mathsf{U}(g)\exp\left(\frac{1}{2}\sum_{k,l}\zeta_{k,l}\mathsf{a}_k^\dagger\mathsf{a}_l^\dagger - C.C.\right)\\
=&\mathsf{U}(g)\mathsf{U}(g')^\dagger\mathsf{U}(g')\exp(\frac{1}{2}\sum_{k,l}\zeta_{k,l}\mathsf{a}_k^\dagger\mathsf{a}_l^\dagger - C.C.)\mathsf{U}(g')^\dagger\mathsf{U}(g')\\
=&\mathsf{U}(g)\mathsf{U}(g')^\dagger\mathsf{S}^r_1(\eta_1)\mathsf{S}^r_2(\eta_2)\cdots\mathsf{S}^r_r(\eta_r)\mathsf{U}(g')\\
=&\mathsf{U}(g)\mathsf{U}(g')^\dagger\mathsf{S}^r_1(\eta_1)\mathsf{U}_{1,2}\mathsf{S}^r_1(\eta_2)\mathsf{U}_{1,2}^\dagger\cdots\mathsf{U}_{1,r}\mathsf{S}^r_1(\eta_r)\mathsf{U}_{1,r}^\dagger\mathsf{U}(g'),
\end{aligned}
$$

which implies that the representation of $\mathrm{Sp}(\check{2r},\mathbb{R})$ can be constructed from the combination of the representations of the squeezing for the first mode and $\mathrm{U}(r)$.

Next, we describe a squeezed vector of the r-mode bosonic system as the generalized coherent vector with respect to $\mathfrak{sp}(2r,\mathbb{R})$ given in Sect. 6.9. Since a generalized coherent vector is written by a complex-valued vector, the generalized coherent vector with respect to $\mathfrak{sp}(2r,\mathbb{R})$ is written with a complex symmetric matrix $\zeta=\{\zeta_{k,l}\}$ as

$$
|-\frac{1}{2}\boldsymbol{v}:\zeta\rangle = C_\zeta\exp\left(\frac{1}{2}\sum_{k,l}\zeta_{k,l}\mathsf{a}_k^\dagger\mathsf{a}_l^\dagger\right)|-\frac{1}{2}\boldsymbol{v};-\frac{1}{2}\boldsymbol{v}\rangle, \tag{7.81}
$$

where C_ζ is a normalized constant. Especially, the state $|-\frac{1}{2}\boldsymbol{v}:\zeta\rangle\langle-\frac{1}{2}\boldsymbol{v}:\zeta|$ is transformed via the representation of the unitary matrix $g\in\mathrm{U}(r)$ as

$$\begin{aligned}
&\mathsf{U}(g)|-\frac{1}{2}\boldsymbol{v}:\zeta\rangle \\
=&C_\zeta \exp\left(\frac{1}{2}\sum_{k,l}\zeta_{k,l}\mathsf{U}(g)\mathsf{a}_k^\dagger\mathsf{U}(g)^\dagger\mathsf{U}(g)\mathsf{a}_l^\dagger\mathsf{U}(g)^\dagger\right)\mathsf{U}(g)|-\frac{1}{2}\boldsymbol{v};-\frac{1}{2}\boldsymbol{v}\rangle \\
=&C_\zeta \exp\left(\frac{1}{2}\sum_{k,k',l,l'}\zeta_{k,l}g_{l',l}g_{k',k}\mathsf{a}_{k'}^\dagger\mathsf{a}_{l'}^\dagger\right)|-\frac{1}{2}\boldsymbol{v};-\frac{1}{2}\boldsymbol{v}\rangle=|-\frac{1}{2}\boldsymbol{v}:g\zeta g^T\rangle.
\end{aligned}$$

When g is chosen as the element g' diagonalizing ζ, the above vector is $C_\zeta \exp(\frac{\eta_1}{2}(\mathsf{a}_1^\dagger)^2)|0\rangle\otimes\exp(\frac{\eta_r}{2}(\mathsf{a}_r^\dagger)^2)|0\rangle$. Hence, the normalized constant C_ζ is calculated to $(1-|\eta_1|^2)^{\frac{1}{4}}\cdots(1-|\eta_r|^2)^{\frac{1}{4}}=(\det(I-\zeta^\dagger\zeta))^{\frac{1}{4}}$. That is, the above vector is the tensor product $|\eta_1,0\rangle\otimes\cdots\otimes|\eta_r,0\rangle$ of the one-mode squeezed vector. The inner product between the above vector and the highest weight vector is calculated to be $|\langle-\frac{1}{2}\boldsymbol{v};-\frac{1}{2}\boldsymbol{v}|-\frac{1}{2}\boldsymbol{v}:\zeta\rangle|^2=(\det(I-\zeta^\dagger\zeta))^{-\frac{1}{2}}$. The highest weight vector of the space $L^2(\mathbb{R})^{\otimes r:\mathrm{odd}}$ is $|1,0,\ldots,0\rangle$, and the highest weight is $-\frac{1}{2}\boldsymbol{v}+\frac{1}{2}\boldsymbol{\beta}_{1,1}$.

Chapter 8
Discretization of Bosonic System

Abstract Chapter 4 addressed the Lie groups $\mathrm{U}(r)$ and $\mathrm{SU}(r)$ composed of unitary matrices on the space $\mathbb{C}^r$. Dependently of the physical situation, all of unitary matrices do not necessarily have the equal possibility to physically realize. When we realize them artificially, it is natural that several specific transformations can be easily realized. Due to such a situation, it is important to address the symmetry of a specific subgroup as well as that of the Lie groups $\mathrm{U}(r)$ and $\mathrm{SU}(r)$. On the other hand, the Heisenberg representation discussed in Chap. 7 is a representation on an infinite-dimensional space, but the Lie group of interest has the finite dimension. Besides, the representation is irreducible. Hence, we might construct a small subgroup of $\mathrm{U}(r)$ or $\mathrm{SU}(r)$ such that it plays a similar role to $\mathrm{U}(r)$ or $\mathrm{SU}(r)$ if we employ a similar structure as the Heisenberg representation. For this purpose, we address the discretizations of the Heisenberg representation and the representation of $\mathrm{Sp}(2r, \mathbb{R})$ that corresponds to the squeezing operation.

8.1 Discrete Heisenberg Representation

8.1.1 Discrete Heisenberg Representation on $\mathbb{Z}_d$

In quantum information, when a d-dimensional system $\mathcal{H}$ is given, we usually fix one CONS to describe our information processing. Such a CONS is called the **computational basis**, and is written as $\{|0\rangle, \ldots, |d-1\rangle\}$ with modulo d. Our information processings, e.g., quantum error correction and quantum computation are designed under the computational basis.

This section investigates the discretization of Heisenberg representation. Since the Heisenberg representation is a projective unitary representation of the Lie group $\mathbb{R}^2$ on an infinite-dimensional space, replacing $\mathbb{R}$ by the commutative algebra $\mathbb{Z}_d := \mathbb{Z}/d\mathbb{Z}$, we deal with a projective unitary representation of $\mathbb{Z}_d^2$ [107]. Since the group $\mathbb{Z}_d^2$ is a commutative group, any irreducible unitary representation is one of the following one-dimensional representations $\mathsf{f}_{s',t'}$ $(s', t' = 0, \ldots, d-1)$.

$$\mathsf{f}_{s',t'}(s, t)|v\rangle := e^{2\pi i (ss'+tt')/d}|v\rangle, \quad (s, t) \in \mathbb{Z}_d^2,$$

M. Hayashi, *Group Representation for Quantum Theory*,
DOI 10.1007/978-3-319-44906-7_8

where $|v\rangle$ is the basis of one of one-dimensional representation spaces. That is, there are d^2 distinct one-dimensional irreducible representations. So, we can easily check that the relation (2.56) holds.

However, there exists a d-dimensional irreducible projective unitary representation of $\mathbb{Z}_d^2$. Firstly, we define the following matrices $\mathsf{X}_{\mathbb{Z}}$ and $\mathsf{Z}_{\mathbb{Z}}$ on the space $\mathcal{H}$ whose CONS is the computational basis $\{|0\rangle, \ldots, |d-1\rangle\}$.

$$\mathsf{X}_{\mathbb{Z}}|x\rangle = |x+1\rangle, \quad \mathsf{Z}_{\mathbb{Z}}|x\rangle = \omega_{\mathbb{Z}}^x |x\rangle, \tag{8.1}$$

$\omega_{\mathbb{Z}} := e^{i2\pi/d}$. This definition is equivalent to the following conditions.

$$\mathsf{X}_{\mathbb{Z}} = \sum_x |x+1\rangle\langle x|, \quad \mathsf{Z}_{\mathbb{Z}} = \sum_x \omega_{\mathbb{Z}}^x |x\rangle\langle x|. \tag{8.2}$$

Then, we define $\tilde{\mathsf{W}}_{\mathbb{Z}}$ as

$$\tilde{\mathsf{W}}_{\mathbb{Z}}(s,t) := \mathsf{X}^s \mathsf{Z}^t = \sum_s \omega_{\mathbb{Z}}^{xt} |x+s\rangle\langle x|, \quad \forall (s,t) \in \mathbb{Z}_d^2. \tag{8.3}$$

Since the relation $\omega_{\mathbb{Z}}\mathsf{XZ} = \mathsf{ZX}$ holds, we have

$$\tilde{\mathsf{W}}_{\mathbb{Z}}(s,t)\tilde{\mathsf{W}}_{\mathbb{Z}}(s',t') = \omega_{\mathbb{Z}}^{s't}\tilde{\mathsf{W}}_{\mathbb{Z}}(s+s',t+t'). \tag{8.4}$$

As the factor system $\{\omega_{\mathbb{Z}}^{s't}\}$ satisfies (2.5), $\tilde{\mathsf{W}}_{\mathbb{Z}}$ is a projective unitary representation of $\mathbb{Z}_d^2$. Since the dimension of the representation space is d and the order of the group $\mathbb{Z}_d^2$ is d^2, any irreducible projective unitary representation with factor system $\{\omega_{\mathbb{Z}}^{s't}\}$ is isomorphic to the representation $\tilde{\mathsf{W}}_{\mathbb{Z}}(s,t)$ because of the projective representation version of (2.56). Also, the commutation relation

$$\tilde{\mathsf{W}}_{\mathbb{Z}}(s,t)\tilde{\mathsf{W}}_{\mathbb{Z}}(s',t') = \omega_{\mathbb{Z}}^{s't-t's}\tilde{\mathsf{W}}_{\mathbb{Z}}(s',t')\tilde{\mathsf{W}}_{\mathbb{Z}}(s,t) \tag{8.5}$$

holds. Here, we define $\tau_{\mathbb{Z}}$ to be $\omega_{\mathbb{Z}}^{(d+1)/2}$ when d is an odd number, and to be $e^{i\pi/d}$ when d is an even number. Then, we define the projective unitary representation $\mathsf{W}_{\mathbb{Z}}(s,t)$ as

$$\mathsf{W}_{\mathbb{Z}}(s,t) := \tau_{\mathbb{Z}}^{st}\tilde{\mathsf{W}}_{\mathbb{Z}}(s,t) = \tau_{\mathbb{Z}}^{st}\sum_s \omega_{\mathbb{Z}}^{xt}|x+s\rangle\langle x|, \tag{8.6}$$

s and t are integers $0, \ldots, d-1$. Thus, we have

$$\begin{aligned} &\mathsf{W}_{\mathbb{Z}}(s,t)\mathsf{W}_{\mathbb{Z}}(s',t') \\ =&\begin{cases} \tau_{\mathbb{Z}}^{s't-t's}\mathsf{W}_{\mathbb{Z}}(s+s',t+t') & \text{when } d \text{ is an odd number} \\ (-1)^{\epsilon}\tau_{\mathbb{Z}}^{s't-t's}\mathsf{W}_{\mathbb{Z}}(s+s',t+t') & \text{when } d \text{ is an even number,} \end{cases} \end{aligned} \tag{8.7}$$

where ϵ is chosen as an integer 0 or 1 dependently of s, t, s', and t'. Hence, $\mathsf{W}_{\mathbb{Z}}(s,t)$ is a projective unitary representation of $\mathbb{Z}_d^2$ and can be regarded as a discrete version of $\mathsf{W}((q+ip)/\sqrt{2})$. So, we call it the **discrete Heisenberg representation** of $\mathbb{Z}_d^2$. To distinguish $\tilde{\mathsf{W}}_{\mathbb{Z}}$ from $\mathsf{W}_{\mathbb{Z}}$, we call $\tilde{\mathsf{W}}_{\mathbb{Z}}$ pre-discrete-Heisenberg representation.

Now, we define the **dual computational basis** for a given the computational basis $\{|x\rangle\}_{x\in\mathbb{Z}_d}$ as

$$|\hat{e}_{\mathbb{Z}}(l)\rangle := \frac{1}{\sqrt{d}}\sum_{k\in\mathbb{Z}_d}\omega_{\mathbb{Z}}^{-kl}|k\rangle, \quad l\in\mathbb{Z}_d. \tag{8.8}$$

Then, the dual computational basis is composed of the eigenvectors of $\mathsf{X}_{\mathbb{Z}}$. The matrix $\mathsf{Z}_{\mathbb{Z}}$ cyclically changes the elements of dual computational basis.

Lemma 8.1 *The relation*

$$(\mathsf{W}_{\mathbb{Z}}(s,t))^d = I \tag{8.9}$$

holds. Hence, when a unitary matrix U, two integers s and t, and a constant c satisfy $U\mathsf{W}_{\mathbb{Z}}(s,t)U^\dagger = c\mathsf{W}_{\mathbb{Z}}(s,t)$, the constant c is limited to be $\omega_{\mathbb{Z}}^n$.

Proof When d is an odd number, the relation (8.7) yields that $(\mathsf{W}_{\mathbb{Z}}(s,t))^d = \mathsf{W}_{\mathbb{Z}}(ds,dt) = I$, which implies the desired argument.

On the other hand, when d is an even number, the relation yields that

$$(\tilde{\mathsf{W}}_{\mathbb{Z}}(s,t))^d = \omega_{\mathbb{Z}}^{st+2st+\cdots+(d-1)st}\tilde{\mathsf{W}}_{\mathbb{Z}}^{(ds,dt)} = \omega_{\mathbb{Z}}^{std(d-1)/2}I = \tau_{\mathbb{Z}}^{std(d-1)}I.$$

Hence,

$$(\mathsf{W}_{\mathbb{Z}}(s,t))^d = \tau_{\mathbb{Z}}^{dst}\tau_{\mathbb{Z}}^{std(d-1)}I = \tau_{\mathbb{Z}}^{d^2st}I.$$

Since d is an even number, $2d$ is a divisor of d^2. So, we obtain (8.9). ∎

For any element $(s,t)\in\mathbb{Z}_d$ and any integer n, matrices on the line $\{\mathsf{W}_{\mathbb{Z}}(ns,nt)\}_{n=0,\ldots,d-1}$ are commutative with each other. When d is a prime p, the number of non-identity matrices is p^2-1. There are $p+1$ lines composed of commutative matrices and each line contains $p-1$ non-identity matrices, as Fig. 8.1. That is, we can make $p+1$ bases by diagonalizing these matrices.

Further, we can show that the central extension by $\mathbb{Z}_d^2$ based on the factor system (8.7) is isomorphic to discrete Heisenberg group $\mathrm{H}(2,\mathbb{Z}_d)$ given in Example 2.4. Hence, we can define the map $\mathsf{W}_{\mathbb{Z},\mathrm{H}}$ as a unitary representation of the group $\mathrm{H}(2,\mathbb{Z}_d)$ by

$$\mathsf{W}_{\mathbb{Z},\mathrm{H}}(s,t,\omega_d^n) := \tau_{\mathbb{Z}}^n\mathsf{W}_{\mathbb{Z}}(s,t) \text{ when } d \text{ is an odd number}$$
$$\mathsf{W}_{\mathbb{Z},\mathrm{H}}(s,t,\omega_{2d}^n) := \tau_{\mathbb{Z}}^n\mathsf{W}_{\mathbb{Z}}(s,t) \text{ when } d \text{ is an even number.}$$

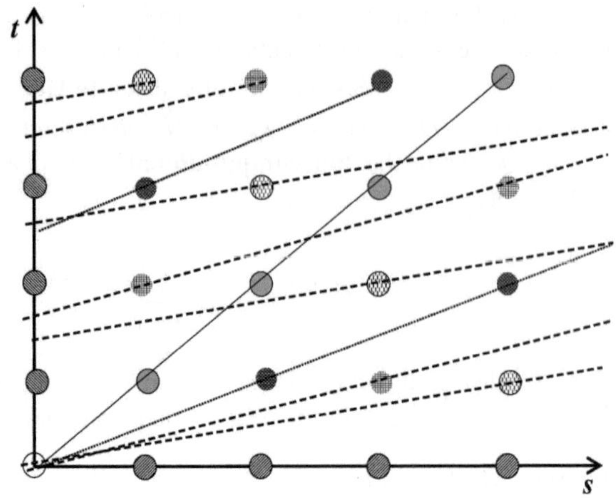

Fig. 8.1 $p+1$ lines in $\mathbb{Z}_p$ with $p=5$

Example 8.1 When $d=2$, the discrete Heisenberg representation $\mathsf{W}_{\mathbb{Z}}$ is given as follows.

$$\mathsf{W}_{\mathbb{Z}}(0,0)=I,\quad \mathsf{W}_{\mathbb{Z}}(1,0)=\begin{pmatrix}0&1\\1&0\end{pmatrix},$$
$$\mathsf{W}_{\mathbb{Z}}(0,1)=\begin{pmatrix}1&0\\0&-1\end{pmatrix},\quad \mathsf{W}_{\mathbb{Z}}(1,1)=\begin{pmatrix}0&-i\\i&0\end{pmatrix}.$$

Further, the set $\left\{|\frac{1}{\sqrt{d}}\mathsf{W}_{\mathbb{Z}}(s,t)\rangle\rangle\right\}_{(s,t)\in\mathbb{Z}_d^2}$ forms a CONS on $\mathbb{C}^d\otimes\mathbb{C}^d$ (see Exercise 8.2). So, the set gives a PVM on the composite system $\mathbb{C}^d\otimes\mathbb{C}^d$. This measurement is called the generalized Bell measurement, and the states $|\frac{1}{\sqrt{d}}\mathsf{W}_{\mathbb{Z}}(s,t)\rangle\rangle$ is called the generalized Bell states.

Exercise 8.1 List up all elements of $\mathsf{W}_{\mathbb{Z}}(k,j)$ when $d=3$.

Exercise 8.2 Show that $\left\{|\frac{1}{\sqrt{d}}\mathsf{W}_{\mathbb{Z}}(s,t)\rangle\rangle\right\}_{(s,t)\in\mathbb{Z}_d^2}$ forms a CONS on $\mathbb{C}^d\otimes\mathbb{C}^d$.

8.1.2 Discrete Heisenberg Representation on $\mathbb{F}_q$

When d is not a prime number, there is an element $s\in\mathbb{Z}_d^2$ such that s has no inverse element. Hence, $\mathbb{Z}_d$ is not a field, which causes inconveniences in the latter discussion. To resolve these inconveniences, we consider another discrete version of $\mathsf{W}((x+iy)/\sqrt{2})$ when d is a power $q=p^m$ of a prime number p. That is, we focus on the Galois extension of the finite field $\mathbb{F}_p:=\mathbb{Z}/p\mathbb{Z}$, which is given as the finite field $\mathbb{F}_q:=\mathbb{F}_p[X]/f(X)$ with an irreducible polynomial f with degree m. For this purposes, we define the following matrix $\mathsf{X}_{\mathbb{F}}, \mathsf{Z}_{\mathbb{F}}$ on the space $\mathcal{H}$ whose CONS is the computational basis $\{|x\rangle\}_{x\in\mathbb{F}_q}$;

$$\mathsf{X}_{\mathbb{F}}(s)|x\rangle = |x+s\rangle, \quad \mathsf{Z}_{\mathbb{F}}(t)|x\rangle = \omega_{\mathbb{F}}^{trtx}|x\rangle, \tag{8.10}$$

where $\omega_{\mathbb{F}}$ is defined to be $e^{i2\pi/p}$ and trz is to be $\mathrm{Tr}\, M_z$ when M_z denotes the matrix representation of the multiplication map $x \mapsto zx$ with identifying the finite filed $\mathbb{F}_q$ with the vector space $\mathbb{F}_p^m$. This definition is equivalent to the following definition.

$$\mathsf{X}_{\mathbb{F}}(s) = \sum_{x\in\mathbb{F}_q} |x+s\rangle\langle x|, \quad \mathsf{Z}_{\mathbb{F}}(t) = \sum_{x\in\mathbb{F}_q} \omega_{\mathbb{Z}}^{trtx}|x\rangle\langle x|.$$

Since the relation $\mathsf{X}_{\mathbb{F}}(s)\mathsf{Z}_{\mathbb{F}}(t)\mathsf{X}_{\mathbb{F}}(-s) = \omega_{\mathbb{F}}^{-trst}\mathsf{Z}_{\mathbb{F}}(t)$ holds, when we define $\tilde{\mathsf{W}}_{\mathbb{F}}(s,t) := \mathsf{X}_{\mathbb{F}}(s)\mathsf{Z}_{\mathbb{F}}(t) = \sum_{x\in\mathbb{F}_q} \omega_{\mathbb{Z}}^{trtx}|x+s\rangle\langle x|$ for any pair $(s,t) \in \mathbb{F}_q^2$, we have $\tilde{\mathsf{W}}_{\mathbb{F}}(s,t)\tilde{\mathsf{W}}_{\mathbb{F}}(s',t') = \omega_{\mathbb{F}}^{trs't}\tilde{\mathsf{W}}_{\mathbb{F}}(s+s',t+t')$. Because the factor system $\{\omega_{\mathbb{F}}^{trs't}\}$ satisfies (2.5), $\tilde{\mathsf{W}}_{\mathbb{F}}$ is a projective unitary representation of $\mathbb{F}_q^2$. Hence, we have the commutation relation

$$\tilde{\mathsf{W}}_{\mathbb{F}}(s,t)\tilde{\mathsf{W}}_{\mathbb{F}}(s',t') = \omega_{\mathbb{F}}^{tr(s't-t's)}\tilde{\mathsf{W}}_{\mathbb{F}}(s',t')\tilde{\mathsf{W}}_{\mathbb{F}}(s,t). \tag{8.11}$$

Since the square of the group $\mathbb{F}_q^2$ is the dimension of the representation space of $\tilde{\mathsf{W}}_{\mathbb{F}}(s,t)$, when the factor system of a projective unitary representation of $\mathbb{F}_q^2$ is equivalent to $\{\omega_{\mathbb{F}}^{trs't}\}$, the projective unitary representation is limited to a representation isomorphic to $\tilde{\mathsf{W}}_{\mathbb{F}}(s,t)$.

Now, we define $\tau_{\mathbb{F}}$ as $\tau_{\mathbb{F}} := \omega_{\mathbb{Z}}^{(p+1)/2}$ when the characteristic p of $\mathbb{F}_q$ is not 2, and we define $\tau_{\mathbb{F}}$ as $\tau_{\mathbb{F}} := i$ when the characteristic p of $\mathbb{F}_q$ is 2. Then, we define the projective representation $\mathsf{W}_{\mathbb{F}}(s,t)$ as

$$\mathsf{W}_{\mathbb{F}}(s,t) := \tau_{\mathbb{F}}^{trts}\tilde{\mathsf{W}}_{\mathbb{F}}(s,t) = \tau_{\mathbb{F}}^{trts}\sum_{x\in\mathbb{F}_q} \omega_{\mathbb{F}}^{trtx}|x+s\rangle\langle x|,$$

where we assume that $trts$ takes values in integers $0,\ldots,p-1$. Then, the relation

$$\mathsf{W}_{\mathbb{F}}(s,t)\mathsf{W}_{\mathbb{F}}(s',t') = \begin{cases} \tau_{\mathbb{F}}^{tr(s't-t's)}\mathsf{W}_{\mathbb{F}}(s+s',t+t') & \text{when } p \neq 2 \\ (-1)^{\epsilon}\tau_{\mathbb{F}}^{tr(s't-t's)}\mathsf{W}_{\mathbb{F}}(s+s',t+t') & \text{when } p = 2 \end{cases} \tag{8.12}$$

holds, where ϵ is chosen as an integer 0 or 1 dependently of s, t, s', and t'. Hence, $\mathsf{W}_{\mathbb{F}}(s,t)$ also can be regarded as a discrete version of $\mathsf{W}((q+ip)/\sqrt{2})$. So, the projective unitary representation of $\mathbb{F}_q^2$ given here is called the **discrete Heisenberg representation** of $\mathbb{F}_q^2$. When q is a prime number p, identifying $\mathbb{F}_p^2$ with $\mathbb{Z}_p^2$, we find that the discrete Heisenberg representation of $\mathbb{F}_p^2$ is the discrete Heisenberg representation of $\mathbb{Z}_p^2$. However, when q is a power p^m of a prime number p, the discrete Heisenberg representation of $\mathbb{F}_q^2$ is different from the discrete Heisenberg representation of $\mathbb{Z}_q^2$. We need to remark that a representation based on the finite field $\mathbb{F}_q$ is different from a representation based on the commutative finite algebra $\mathbb{Z}_q$.

Here, based on the computational basis $\{|x\rangle\}_{x\in\mathbb{F}_q}$, we define the **dual computational basis**

$$|\hat{e}_{\mathbb{F}}(l)\rangle := \frac{1}{\sqrt{q}}\sum_{k\in\mathbb{F}_q}\omega_{\mathbb{F}}^{-trkl}|k\rangle, \quad l\in\mathbb{F}_q. \tag{8.13}$$

Then, the dual computational basis forms the set of eigenvectors of $\mathsf{X}_{\mathbb{F}}$. The operator $\mathsf{Z}_{\mathbb{F}}$ maps the dual computational basis cyclically.

Further, similar to Lemma 8.1, we obtain the following Lemma.

Lemma 8.2 *The equation* $(\mathsf{W}_{\mathbb{F}}(s,t))^p = I$ *holds. Hence, when a unitary matrix* U*, two integers* s *and* t*, and a constant* c *satisfy* $U\mathsf{W}_{\mathbb{F}}(s,t)U^\dagger = c\mathsf{W}_{\mathbb{F}}(s,t)$*, the constant* c *is limited to be* $\omega_{\mathbb{F}}^n$.

Further, we find that the central extension by $\mathbb{F}_q^2$ based on the factor system (8.12) is isomorphic to the discrete Heisenberg group $\mathrm{H}(2,\mathbb{F}_q)$ given in Example 2.6.

Then, we define the unitary representation $\mathsf{W}_{\mathbb{F},\mathrm{H}}$ of the group $\mathrm{H}(2,\mathbb{F}_q)$ as

$$\begin{aligned}\mathsf{W}_{\mathbb{F},\mathrm{H}}(s,t,\omega_p^n) &:= \tau_{\mathbb{F}}^n\mathsf{W}_{\mathbb{F}}(s,t) \text{ when } p \text{ is a odd number}\\ \mathsf{W}_{\mathbb{F},\mathrm{H}}(s,t,\omega_4^n) &:= \tau_{\mathbb{F}}^n\mathsf{W}_{\mathbb{F}}(s,t) \text{ when } p=2.\end{aligned} \tag{8.14}$$

8.1.3 Multi-mode Discrete Heisenberg Representation

Next, based on the discrete Heisenberg representation of $\mathbb{Z}_d^2$ or $\mathbb{F}_q^2$, we define a projective unitary representation of the group $\mathbb{Z}_d^{2r}$ or the vector space $\mathbb{F}_q^{2r}$ on $\mathcal{H}^{\otimes r} = \mathcal{H}_1\otimes\cdots\otimes\mathcal{H}_r$ as follows. Here, $\mathcal{H}_i$ is a space isomorphic to $\mathcal{H}$. Then, the computational basis of the space $\mathcal{H}^{\otimes r}$ can be written as $|s\rangle$ by using the element s of $\mathbb{Z}_d^r$ or $\mathbb{F}_q^r$. In the following, to treat the two cases with the finite field $\mathbb{F}_q$ and the finite commutative algebra $\mathbb{Z}_d$ unifiedly, we denote them by $\mathbb{X}$. That is, we can substitute $\mathbb{F}_q$ and $\mathbb{Z}_d$ into $\mathbb{X}$ in this notation, and $\mathbb{X}$ has the structure of a finite commutative algebra, at least. Then, $\mathsf{W}_{\mathbb{X},i}(s,t)$ is defined to be the action $\mathsf{W}_{\mathbb{X}}(s,t)$ on the i-th space $\mathcal{H}_i$. So, we can define the projective unitary representation $\mathsf{W}_{\mathbb{X}}^r$ as

$$\begin{aligned}\mathsf{W}_{\mathbb{X}}^r(\vec{s}) &:= \mathsf{W}_{\mathbb{X},1}(s_1,t_1)\otimes\cdots\otimes\mathsf{W}_{\mathbb{X},r}(s_r,t_r),\\ \tilde{\mathsf{W}}_{\mathbb{X}}^r(\vec{s}) &:= \tilde{\mathsf{W}}_{\mathbb{X},1}(s_1,t_1)\otimes\cdots\otimes\tilde{\mathsf{W}}_{\mathbb{X},r}(s_r,t_r)\end{aligned}$$

for any vector $\vec{s} = (s_1,\ldots s_r,t_1,\ldots,t_r)\in\mathbb{X}^{2r}$. Then, for $\boldsymbol{s} = (s_1,\ldots s_r)$, $\boldsymbol{t} = (t_1,\ldots t_r)\in\mathbb{X}^r$, we define $(\boldsymbol{s},\boldsymbol{t})_{\mathbb{Z}} := s_1t_1+\cdots+s_rt_r$ and $(\boldsymbol{s},\boldsymbol{t})_{\mathbb{F}} := tr(s_1t_1+\cdots+s_rt_r)$. So, we have $\tilde{\mathsf{W}}_{\mathbb{X}}^r(\boldsymbol{s},\boldsymbol{t})\tilde{\mathsf{W}}_{\mathbb{X}}^r(\boldsymbol{s}',\boldsymbol{t}') = \omega_{\mathbb{Z}}^{(\boldsymbol{s}',\boldsymbol{t})_{\mathbb{X}}}\tilde{\mathsf{W}}_{\mathbb{X}}^r(\boldsymbol{s}+\boldsymbol{s}',\boldsymbol{t}+\boldsymbol{t}')$. Also, for $\vec{s} = (\boldsymbol{s},\boldsymbol{t})$, $\vec{s}' = (\boldsymbol{s}',\boldsymbol{t}')\in\mathbb{X}^{2r}$, we define $\langle\vec{s},\vec{s}'\rangle_{\mathbb{X}} := (\boldsymbol{s}',\boldsymbol{t})_{\mathbb{X}} - (\boldsymbol{s},\boldsymbol{t}')_{\mathbb{X}}$. Thus,

$$\mathsf{W}^r_{\mathbb{X}}(\vec{s})\mathsf{W}^r_{\mathbb{X}}(\vec{s}') = \begin{cases} \tau_{\mathbb{X}}^{\langle \vec{s},\vec{s}'\rangle_{\mathbb{X}}} \mathsf{W}^r_{\mathbb{X}}(\vec{s}+\vec{s}') & \text{when } d \text{ or } p \text{ is an odd number} \\ (-1)^{\epsilon}\tau_{\mathbb{X}}^{\langle \vec{s},\vec{s}'\rangle_{\mathbb{X}}} \mathsf{W}^r_{\mathbb{X}}(\vec{s}+\vec{s}') & \text{when } d \text{ or } p \text{ is an even number,} \end{cases} \tag{8.15}$$

where ϵ is an integer 0 or 1 dependently of s, t, s', and t'.

Hence, $\tilde{\mathsf{W}}^r_{\mathbb{Z}}$, $\mathsf{W}^r_{\mathbb{Z}}$, $\tilde{\mathsf{W}}^r_{\mathbb{F}}$, and $\mathsf{W}^r_{\mathbb{F}}$ are projective unitary representations of $\mathbb{Z}_d^{2r}$ or $\mathbb{F}_q^{2r}$. The projective unitary representation $\mathsf{W}^r_{\mathbb{Z}}$ ($\mathsf{W}^r_{\mathbb{F}}$) is called the **discrete Heisenberg representation** of $\mathbb{Z}_d^{2r}$ ($\mathbb{F}_q^{2r}$). Dependently of the situation, we use one of two types of discrete Heisenberg representations in the following discussion. Also, the commutation relation $\mathsf{W}^r_{\mathbb{X}}(\vec{s})\mathsf{W}^r_{\mathbb{X}}(\vec{s}') = \omega_{\mathbb{X}}^{\langle \vec{s},\vec{s}'\rangle_{\mathbb{X}}}\mathsf{W}^r_{\mathbb{X}}(\vec{s}')\mathsf{W}^r_{\mathbb{X}}(\vec{s})$ holds, and a similar commutation relation holds for another projective unitary representation $\tilde{\mathsf{W}}_{\mathbb{X}}$.

We have the following lemma for the representation given here.

Lemma 8.3 *The projective unitary representation $\mathsf{W}^r_{\mathbb{X}}$ is irreducible. Further, when an irreducible projective unitary representation of $\mathbb{X}^{2r}$ has the same factor system as $\mathsf{W}^r_{\mathbb{X}}$, it is isomorphic to $\mathsf{W}^r_{\mathbb{X}}$. The same fact holds for $\tilde{\mathsf{W}}^r_{\mathbb{X}}$.*

Proof When $\vec{s} \neq 0$, $\mathrm{Tr}\,\mathsf{W}^r_{\mathbb{X}}(\vec{s})$ is 0, and $\mathrm{Tr}\,\mathsf{W}^r_{\mathbb{X}}(0)$ equals q^r or d^r, which is the dimension of the representation space. Hence, Lemma 2.15 with the case of projective unitary representation guarantees that $\mathsf{W}^r_{\mathbb{X}}$ is irreducible. Further, due to the relation (2.56) with the case of projective unitary representation, a projective unitary representation is isomorphic to $\mathsf{W}^r_{\mathbb{X}}(\vec{s})$ when its factor system equals that of $\mathsf{W}^r_{\mathbb{X}}(\vec{s})$.

The same fact can be shown for $\tilde{\mathsf{W}}^r_{\mathbb{X}}$ in the same way. ■

Further, when d or p is an odd number and the positive integer k is invertible as an element of $\mathbb{X}$, a projective unitary representation whose factor system is

$$e^{i\phi(\vec{s},\vec{s}')} = \tau_{\mathbb{X}}^{k\langle \vec{s},\vec{s}'\rangle_{\mathbb{X}}} \tag{8.16}$$

satisfies the same fact as Lemma 8.3. That is, when we define the projective unitary representation $\mathsf{W}^r_{\mathbb{X}:k}$ by replacing $\omega_{\mathbb{X}}$ in the definition of $\mathsf{W}^r_{\mathbb{X}}$ by $\omega_{\mathbb{X}}^k$, $\mathsf{W}^r_{\mathbb{X}:k}$ is an irreducible projective unitary representation whose factor system is (8.16). Then, in the same way as the proof of Lemma 8.3, we can show that any irreducible projective unitary representation of $\mathbb{X}^r$ with the factor system (8.16) is isomorphic to $\mathsf{W}^r_{\mathbb{X}:k}$.

From Lemmas 8.1 and 8.2, we obtain the following lemma.

Lemma 8.4 *When a unitary matrix U and a constant c satisfy $U\mathsf{W}^r_{\mathbb{X}}(\vec{s})U^\dagger = c\mathsf{W}^r_{\mathbb{X}}(\vec{s})$ for a given $\vec{s}$, the constant c is limited to $\omega_{\mathbb{X}}^n$.*

Let $\mathsf{W}^r_{\mathbb{X}} \otimes \overline{\mathsf{W}^r}_{\mathbb{X}}$ be a projective unitary representation on the space composed of matrices on $\mathcal{H}^{\otimes r}$. Then, all of its irreducible subspaces are one-dimensional spaces. So, the irreducible decomposition is given as $\oplus_{\vec{s}\in\mathbb{X}^{2r}} < \mathsf{W}^r_{\mathbb{X}}(\vec{s}) >$. So, the representation is characterized as

$$\mathsf{W}^r_{\mathbb{X}} \otimes \overline{\mathsf{W}^r}_{\mathbb{X}}(\vec{s}) : \mathsf{W}^r_{\mathbb{X}}(\vec{s}') \mapsto \omega_{\mathbb{X}}^{\langle \vec{s},\vec{s}'\rangle_{\mathbb{X}}}\mathsf{W}^r_{\mathbb{X}}(\vec{s}'). \tag{8.17}$$

Hence, we obtain the following lemma.

Lemma 8.5 *Assume that a unitary matrix U satisfies*

$$\mathsf{W}^r_{\mathbb{X}}(\vec{s})U = e^{i\theta(U,\vec{s})}U\mathsf{W}^r_{\mathbb{X}}(\vec{s}), \quad \forall \in \mathbb{X}^{2r}, \tag{8.18}$$

where $\theta(U,\vec{s})$ is a constant depending on U and θ. Then, there exists an element $\vec{s}' \in \mathbb{X}^{2r}$ such that the unitary U is a constant times of $\mathsf{W}^r_{\mathbb{X}}(\vec{s}')$ and $e^{i\theta(U,\vec{s})} = \omega_{\mathbb{X}}^{\langle \vec{s},\vec{s}'\rangle_{\mathbb{X}}}$.

Proof Modifying (8.18), we obtain $\mathsf{W}^r_{\mathbb{X}}(\vec{s})U(\mathsf{W}^r_{\mathbb{X}}(\vec{s}))^\dagger = e^{i\theta(U,\vec{s})}U$ for an arbitrary element $\vec{s} \in \mathbb{X}^{2r}$. So, the above discussion yields this lemma. ∎

Further, we can show that the central extension by $\mathbb{Z}_d^{2r}$ based on the factor system given in (8.15) is isomorphic to the discrete Heisenberg group $\mathrm{H}(2r, \mathbb{Z}_d)$ defined in Example 2.5. Now, we define the unitary representation $\mathsf{W}_{\mathbb{Z},\mathrm{H}}$ of the group $\mathrm{H}(2r, \mathbb{Z}_d)$ as

$$\begin{aligned} \mathsf{W}_{\mathbb{Z},\mathrm{H}}(\boldsymbol{s}, \boldsymbol{t}, \omega_d^n) &:= \tau_{\mathbb{Z}}^n \mathsf{W}_{\mathbb{Z}}(\boldsymbol{s}, \boldsymbol{t}) \text{ when } d \text{ is an odd number} \\ \mathsf{W}_{\mathbb{Z},\mathrm{H}}(\boldsymbol{s}, \boldsymbol{t}, \omega_{2d}^n) &:= \tau_{\mathbb{Z}}^n \mathsf{W}_{\mathbb{Z}}(\boldsymbol{s}, \boldsymbol{t}) \text{ when } d \text{ is an even number.} \end{aligned} \tag{8.19}$$

Similarly, the central extension by $\mathbb{F}_q^{2r}$ based on the factor system given in (8.15) is isomorphic to the discrete Heisenberg group $\mathrm{H}(2r, \mathbb{F}_q)$ defined in Example 2.7. Now, we define the unitary representation $\mathsf{W}_{\mathbb{F},\mathrm{H}}$ of the group $\mathrm{H}(2r, \mathbb{F}_q)$ as

$$\begin{aligned} \mathsf{W}_{\mathbb{F},\mathrm{H}}(\boldsymbol{s}, \boldsymbol{t}, \omega_p^n) &:= \tau_{\mathbb{F}}^n \mathsf{W}_{\mathbb{Z}}(\boldsymbol{s}, \boldsymbol{t}) \text{ when } p \text{ is an odd number} \\ \mathsf{W}_{\mathbb{F},\mathrm{H}}(\boldsymbol{s}, \boldsymbol{t}, \omega_4^n) &:= \tau_{\mathbb{F}}^n \mathsf{W}_{\mathbb{Z}}(\boldsymbol{s}, \boldsymbol{t}) \text{ when } p = 2. \end{aligned} \tag{8.20}$$

In the above discussion, we have given the unitary representations of the discrete Heisenberg groups $\mathrm{H}(2r, \mathbb{Z}_d)$ and $\mathrm{H}(2r, \mathbb{F}_q)$ from the projective unitary representations of $\mathbb{Z}_d^{2r}$ and $\mathbb{F}_q^{2r}$. Conversely, it is possible to derive projective unitary representations of $\mathbb{Z}_d^{2r}$ and $\mathbb{F}_q^{2r}$ from unitary representations of discrete Heisenberg groups $\mathrm{H}(2r, \mathbb{Z}_d)$ and $\mathrm{H}(2r, \mathbb{F}_q)$.

Related to this relation, we have the following lemma by using Lemma 8.3.

Lemma 8.6 *The unitary representation $\mathsf{W}^r_{\mathbb{X},\mathrm{H}}$ is irreducible. Further, we assume that when an irreducible unitary representation of the discrete Heisenberg group $\mathrm{H}(2r, \mathbb{X})$ is restricted in the center, the restricted representation is isomorphic to the representation restricted in the center of $\mathsf{W}^r_{\mathbb{X},\mathrm{H}}$. Then, the original irreducible unitary representation of $\mathrm{H}(2r, \mathbb{X})$ is isomorphic to $\mathsf{W}^r_{\mathbb{X},\mathrm{H}}$.*

Proof Since the representation $\mathsf{W}^r_{\mathbb{X},\mathrm{H}}$ is irreducible as a projective representation of the subgroup $\mathbb{X}^{2r}$, the representation $\mathsf{W}^r_{\mathbb{X},\mathrm{H}}$ is irreducible. Given an irreducible unitary representation of $\mathrm{H}(2r, \mathbb{X})$, when it is regarded as a representation of the center of $\mathrm{H}(2r, \mathbb{X})$, it gives the factor system of the projective unitary representation of the subgroup $\mathbb{X}^{2r}$. Hence, the latter argument follows from Lemma 8.3. ∎

Further, similar to $\mathsf{W}^r_{\mathbb{X}:k}$, we can define the representation $\mathsf{W}^r_{\mathbb{X},\mathrm{H}:k}$ of the discrete Heisenberg group $\mathrm{H}(2r, \mathbb{X})$. The unitary representation satisfies a lemma similar to Lemma 8.6.

Exercise 8.3 Consider the algebraic extension $\mathbb{F}_4$ of $\mathbb{F}_2$ by using $\alpha^2 = \alpha + 1$. Then, we define $\mathbb{F}_2$ morphism F from $\mathbb{F}_2^2$ to $\mathbb{F}_4$ as

$$(0,0) \mapsto 0,\ (1,0) \mapsto \alpha,\ (0,1) \mapsto \alpha+1,\ (1,1) \mapsto 1.$$

Show the following relation;

$$\mathsf{W}_{\mathbb{F}_4}(F(x), F(y))|F(z)\rangle = \mathsf{W}^2_{\mathbb{F}_2}(x, y)|z\rangle \tag{8.21}$$

for $x, y, z \in \mathbb{F}_2^2$.

8.1.4 *Irreducible Representation of Subgroup*

We call a subgroup N of $\mathbb{X}^{2r}$ an $\mathbb{X}$-subgroup when the subgroup N is closed with respect to the multiplication of the finite commutative algebra $\mathbb{X}$. We choose an $\mathbb{X}$-subgroup N such that $\langle \vec{s}, \vec{s}'\rangle_{\mathbb{X}} = 0$ for any elements $\vec{s}, \vec{s}' \in N$. Such an $\mathbb{X}$-subgroup is called a **self-orthogonal subgroup**. In the following, we assume that N is a self-orthogonal subgroup. Then, by defining $N^\perp := \{\vec{s} \in \mathbb{X}^{2r} | \langle \vec{s}, \vec{s}'\rangle_{\mathbb{X}} = 0, \quad \forall \vec{s}' \in N\}$, an $\mathbb{X}$-subgroup N is self-orthogonal if and only if $N \subset N^\perp$. Especially, when $N = N^\perp$, it is called a **strictly self-orthogonal subgroup**. We define the dual space of N as $N^* := \mathbb{X}^{2r}/N^\perp$. For an element $x \in N^*$, using its representative $\vec{s}_x$, we define $x(\vec{s}) := \langle \vec{s}, \vec{s}_x\rangle_{\mathbb{X}}$, which does not depend on the choice of the representative of x.

Under the projective unitary representation $\mathsf{W}^r_{\mathbb{X}}$, the representations of elements of N are commutative with each other. So, $\mathsf{W}^r_{\mathbb{X}}$ is a representation of the commutative group N. Hence, the irreducible components are one-dimensional. Thus, the matrices $\{\mathsf{W}_{\mathbb{X}}(\vec{s})\}_{\vec{s}\in N}$ can be simultaneously diagonalized. The eigenvalue is given by a homomorphism from the group N to the cyclic group $\{\omega^n_{\mathbb{X}}\}_n$, which is the center of the discrete Heisenberg group $\mathrm{H}(2r, \mathbb{X})$. The homomorphism is given as $N \ni \vec{s} \mapsto \omega_{\mathbb{X}}^{x(\vec{s})}$ by choosing a suitable element $x \in N^*$. Hence, we denote the simultaneous eigenspace of the matrices $\{\mathsf{W}^r_{\mathbb{X}}(\vec{s})\}_{\vec{s}\in N}$ corresponding to the eigenvalues $\omega_{\mathbb{X}}^{x(\vec{s})}$ by $\mathcal{H}_x$. Then, we let the projection to $\mathcal{H}_x$ be P_x, i.e.,

$$\mathsf{W}^r_{\mathbb{X}}(\vec{s}) = \sum_{x\in N^*} \omega_{\mathbb{X}}^{x(\vec{s})} P_x. \tag{8.22}$$

Since any element $\vec{s}' \in \mathbb{X}^{2r}$ satisfies

$$\mathsf{W}^r_{\mathbb{X}}(\vec{s})\mathsf{W}^r_{\mathbb{X}}(\vec{s}') = \mathsf{W}^r_{\mathbb{X}}(\vec{s}')\mathsf{W}_{\mathbb{X}}(\vec{s}')^{-1}\mathsf{W}^r_{\mathbb{X}}(\vec{s})\mathsf{W}^r_{\mathbb{X}}(\vec{s}') \tag{8.23}$$

$$= \mathsf{W}^r_{\mathbb{X}}(\vec{s}')\omega_{\mathbb{X}}^{\langle \vec{s}, \vec{s}'\rangle_{\mathbb{X}}}\mathsf{W}^r_{\mathbb{X}}(\vec{s}) = \mathsf{W}^r_{\mathbb{X}}(\vec{s}')\omega_{\mathbb{X}}^{[\vec{s}'](\vec{s})}\mathsf{W}^r_{\mathbb{X}}(\vec{s}), \quad \vec{s} \in \mathbb{X}^{2r}, \tag{8.24}$$

we have $\mathsf{W}^r_{\mathbb{X}}(\vec{s}')\mathcal{H}_x = \mathcal{H}_{x+[\vec{s}']}$.

Thus, we have the following lemma.

Lemma 8.7 *Assume that N is a self-orthogonal subgroup. Under the projective unitary representation $\mathsf{W}^r_{\mathbb{X}}$, the representation of $N^\perp$ on the space $\mathcal{H}_x$ is an irreducible space and is not isomorphic to that on another space $\mathcal{H}_{x'}$. Further, the dimension of $\mathcal{H}_x$ is $|\mathbb{X}|^r/|N^*|$. An irreducible projective unitary representations of $N^\perp$ is given on $\mathcal{H}_x$ with a suitable choice of x when its factor system equals that of $\mathsf{W}^r_{\mathbb{X}}$.*

Especially, when $\mathbb{X} = \mathbb{F}_q$ and N is an r-dimensional vector subspace, the dimension of $\mathcal{H}_x$ is q^{r-k}.

Proof For an element $x \in N^*$, we assume that the space $\mathcal{H}_x$ is not irreducible with respect to the representation of $N^\perp$. Let $\mathcal{H}'_x$ be an irreducible subspace of $\mathcal{H}_x$. Since the above discussion does not depend on the choice of the representative of an element of $\mathbb{F}_q^{2r}/N^\perp$, an element of $(N^\perp)^c$ maps the space $\mathcal{H}'_x$ to $\oplus_{y \in N^* \setminus \{0\}} \mathcal{H}_{x+y}$. That is, the subspace $\mathcal{H}'_x \oplus (\oplus_{y \in N^* \setminus \{0\}} \mathcal{H}_{x+y})$ is closed with respect to the action of $\mathbb{X}^{2r}$, which contradicts the fact that $\mathcal{H}$ is irreducible. Hence, $\mathcal{H}_x$ is irreducible. Further, the representation space $\mathcal{H}_x$ of the representation of N is not isomorphic to the representation space $\mathcal{H}_{x'}$ with another x'. So, the same fact holds for the subgroup $N^\perp$.

Due to the above discussion, since $\mathcal{H}_x$ and $\mathcal{H}_y$ are mapped to each other by an application of a unitary matrix, they have the same dimension. Since the dimension of the space $\mathcal{H}$ is $|\mathbb{X}|^r$, the dimension of $\mathcal{H}_y$ is $|\mathbb{X}|^r/|N^*|$. We also have the relation $|N^\perp| = |\mathbb{X}|^{2r}/|N^*|$. This value equals $(|\mathbb{X}|^r/|N^*|)^2 \cdot |N^*|$, which is the product of the square of the dimension of the irreducible space and the number of irreducible spaces in $\mathcal{H}$. Hence, due to the relation (2.56) for the projective unitary representation, the space $\mathcal{H}$ contains all of irreducible projective unitary representations with the same factor system as $\mathsf{W}^r_{\mathbb{X}}$. ■

Next, we discuss the same discussion as the above when N is an $\mathbb{X}$-subgroup of the discrete Heisenberg group $\mathrm{H}(2r, \mathbb{X})$. In this case, we can define a self-orthogonal subgroup in the same way. In the following, we assume that N is a self-orthogonal subgroup of $\mathrm{H}(2r, \mathbb{X})$. A subset $N^\perp \cdot Z$ of discrete Heisenberg group $\mathrm{H}(2r, \mathbb{X})$ is a normal subgroup, where Z is the center of $\mathrm{H}(2r, \mathbb{X})$. Also, N^* is an $\mathbb{X}$-subgroup, and $\mathrm{H}(2r, \mathbb{X})$ equals $(N^\perp \cdot Z) \rtimes N^*$. Then, we have a lemma similar to Lemma 8.7 for a self-orthogonal subgroup N of $\mathrm{H}(2r, \mathbb{X})$. Hence, the inertia group of N^* given in Sect. 2.5 is an $\mathbb{X}$-subgroup composed of the identify element. Now, we denote the restriction of $\mathsf{W}^r_{\mathbb{X},\mathrm{H}}|_{N^\perp \cdot Z}$ on $\mathcal{H}_0$ by $\mathsf{W}^r_{\mathbb{X},\mathrm{H}}|_{N^\perp \cdot Z, 0}$. Then, the irreducible unitary representation $\mathsf{W}_{\mathbb{X},\mathrm{H}}$ of discrete Heisenberg group $\mathrm{H}(2r, \mathbb{X})$ equals $\mathsf{W}^r_{\mathbb{X},\mathrm{H}}|_{N^\perp \cdot Z, 0} \tilde{\rtimes} 1$ given in **Construction 2**.

Now, we apply Lemma 2.6 to $N^\perp \cdot Z \lhd \mathrm{H}(2r, \mathbb{X})$. Letting $\chi_{\mathrm{H},x}$ be the character of the irreducible unitary representation $\mathsf{W}^r_{\mathbb{X},H}|_{\mathcal{H}_x}$ of $N^\perp \cdot Z$, we have the relation $\chi_{\mathrm{H},x}(\vec{s}) = 0$ for any element $\vec{s} \in N^\perp \cdot Z \setminus N \cdot Z$ because $C(N^\perp \cdot Z) = N \cdot Z$. Applying this fact to the character χ_x of the projective unitary representation on $\mathcal{H}_x$ of the $\mathbb{X}$-subgroup $N^\perp$ of $\mathbb{X}^{2r}$, we have

$$\chi_x(\vec{s}) = 0, \quad \vec{s} \in N^\perp \setminus N. \tag{8.25}$$

Exercise 8.4 Give a non-trivial example of a strictly self-orthogonal subgroup of $\mathbb{F}_2^4$.

8.2 Discrete Symplectic Group and Clifford Group

8.2.1 *Definitions of Discrete Symplectic Group and Clifford Group*

In this subsection, to address the operation corresponding to squeezing operation over the finite field $\mathbb{F}_q$ and the algebra $\mathbb{Z}_d$, we deal with the symplectic group (**discrete symplectic group**) $\mathrm{Sp}(2r, \mathbb{F}_q)$ and $\mathrm{Sp}(2r, \mathbb{Z}_d)$. Similar to the symplectic group $\mathrm{Sp}(2r, \mathbb{R})$ on $\mathbb{R}$, these groups are given as

$$\mathrm{Sp}(2r, \mathbb{F}_q) := \{g \in \mathrm{GL}(2r, \mathbb{F}_q) | g^T J_r g = J_r\}$$
$$\mathrm{Sp}(2r, \mathbb{Z}_d) := \{g \in \mathrm{GL}(2r, \mathbb{Z}_d) | g^T J_r g = J_r\},$$

where $\mathrm{GL}(2r, \mathbb{F}_q)$ and $\mathrm{GL}(2r, \mathbb{Z}_d)$ are the groups composed of invertible square matrices of oder $2r$ on $\mathbb{F}_q$ and $\mathbb{Z}_d$, respectively. Now, we denote the group composed of square matrices of order 2 with determinant 1 (2-dimensional special linear group) by $\mathrm{SL}(2, \mathbb{F}_q)$ and $\mathrm{SL}(2, \mathbb{Z}_d)$, respectively. So, similar to the case of real numbers, we have $\mathrm{SL}(2, \mathbb{F}_q) = \mathrm{Sp}(2, \mathbb{F}_q)$ and $\mathrm{SL}(2, \mathbb{Z}_d) = \mathrm{Sp}(2, \mathbb{Z}_d)$. The order $|\mathrm{Sp}(2r, \mathbb{F}_q)|$ of the group $\mathrm{Sp}(2r, \mathbb{F}_q)$ is known to be $q^{r^2} \prod_{k=1}^{r} (q^{2k} - 1)$ [2, p. 147]. When $q = p^2$, $|\mathrm{Sp}(2, \mathbb{F}_{p^2})| = p^2(p^4 - 1)$ while $|\mathrm{Sp}(4, \mathbb{F}_p)| = p^4(p^2 - 1)(p^4 - 1)$. So, we find that $\mathrm{Sp}(4, \mathbb{F}_p)$ is much larger than $\mathrm{Sp}(2, \mathbb{F}_{p^2})$ (see Exercise 8.8).

On the other hand, in the case of $\mathbb{Z}_d$, when d is a product of powers of distinct prime numbers $p_1 \ldots, p_l$, we have $|\mathrm{Sp}(2, \mathbb{Z}_d)| = d^3(1 - p_1^{-2}) \cdots (1 - p_l^{-2})$ [81, Lemma 3.2], which can be checked as follows. Firstly, we consider $|\mathrm{Sp}(2, \mathbb{Z}_{p^k})|$. Assume that we denote the entries of an element of $\mathrm{Sp}(2, \mathbb{Z}_{p^k})$ by $a_{i,j}$. The number of elements with invertible $a_{1,1}$ is $p^k p^k (p^k - p^{k-1})$. When $a_{1,1}$ is not invertible, $a_{2,1}$ must be invertible. The number of such elements is $p^k p^{k-1}(p^k - p^{k-1})$. So, the total number is

$$|\mathrm{Sp}(2, \mathbb{Z}_{p^k})| = (p^k)^2(p^k - p^{k-1} + p^{k-1} - p^{k-2}) = (p^k)^3(1 - p^{-2}). \tag{8.26}$$

Next, we assume that the prime factorization of d is given as $d = p_1^{k_1} \cdots p_l^{k_l}$. Then, $\mathbb{Z}_d$ is isomorphic to $\mathbb{Z}_{p_1^{k_1}} \times \cdots \times \mathbb{Z}_{p_l^{k_l}}$. So, $\mathrm{Sp}(2, \mathbb{Z}_d)$ is isomorphic to $\mathrm{Sp}(2, \mathbb{Z}_{p_1^{k_1}}) \times \cdots \times \mathrm{Sp}(2, \mathbb{Z}_{p_l^{k_l}})$. Hence, using (8.26), we have [81, Lemma 3.2]

$$|\mathrm{Sp}(2, \mathbb{Z}_{p_1^{k_1} \cdots p_l^{k_l}})| = (p_1^{k_1} \cdots p_l^{k_l})^3(1 - p_1^{-2}) \cdots (1 - p_l^{-2}). \tag{8.27}$$

In the following, $\mathbb{X}^{2r}$ and $\mathrm{Sp}(2r, \mathbb{X})$ express $\mathbb{F}_q^{2r}$ and $\mathrm{Sp}(2r, \mathbb{F}_q)$ for $\mathbb{X} = \mathbb{F}_q$ and do $\mathbb{Z}_d^{2r}$ and $\mathrm{Sp}(2r, \mathbb{Z}_d)$ for $\mathbb{X} = \mathbb{Z}_d$.

Further, the symplectic group $\mathrm{Sp}(2r, \mathbb{X})$ naturally acts on the commutative group $\mathbb{X}^{2r}$. Using this action, we can naturally define the semi direct product $\mathbb{X}^{2r} \rtimes \mathrm{Sp}(2r, \mathbb{X})$, which is called the **Clifford group**.[1] An element of Clifford group $(\vec{s}, g) \in \mathbb{X}^{2r} \rtimes \mathrm{Sp}(2r, \mathbb{X})$ naturally acts on $\vec{s}' \in \mathbb{X}^2$ as

$$(\vec{s}, g) : \vec{s}' \mapsto (g\vec{s}') + \vec{s}.$$

Example 8.2 When $q = 2$ and $r = 1$, $\mathbb{F}_2^2$ is isometric to Klein four-group $\mathbb{Z}_2 \times \mathbb{Z}_2$. Also, the order of $\mathrm{Sp}(2, \mathbb{F}_2)$ is 6. The all of elements of the group are listed as

$$\begin{pmatrix} 1 & 0 \\ 0 & 1 \end{pmatrix}, \begin{pmatrix} 1 & 1 \\ 1 & 0 \end{pmatrix}, \begin{pmatrix} 0 & 1 \\ 1 & 1 \end{pmatrix}$$
$$\begin{pmatrix} 0 & 1 \\ 1 & 0 \end{pmatrix}, \begin{pmatrix} 1 & 1 \\ 0 & 1 \end{pmatrix}, \begin{pmatrix} 1 & 0 \\ 1 & 1 \end{pmatrix}.$$

The orders of the second and third elements are 3. The orders of the 4th–6th elements are 2. Hence, this group is isomorphic to the permutation group of degree 3 (see Exercise 2.1).

The action of $\mathrm{Sp}(2, \mathbb{F}_2)$ on $\mathbb{F}_2^2$ is the same as the action of S_3 on $\mathbb{Z}_2 \times \mathbb{Z}_2$ given in Example 2.2. Hence, the Clifford group $\mathbb{F}_2^2 \rtimes \mathrm{Sp}(2, \mathbb{F}_2)$ is isomorphic to $(\mathbb{Z}_2 \times \mathbb{Z}_2) \rtimes S_3$, i.e., S_4. In Example 8.3, we will see that the Clifford group $\mathbb{F}_2^2 \rtimes \mathrm{Sp}(2, \mathbb{F}_2)$ is isomorphic to the regular hexahedron group, which is also isometric to S_4, in a different way. For the regular hexahedron group, see Example 2.2 and Fig. 2.13.

Exercise 8.5 Show that $\mathrm{Sp}(2, \mathbb{F}_q) = \mathrm{SL}(2, \mathbb{F}_q)$ and $\mathrm{Sp}(2, \mathbb{Z}_d) = \mathrm{SL}(2, \mathbb{Z}_d)$.

Exercise 8.6 List up all elements of $\mathrm{Sp}(2, \mathbb{F}_3)$.

Exercise 8.7 Compare two groups $\mathrm{GL}(2, \mathbb{F}_2)$ and $\mathbb{F}_4^\times$ by using the $\mathbb{F}_2$-morphism F defined in Exercise 8.3. For this comparison, give an element of $\mathrm{GL}(2, \mathbb{F}_2)$ that does not belong to $\mathbb{F}_4^\times$ via the $\mathbb{F}_2$-morphism F.

Exercise 8.8 Compare two groups $\mathrm{Sp}(2, \mathbb{F}_4)$ and $\mathrm{Sp}(4, \mathbb{F}_2)$ by using the $\mathbb{F}_2$-morphism F defined in Exercise 8.3. For this comparison, give an element of $\mathrm{Sp}(4, \mathbb{F}_2)$ that does not belong to $\mathrm{Sp}(2, \mathbb{F}_4)$ via the $\mathbb{F}_2$-morphism F.

8.2.2 Generator

In the following, we investigate the generators of $\mathrm{Sp}(2, \mathbb{F}_q)$. When all of diagonal entries of an element g of $\mathrm{Sp}(2, \mathbb{F}_q)$ are not zero, there exist elements $a, a', b \in \mathbb{F}_q$ such that its Gaussian decomposition like (4.2) is available as

[1] Appleby [3] extended the Clifford group.

$$g = \begin{pmatrix} 1 & a \\ 0 & 1 \end{pmatrix} \begin{pmatrix} b^{-1} & 0 \\ 0 & b \end{pmatrix} \begin{pmatrix} 1 & 0 \\ a' & 1 \end{pmatrix}. \tag{8.28}$$

That is, choosing

$$M_b := \begin{pmatrix} b^{-1} & 0 \\ 0 & b \end{pmatrix}, \; Q_a := \begin{pmatrix} 1 & 0 \\ a & 1 \end{pmatrix}, \; P_a := \begin{pmatrix} 1 & a \\ 0 & 1 \end{pmatrix}, \tag{8.29}$$

we have $g = P_a M_b Q_{a'}$. Now, we denote the entries of g by $a_{k,l}$. On the other hand, when $a_{1,1}$ is 0, all of diagonal entries of gJ are invertible. Hence, any element of $\mathrm{Sp}(2, \mathbb{F}_q)$ can be expressed by the products of Q_a, $P_{a'}$, M_b, J. Especially, since $P_a = -J Q_a J$, the generators of $\mathrm{Sp}(2, \mathbb{F}_q)$ can be restricted to $\{Q_a\}_a \cup \{M_b\}_b \cup \{J\}$.

Now, we assume that $p \neq 2$. We choose a specific element $a_0 \in \mathbb{F}_q$ such that there does not exist $b \in \mathbb{F}_q$ satisfying $a_0 = b^2$. Then, any element of $\mathbb{F}_q$ except for 0 can be written as b^2 or $b^2 a_0$. Hence, due to the relation $M_b Q_a M_{b^{-1}} = Q_{b^2 a}$, the generators of $\mathrm{Sp}(2, \mathbb{F}_q)$ can be restricted to $\{Q_1\} \cup \{Q_{a_0}\} \cup \{M_b\} \cup \{J\}$. Further, since $M_{b'} M_b = M_{b'b}$, the generators of $\mathrm{Sp}(2, \mathbb{F}_q)$ can be restricted to $\{Q_1\} \cup \{Q_{a_0}\} \cup \{M_b\}_{b \in G_\times(\mathbb{F}_q)} \cup \{J\}$, where $G_\times(\mathbb{F}_q)$ is the generators of $\mathbb{F}_q$ with respect to multiplication.

Now, we assume that $p = 2$. Since $x \mapsto x^2$ is bijective on $\mathbb{F}_q$, any element of $\mathbb{F}_q$ except for 0 can be written as b^2. Hence, the generators of $\mathrm{Sp}(2, \mathbb{F}_q)$ can be restricted to $\{Q_1\} \cup \{M_b\}_{b \in G_\times(\mathbb{F}_q)} \cup \{J\}$.

Then, the generators satisfy the following relations

$$Q_{a'} Q_a = Q_{a'+a}, \; M_b Q_a = Q_{ba} M_b, \; J M_b = M_{-b} J, \; J^2 = -I. \tag{8.30}$$

When all of diagonal entries of an element of $\mathrm{Sp}(2, \mathbb{Z}_d)$ are invertible, the decomposition (8.28) is available. Hence, the above discussion can be applied to the case of $\mathrm{Sp}(2, \mathbb{Z}_d)$.

Next, we focus on the group $\mathrm{Sp}(2r, \mathbb{F}_q)$. We choose two $r \times r$ matrices ζ and γ on $\mathbb{F}_q$ such that $\zeta^T = \zeta$ and $\det \gamma \neq 0$. That is, ζ is a symmetric matrix and γ is an invertible matrix. We define three elements M_γ, Q_ζ, and P_ζ of $\mathrm{Sp}(2r, \mathbb{F}_q)$ as

$$M_\gamma := \begin{pmatrix} \gamma^{-1} & 0 \\ 0 & \gamma^T \end{pmatrix}, \quad Q_\zeta := \begin{pmatrix} I_r & 0 \\ \zeta & I_r \end{pmatrix}, \quad P_\zeta := \begin{pmatrix} I_r & \zeta \\ 0 & I_r \end{pmatrix}.$$

Then, the matrices J_r, Q_ζ, P_ζ, and M_γ satisfy the above relations. In the following, we denote the basis of the vector space $\mathbb{F}_q^r$ on the finite field $\mathbb{F}_q$ by $\{|k)\}_{k=1}^r$. Then, for $a \in \mathbb{F}_q$, we define a matrix $\gamma_{k,l:a} := a|k)(l| + \sum_{k'=1}^r |k')(k'|$ as an element of the general linear group $\mathrm{GL}(\mathbb{F}_q, r)$ on the finite field $\mathbb{F}_q$.

Consider an element of $\mathrm{Sp}(2r, \mathbb{F}_q)$ that has non-zero entries only on (j, j), $(j, j + r)$, $(j + r, j)$, $(j + r, j + r)$ entries. The set of such elements forms a subgroup of $\mathrm{Sp}(2r, \mathbb{F}_q)$ that is isomorphic to $\mathrm{Sp}(2, \mathbb{F}_q)$. So, $\mathrm{Sp}(2r, \mathbb{F}_q)$ can be regarded as a subgroup of $\mathrm{Sp}(2r, \mathbb{F}_q)$. Hence, the direct product group $\underbrace{\mathrm{Sp}(2, \mathbb{F}_q) \times \cdots \times \mathrm{Sp}(2, \mathbb{F}_q)}_{r}$ is also a subgroup of $\mathrm{Sp}(2r, \mathbb{F}_q)$. Similarly, $\underbrace{\mathbb{F}_q^\times \times \cdots \times \mathbb{F}_q^\times}_{r}$ is a subgroup of $\mathrm{GL}(\mathbb{F}_q, r)$.

Then, we have the following theorem.

Theorem 8.1 *When $p \neq 2$, the group $\mathrm{Sp}(2r, \mathbb{F}_q)$ is generated by the subgroup $\underbrace{\mathrm{Sp}(2, \mathbb{F}_q) \times \cdots \times \mathrm{Sp}(2, \mathbb{F}_q)}_{r}$ and the set $\{M_{\gamma_{j,l:1}}\}_{j,l} \cup \{M_{\gamma_{j,l:a_0}}\}_{j,l}$, where $a_0 \in \mathbb{F}_q$ is defined above. When a_0 is given as a finite sum of elements of $\{a^2\}_{a\in\mathbb{F}_q}$, the group $\mathrm{Sp}(2r, \mathbb{F}_q)$ is generated by $\underbrace{\mathrm{Sp}(2, \mathbb{F}_q) \times \cdots \times \mathrm{Sp}(2, \mathbb{F}_q)}_{r}$ and $\{M_{\gamma_{j,l:1}}\}_{j,l}$.*

When $p = 2$, the group $\mathrm{Sp}(2r, \mathbb{F}_q)$ is generated by its subgroup $\underbrace{\mathrm{Sp}(2, \mathbb{F}_q) \times \cdots \times \mathrm{Sp}(2, \mathbb{F}_q)}_{r}$ and $\{M_{\gamma_{j,l:1}}\}_{j,l}$.

To show this Theorem, we prepare the following Lemma.

Lemma 8.8 *(1) When $p \neq 2$, the group $\mathrm{GL}(\mathbb{F}_q, r)$ is generated by the subgroup $\underbrace{\mathbb{F}_q^\times \times \cdots \times \mathbb{F}_q^\times}_{r}$ and $\{\gamma_{j,l:1}\}_{j,l} \cup \{\gamma_{j,l:a_0}\}_{j,l}$. (2) When a_0 is given as a finite sum of elements of $\{a^2\}_{a\in\mathbb{F}_q}$, it is generated by the subgroup $\underbrace{\mathbb{F}_q^\times \times \cdots \times \mathbb{F}_q^\times}_{r}$ and $\{\gamma_{j,l:1}\}_{j,l}$.*

(3) When $p = 2$, the group $\mathrm{GL}(\mathbb{F}_q, r)$ is generated by the subgroup its subgroup $\underbrace{\mathbb{F}_q^\times \times \cdots \times \mathbb{F}_q^\times}_{r}$ and $\{\gamma_{j,l:1}\}_{j,l}$.

Proof of Lemma 8.8 Elementary operations for matrices show that $\mathrm{GL}(\mathbb{F}_q, r)$ is generated by its subgroup $\underbrace{\mathbb{F}_q^\times \times \cdots \times \mathbb{F}_q^\times}_{r}$, $\{M_{\gamma_{j,l:a}}\}_{j,l,a\in\mathbb{F}_q}$, and the set of permutation matrices $T_{j,l} := |j\rangle\langle l| + |l\rangle\langle j|$. Let $D_j(x)$ be the diagonal matrix whose j-th diagonal entry is x and whose other diagonal entries are 1. Then, we have $T_{j,l} = D_j(-1)\gamma_{j,l:-1}\gamma_{l,j:1}\gamma_{j,l:-1}$. Hence, it is sufficient to show that $\gamma_{j,l:-1}$ can be generated by the proposed subset. Now, we have $D_j(b)D_l(b^{-1})\gamma_{l,j:a}D_j(b^{-1})D_l(b) = \gamma_{l,j:b^2a}$ for any $a \in \mathbb{F}_q$.

For (1), the above relation with $a = -1$ and the choice of a_0 yield that $\gamma_{j,l:-1}$ can be generated by the proposed subset. For (2), the relation $Q_{a'}Q_a = Q_{a'+a}$ guarantees the desired argument. For (3), there is an element b such that $b^2 \cdot (-1) = 1$ when $p \neq 2$. So, the above relation with $a = -1$ guarantees the desired argument. ■

Proof of Theorem 8.1 It is known that $\mathrm{Sp}(2r, \mathbb{F}_q)$ is generated by $\{J_r\} \cup \{Q_\zeta\} \cup \{M_\gamma\}$ [80, Theorem 3.3].[2] Let $\{|k\rangle\}_{k=1}^{2r}$ be the basis of the vector space $\mathbb{F}_q^{2r}$ on the finite field $\mathbb{F}_q$. Then, we define the matrix $J_{r,j}$ as

[2]Since the paper [80] assumes the condition $p \neq 2$, it shows this fact in this case. However, the proof is still valid even when $p = 2$.

$$J_{r,j} = -|j)(j+r| + |j+r)(j| + \sum_{k \neq j} (|k)(k| + |k+r)(k+r|), \tag{8.31}$$

which is contained in the subgroup $\underbrace{\mathrm{Sp}(2, \mathbb{F}_q) \times \cdots \times \mathrm{Sp}(2, \mathbb{F}_q)}_{r}$. Since the equation $J_r = J_{r,1} \cdots J_{r,r}$ holds, J_r is also contained in the above subgroup. Further, since $M_\gamma Q_\zeta M_{\gamma^{-1}} = Q_{\gamma^T \zeta \gamma}$, any symmetric matrix ζ can be diagonalized by applying the matrix M_γ from both sides. Let $\zeta_j(x)$ be the matrix whose jth diagonal entry is x and whose other diagonal entry is 0. Then, the matrix $\zeta_j(x)$ is contained in the above subgroup. Hence, we obtain the desired argument from these facts and Lemma 8.8. ∎

8.2.3 Structure of Self-orthogonal Subgroup

In the following, we consider the expression of the self-orthogonal subgroup N on the finite filed $\mathbb{F}_q^{2r}$ based on an element of the symplectic group $\mathrm{Sp}(2r, \mathbb{F}_q)$. Let e_j be the element of $\mathbb{F}_q^{2r}$ whose jth entry is 1 and whose other entries are 0. When the self-orthogonal subgroup N is isomorphic to $\mathbb{F}_q^k$, i.e., the dimension of N is k, we have $k \leq r$. As a typical case, we define the $\mathbb{F}_q$-subgroup $\mathbb{F}_q^{(k)}$ generated by $e_1, \ldots, e_k$. Hence, we can choose an element $g \in \mathrm{Sp}(2r, \mathbb{F}_q)$ such that any k-dimensional self-orthogonal subgroup N is the image of $\mathbb{F}_q^{(k)}$ by the map g. Especially, the image of $\mathbb{F}_q^{(k)}$ by the map g is the $\mathbb{F}_q$-subgroup generated by the vectors $e(g)_1, \ldots, e(g)_k$, where $e(g)_j$ is the jth column vector of g. Now, we call a $2r \times k$ matrix a **generating matrix** of the self-orthogonal subgroup N when its image is the self-orthogonal subgroup N. Hence, defining the $2r \times k$ matrix $A(g)$ as the matrix composed of k vectors $e(g)_1, \ldots, e(g)_k$, we find that $A(g)$ is a generating matrix of the self-orthogonal subgroup N. Also, for any element $h \in \mathrm{GL}(k, \mathbb{F}_q)$, $A(g)h$ is a generating matrix of the self-orthogonal subgroup N.

Especially, when N is a strictly self-orthogonal subgroup, we have $k = r$. So, it is composed of two square matrices of order r, the upper half and the lower half. When the upper half is invertible, by multiplying the inverse of the upper matrix from the left side, the generating matrix $A(g)$ is transformed to a matrix whose upper half is the identity matrix. The lower half of the transformed matrix is an alternative matrix ζ because N is a strictly self-orthogonal subgroup. Hence, unless the upper half of the generating matrix is non-invertible, any strictly self-orthogonal subgroup can be described by an $r \times r$ alternative matrix ζ on the finite filed $\mathbb{F}_q^r$. Note that when $p = 2$, a matrix is alternative if and only if it is symmetric. So, an alternative matrix may have non-zero diagonal element in this case.

Exercise 8.9 Give a generating matrix of a strictly self-orthogonal subgroup given in Exercise 8.4.

8.3 Metaplectic Representation

8.3.1 Existence of Projective Unitary Representation

Next, we give a projective unitary representation of the symplectic group $\mathrm{Sp}(2r, \mathbb{X})$ on $\mathcal{H}^{\otimes r}$. Firstly, we consider the case when d or p is an odd number. Then, for two elements g, g' of $\mathrm{Sp}(2r, \mathbb{X})$, we have

$$\langle \vec{s}, \vec{s}' \rangle_{\mathbb{X}} = \langle g(\vec{s}), g(\vec{s}') \rangle_{\mathbb{X}}. \tag{8.32}$$

Due to (8.15), this fact shows that the projective representation $\vec{s} \mapsto \mathsf{W}^r_{\mathbb{X}} g(\vec{s})$ is isomorphic to the projective representation $\vec{s} \mapsto \mathsf{W}^r_{\mathbb{X}}(\vec{s})$. Since Lemma 8.3 guarantees that the projective representation $\vec{s} \mapsto \mathsf{W}^r_{\mathbb{X}}(\vec{s})$ is irreducible, there exists a unitary matrix $\mathsf{S}^r_{\mathbb{X}}(g)$ such that

$$\mathsf{S}^r_{\mathbb{X}}(g) \mathsf{W}^r_{\mathbb{X}}(\vec{s}) \mathsf{S}^r_{\mathbb{X}}(g)^\dagger = \mathsf{W}^r_{\mathbb{X}}(g(\vec{s})). \tag{8.33}$$

Further, since

$$\mathsf{S}^r_{\mathbb{X}}(g)\mathsf{S}^r_{\mathbb{X}}(g')\mathsf{W}^r_{\mathbb{X}}(\vec{s})\mathsf{S}^r_{\mathbb{X}}(g')^\dagger\mathsf{S}^r_{\mathbb{X}}(g)^\dagger = \mathsf{W}^r_{\mathbb{X}}(gg'(\vec{s})) = \mathsf{S}^r_{\mathbb{X}}(gg')\mathsf{W}^r_{\mathbb{X}}(\vec{s})\mathsf{S}^r_{\mathbb{X}}(gg')^\dagger,$$

there exists a real number $\theta(g, g')$ such that $\mathsf{S}^r_{\mathbb{X}}(g)\mathsf{S}^r_{\mathbb{X}}(g') = e^{i\theta(g,g')}\mathsf{S}^r_{\mathbb{X}}(gg')$. Hence, $g \mapsto \mathsf{S}^r_{\mathbb{X}}(g)$ is a projective unitary representation. The projective unitary representation is called the **Metaplectic representation** of the symplectic group $\mathrm{Sp}(2r, \mathbb{X})$.

On the other hand, when d or p is an even number, the similar fact can be shown in the same way basically. However, we need some additional ideas. For two elements $g, g' \in \mathrm{Sp}(2r, \mathbb{X})$, the relation (8.32) holds. However, the projective representation $\vec{s} \mapsto \mathsf{W}^r_{\mathbb{X}}(g(\vec{s}))$ is not necessarily isomorphic to the projective representation $\vec{s} \mapsto \mathsf{W}^r_{\mathbb{X}}(\vec{s})$.

Here, we choose generators of $\mathbb{X}^{2r} = \mathbb{F}_q^{2r}, \mathbb{Z}_d^{2r}$ as follows. When $\mathbb{X} = \mathbb{Z}_d$, we consider the vector e_j whose jth entry is 1 and whose other entries are 0. Then, $\mathbb{Z}_d^{2r}$ is a commutative group generated by $\{e_j\}_{j=1}^{2r}$. When $\mathbb{X} = \mathbb{F}_q$, we regard $\mathbb{F}_q$ as a vector space on the finite field $\mathbb{F}_p$, and focus on the basis $\{f_k\}$ of $\mathbb{F}_q$. Let $f_{k,j}$ be the vector in $\mathbb{F}_q^{2r}$ whose jth entry is f_k and whose other entries are 0. Then, $\mathbb{F}_q^{2r}$ is a commutative group generated by $\{f_{k,j}\}$. We denote the generators by $\{e_j\}$ with a proper reordering.

The above given discrete Heisenberg representation can be given by the representations $\mathsf{W}^r_{\mathbb{X}}(e_j)$ of the generators $\{e_j\}$ because the representation of another element can be given from the products of those of the generators in the same way as Sect. 8.1. Here, we define another projective representation $\vec{s} \mapsto \mathsf{W}'^r_{\mathbb{X}}(\vec{s})$ by $\mathsf{W}'^r_{\mathbb{X}}(e_j) = \mathsf{W}'^r_{\mathbb{X}}(g(e_j))$. Here, the representation of another element is defined by the above method. Then, since the commutation relation among the representations of the generators for $\mathsf{W}'^r_{\mathbb{X}}$ is that of $\mathsf{W}^r_{\mathbb{X}}$, there exists an integer $\epsilon_{g,\vec{s}}$ for $\vec{s}$ such that

$\mathsf{W}'^r_{\mathbb{X}}(\vec{s}) = (-1)^{\epsilon_{g,\vec{s}}}\mathsf{W}^r_{\mathbb{X}}(g(\vec{s}))$. Thus, because $\mathsf{W}'^r_{\mathbb{X}}(\vec{s})$ is irreducible and is isomorphic to $\mathsf{W}^r_{\mathbb{X}}(\vec{s})$, there exists a unitary matrix $\mathsf{S}^r_{\mathbb{X}}(g)$ such that

$$\mathsf{S}^r_{\mathbb{X}}(g)\mathsf{W}^r_{\mathbb{X}}(\vec{s})\mathsf{S}^r_{\mathbb{X}}(g)^\dagger = (-1)^{\epsilon_{g,\vec{s}}}\mathsf{W}^r_{\mathbb{X}}(g(\vec{s})). \tag{8.34}$$

Further, since

$$\begin{aligned}&(-1)^{\epsilon_{g',\vec{s}}+\epsilon_{g,g(\vec{s})}}\mathsf{S}^r_{\mathbb{X}}(g)\mathsf{S}^r_{\mathbb{X}}(g')\mathsf{W}^r_{\mathbb{X}}(\vec{s})\mathsf{S}^r_{\mathbb{X}}(g')^\dagger\mathsf{S}^r_{\mathbb{X}}(g)^\dagger\\ =&\mathsf{W}^r_{\mathbb{X}}(gg'(\vec{s})) = (-1)^{\epsilon_{gg',\vec{s}}}\mathsf{S}^r_{\mathbb{X}}(gg')\mathsf{W}^r_{\mathbb{X}}(\vec{s})\mathsf{S}^r_{\mathbb{X}}(gg')^\dagger,\end{aligned}$$

we have $\epsilon_{g',\vec{s}} + \epsilon_{g,g'(\vec{s})} = \epsilon_{gg',\vec{s}}$ (mod) 2. So, there exists a real number $\theta(g, g')$ such that $\mathsf{S}^r_{\mathbb{X}}(g)\mathsf{S}^r_{\mathbb{X}}(g') = e^{i\theta(g,g')}\mathsf{S}^r_{\mathbb{X}}(gg')$. Hence, we find that $g \mapsto \mathsf{S}^r_{\mathbb{X}}(g)$ is a projective unitary representation. Similar to the odd number case, this representation is called the **Metaplectic representation** of the symplectic group Sp($2r$, $\mathbb{X}$). The properties (8.33) and (8.34) guarantee that the representation is faithful.

8.3.2 Concrete Construction of Representation

In this subsection, for the case of $X = \mathbb{F}_q$, we concretely construct the Metaplectic representation of Sp($2r$, $\mathbb{F}_q$). It suffices for this purpose to give a matrix $\mathsf{S}^r_{\mathbb{F}}(g)$ for a generator g such that the relations (8.33) and (8.34) hold. When $p \neq 2$, the representations of the generators are given as [80, Theorem 4.1]

$$\mathsf{S}^r_{\mathbb{F}}(J_{r,j}) := I_r^{\otimes(j-1)} \otimes \Big(\frac{1}{q^{1/2}}\sum_{x,y\in\mathbb{F}_q}\omega_{\mathbb{F}}^{\mathrm{tr}xy}|x\rangle\langle y|\Big)\otimes I_r^{\otimes(r-j)} \tag{8.35}$$

$$\mathsf{S}^r_{\mathbb{F}}(M_\gamma) := \sum_{\boldsymbol{x}\in\mathbb{F}_q^r}|\gamma\boldsymbol{x}\rangle\langle\boldsymbol{x}| \tag{8.36}$$

$$\mathsf{S}^r_{\mathbb{F}}(Q_\zeta) := \sum_{\boldsymbol{x}\in\mathbb{F}_q^r}\tau_{\mathbb{F}}^{(\boldsymbol{x},\zeta\boldsymbol{x})_{\mathbb{F}}}|\boldsymbol{x}\rangle\langle\boldsymbol{x}|. \tag{8.37}$$

When $r = 1$, $\mathsf{S}^r_{\mathbb{F}}(J_r)$ is the discrete Fourier transform. Since $\{M_\gamma\}$ is the subgroup GL(r, $\mathbb{F}_q$), the relation (8.36) can be regarded as the definition of a projective representation of the subgroup GL(r, $\mathbb{F}_q$), which is an invertible multiplication of γ on the computational basis. Then, we can show that

$$\mathsf{S}^r_{\mathbb{F}}(M_\gamma)|\hat{e}_{\mathbb{F}}(l)\rangle = |\hat{e}_{\mathbb{F}}(\gamma^{-1}l)\rangle.$$

Also, the above defined matrix $\mathsf{S}^r_{\mathbb{F}}(g)$ satisfies

$$\mathsf{S}^r_{\mathbb{F}}(J_{r,j})\mathsf{W}^r_{\mathbb{X}}(s,t)\mathsf{S}^r_{\mathbb{F}}(J_{r,j})^\dagger = \mathsf{W}^r_{\mathbb{X}}(J_{r,j}(s,t))$$
$$\mathsf{S}^r_{\mathbb{F}}(M_\gamma)\mathsf{W}^r_{\mathbb{X}}(s,t)\mathsf{S}^r_{\mathbb{F}}(M_\gamma)^\dagger = \mathsf{W}^r_{\mathbb{X}}(\gamma s,\gamma^{-1}t)$$
$$\mathsf{S}^r_{\mathbb{F}}(Q_\varsigma)\mathsf{W}^r_{\mathbb{X}}(s,t)\mathsf{S}^r_{\mathbb{F}}(Q_\varsigma)^\dagger = \mathsf{W}^r_{\mathbb{X}}(s,t+\varsigma s).$$

Hence, Theorem 8.1 guarantees that the relations (8.35), (8.36), and (8.37) give a projective unitary representation satisfying (8.33). That is, due to Theorem 8.1, we can implement the representation $\mathsf{S}^r_{\mathbb{F}}$ physically by implementing the unitary matrices corresponding to the projective representation of $\mathrm{Sp}(2,\mathbb{F}_q)$ on each system $\mathcal{H}_j$ and the unitary matrices $\{\mathsf{S}^r_{\mathbb{F}}(M_{\gamma_{j,l:1}})\}_{j,l}$ and $\{\mathsf{S}^r_{\mathbb{F}}(M_{\gamma_{j,l:a}})\}_{j,l}$ across two quantum systems $\mathcal{H}_j$ and $\mathcal{H}_l$. Further, it is known that, given a projective unitary representation $\mathsf{S}^r_{\mathbb{F}}(g)$, there exist scalars $c(g)$ such that $c(g)\mathsf{S}^r_{\mathbb{F}}(g)$ is a unitary representation [80, Theorem 4.3].

On the other hand, when $p=2$, we can define $\mathsf{S}^r_{\mathbb{F}}(J_{r,j})$ and $\mathsf{S}^r_{\mathbb{F}}(M_\gamma)$ in the same way. However, we need to define $\mathsf{S}^r_{\mathbb{F}}(Q_\varsigma)$ in a different way from the above. Since the given generators contain only $Q_{\varsigma_j(y)}$ among the elements $\{Q_\varsigma\}$, we need to construct only unitary matrices $\mathsf{S}^r_{\mathbb{F}}(Q_{\varsigma_j(y)})$. For this purpose, we regard the finite field $\mathbb{F}_q$ as a vector space $\mathbb{F}_2^m$ on the finite field $\mathbb{F}_2$, and choose the CONS $\{e_1,\ldots,e_m\}$ such that $tre_je_k=\delta_{j,k}$. Then, we define the integer $\mathrm{wgt}(x)$ to be the number of elements $\#\{j\in\{1,\ldots,m\}|tr(xe_j)\neq 0\}$. So, for two elements $x,y\in\mathbb{F}_q$, we have $\mathrm{wgt}(x+y)=\mathrm{wgt}(x)+\mathrm{wgt}(y)-2\#\{j|x_j=y_j=1\}$, where $x_j:=tr(xe_j)\in\mathbb{F}_2$. Further, since $x=\sum_j x_je_j$, we have $tr(xy)=\sum_j x_jy_j=\#\{j|x_j=y_j=1 \pmod 2\}$. Hence,

$$\tau_{\mathbb{F}}^{\mathrm{wgt}(x+y)} = \tau_{\mathbb{F}}^{\mathrm{wgt}(x)+\mathrm{wgt}(y)}\omega_{\mathbb{F}}^{tr(xy)}. \tag{8.38}$$

Because the map $x\mapsto x^2$ is bijective on $\mathbb{F}_q$, for an arbitrary element $y\in\mathbb{F}_q$, there exists its square root $y'\in\mathbb{F}_q$. Using this fact, we can define [46]

$$\mathsf{S}^r_{\mathbb{F}}(Q_{\varsigma_j(y)}) := I^{\otimes j-1}\left(\sum_{x\in\mathbb{F}_q}\tau_{\mathbb{F}}^{\mathrm{wgt}(y'x)}|x\rangle\langle x|\right)I^{\otimes r-j}, \tag{8.39}$$

which satisfies (8.34), as will be shown later. Hence, due to Theorem 8.1, the relations (8.35), (8.36), and (8.39) give a projective unitary representation satisfying (8.33). That is, due to Theorem 8.1, we can implement the representation $\mathsf{S}^r_{\mathbb{F}}$ physically by implementing the unitary matrices corresponding to the projective representation of $\mathrm{Sp}(2,\mathbb{F}_q)$ on each system $\mathcal{H}_j$ and the unitary matrix $\mathsf{S}^r_{\mathbb{F}}(M_{\gamma_{j,l:1}})$ across two systems $\mathcal{H}_j$ and $\mathcal{H}_l$. Especially, in this case, the unitary matrix $\mathsf{S}^r_{\mathbb{F}}(M_{\gamma_{j,l:1}})$ is called the Control NOT (**C-NOT**) operation with the control system $\mathcal{H}_l$ and the target system $\mathcal{H}_j$. Indeed, any quantum circuit can be constructed from the combinations of C-NOT and local operations as Fig. 8.2. Hence, it is an important topic in physics to implement the C-NOT operation physically.

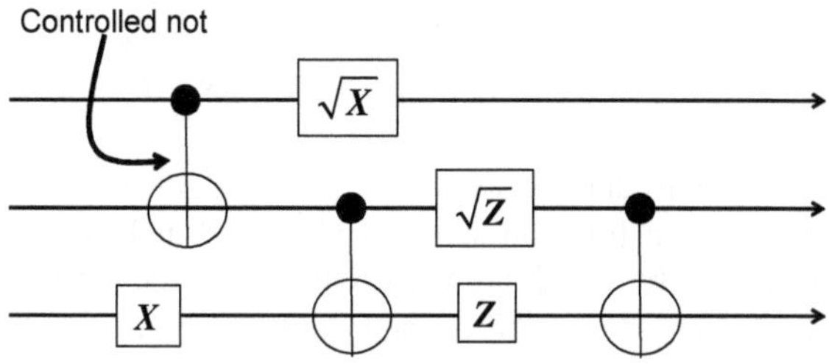

Fig. 8.2 Quantum circuit

Proof of (8.34) In the following, we show that the unitary matrix defined in (8.39) satisfies the condition (8.34) for the projective unitary representation $\mathsf{S}_{\mathbb{F}}^r$. For this purpose, it is enough to discuss only the j-th Hilbert space. That is, the problem can be reduced to the case when $r = 1$, i.e., it is enough to show the condition (8.34) for the case of $r = 1$. Now, we abbreviate $\mathsf{S}_{\mathbb{F}}^1$ to $\mathsf{S}_{\mathbb{F}}$ in this case. Due to (8.38), we can check that

$$\mathsf{S}_{\mathbb{F}}(Q_y)\mathsf{W}_{\mathbb{F}}(s,t)\mathsf{S}_{\mathbb{F}}(Q_y)^\dagger = \tau_{\mathbb{F}}^{\mathrm{wgt}(y's)-tr(y's)^2}\mathsf{W}_{\mathbb{F}}(s,t+ys).$$

Here, due to (8.34), there exists a unitary matrix U' such that

$$U'\mathsf{W}_{\mathbb{F}}(s,t)U'^\dagger = (-1)^\epsilon \mathsf{W}_{\mathbb{F}}(s,t+ys).$$

Hence,

$$(-1)^\epsilon U'^\dagger \mathsf{W}_{\mathbb{F}}(s,t+ys)U' = \tau_{\mathbb{F}}^{\mathrm{wgt}(y's)-tr(y's)^2}\mathsf{S}_{\mathbb{F}}(Q_y)^\dagger \mathsf{W}_{\mathbb{F}}(s,t+ys)\mathsf{S}_{\mathbb{F}}(Q_y).$$

Thus, Lemma 8.4 guarantees that $\tau_{\mathbb{F}}^{\mathrm{wgt}(y's)-tr(y's)^2}$ is 1 or -1. That is, $\mathrm{wgt}(y's) - tr(y's)^2 = 0 \bmod 2$. Therefore, we have shown that the unitary matrix defined in (8.39) satisfies the condition (8.34). ∎

The above discussion can be applied to the group $\mathrm{Sp}(2r, \mathbb{Z}_d)$. That is, the representations of J_r, M_γ, and Q_ζ can be given as the above way. Further, since these elements generate the group $\mathrm{Sp}(2, \mathbb{Z}_d)$, its projective representation can be concretely constructed from the above method.

8.3.3 Properties of Irreducible Decomposition

When $X = \mathbb{F}_q$ and $p \neq 2$, the representation space $\mathcal{H}^{\otimes r}$ of the Metaplectic representation is divided into two irreducible subspaces as [80, Theorem 4.2]

$$\mathcal{H}^{\otimes r:\text{even}} := \{|\phi\rangle \in \mathcal{H}^{\otimes r} | \forall \boldsymbol{s} \in \mathbb{F}_q^r,\quad \langle \boldsymbol{s}|\phi\rangle = \langle -\boldsymbol{s}|\phi\rangle\}$$
$$\mathcal{H}^{\otimes r:\text{odd}} := \{|\phi\rangle \in \mathcal{H}^{\otimes r} | \forall \boldsymbol{s} \in \mathbb{F}_q^r,\quad \langle \boldsymbol{s}|\phi\rangle = -\langle -\boldsymbol{s}|\phi\rangle\}.$$

The two irreducible spaces $\mathcal{H}^{\otimes r:\text{even}}$ and $\mathcal{H}^{\otimes r:\text{odd}}$ correspond to the discrete versions of the even function space $L^2(\mathbb{R})^{\otimes r:\text{even}}$ and the odd function space $L^2(\mathbb{R})^{\otimes r:\text{odd}}$, respectively, which are the irreducible spaces of the projective unitary representation given in Sect. 7.8 of the symplectic group $\mathrm{Sp}(2r, \mathbb{R})$. The dimensions of these subspaces are given as

$$\dim \mathcal{H}^{\otimes r:\text{even}} = \frac{q^r + 1}{2}, \quad \dim \mathcal{H}^{\otimes r:\text{odd}} = \frac{q^r - 1}{2}.$$

On the other hand, we have the following theorem in the case of $p = 2$.

Lemma 8.9 *When $p = 2$, the Metaplectic representation $\mathsf{S}_{\mathbb{F}}^r$ is irreducible.*

Proof The symplectic group $\mathrm{Sp}(2r, \mathbb{F}_q)$ contains the direct product group $\underbrace{\mathrm{Sp}(2, \mathbb{F}_q) \times \cdots \times \mathrm{Sp}(2, \mathbb{F}_q)}_{r}$ as a subgroup. Hence, due to Lemma 2.8, it is enough to show only the case of $r = 1$. When we focus on the subgroup generated by Q_y, any irreducible subspaces of the representation of the subgroup are given as 1-dimensional spaces generated by $|x\rangle$. These 1-dimensional representation spaces are not isometric to each other.

Next, we consider the action of element M_γ. So, we find that $|x\rangle$ and $|x'\rangle$ can be mapped to each other for two non-zero elements x and x' of $\mathbb{F}_q$. Since the element J_r maps the vector $|0\rangle$, we find that the Metaplectic representation $\mathsf{S}_{\mathbb{F}}^1$ is irreducible. ■

Further, since the discrete Heisenberg representation and the Metaplectic representation satisfy the condition (2.39), as mentioned in (8.33) and (8.34), we can define the semi direct product representation $\mathsf{V}_{\mathbb{X}}^r$ of the semi direct product $\mathbb{X}^{2r} \rtimes \mathrm{Sp}(2r, \mathbb{X})$. Hence, $\mathsf{V}_{\mathbb{X}}^r(\vec{s}, g)\mathsf{W}_{\mathbb{X}}^r(\vec{s}\,')\mathsf{V}_{\mathbb{X}}^r(\vec{s}, g)^\dagger = \omega_{\mathbb{X}}^{n(g,\vec{s},\vec{s}\,')}\mathsf{W}_{\mathbb{X}}^r(g(\vec{s}\,') + \vec{s})$, where $n(g, \vec{s}, \vec{s}\,')$ is an integer depending on $g, \vec{s}, \vec{s}\,'$. Thus, the semi direct product representation $\mathsf{V}_{\mathbb{X}}^r$ is faithful.

Next, we define the group

$$C_{\mathbb{X}}(\mathcal{H}^{\otimes r})$$
$$:= \{U \in \mathrm{U}(\mathcal{H}^{\otimes r}) | \forall \vec{s} \in \mathbb{X}^{2r}, \exists \theta(\vec{s}, g), \exists \vec{s}\,' \in \mathbb{X}^{2r}, U\mathsf{W}_{\mathbb{X}}^r(\vec{s})U^\dagger = e^{i\theta(\vec{s},U)}\mathsf{W}_{\mathbb{X}}^r(\vec{s}\,')\}.$$

Then, the group $C_{\mathbb{X}}(\mathcal{H}^{\otimes r})$ is a subgroup of $\mathrm{U}(\mathcal{H}^{\otimes r})$. The subgroup naturally contains the group $\mathrm{U}(1)$ as its subgroup of the scalar multiplication. Then, the following theorem holds.

Theorem 8.2 *The semi direct product representation $\mathsf{V}_{\mathbb{Z}}^r$ of the discrete Heisenberg representation and the Metaplectic representation of the Clifford group $\mathbb{Z}_d^{2r} \rtimes$*

$\mathrm{Sp}(2r, \mathbb{Z}_d)$ *gives the isomorphism between the Clifford group* $\mathbb{Z}_d^{2r} \rtimes \mathrm{Sp}(2r, \mathbb{Z}_d)$ *and the group* $C_{\mathbb{Z}}(\mathcal{H}^{\otimes r})/\mathrm{U}(1)$.

Proof Since the semi direct product representation is faithful, the semi direct product representation is an injective map from the group $\mathbb{Z}_d^{2r} \rtimes \mathrm{Sp}(2r, \mathbb{Z}_d)$ to $\mathrm{U}(\mathcal{H}^{\otimes r})$. On the other hand, we can show that the semi direct product representation is a surjective map from the group $\mathbb{Z}_d^{2r} \rtimes \mathrm{Sp}(2r, \mathbb{Z}_d)$ to $C_{\mathbb{Z}}(\mathcal{H}^{\otimes r})/\mathrm{U}(1)$ by the following way. For an element $U \in \hat{C}_{\mathbb{Z}}(\mathcal{H}^{\otimes r})$, we choose a vector $u(\vec{s})$ such that $g'\mathsf{W}_{\mathbb{Z}}^r(\vec{s})g'^{\dagger} = e^{i\theta(\vec{s},U)}\mathsf{W}_{\mathbb{Z}}^r(u(\vec{s}))$. Since the commutation relation between $\mathsf{W}_{\mathbb{Z}}^r(\vec{s}')$ and $\mathsf{W}_{\mathbb{Z}}^r(\vec{s})$ is the commutation relation between $\mathsf{W}_{\mathbb{Z}}^r(u(\vec{s}'))$ and $\mathsf{W}_{\mathbb{Z}}^r(u(\vec{s}))$, we obtain

$$\langle \vec{s}', \vec{s} \rangle_{\mathbb{Z}} = \langle u(\vec{s}'), u(\vec{s}) \rangle_{\mathbb{Z}}. \tag{8.40}$$

Hence,

$$\begin{aligned} e^{i\theta(\vec{s}'+\vec{s},g)}\mathsf{W}_{\mathbb{Z}}^r(u(\vec{s}'+\vec{s})) &= g'\mathsf{W}_{\mathbb{Z}}^r(\vec{s})\mathsf{W}_{\mathbb{Z}}^r(\vec{s}')g'^{\dagger} = g'\mathsf{W}_{\mathbb{Z}}^r(\vec{s})g'^{\dagger}g'\mathsf{W}_{\mathbb{Z}}^r(\vec{s}')g'^{\dagger} \\ &= e^{i(\theta(\vec{s}',g)+\theta(\vec{s},g))}\mathsf{W}_{\mathbb{Z}}^r(u(\vec{s}'))\mathsf{W}_{\mathbb{Z}}^r(u(\vec{s})). \end{aligned}$$

This fact indicates that $u(\vec{s}'+\vec{s}) = u(\vec{s}') + u(\vec{s})$. In particular, since $u(0) = 0$, we have $u(n'\vec{s}' + n\vec{s}) = n'(\vec{s}') + nu(\vec{s})$ for any two integers n, n'. Because this fact shows the linearity of u, the map u can be written as a matrix. Further, the relation (8.40) guarantees that the map u is an element of $\mathrm{Sp}(2r, \mathbb{Z}_d)$. Hence, we have $\mathsf{S}_{\mathbb{Z}}^r(u)^{\dagger}U\mathsf{W}_{\mathbb{Z}}^r(\vec{s})U^{\dagger}\mathsf{S}_{\mathbb{Z}}^r(u) = (-1)^{\epsilon}e^{i\theta'(\vec{s},U)}\mathsf{W}_{\mathbb{Z}}^r(\vec{s})$. Due to Lemma 8.5, there exists an element $\vec{s}' \in \mathbb{Z}_d^{2r}$ such that $\mathsf{W}_{\mathbb{Z}}^r(\vec{s}')$ is a constant times of $\mathsf{S}_{\mathbb{Z}}^r(u)^{\dagger}U$. Therefore, we obtain the desired argument. ■

Thus, we obtain the following lemma.

Lemma 8.10 *The representation space of the tensor product representation between the projective unitary representation* $\mathsf{V}_{\mathbb{F}}^r$ *of the Clifford group* $\mathbb{F}_q^{2r} \rtimes \mathrm{Sp}(2r, \mathbb{F}_q)$ *on the space* $\mathcal{H}^{\otimes r}$ *and its complex conjugate representation* $\overline{\mathsf{V}_{\mathbb{F}}^r}$ *can be regarded as the space composed of the linear maps form* $\mathcal{H}^{\otimes r}$ *to* $\mathcal{H}^{\otimes r}$. *Then, the representation space is irreducibly decomposed to the one-dimensional space spanned by the identity matrix and its orthogonal complement.*

Proof Firstly, we focus on the representation of the subgroup $\mathbb{F}_q^{2r}$. Then, each irreducible subspace is a one-dimensional subspace spanned by $\mathsf{W}_{\mathbb{F}}^r(\vec{s})$. For an arbitrary non-zero element $\vec{s}, \vec{s}' \in \mathbb{F}_q^{2r}$, there exists an element of $\mathrm{Sp}(2r, \mathbb{F}_q)$ that maps $\mathsf{W}_{\mathbb{F}}^r(\vec{s})$ to $\mathsf{W}_{\mathbb{F}}^r(\vec{s}')$. Hence, $< \mathsf{W}_{\mathbb{F}}^r(0) >$ and $< \{\mathsf{W}_{\mathbb{F}}^r(\vec{s})\}_{\vec{s}\neq 0} >$ are invariant subspaces, which shows the desired argument. ■

Example 8.3 When q=2 and r=1, we consider the Clifford group $\mathbb{F}_q^{2r} \rtimes \mathrm{Sp}(2r, \mathbb{F}_q)$ and the representation $\mathsf{V}_{\mathbb{F}}^r \otimes \overline{\mathsf{V}_{\mathbb{F}}^r}$. Now, we restrict this representation to the subspace $< \mathsf{W}_{\mathbb{F}}(1, 0), \mathsf{W}_{\mathbb{F}}(0, 1), \mathsf{W}_{\mathbb{F}}(1, 1) >$. Due to Example 8.1, this space is composed of all of Hermitian traceless matrices. The space is isomorphic to $\mathbb{R}^3$. The concrete form of $\mathsf{W}_{\mathbb{F}}$ given in Example 8.1 shows that $\epsilon_{g,\vec{s}}$ in (8.34) is 0 for any g and $\vec{s}$. Hence, by regarding $\mathsf{W}_{\mathbb{F}}(1,0)$, $\mathsf{W}_{\mathbb{F}}(0,1)$, $\mathsf{W}_{\mathbb{F}}(1,1)$ as the axis of the space $\mathbb{R}^3$, the action of the Clifford group $\mathbb{F}_2^2 \rtimes \mathrm{Sp}(2,\mathbb{F}_2)$ on the space $\mathbb{R}^3$ is the natural action of the regular hexahedron group on the space $\mathbb{R}^3$.

When the Clifford group $\mathbb{F}_q^{2r} \rtimes \mathrm{Sp}(2r,\mathbb{F}_q)$ is regarded as a subset of $\mathrm{U}(r)$ via the above representation based on Lemma 8.10, the Clifford group $\mathbb{F}_q^{2r} \rtimes \mathrm{Sp}(2r,\mathbb{F}_q)$ is a $(1,1)$-design. Hence, Theorem 4.6 guarantees that it is also a $(2,0)$-design. As shown in Lemma 4.3, the tensor product representation $(\mathsf{V}_{\mathbb{F}}^r)^{\otimes 2}$ is irreducibly decomposed to symmetric tensor product space $\mathcal{H}_s$ and alternative tensor product space $\mathcal{H}_a$. In the following, when q is an odd number, using Theorem 2.2, we investigate the projective unitary representation $(\mathsf{V}_{\mathbb{F}}^r)^{\otimes 2}$ of the Clifford group $\mathbb{F}_q^{2r} \rtimes \mathrm{Sp}(2r,\mathbb{F}_q)$ on $\mathcal{H}^{\otimes 2r}$. When we consider the representation only of $\mathbb{F}_q^{2r}$, the factor system of the representation is the square of that of the irreducible projective unitary representation $\mathsf{W}_{\mathbb{F}}^r$. As explained in the sentence before Lemma 8.4, when a projective unitary representation has the same factor system, it is isometric to the projective unitary representation $\mathsf{W}_{\mathbb{F}:2}^r$. We denote the representation space by $\mathcal{U}_{\mathbb{F}_q^r,2}$.

Next, in order to apply Theorem 2.2, we discuss $\mathrm{H}(2r,\mathbb{F}_q)$ instead of $\mathbb{F}_q^{2r}$. That is, we focus on the unitary representation $(\mathsf{W}_{\mathbb{F},\mathrm{H}}^r \rtimes \mathsf{S}_{\mathbb{F}}^r)^{\otimes 2}$ of $\mathrm{H}(2r,\mathbb{F}_q) \rtimes \mathrm{Sp}(2r,\mathbb{F}_q)$. Here, we change the phase factor of $\mathsf{S}_{\mathbb{F}}^r$ so that it is a unitary representation. Then, due to the above discussion, the irreducible unitary representation component of $\mathcal{H}^{\otimes 2r}$ with respect to the normal subgroup $\mathrm{H}(2r,\mathbb{F}_q)$ is isometric to $\mathsf{W}_{\mathbb{F},\mathrm{H}:2}^r$. Hence,

$$\mathcal{H}_s = \mathcal{U}_{\mathbb{F}_q^r,2} \otimes \mathbb{C}^{(q^r+1)/2} \tag{8.41}$$

$$\mathcal{H}_a = \mathcal{U}_{\mathbb{F}_q^r,2} \otimes \mathbb{C}^{(q^r-1)/2}. \tag{8.42}$$

Since the unitary representation $\mathsf{W}_{\mathbb{F},\mathrm{H}:2}^r$ can be constructed from the unitary representation $\mathsf{W}_{\mathbb{F},\mathrm{H}}^r$ by replacing $\omega_{\mathbb{F}}$ by $\omega_{\mathbb{F}}^2$, the associated representation of $\mathrm{Sp}(2r,\mathbb{F}_q)$ with respect to the unitary representation $\mathsf{W}_{\mathbb{F},\mathrm{H}:2}^r$ can be chosen as a unitary representation. This unitary representation is denoted by $\mathsf{S}_{\mathbb{F}:2}^r$. Since the unitary representations of $\mathrm{H}(2r,\mathbb{F}_q) \rtimes \mathrm{Sp}(2r,\mathbb{F}_q)$ on $\mathcal{H}_s$ and $\mathcal{H}_a$ are irreducible, due to Theorem 2.2, there exist irreducible unitary representations of $\mathrm{Sp}(2r,\mathbb{F}_q)$ on $\mathbb{C}^{(q^r+1)/2}$ and $\mathbb{C}^{(q^r-1)/2}$ given the right hand sides of (8.41) and (8.42). We denote these unitary representations by $\mathsf{S}_{\mathbb{F}:s}^r$ and $\mathsf{S}_{\mathbb{F}:a}^r$. So, $(\mathsf{W}_{\mathbb{F},\mathrm{H}}^r \rtimes \mathsf{S}_{\mathbb{F}}^r)^{\otimes 2}$ can be described as

$$(\mathsf{W}_{\mathbb{F},\mathrm{H}}^r \rtimes \mathsf{S}_{\mathbb{F}}^r)^{\otimes 2} = \mathsf{W}_{\mathbb{F},\mathrm{H}:2}^r \tilde{\rtimes} \mathsf{S}_{\mathbb{F}:2}^r \otimes \mathsf{S}_{\mathbb{F}:s}^r \bigoplus \mathsf{W}_{\mathbb{F},\mathrm{H}:2}^r \tilde{\rtimes} \mathsf{S}_{\mathbb{F}:2}^r \otimes \mathsf{S}_{\mathbb{F}:a}^r.$$

Next, we consider the case when $\mathbb{X} = \mathbb{Z}_d$. The order of an element of $\mathbb{Z}_d^{2r}$ is given as a divisor of d. For a divisor m of d, let N_m be the set of elements of $\mathbb{Z}_d^{2r}$ whose order is m. Then, the number of elements of N_m is given as $|N_m| = m^2 - \sum_{m'|m} |N_{m'}|$. Especially, when $d = p^k$, we have $|N_{p^l}| = p^{2l} - p^{2(l-1)} = p^{2(l-1)}(p^2-1)$.

Lemma 8.11 *The representation space of the tensor product representation between the projective unitary representation* $\mathsf{V}_{\mathbb{Z}}^r$ *of the Clifford group* $\mathbb{Z}_d^{2r} \rtimes \mathrm{Sp}(2r,\mathbb{Z}_d)$ *on* *the space $\mathcal{H}^{\otimes r}$ and its complex conjugate representation $\overline{\mathsf{V}^r_{\mathbb{Z}}}$ can be regarded as the space of liner maps from $\mathcal{H}^{\otimes r}$ to $\mathcal{H}^{\otimes r}$. Then, the irreducible subspace of this representation is given as $< \{\mathsf{W}^r_{\mathbb{Z}}(\vec{s})\}_{\vec{s}\in N_m} >$.*

Proof Any irreducible subspace of the representation of the subgroup $\mathbb{Z}_d^{2r}$ is given as the 1-dimensional subspace spanned by $\mathsf{W}^r_{\mathbb{Z}}(\vec{s})$. For any two elements $\vec{s}$ and $\vec{s}'$ of order m, there exists an element g of $\mathrm{Sp}(2r, \mathbb{Z}_d)$ such that $g(\vec{s}') = \vec{s}$. That is, $\mathsf{W}^r_{\mathbb{Z}}(\vec{s}) = \mathsf{S}^r_{\mathbb{F}}(g)\mathsf{W}^r_{\mathbb{Z}}(\vec{s}')\mathsf{S}^r_{\mathbb{F}}(g)^\dagger$. Conversely, when $\vec{s}'$ has a different order from $\vec{s}$, there is no unitary matrix U such that $\mathsf{W}^r_{\mathbb{Z}}(\vec{s}) = U\mathsf{W}^r_{\mathbb{Z}}(\vec{s}')U^\dagger$. So, we have the desired argument. ∎

Exercise 8.10 Give a CONS of $\mathcal{H}^{\otimes 1:\mathrm{even}}$ and $\mathcal{H}^{\otimes 1:\mathrm{odd}}$ when $d = 3$.

8.4 MUB and SIC Vectors*

For a CONS $B_0 := \{|0\rangle, \ldots, |d-1\rangle\}$, we define another CONS $B_1 := \{|e_0\rangle, \ldots, |e_{d-1}\rangle\}$ by (2.51) via the discrete Fourier transform DFT. Then, the condition

$$|\langle e_l|j\rangle| = \frac{1}{\sqrt{d}}, \forall l, j \in \mathbb{Z}_d \tag{8.43}$$

holds. Since any base of the CONS B_1 has a frequency of the same size based on the other CONS B_0, it seems to have no bias on the other CONS B_0. Hence, two CONSs B_0 and B_1 are called mutually unbiased to each other when they satisfy (8.43). So, they are called mutually unbiased bases (**MUB**). For example, the CONS

$$|\hat{e}_l\rangle := \frac{1}{\sqrt{d}}\sum_{k\in\mathbb{F}_q} e^{-2\pi i lk/d}|k\rangle = \mathsf{DFT}^\dagger|l\rangle$$

is given by the discrete Fourier transform for the computational basis $\{|x\rangle\}_{x=0}^{d-1}$ on $\mathbb{C}^d$. Then, these CONSs are MUB. As will be explained later, when $d+1$ bases mutually unbiased to each other, we can construct a highly symmetric POVM, which enables us to construct a useful protocol. MUB can also be applied to signal processing is given in [1].

A MUB can be constructed as follows.

Theorem 8.3 ([110]) *When the dimension d of the Hilbert space $\mathcal{H}$ is the power $d = p^m$ of the prime number p, there exist $d+1$ bases mutually unbiased to each other.*

When $d = 2$, three MUBs can be constructed in Riemann sphere as Fig. 8.3. This construction of MUB is called the **canonical MUB**. An alternative construction was given based on strictly self-orthogonal subgroup in the sense of Subsect. 8.1.4 under the same condition [14, 70]. It is known that the number of MUB is at most $d+1$ for

Fig. 8.3 MUBs in Riemann sphere

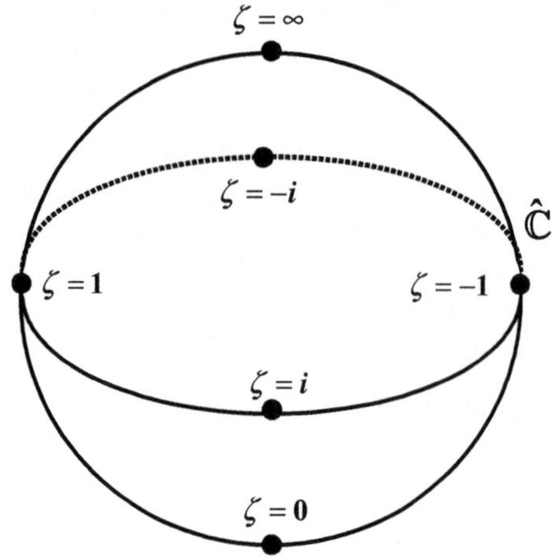

any dimension d [14]. Also, in any prime power dimension, the canonical MUB is uniquely determined by an extremal orbit of the (restricted) Clifford group except in dimension 3 [119]. On the other hand, there are several MUBs that are not unitarily equivalent to each other. Inequivalent MUBs are classified in [62]. A general review for MUB is available in [28].

Proof When $q = p^m$, we employ the discrete Heisenberg representation on the finite field $\mathbb{F}_q$ so that we can construct $d + 1$ bases mutually unbiased to each other as follows.

We can choose a pair $(s, t) \neq (0, 0)$ of elements of $\mathbb{F}_q$ such that the set $\{\mathsf{W}^r_{\mathbb{F}}(xs, xt)\}_{x \in \mathbb{F}_q}$ is composed of matrices commutative with each other. For $p = 5$, see Fig. 8.1. Then, we can choose the CONS $B_{(s,t)}$ that diagonalizes these matrices simultaneously. Here, the CONS $B_{(0,1)}$ corresponds to the computational basis, and an element of the CONS $B_{(1,0)}$ is given as $\sum_{x \in \mathbb{F}_q} \frac{1}{q^{1/2}} \omega_{\mathbb{F}}^{trxt} |x\rangle$. So, they are mutually unbiased with each other.

Generally, for two pairs $(s, t), (s', t') \in \mathbb{F}_q^2 \setminus (0, 0)$, we say that (s, t) is equivalent to (s', t'), i.e., $(s, t) \sim (s', t')$ when there exists a scalar $x \in \mathbb{F}_q$ such that $(s, t) = (xs', xt')$. When two pairs $(s, t), (s', t') \in \mathbb{F}_q^2$ are not equivalent to each other, there exists an element $g \in \mathrm{Sp}(2, \mathbb{F}_q)$ such that $g(s, t) = (0, 1), g(s', t') = (1, 0)$.

Since the unitary matrix $\mathsf{S}^r_{\mathbb{F}}(g)$ maps the bases $B_{(s,t)}$ and $B_{(s',t')}$ to the bases $B_{(0,1)}$ and $B_{(1,0)}$, the bases $B_{(s,t)}$ and $B_{(s',t')}$ are mutually unbiased to each other. Further, because $|(\mathbb{F}_q^2 \setminus (0, 0))/ \sim | = q + 1$, we obtain the desired argument. ∎

MUB is a 2-design as follows.

Theorem 8.4 *When $d + 1$ bases $B_1, \cdots, B_{d+1}$ are mutually unbiased to each other, the set $B_1 \cup \cdots \cup B_{d+1}$ is a $(1, 1)$-design. So, it is also a 2-design.*

Further, it is known that the canonical MUB is the unique minimal 2-design covariant with respect to the Clifford group except in dimension 3 [119].

Proof We denote the basis of B_j by $u_{1,j},\ldots,u_{d,j}$, and the vector $|\frac{1}{\sqrt{d}}I\rangle\rangle$ on the composite system $\mathbb{C}^d\otimes\mathbb{C}^d$ by Φ. Then, the subspace $\mathcal{B}'_j =< u_{1,j}\otimes\overline{u_{1,j}}-\frac{1}{\sqrt{d}}\Phi,\ldots,u_{d-1,j}\otimes\overline{u_{d-1,j}}-\frac{1}{\sqrt{d}}\Phi >$ of the composite system $\mathbb{C}^d\otimes\mathbb{C}^d$ is a subspace of the orthogonal complement to Φ, and has the dimension $d-1$. Since

$$\langle u_{i,j}\otimes\overline{u_{i,j}}-\frac{1}{\sqrt{d}}\Phi|u_{i',j'}\otimes\overline{u_{i',j'}}-\frac{1}{\sqrt{d}}\Phi\rangle=0,\quad j\neq j',$$

the subspaces $\mathcal{B}'_1,\ldots,\mathcal{B}'_{d+1}$ are orthogonal to each other. Because the orthogonal complement to Φ has the dimension d^2-1, the subset $\mathcal{B}'_1\cup\ldots\cup\mathcal{B}'_{d+1}$ spans the orthogonal complement to Φ. Hence, an arbitrary vector on the composite system $\mathbb{C}^d\otimes\mathbb{C}^d$ can be written as $\sum_{j=1}^{d+1}\sum_{i=1}^{d}a_{i,j}u_{i,j}\otimes\overline{u_{i,j}}$.

Since

$$\left\|\sum_{j,i}a_{i,j}u_{i,j}\otimes\overline{u_{i,j}}\right\|^2=\frac{1}{d}\left|\sum_{j,i}a_{i,j}\right|^2-\frac{1}{d}\sum_j|\sum_i a_{i,j}|^2+\sum_{j,i}|a_{i,j}|^2$$

$$\left|\left\langle\phi^0_{A,B}\left|\sum_{j,i}a_{i,j}u_{i,j}\otimes\overline{u_{i,j}}\right.\right\rangle\right|^2=\frac{1}{d}\left|\sum_{j,i}a_{i,j}\right|^2,$$

we have

$$\begin{aligned}
&\left\langle\sum_{j,i}a_{i,j}u_{i,j}\otimes\overline{u_{i,j}}\right|\left(\sum_{l,k}\frac{1}{d+1}|u_{k,l}\otimes\overline{u_{k,l}}\rangle\langle u_{k,l}\otimes\overline{u_{k,l}}|\right)\\
&\left|\sum_{j',i'}a_{i',j'}u_{i',j'}\otimes\overline{u_{i',j'}}\right\rangle\\
&=\frac{1}{d+1}\sum_{l,k}\left|\sum_{j,i}\langle u_{k,l}\otimes\overline{u_{k,l}}||a_{i,j}u_{i,j}\otimes\overline{u_{i,j}}\rangle\right|^2\\
&=\frac{1}{d}\left|\sum_{j,i}a_{i,j}\right|^2-\frac{1}{d(d+1)}\sum_j|\sum_i a_{i,j}|^2+\frac{1}{d+1}\sum_{j,i}|a_{i,j}|^2\\
&=\left\langle\sum_{j,i}a_{i,j}u_{i,j}\otimes\overline{u_{i,j}}\right|\frac{d}{d+1}|\Phi\rangle\langle\Phi|+\frac{1}{d+1}I\left|\sum_{j',i'}a_{i',j'}u_{i',j'}\otimes\overline{u_{i',j'}}\right\rangle.
\end{aligned}$$

Thus, we obtain $\sum_{l,k}\frac{1}{d+1}|u_{k,l}\otimes\overline{u_{k,l}}\rangle\langle u_{k,l}\otimes\overline{u_{k,l}}|=\frac{d}{d+1}|\Phi\rangle\langle\Phi|+\frac{1}{d+1}I$, which implies that $B_1\cup\cdots\cup B_{d+1}$ is a (1, 1)-design. Therefore, Theorem 4.6 guarantees that $B_1\cup\cdots\cup B_{d+1}$ is a (2, 0)-design. ■

Fig. 8.4 SIC vectors in Riemann sphere

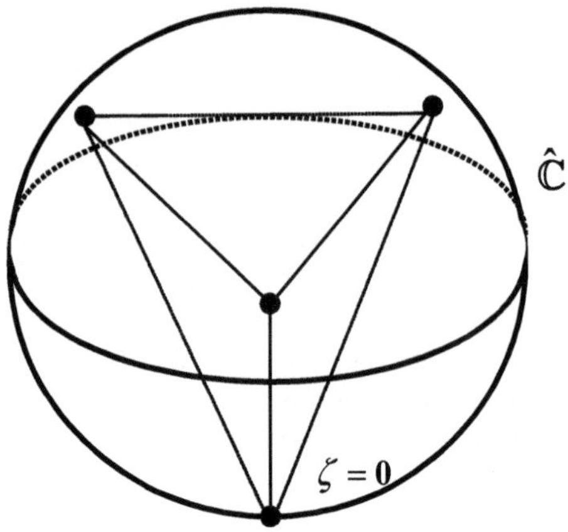

Next, we focus on d^2 vectors $\{|s_j\rangle\}_{j=0}^{d^2-1}$ satisfying the symmetry $|\langle s_l|s_j\rangle| = \frac{1}{d+1}$, where $j \neq l$. Such vectors are called symmetric informationally complete (**SIC**) vectors. When $d = 2$, SIC vectors can be constructed from the vertexes of tetrahedron on Riemann sphere by vectors $\Big\{|1\rangle, \sqrt{\frac{2}{3}}|0\rangle + \sqrt{\frac{1}{3}}|1\rangle, \sqrt{\frac{2}{3}}|0\rangle + e^{2\pi i/3}\sqrt{\frac{1}{3}}|1\rangle, \sqrt{\frac{2}{3}}|0\rangle$ $+e^{4\pi i/3}\sqrt{\frac{1}{3}}|1\rangle\Big\}$ (see Fig. 8.4).

SIC vectors satisfy the following properties.

Theorem 8.5 *SIC vectors* $\{|s_j\rangle\}_{j=0}^{d^2-1}$ *form a POVM* $\frac{1}{d}\sum_{j=0}^{d^2-1}|s_j\rangle\langle s_j| = I$, *and form a* $(1, 1)$*-design, i.e., a* $(2, 0)$*-design.*

The literatures [1, 90, 111, 112] are basic for SIC vectors. The existence of SIC vectors has been analytically shown when the dimension d is one of integers from 2 to 16, or one of 19, 24, 28, 31, 35, 37, 43, and 48 [7, 97]. Also, when $d \leq 67$, SIC vectors has been approximately given by a numerical analysis [42, 90, 97]. Since, in fact, SIC vectors are closely related to Lie algebra, Lie groups [5, 7, 115] and Galois field [6] and has interesting symmetric properties [4, 113, 114, 120], they are interesting mathematical objects. Due to the symmetry, SIC can be applied to state estimation [95, 121, 115]. It was also applied to Bayesian statistics with quantum setting [33].

Further, even when t is larger than 2, a construction of a $(t, 0)$-design has been discussed based on SIC vectors [16, 95, 17, 42, 56, 97]. Also, SIC vectors can be applied to equiangular lines, which are collections of vectors with only one angle between them [25, 36, 41, 64, 71, 93, 103].

Proof Firstly, we show that the vectors $s_1 \otimes \overline{s_1}, \ldots, s_{d^2} \otimes \overline{s_{d^2}}$ are linearly independent on $\mathbb{C}^d \otimes \mathbb{C}^d$. For this purpose, we choose complex numbers $a_1, \ldots, a_{d^2}$ such that $\sum_i a_i s_i \otimes \overline{s_i} = 0$, which implies that

$$0 = \sqrt{d}\langle \Phi | \sum_i a_i s_i \otimes \overline{s_i} \rangle = a_1 + \sum_{i \neq 1} a_i$$

$$0 = \langle u_1 \otimes \overline{u_1} | \sum_i a_i u_i \otimes \overline{u_i} \rangle = a_1 + \frac{1}{d+1} \sum_{i \neq 1} a_i .$$

These two equations show that $a_1 = 0$. Similarly, we have $a_i = 0$. So, $s_1 \otimes \overline{s_1}, \ldots, s_{d^2} \otimes \overline{s_{d^2}}$ are linearly independent. Since the composite system $\mathbb{C}^d \otimes \mathbb{C}^d$ has the dimension d^2, any vector on $\mathbb{C}^d \otimes \mathbb{C}^d$ is written as $\sum_j a_i u_i \otimes \overline{u_i}$.

Because

$$\left\| \sum_j a_j s_j \otimes \overline{s_j} \right\|^2 = \frac{1}{d+1} \left| \sum_k a_k \right|^2 + \frac{d}{d+1} \sum_k |a_k|^2$$

$$\left| \langle \phi^0_{A,B} | \sum_j a_j s_j \otimes \overline{s_j} \rangle \right|^2 = \frac{1}{d} \left| \sum_j a_j \right|^2 ,$$

we have

$$\begin{aligned} &\left\langle \sum_i a_i s_i \otimes \overline{s_i} \right| \left(\sum_k \frac{1}{d} | s_k \otimes \overline{s_k} \rangle \langle s_k \otimes \overline{s_k} | \right) \left| \sum_j a_j s_j \otimes \overline{s_j} \right\rangle \\ =& \frac{d+2}{(d+1)^2} \left| \sum_k a_k \right|^2 + \frac{d}{(d+1)^2} \sum_k |a_k|^2 \\ =& \left\langle \sum_i a_i u_i \otimes \overline{s_i} \right| \left(\frac{d}{d+1} |\Phi\rangle\langle\Phi| + \frac{1}{d+1} I \right) \left| \sum_j a_j s_j \otimes \overline{s_j} \right\rangle , \end{aligned}$$

which implies that

$$\sum_k \frac{1}{d} |s_k \otimes \overline{s_k}\rangle\langle s_k \otimes \overline{s_k}| = \frac{d}{d+1} |\Phi\rangle\langle\Phi| + \frac{1}{d+1} I. \tag{8.44}$$

That is, $\{|s_j\rangle\}_{j=0}^{d^2-1}$ is a (1, 1)-design. Further, Theorem 4.6 guarantees that $\{|s_j\rangle\}_{j=0}^{d^2-1}$ is a (2, 0)-design. Taking the partial trace in (8.44), we obtain $\sum_k \frac{1}{d} |s_k\rangle\langle s_k| = I$, which shows that $\{\frac{1}{d} |s_k\rangle\langle s_k|\}$ is a POVM on $\mathbb{C}^d$. ∎

Exercise 8.11 Show that vectors $\left\{ |1\rangle, \sqrt{\frac{2}{3}}|0\rangle + \sqrt{\frac{1}{3}}|1\rangle, \sqrt{\frac{2}{3}}|0\rangle + e^{2\pi i/3}\sqrt{\frac{1}{3}}|1\rangle, \sqrt{\frac{2}{3}}|0\rangle + e^{4\pi i/3}\sqrt{\frac{1}{3}}|1\rangle \right\}$ form SIC vectors.

Appendix A
Solutions of Exercises

Exercise 1.1
$P^{\rho}_{M}(1) = \frac{5}{16}$, $P^{\rho}_{M}(2) = \frac{5}{16}$, and $P^{\rho}_{M}(3) = \frac{3}{8}$.

Exercise 1.2
$A = E_1 - E_2$ and $B = F_1 - F_2$, where $E_1 = \frac{1}{2}\begin{pmatrix} 1 & i \\ -i & 1 \end{pmatrix}$, $E_2 = \frac{1}{2}\begin{pmatrix} 1 & -i \\ i & 1 \end{pmatrix}$, $F_1 = \frac{1}{2}\begin{pmatrix} 1 & 1 \\ 1 & 1 \end{pmatrix}$ and $F_2 = \frac{1}{2}\begin{pmatrix} 1 & -1 \\ 1 & 1 \end{pmatrix}$.

Exercise 1.3
The expectation can be calculated in two ways. (1) $1 \times \frac{1}{2} + (-1) \times \frac{1}{2} = 0$. (2) $\operatorname{Tr} \rho A = 0$.

Exercise 1.4
The variance $\Delta_\rho A$ can be calculated in two ways. (1) $1^2 \times \frac{1}{2} + (-1)^2 \times \frac{1}{2} = 1$. (2) $\operatorname{Tr} \rho A^2 = 1$.

Exercise 1.5
Since $\operatorname{Tr} B\rho = 0$, the value $\Delta_\rho A \circ B$ is $\operatorname{Tr} \frac{1}{2}(AB + BA)\rho = \operatorname{Tr} \frac{1}{2}\left(\begin{pmatrix} i & 0 \\ 0 & -i \end{pmatrix} + \begin{pmatrix} -i & 0 \\ 0 & i \end{pmatrix}\right)\rho = 0$.

Exercise 1.6
The state $|X\rangle\rangle_{A,B}\ {}_{A,B}\langle\langle X|$ is entangled if and only if the vector $|X\rangle\rangle_{A,B}$ is in a tensor product form. This condition is equivalent to $\det X \neq 0$.

Exercise 1.7
The orthogonality can be shown by using (1.16).

M. Hayashi, *Group Representation for Quantum Theory*,
DOI 10.1007/978-3-319-44906-7

Exercise 1.8
The spectral decomposition of $\rho^{\otimes n}$ is $\rho^{\otimes n} = \sum_{k=0}^{n} \frac{2^k}{3^n} F_k$, where $F_k := \sum_{i_1,i_2,\ldots,i_n:i_1+i_2+\cdots+i_n=k} E_{i_1} \otimes E_{i_2} \otimes \cdots \otimes E_{i_n}$.

Exercise 1.9
Since $\lim_{T\to\infty} \frac{1}{T} \int_0^T e^{it(h_j-h_{j'})} dt = 0$ for $h_j \neq h_{j'}$, we have

$$\lim_{T\to\infty} \frac{1}{T} \int_0^T e^{itH} \rho e^{-itH} dt = \lim_{T\to\infty} \frac{1}{T} \int_0^T e^{it(h_j-h_{j'})} dt E_j \rho E_{j'} = \sum_j E_j \rho E_j.$$

Exercise 1.10
The PVM $\{E_i \otimes F_j\}_{i,j}$ gives the simultaneous diagonalization of $A \otimes I$ and $I \otimes B$, where E_i and F_j are given in Exercise 1.2.

Exercise 1.11
For simplicity, we show the case with $d = 1$. Given two non-negative integers $\alpha, \beta \in \mathbb{Z}_+$, we assume that $\beta \geq 1$. Then, we have

$$\begin{aligned}
&\sup_x \left|x^\alpha \frac{d^\beta}{dx^\beta} x f(x)\right| = \sup_x \left|x^\alpha \frac{d^{\beta-1}}{dx^{\beta-1}} f(x) + x^{\alpha+1} \frac{d^\beta}{dx^\beta} f(x)\right| \\
\leq &\sup_x \left|x^\alpha \frac{d^{\beta-1}}{dx^{\beta-1}} f(x)\right| + \sup_x \left|x^{\alpha+1} \frac{d^\beta}{dx^\beta} f(x)\right| < \infty.
\end{aligned}$$

When $\beta = 0$,

$$\sup_x \left|x^\alpha \frac{d^\beta}{dx^\beta} x f(x)\right| = \sup_x \left|x^{\alpha+1} f(x)\right| < \infty.$$

So, he function $xf(x)$ belongs to $\mathscr{S}(\mathbb{R}^d)$.

For a given $\alpha, \beta \in \mathbb{Z}_+$, we assume that $\beta \geq 1$. Then, we have $\sup_x \left|x^\alpha \frac{d^\beta}{dx^\beta} \frac{d}{dx} f(x)\right| = \sup_x \left|x^\alpha \frac{d^{\beta+1}}{dx^{\beta+1}} f(x)\right| < \infty$. So, the function $\frac{d}{dx} f(x)$ belongs to $\mathscr{S}(\mathbb{R}^d)$.

Exercise 1.12
For simplicity, we show the case with $d = 1$. Given two non-negative integers $\alpha, \beta \in \mathbb{Z}_+$, we assume that $\beta \geq 1$. Then, we have

$$\begin{aligned}
&\sup_x \left|x^\alpha \frac{d^\beta}{dx^\beta} g(x) f(x)\right| = \sup_x \left|x^\alpha \sum_{k=0}^{\beta} \binom{\beta}{k} \frac{d^k}{dx^k} g(x) \frac{d^{\beta-k}}{dx^{\beta-k}} f(x)\right| \\
\leq &\sum_{k=0}^{\beta} \binom{\beta}{k} \sup_x \left|x^\alpha \frac{d^k}{dx^k} g(x)\right| \cdot \left|\frac{d^{\beta-k}}{dx^{\beta-k}} f(x)\right| < \infty.
\end{aligned}$$

So, the product $f(x)g(x)$ belongs to $\mathscr{S}(\mathbb{R}^d)$.

Exercise 2.1
Let (a, b, c) be the cyclic permutation among a, b, c. Then, the permutation group S_3 is composed of (1, 2, 3), (3, 1, 2), (2, 3, 1), (1, 2), (2, 3), and (3, 1). The orders of (1, 2, 3), (3, 1, 2), (2, 3, 1) are 3. The orders of (1, 2), (2, 3), (3, 1) are 2.

Exercise 2.2
This statement can be shown by the following relation. $(1, 2)(3, 4)(1, 3)(2, 4) = (1, 3)(2, 4)(1, 2)(3, 4) = (1, 4)(2, 3)$, $(1, 3)(2, 4)(1, 4)(2, 3) = (1, 4)(2, 3)(1, 3)(2, 4) = (1, 2)(3, 4)$, $(1, 4)(2, 3)(1, 2)(3, 4) = (1, 2)(3, 4)(1, 4)(2, 3) = (1, 3)(2, 4)$, $((1, 2)(3, 4))^2 = ((1, 3)(2, 4))^2 = ((1, 4)(2, 3))^2 = e$.

Exercise 2.3
The following map gives the isometric relation. $e \mapsto e$, $(1, 2)(3, 4) \mapsto i$, $(1, 3)(2, 4) \mapsto j$, $(1, 4)(2, 3) \mapsto k$.

Exercise 2.4
It is enough to show that $(a, b)\{e, (1, 2)(3, 4), (1, 3)(2, 4), (1, 4)(2, 3)\}(a, b) = \{e, (1, 2)(3, 4), (1, 3)(2, 4), (1, 4)(2, 3)\}$ because any permutation is given as the products of transpositions. Consider the case of $(a, b) = (1, 2)$ because other cases can be shown in the same way. $(1, 2)(1, 2)(3, 4)(1, 2) = (1, 2)(1, 2)(1, 2)(3, 4) = (1, 2)(3, 4)$. $(1, 2)(1, 3)(2, 4)(1, 2) = (2, 3)(1, 2)(1, 2)(1, 4) = (2, 3)(1, 4) = (1, 4)(2, 3)$. $(1, 2)(1, 4)(2, 3)(1, 2) = (2, 4)(1, 2)(1, 2)(1, 3) = (2, 4)(1, 3) = (1, 3)(2, 4)$.

Exercise 2.5
Consider the subgroup S_3 of S_4, which permutes the letters $\{1, 2, 3\}$. For any different element $g \neq e \in S_3$, the set $g(\mathbb{Z}_2 \times \mathbb{Z}_2)$ is different from $\mathbb{Z}_2 \times \mathbb{Z}_2$. Hence, for any different elements $g_1 \neq g_2 \in S_3$, the set $g_1(\mathbb{Z}_2 \times \mathbb{Z}_2)$ is different from $g_2(\mathbb{Z}_2 \times \mathbb{Z}_2)$. So, each element of S_3 gives different residue class. Since the order of S_3 is 6 and the order of $S_4/\mathbb{Z}_2 \times \mathbb{Z}_2$ is $24/4 = 6$, all of elements of $S_4/\mathbb{Z}_2 \times \mathbb{Z}_2$ can be written by S_3. Since they have the same products, the quotient group $S_4/\mathbb{Z}_2 \times \mathbb{Z}_2$ is isometric to S_3.

Exercise 2.6
It is trivial that the subset $\{g \in G|T(g)\theta_0 = \theta_0\}$ contains the identity element. When $g_1, g_2 \in \{g \in G|T(g)\theta_0 = \theta_0\}$, we have $T(g_1g_2)\theta_0 = T(g_1)T(g_2)\theta_0T(g_1)\theta_0 = \theta_0$. Hence, $\{g \in G|T(g)\theta_0 = \theta_0\}$ is a subgroup.

Exercise 2.7
Firstly, we fix a point θ_0 in Θ. For an element $\theta \in \Theta$, we choose $g_\theta \in G$ such that $g_\theta \cdot \theta_0 = \theta$. Then, $\theta \mapsto [g_\theta] \in G/H$ is an isomorphism for homogeneous spaces.

Exercise 2.8
Let X be the set $\{g^{-1}h^{-1}gh|g, h \in G\}$ of generators of the commutator subgroup $[G, G]$. For elements $z \in G$ and $x, y \in G$, we have $z(xyx^{-1}y^{-1})z^{-1} = (zxz^{-1})(zyz^{-1})(zxz^{-1})^{-1}(zyz^{-1})^{-1} \in X$. That is, $zXz^{-1} \subset X$. Since X generates $[G, G]$, we have $z[G, G]z^{-1} \subset [G, G]$. Especially, considering the case when z is an element of $[G, G]$, we find that $[G, G]$ is a normal subgroup.

Exercise 2.9
The desired statement follows from the calculation given in Exercise 2.4.

Exercise 2.10
Since

$$\begin{aligned}
&((h,k)(h',k'))(h'',k'') = (hT(k)(h')\phi(k,k'), kk')(h'',k'')\\
=&(hT(k)(h')\phi(k,k')T(kk')(h'')\phi(kk',k''), kk'k''),\\
&(h,k)((h',k')(h'',k'')) = (h,k)(h'T(k')(h'')\phi(k',k''), k'k'')\\
=&(hT(k)(h'T(k')(h'')\phi(k',k''))\phi(k,k'k''), kk'k''),
\end{aligned}$$

it is sufficient to show that

$$\begin{aligned}
&hT(k)(h')\phi(k,k')T(kk')(h'')\phi(kk',k'')\\
=&hT(k)(h'T(k')(h'')\phi(k',k''))\phi(k,k'k'').
\end{aligned} \tag{A.1}$$

This equation can be shown as follows.

$$\begin{aligned}
&T(k)(h'T(k')(h'')\phi(k',k''))\phi(k,k'k'')\\
=&T(k)(h')T(k)(T(k')(h''))T(k)(\phi(k',k''))\phi(k,k'k'')\\
\stackrel{(a)}{=}&T(k)(h')\phi(k,k')T(kk')(h'')\phi(k,k')^{-1}\phi(k,k')\phi(kk',k'')\phi(k,k'k'')^{-1}\\
&\cdot\phi(k,k'k'')\\
=&T(k)(h')\phi(k,k')T(kk')(h'')\phi(kk',k''),
\end{aligned} \tag{A.2}$$

where (a) follows from (2.7) and (2.8).

Exercise 2.11
$hk = (h,e)(e,k) = (h,k)$. $khk^{-1} = (e,k)(h,e)(e,k^{-1}) = (e,k)(h,k^{-1}) = (T(k)h, kk^{-1}) = (T(k)h, e) = T(k)(h)$.

Exercise 2.12
The desired argument can be checked by the relations $(h,k)(h',e)(h,k)^{-1} = (h,k)(h',e)(h^{-1},k^{-1}) = (h,k)(h'h^{-1},k^{-1}) = (h, T(k)h'h^{-1}, kk^{-1}) = (h, T(k)h'h^{-1}, e) \in H$.

Exercise 2.13
When $T(k)$ is the identity map, the discussions in Sect. 2.2.2 give the central extension.

Conversely, we assume that the given extension is a central extension. Since any factor system can be replaced by a normalized factor system, we treat only the case with a normalized factor system. We consider the case with $(h', e) = (e, k)(h', e)(e, k)^{-1} = (e, k)(h', k^{-1}) = (T(k)h'\phi(k, k^{-}), kk^{-1}) = (T(k)h'\phi(k, k^{-}), e)$ $h' = e$. So, we obtain $\phi(k, k^{-}) = e$. Hence, we have $T(k)h' = h'$, which implies that $T(k)$ is the identity map.

Exercise 2.14
The desired argument can be shown by focusing on the simultaneous diagonalization.

Exercise 2.15
When $\mathfrak{f}_\lambda$ is a self-conjugate representation, we have $\overline{\mathfrak{f}_\lambda(g)} = \mathfrak{f}_{\lambda^*}(g) = \mathfrak{f}_\lambda(g)$. So, the character takes real numbers.

Exercise 2.16
The representation is complexifiable if and only if it is commutative with the multiplication of the imaginary number i, which is equivalent to (2.44).

Exercise 2.17
It is enough to consider the case when $j = 2$ and $i = 1$ in (2.72). So, (2.72) implies that

$$
\begin{aligned}
d_{S_n,(\frac{n+k}{2},\frac{n-k}{2})} =& \frac{n!}{(\frac{n+k}{2}+1)!\frac{n-k}{2}!}\Big(\frac{n+k}{2} - \frac{n-k}{2} + 2 - 1\Big) \\
=& \frac{(n+1)!}{(n+1)(\frac{n+k}{2}+1)!\frac{n-k}{2}!}(k+1) = \frac{k+1}{n+1}\binom{n+1}{\frac{n+k}{2}+1}.
\end{aligned}
$$

Exercise 3.1
Any element g of SO(2, $\mathbb{R}$) is written as $\begin{pmatrix} \cos\theta & \sin\theta \\ -\sin\theta & -\cos\theta \end{pmatrix}$. So, any element connected to the connected components including I. Thus, SO(2, $\mathbb{R}$) is connected.

Exercise 3.2
Since $gg^\dagger = g^\dagger g$, there are d eigenvectors $v_1, \ldots, v_d$ of g. Since g preserves the inner product, the eigenvalue λ_j corresponding to v_j has the absolute value 1. Taking the complex conjugate of the eigenvalue equation $\lambda_j v_j = g v_j$, we have $\overline{\lambda_j} v_j^* = g v_j^*$. This fact shows that the number of non-real eigenvectors is even, which is denoted by $2r$. We sort the eigenvectors v_i such that initial $2r$ eigenvectors are non-real, and v_{2j} is the complex conjugate of v_{2j-1}. Now, we define two vectors $w_{2j-1} := \frac{1}{\sqrt{2}}(v_{2j-1} + v_{2j})$ and $w_{2j} := \frac{i}{\sqrt{2}}(v_{2j-1} - v_{2j})$ for $j = 1, \ldots, r$. For

$j \geq 2r+1$, we define $w_j := v_j$. Since $|\lambda_j| = 1$, λ_j is written as $\cos\theta + i\sin\theta$. So, we have $gw_{2j-1} = \cos\theta w_{2j-1} + \sin\theta w_{2j}$ and $gw_{2j} = -\sin\theta w_{2j-1} + \cos\theta w_{2j}$ for $j = 1, \ldots, r$. Arranging the d vectors $w_1, \ldots, w_d$, we define orthogonal matrix g'. Then, $g'gg'^{-1}$ satisfies the required condition.

Exercise 3.3
Due to Exercises 3.1 and 3.2, Any element $g \in \mathrm{SO}(d, \mathbb{R})$ is connected to the connected components including I. So, $\mathrm{SO}(d, \mathbb{R})$ is connected.

Exercise 3.4
We choose one-dimensional differentiable subsets $g_i(t)$ of G_i. Then, $g_1(t)$ and $g_2(t)$ are commutative with each other as elements of $G_1 \times G_2$. This fact shows that $\mathfrak{g}(G_1) \oplus \mathfrak{g}(G_2) = \mathfrak{g}(G_1 \times G_2)$.

Exercise 3.5
The desired argument can be checked by expanding both sides of (3.20) and (3.21) up to the order t^2.

Exercise 3.6
It is sufficient to show that (3.41) satisfies Jacobi law, which can be shown as follows.

$$
\begin{aligned}
&[(X_1, X_2), [(Y_1, Y_2), (Z_1, Z_2)]] + [(Y_1, Y_2), [(Z_1, Z_2), (X_1, X_2)]] \\
=&[(X_1, X_2), (T(Y_2)Z_1 - T(Z_2)Y_1 + [Y_1, Z_1], [Y_2, Z_2])] \\
&+ [(Y_1, Y_2), (T(Z_2)X_1 - T(X_2)Z_1 + [Z_1, X_1], [Z_2, X_2])] \\
=&(T(X_2)(T(Y_2)Z_1 - T(Z_2)Y_1 + [Y_1, Z_1]) - T([Y_2, Z_2])X_1 \\
&+ [X_1, (T(Y_2)Z_1 - T(Z_2)Y_1 + [Y_1, Z_1]], [X_2, [Y_2, Z_2]]) \\
&+ (T(Y_2)(T(Z_2)X_1 - T(X_2)Z_1 + [Z_1, X_1]) - T([Z_2, X_2])Y_1 \\
&+ [Y_1, (T(Z_2)X_1 - T(X_2)Z_1 + [Z_1, X_1]], [Y_2, [Z_2, X_2]]) \\
=&(T(X_2)(T(Y_2)Z_1 - T(Z_2)Y_1 + [Y_1, Z_1]) - T([Y_2, Z_2])X_1 \\
&+ T(Y_2)(T(Z_2)X_1 - T(X_2)Z_1 + [Z_1, X_1]) - T([Z_2, X_2])Y_1 \\
&+ [X_1, (T(Y_2)Z_1 - T(Z_2)Y_1 + [Y_1, Z_1]] \\
&+ [Y_1, (T(Z_2)X_1 - T(X_2)Z_1 + [Z_1, X_1]], \\
&[X_2, [Y_2, Z_2]] + [Y_2, [Z_2, X_2]]) \\
=&([T(X_2), T(Y_2)]Z_1 + T(Z_2)T(Y_2)X_1 - T(Z_2)T(X_2)Y_1 \\
&+ T(Z_2)[X_1, Y_1] + [Z_1, (T(X_2)Y_1 - T(Y_2)X_1] + [Z_1, [X_1, Y_1]], \\
&[Z_2, [X_2, Y_2]]) \\
=&[(Z_1, Z_2), (T(X_2)Y_1 - T(Y_2)X_1 + [X_1, Y_1], [X_2, Y_2])] \\
=&[(Z_1, Z_2), [(X_1, X_2), (Y_1, Y_2)]].
\end{aligned}
$$

Exercise 3.7

$$
\begin{aligned}
&[\mathfrak{f}((X_1, X_2)), \mathfrak{f}((Y_1, Y_2))] \\
=&(\mathfrak{f}_1(X_1) + \mathfrak{f}_2(Y_1))(\mathfrak{f}_1(X_2) + \mathfrak{f}_2(Y_2)) - (\mathfrak{f}_1(X_2) + \mathfrak{f}_2(Y_2))(\mathfrak{f}_1(X_1) + \mathfrak{f}_2(Y_1)) \\
=&[\mathfrak{f}_1(X_1), \mathfrak{f}_1(Y_1)] + [\mathfrak{f}_2(X_2), \mathfrak{f}_2(Y_2)] + [\mathfrak{f}_1(X_1), \mathfrak{f}_2(Y_2)] + [\mathfrak{f}_2(X_2), \mathfrak{f}_1(Y_1)] \\
=&\mathfrak{f}_1([X_1, Y_1]) + \mathfrak{f}_2([X_2, Y_2]) - \mathfrak{f}_1(T(Y_2)X_1) + \mathfrak{f}_1(T(X_2)Y_1) \\
=&\mathfrak{f}([(X_1, X_2), (Y_2, Y_2)]).
\end{aligned}
$$

Exercise 3.8
Property (3.44) shows that the matrix $\mathsf{ad}(Y)$ is represented as an alternative matrix. Hence, $\exp(\mathsf{ad}(Y))$ is an orthogonal matrix. When we choose Y as $g = \exp(Y)$, we have $(\mathsf{ad}(g)X, \mathsf{ad}(g)Z)_{\mathfrak{g}} = (X, Z)_{\mathfrak{g}}$.

Exercise 3.9
Since

$$
\mathcal{F}_{\mathrm{U}(1)} \circ \mathcal{F}_{\mathrm{U}(1)}^{-1}[\{a_n\}](n') = \frac{1}{2\pi}\int_0^{2\pi} \sum_{n\in\mathbb{Z}} a_n e^{i\theta n} e^{-i\theta n'} d\theta = \sum_{n\in\mathbb{Z}} a_n \delta_{n,n'} = a_n,
$$

the map $\mathcal{F}_{\mathrm{U}(1)} \circ \mathcal{F}_{\mathrm{U}(1)}^{-1}$ is the identity map.

For any function $\phi(\neq 0) \in \mathrm{Ł}^2(\mathrm{U}(1))$, $\mathcal{F}_{\mathrm{U}(1)}[\phi]$ is not zero. So, the kernel of $\mathcal{F}_{\mathrm{U}(1)}$ is $\{0\}$. Thus, the map $\mathcal{F}_{\mathrm{U}(1)}^{-1} \circ \mathcal{F}_{\mathrm{U}(1)}$ is also the identity map.

Exercise 3.10

$$
\begin{aligned}
\|\mathcal{F}_{\mathrm{U}(1)}^{-1}[\{a_n\}]\|^2 &= \frac{1}{2\pi}\int_0^{2\pi} \sum_{n,m} \overline{a_n} a_m e^{-in\theta} e^{im\theta} d\theta \\
=\sum_{n,m} \overline{a_n} a_m \frac{1}{2\pi}\int_0^{2\pi} & e^{i(m-n)\theta} d\theta = \sum_{n,m} \overline{a_n} a_m \delta_{n,m} = \|\{a_n\}\|^2.
\end{aligned}
$$

Thus, $\mathcal{F}_{\mathrm{U}(1)}^{-1}$ is a unitary map. Since $\mathcal{F}_{\mathrm{U}(1)}^{-1} \circ \mathcal{F}_{\mathrm{U}(1)}$ is the identity map, $\mathcal{F}_{\mathrm{U}(1)}$ is a unitary map.

Exercise 3.11

$$
\begin{aligned}
&-i\mathcal{F}_d[\frac{df}{dx_j}](\lambda) = -i\frac{1}{(2\pi)^{\frac{d}{2}}}\int_{\mathbb{R}^d}\frac{df}{dx_j}\phi(x)e^{-ix\cdot\lambda}dx\\
&=-\frac{i}{(2\pi)^{\frac{d}{2}}}\int_{\mathbb{R}^{d-1}}[\phi(x)e^{-ix\cdot\lambda}]_{x_j=-\infty}^{\infty}dx_1\ldots dx_{j-1}dx_{j+1}\ldots dx_d\\
&+\frac{i}{(2\pi)^{\frac{d}{2}}}\int_{\mathbb{R}^{d-1}}\phi(x)\frac{df}{dx_j}e^{-ix\cdot\lambda}dx\\
&=\frac{1}{(2\pi)^{\frac{d}{2}}}\int_{\mathbb{R}^{d-1}}\phi(x)\lambda_j e^{-ix\cdot\lambda}dx = \lambda_j\mathcal{F}_d[f](\lambda).
\end{aligned}
$$

Exercise 3.12

$$
\begin{aligned}
&\mathcal{F}_d^{-1}[\phi]^*(\lambda) = \frac{1}{(2\pi)^{\frac{d}{2}}}\left(\int_{\mathbb{R}^d}\phi(x)e^{ix\cdot\lambda}dx\right)^*\\
&=\frac{1}{(2\pi)^{\frac{d}{2}}}\int_{\mathbb{R}^d}\phi(x)^*e^{-ix\cdot\lambda}dx = \mathcal{F}_d[\phi^*](\lambda).
\end{aligned}
$$

Exercise 3.13

$$
\begin{aligned}
&\mathcal{F}_d[\phi](-\lambda) = \frac{1}{(2\pi)^{\frac{d}{2}}}\left(\int_{\mathbb{R}^d}\phi(x)e^{-ix\cdot(-\lambda)}dx\right)^*\\
&=\frac{1}{(2\pi)^{\frac{d}{2}}}\left(\int_{\mathbb{R}^d}\phi(x)e^{ix\cdot\lambda}dx\right)^* = \mathcal{F}_d^{-1}[\phi].
\end{aligned}
$$

Exercise 3.14
For simplicity, we show the case with $d = 1$. Using Exercise 3.11, we have

$$
\begin{aligned}
&\sup_\lambda\left|\lambda^\alpha\frac{d^\beta}{d\lambda^\beta}\mathcal{F}[f](\lambda)\right| = \sup_\lambda\left|\lambda^\alpha\frac{d^\beta}{d\lambda^\beta}\frac{1}{\sqrt{2\pi}}\int_{\mathbb{R}}\phi(x)e^{-ix\lambda}dx\right|\\
&=\sup_\lambda\left|\lambda^\alpha\frac{1}{\sqrt{2\pi}}\int_{\mathbb{R}}\frac{d^\beta}{d\lambda^\beta}\phi(x)e^{-ix\lambda}dx\right|\\
&=\sup_\lambda\left|\lambda^\alpha\frac{1}{\sqrt{2\pi}}\int_{\mathbb{R}}(-ix)^\beta\phi(x)e^{-ix\lambda}dx\right|\\
&=\sup_\lambda\left|\frac{1}{\sqrt{2\pi}}\int_{\mathbb{R}}\frac{d^\alpha}{dx^\alpha}(-ix)^\beta\phi(x)e^{-ix\lambda}dx\right|.
\end{aligned}
$$

Exercise 1.11 guarantees that $\frac{d^\alpha}{dx^\alpha}(-ix)^\beta\phi(x)$ belongs to $\mathscr{S}(\mathbb{R})$.

For a function f in $\mathscr{S}(\mathbb{R}^d)$, the values $\sup_x |x|^2|f(x)|$ and $\sup_x |f(x)|$ have finite values C_1 and C_2, we have $|\phi(x)| \le \frac{C_1}{|x^2|}$ and $|\phi(x)| \le C_2$. We apply this fact to $\frac{d^\alpha}{dx^\alpha}(-ix)^\beta\phi(x)$. So, we have

$$\sup_\lambda \Big|\frac{1}{\sqrt{2\pi}}\int_{\mathbb{R}} \frac{d^\alpha}{dx^\alpha}(-ix)^\beta\phi(x)e^{-ix\lambda}dx\Big|$$

$$\le \frac{1}{\sqrt{2\pi}}\int_{\mathbb{R}}\Big|\frac{d^\alpha}{dx^\alpha}(-ix)^\beta\phi(x)\Big|dx \le \frac{1}{\sqrt{2\pi}}\left(\int_{|x|>1}\frac{C_1}{|x^2|}dx + \int_{|x|\le 1} C_2 dx\right) < \infty.$$

So, $\mathcal{F}_d[\phi]$ belongs to $\mathscr{S}(\mathbb{R}^d)$.

Exercise 3.15

$$\langle\phi_1|\mathcal{F}_d[\phi_2]\rangle = \int_{\mathbb{R}^d}\overline{\phi_1(x)}\mathcal{F}_d[\phi_2](x)dx$$

$$= \int_{\mathbb{R}^d}\overline{\phi_1(x)}\left(\int_{\mathbb{R}^d}\frac{1}{(2\pi)^{\frac{d}{2}}}\phi_2(y)e^{-iy\cdot x}dy\right)dx$$

$$= \int_{\mathbb{R}^d}\left(\frac{1}{(2\pi)^{\frac{d}{2}}}\int_{\mathbb{R}^d}e^{-iy\cdot x}\overline{\phi_1(x)}dx\right)\phi_2(y)dy$$

$$= \int_{\mathbb{R}^d}\overline{\left(\frac{1}{(2\pi)^{\frac{d}{2}}}\int_{\mathbb{R}^d}e^{iy\cdot x}\phi_1(x)dx\right)}\phi_2(y)dy = \langle\mathcal{F}_d^{-1}[\phi_1]|\phi_2\rangle.$$

Exercise 3.16

a Firstly, considering the complex integral in the path $-R + i\frac{y}{\sqrt{\epsilon}} \to R + i\frac{y}{\sqrt{\epsilon}} \to R \to -R \to -R + i\frac{y}{\sqrt{\epsilon}}$. Then, we have

$$\frac{\sqrt{\epsilon}}{\sqrt{2\pi}}\int_{-R}^{R} e^{-\frac{\epsilon}{2}(x-i\frac{y}{\sqrt{\epsilon}})^2}dy = \frac{\sqrt{\epsilon}}{\sqrt{2\pi}}\int_{-R+i\frac{y}{\sqrt{\epsilon}}}^{R+i\frac{y}{\sqrt{\epsilon}}} e^{-\frac{\epsilon}{2}x^2}dy = \frac{\sqrt{\epsilon}}{\sqrt{2\pi}}\int_{-R}^{R} e^{-\frac{\epsilon}{2}x^2}dy.$$

Taking the limit $R \to \infty$, we have

$$\frac{\sqrt{\epsilon}}{\sqrt{2\pi}}\int_{-\infty}^{\infty} e^{-\frac{\epsilon}{2}(x-i\frac{y}{\sqrt{\epsilon}})^2}dy = \frac{\sqrt{\epsilon}}{\sqrt{2\pi}}\int_{-\infty}^{\infty} e^{-\frac{\epsilon}{2}x^2}dy = 1.$$

Thus,

$$\mathcal{F}[\varphi_{1/\epsilon}](y) = \frac{1}{\sqrt{2\pi}}\int_{-\infty}^{\infty} e^{-ixy}\left(\frac{\epsilon}{2\pi}\right)^{\frac{1}{4}} e^{-\frac{\epsilon x^2}{2}}dy = \frac{1}{\sqrt{2\pi}}\left(\frac{\epsilon}{2\pi}\right)^{\frac{1}{4}}\int_{-\infty}^{\infty} e^{-\frac{\epsilon x^2}{2}-ixy}dy$$

$$= \frac{1}{\sqrt{2\pi}}\left(\frac{\epsilon}{2\pi}\right)^{\frac{1}{4}}\int_{-\infty}^{\infty} e^{-\frac{\epsilon}{2}(x-i\frac{y}{\sqrt{\epsilon}})^2-\frac{x^2}{2\epsilon}}dy = \left(\frac{1}{2\pi\epsilon}\right)^{\frac{1}{4}} e^{-\frac{x^2}{2\epsilon}} = \varphi_\epsilon(x).$$

b

$$
\begin{aligned}
&\left(\frac{1}{2\pi\epsilon}\right)^{\frac{1}{4}}\int_{-\infty}^{\infty}\phi(x-y)\varphi_\epsilon(y)dy=(\frac{1}{2\pi\epsilon})^{\frac{1}{4}}\langle\overline{\phi(x-\cdot)}|\mathcal{F}[\varphi_{1/\epsilon}]\rangle\\
=&\left(\frac{1}{2\pi\epsilon}\right)^{\frac{1}{4}}\langle\mathcal{F}^{-1}[\overline{\phi(x-\cdot)}]|\varphi_{1/\epsilon}\rangle=\left(\frac{1}{2\pi\epsilon}\right)^{\frac{1}{4}}\langle\mathcal{F}[\overline{\phi(x+\cdot)}]|\varphi_{1/\epsilon}\rangle\\
=&\left(\frac{1}{2\pi\epsilon}\right)^{\frac{1}{4}}\langle\overline{\mathcal{F}^{-1}[\phi(x+\cdot)]}|\varphi_{1/\epsilon}\rangle=\left(\frac{1}{2\pi\epsilon}\right)^{\frac{1}{4}}\int_{-\infty}^{\infty}\mathcal{F}^{-1}[\phi(x+\cdot)](y)\varphi_{1/\epsilon}(y)dy\\
=&\frac{1}{\sqrt{2\pi}}\int_{-\infty}^{\infty}\mathcal{F}[\phi](y)e^{ixy}e^{-\epsilon y^2/2}dy.
\end{aligned}
$$

c Since $(\frac{1}{2\pi\epsilon})^{\frac{1}{4}}\varphi_\epsilon(y)$ is the Gaussian distribution with variance ϵ, taking the limit $\epsilon\to 0$, we have

$$
\phi(x)=\frac{1}{\sqrt{2\pi}}\int_{-\infty}^{\infty}\mathcal{F}[\phi](y)e^{ixy}dy=\mathcal{F}^{-1}[\mathcal{F}[\phi]].
$$

So, $\mathcal{F}^{-1}\circ\mathcal{F}$ is the identity map on $\mathcal{S}(\mathbb{R})$. Since

$$
\mathcal{F}^{-1}[\mathcal{F}[\phi]]=\frac{1}{\sqrt{2\pi}}\int_{-\infty}^{\infty}\mathcal{F}[\phi](y)e^{ixy}dy=\frac{1}{\sqrt{2\pi}}\int_{-\infty}^{\infty}\mathcal{F}^{-1}[\phi](-z)e^{-ixz}dz=\mathcal{F}[\mathcal{F}^{-1}[\phi]],
$$

$\mathcal{F}\circ\mathcal{F}^{-1}$ is the identity map on $\mathcal{S}(\mathbb{R})$.
d Since $\mathcal{S}(\mathbb{R})$ is a dense subset of $L^2(\mathbb{R})$, $\mathcal{F}^{-1}\circ\mathcal{F}$ and $\mathcal{F}\circ\mathcal{F}^{-1}$ are the identity map on $L^2(\mathbb{R})$.
e Since $\mathcal{F}_d=\mathcal{F}^{\otimes d}$ and $\mathcal{F}_d^{-1}=(\mathcal{F}^{-1})^{\otimes d}$, $\mathcal{F}_d^{-1}\circ\mathcal{F}_d$ and $\mathcal{F}_d\circ\mathcal{F}_d^{-1}$ are the identity map on $L^2(\mathbb{R}^d)$.

Exercise 3.17
Substituting $\mathcal{F}_d[\phi_2]$ into ϕ_1 in Exercise 3.15, we have

$$
\langle\mathcal{F}_d[\phi_2]|\mathcal{F}_d[\phi_2]\rangle=\langle\mathcal{F}_d^{-1}[\mathcal{F}_d[\phi_2]]|\phi_2\rangle.=\langle\phi_2]|\phi_2\rangle.
$$

So, $\mathcal{F}_d$ preserves the norm, i.e., $\mathcal{F}_d$ is a unitary map.

Exercise 3.18
Since $\mathcal{S}(\mathbb{R}^d)$ is a dense subset of $L^2(\mathbb{R}^d)$, Exercise 3.11 implies (3.70), which contains (3.61) as a special case.

Exercise 3.19

$$\mathcal{F}[e^{ip\mathsf{Q}}\phi](\lambda) = \frac{1}{\sqrt{2\pi}} \int_{-\infty}^{\infty} e^{-i\lambda x} e^{ipx} \phi(x) dx$$
$$= \frac{1}{\sqrt{2\pi}} \int_{-\infty}^{\infty} e^{-i(\lambda-p)x} = \mathcal{F}[\phi](\lambda - p).$$

Exercise 3.20
When we replace λ by $-\lambda$ in (3.61), we have

$$\mathcal{F}^{-1}[\mathsf{P}\psi](\lambda) = -\lambda \mathcal{F}^{-1}[\psi](\lambda),$$

which implies that $\mathcal{F}^{-1}\mathsf{P} = -\mathsf{Q}\mathcal{F}^{-1}$, i.e., $\mathsf{P} = -\mathcal{F}^{-1}\mathsf{Q}\mathcal{F}^{-1}$. Substitute $\mathcal{F}^{-1}[\phi]$ into ϕ in (3.64), we have

$$e^{-ip\mathsf{P}}[\phi](\lambda) e^{ip\mathcal{F}\mathsf{Q}\mathcal{F}^{-1}}[\phi](\lambda) = \mathcal{F} e^{ip\mathsf{Q}} \mathcal{F}^{-1}[\phi](\lambda)$$
$$= \mathcal{F}[e^{ip\mathsf{Q}} \mathcal{F}^{-1}[\phi]](\lambda) = \mathcal{F}[\mathcal{F}^{-1}[\phi]](\lambda - p) = \phi(\lambda - p).$$

Exercise 3.21
Consider the inverse Fourier transform of the given function. Exercise 3.22 shows that the product $\mathcal{F}_d^{-1}[\phi_1] \cdot \mathcal{F}_d^{-1}[\phi_2]$ of the inverse Fourier transforms of ϕ_1, ϕ_2. Exercise 3.14 shows that these inverse Fourier transforms $\mathcal{F}_d^{-1}[\phi_1]$ and $\mathcal{F}_d^{-1}[\phi_2]$ belong to $\mathscr{S}(\mathbb{R}^d)$. So, Exercise 1.12 shows that their product $\mathcal{F}_d^{-1}[\phi_1] \cdot \mathcal{F}_d^{-1}[\phi_2]$ also belongs to $\mathscr{S}(\mathbb{R}^d)$. Since Exercise 3.14 shows that the Fourier transform $\mathcal{F}_d[\mathcal{F}_d^{-1}[\phi_1] \cdot \mathcal{F}_d^{-1}[\phi_2]]$ belongs to $\mathscr{S}(\mathbb{R}^d)$, the given function belongs to $\mathscr{S}(\mathbb{R}^d)$.

Exercise 3.22
We can show this relation as follows.

$$\frac{1}{(2\pi)^{\frac{d}{2}}} \int_{\mathbb{R}^d} \mathcal{F}_d[\phi_1](\lambda') \mathcal{F}_d[\phi_2](\lambda - \lambda') d\lambda'$$
$$= \frac{1}{(2\pi)^d} \int_{\mathbb{R}^d} \int_{\mathbb{R}^d} \mathcal{F}_d[\phi_1](\lambda') \phi_2(x) e^{-ix\cdot(\lambda-\lambda')} dx d\lambda'$$
$$= \frac{1}{(2\pi)^d} \int_{\mathbb{R}^d} \int_{\mathbb{R}^d} \mathcal{F}_d[\phi_1](\lambda') e^{ix\cdot\lambda'} d\lambda' \phi_2(x) e^{-ix\cdot\lambda} dx$$
$$= \frac{1}{(2\pi)^{\frac{d}{2}}} \int_{\mathbb{R}^d} \mathcal{F}_d^{-1} \circ \mathcal{F}_d[\phi_1](x) \phi_2(x) e^{-ix\cdot\lambda} dx$$
$$= \frac{1}{(2\pi)^{\frac{d}{2}}} \int_{\mathbb{R}^d} \phi_1(x) \phi_2(x) e^{-ix\cdot\lambda} dx = \mathcal{F}_d[\phi_1 \cdot \phi_2](\lambda).$$

Exercise 4.1

a The set $\mathcal{M}_{1,1}$ is the set of Hermitian matrices X satisfying the condition $\operatorname{Tr} X \begin{pmatrix} 1 & 0 \\ 0 & -1 \end{pmatrix} = 0$. Since $\operatorname{Tr} gXg^\dagger \begin{pmatrix} 1 & 0 \\ 0 & -1 \end{pmatrix} = \operatorname{Tr} Xg^\dagger \begin{pmatrix} 1 & 0 \\ 0 & -1 \end{pmatrix} g = \operatorname{Tr} X \begin{pmatrix} 1 & 0 \\ 0 & -1 \end{pmatrix}$ and the image of ι is limited to the connected component, we obtain the desired statement.

b $\det gXg^\dagger = \det g \det X \det g^\dagger = \det X$.

c Since $\det \begin{pmatrix} z & x - yi \\ x + yi & z \end{pmatrix} = z^2 - x^2 - y^2$, any element $g \in \mathrm{SU}(1,1)$ gives a 3×3 matrix $\iota(g) \in \mathrm{SO}(1,2)$ based on the coordinate $(x, y, z) \in \mathbb{R}^3$ and the above relation. The map ι is a homomorphism ι from SU(2) to $\mathrm{SO}(3, \mathbb{R})$, and is not an isomorphism because the matrices $I, -I \in \mathrm{SU}(1,1)$ are mapped to the unit matrix on $\mathbb{R}^3$. In particular, we find that $\iota^{-1}(e) = \{I, -I\}$. So, we obtain the relation $\mathrm{SO}(1,2)_0 \cong \mathrm{SU}(1,1)/\mathbb{Z}_2$.

Exercise 4.2

Using (4.13), we have

$$\begin{aligned}
&\mathsf{K}_{+,s}\mathsf{K}_{-,s}v_{n-1} = (\mathsf{K}_{-,s}^n \mathsf{K}_{+,s} + [\mathsf{K}_{+,s}, \mathsf{K}_{-,s}^n])v_0 \\
=&(\mathsf{K}_{-,s}^n \mathsf{K}_{+,s} + \mathsf{K}_{-,s}^{n-1} ns(2\mathsf{E}_{0,s} - (n-1)I))v_0 \\
=&(\mathsf{K}_{-,s}^{n-1} ns(2\lambda - (n-1)))v_0 = ns(2\lambda - (n-1))v_{n-1}.
\end{aligned}$$

Exercise 4.3

Since

$$[\mathsf{E}_{0,s}, \mathsf{K}_{-,s}^n] = \sum_{j=0}^{n-1} \mathsf{K}_{-,s}^j [\mathsf{E}_{0,s}, \mathsf{K}_{-,s}] \mathsf{K}_{-,s}^{n-1-j} = -n\mathsf{K}_{-,s}^n,$$

we obtain the first equation in (4.14). The second equation in (4.14) follows from the relations

$$\begin{aligned}
&[\mathsf{K}_{+,s}, \mathsf{K}_{-,s}^n] = \sum_{j=0}^{n-1} \mathsf{K}_{-,s}^j [\mathsf{K}_{+,s}, \mathsf{K}_{-,s}] \mathsf{K}_{-,s}^{n-1-j} = \sum_{j=0}^{n-1} \mathsf{K}_{-,s}^j 2s\mathsf{E}_{0,s} \mathsf{K}_{-,s}^{n-1-j} \\
=&\sum_{j=0}^{n-1} (\mathsf{K}_{-,s}^{n-1} 2s\mathsf{E}_{0,s} + \mathsf{K}_{-,s}^j [2s\mathsf{E}_{0,s}, \mathsf{K}_{-,s}^{n-1-j}]) \\
=&\sum_{j=0}^{n-1} (\mathsf{K}_{-,s}^{n-1} 2s\mathsf{E}_{0,s} + \mathsf{K}_{-,s}^j \cdot -2s(n-1-j)\mathsf{K}_{-,s}^{n-1-j}) \\
=&\mathsf{K}_{-,s}^{n-1} ns(2\mathsf{E}_{0,s} - (n-1)I).
\end{aligned}$$

Exercise 4.4

Equation (4.25) is changed to

$$m(\alpha^2 - \beta^2) + 2\sqrt{(\lambda_1 + \frac{1}{2})^2 - m^2\alpha\beta} = (\lambda - \lambda_1)(\lambda + \lambda_1 + 1) + \frac{1}{4}$$
$$= \begin{cases} \lambda + \frac{1}{2} & \text{when } \lambda = \lambda_1 + \frac{1}{2} \\ -\lambda - \frac{1}{2} & \text{when } \lambda = \lambda_1 - \frac{1}{2}, \end{cases}$$

which implies 3.28.

Exercise 4.5

Due to the invariance of the LHS of (4.32), it is sufficient to consider the case with the state $|\lambda;\lambda\rangle$. Since $|\langle\lambda;\lambda|\lambda:\zeta\rangle|^2 = (1+|\zeta|^2)^{-2\lambda} = (1+(x^2+y^2)/2)^{-2\lambda}$, using $r = \sqrt{x^2+y^2}$ and $t = r^2/2$, we have

$$\begin{aligned}
&\langle\lambda;\lambda|\int_{\hat{\mathbb{C}}} |\lambda:\zeta\rangle\langle\lambda:\zeta|\frac{dxdy}{2\pi(1+|\zeta|^2)^2}|\lambda;\lambda\rangle \\
=&\int_{\hat{\mathbb{C}}} |\langle\lambda;\lambda|\lambda:\zeta\rangle|^2 \frac{dxdy}{2\pi(1+|\zeta|^2)^2} \\
=&\int_{\hat{\mathbb{C}}} (1+(x^2+y^2)/2)^{-2\lambda}\frac{dxdy}{2\pi(1+(x^2+y^2)/2)^2} \\
=&\int_0^\infty \int_0^{2\pi} (1+(x^2+y^2)/2)^{-2\lambda-2}\frac{d\theta r dr}{2\pi} \\
=&\int_0^\infty (1+t)^{-2\lambda-2}dt \\
=&\frac{1}{-2\lambda-1}[(1+t)^{-2\lambda-1}]_0^\infty = \frac{1}{2\lambda+1}.
\end{aligned}$$

Exercise 4.6

Due to the invariance of the LHS of (4.33), it is sufficient to consider the case with the state $|\lambda;\lambda\rangle$. Since $|\langle\lambda;\lambda|\lambda:\zeta\rangle|^2 = (1-|\zeta|^2)^{-2\lambda} = (1-(x^2+y^2)/2)^{-2\lambda}$, using $r = \sqrt{x^2+y^2}$ and $t = r^2/2$, we have

$$\begin{aligned}
&\langle\lambda;\lambda|\int_D |\lambda:\zeta\rangle\langle\lambda:\zeta|\frac{dxdy}{2\pi(1-|\zeta|^2)^2}|\lambda;\lambda\rangle \\
=&\int_D |\langle\lambda;\lambda|\lambda:\zeta\rangle|^2 \frac{dxdy}{2\pi(1-|\zeta|^2)^2} \\
=&\int_D (1-(x^2+y^2)/2)^{-2\lambda}\frac{dxdy}{2\pi(1-(x^2+y^2)/2)^2} \\
=&\int_0^{\sqrt{2}} \int_0^{2\pi} (1-(x^2+y^2)/2)^{-2\lambda-2}\frac{d\theta r dr}{2\pi}
\end{aligned}$$

$$
\begin{aligned}
&=\int_0^1 (1-t)^{-2\lambda-2}dt \\
&=-\frac{1}{-2\lambda-1}[(1-t)^{-2\lambda-1}]_0^1=-\frac{1}{2\lambda+1}.
\end{aligned}
$$

Exercise 4.7

$$
|\langle\lambda;\lambda|\lambda:\zeta\rangle|^2\frac{dxdy}{2\pi(1+|\zeta|^2)^2}=(1+(x^2+y^2)/2)^{-2\lambda-2}\frac{dxdy}{2\pi}.
$$

Exercise 4.8

$$
|\langle\lambda;\lambda|\lambda:\zeta\rangle|^2\frac{dxdy}{2\pi(1-|\zeta|^2)^2}=(1-(x^2+y^2)/2)^{-2\lambda-2}\frac{dxdy}{2\pi}.
$$

Exercise 4.9
Since the basis of $\mathcal{U}_{[n,0,\ldots,0]}(\mathrm{U}(r))$ are identified with n-combination with repetition of r things, which is $\frac{(n+r-1)!}{(r-1)!n!}$.

Exercise 4.10
Substituting 1 into both of λ_1 and λ_2 in (4.48), we obtain the dimension $\frac{(r+1)r(r-1)}{3}$.

Exercise 4.11
Counting the dimension of both sides in (4.11), we have $r^2=1+\dim\mathcal{U}_{[1,0,\ldots,0,1]}$ $(\mathrm{U}(r))$.

Exercise 4.12
Taking the complex conjugate in (4.65), we obtain (4.68).

Exercise 5.1
We show only the case with $t=1, l=2$. Since

$$
\begin{aligned}
&(x_2\partial_3-x_3\partial_2)(x_3\partial_1-x_1\partial_3) \\
=&x_2\partial_1+(x_2x_3\partial_3\partial_1-x_2x_1\partial_3\partial_3-x_3^2\partial_2\partial_1+x_3x_1\partial_2\partial_3), \\
&(x_3\partial_1-x_1\partial_3)(x_2\partial_3-x_3\partial_2) \\
=&x_1\partial_2+(x_2x_3\partial_3\partial_1-x_2x_1\partial_3\partial_3-x_3^2\partial_2\partial_1+x_3x_1\partial_2\partial_3),
\end{aligned}
$$

we have

$$
[L_1,L_2]=-\hbar^2[(x_2\partial_3-x_3\partial_2),(x_3\partial_1-x_1\partial_3)]=\hbar^2(x_1\partial_2-x_2\partial_1)=i\hbar L_3.
$$

Exercise 5.2
For $k > l$, we have

$$F_{k,l}^2 = Q_k^2\partial_l^2 + Q_l^2\partial_k^2 - 2Q_kQ_l\partial_k\partial_l - Q_k\partial_k - Q_l\partial_l.$$

Taking the sum for $k \neq l$, we have

$$\begin{aligned}\sum_{k>l} F_{k,l}^2 &= \sum_{k>l} Q_k^2\partial_l^2 + Q_l^2\partial_k^2 - 2Q_kQ_l\partial_k\partial_l - Q_k\partial_k - Q_l\partial_l \\ &= \sum_{k\neq l} Q_k^2\partial_l^2 - \sum_{k\neq l} Q_kQ_l\partial_k\partial_l - \sum_{k\neq l} Q_k\partial_k \\ &= \sum_{k\neq l} Q_k^2\partial_l^2 - \sum_{k\neq l} Q_kQ_l\partial_k\partial_l - (d-1)\sum_{k} Q_k\partial_k,\end{aligned}$$

which implies (5.40). Also, we have

$$\begin{aligned}\left(\sum_k Q_k\partial_k\right)^2 &= \sum_k Q_k^2\partial_k^2 + \sum_{k\neq l} Q_k\partial_kQ_l\partial_l \\ &= \sum_k Q_k^2\partial_k^2 + \sum_k Q_k\partial_k + \sum_{k\neq l} Q_kQ_l\partial_k\partial_l,\end{aligned}$$

which implies (5.41).

Exercise 5.3
For $k = 1, \ldots, d-1$, $j > k$, we have

$$\frac{\partial r}{\partial x_k} = \frac{\frac{1}{2}\cdot 2x_k}{\sqrt{\sum_{k'=1}^d x_{k'}^2}} = \frac{w_k}{r},$$

which implies (5.50). Using the formula $\frac{d}{dx}\cos^{-1} f(x) = \frac{-1}{\sqrt{1-f(x)^2}}\frac{df}{dx}$, we have

$$\begin{aligned}\frac{\partial\theta_k}{\partial x_k} &= -\sqrt{\frac{\sum_{k'=k}^d x_{k'}^2}{\sum_{k'=k+1}^d x_{k'}^2}}\frac{\sum_{k'=k+1}^d x_{k'}^2}{(\sum_{k'=k}^d x_{k'}^2)^{3/2}} = \frac{\sqrt{\sum_{k'=k+1}^d x_{k'}^2}}{\sum_{k'=k}^d x_{k'}^2} \\ &= \frac{\sqrt{\sum_{k'=k+1}^d w_{k'}^2}}{r\sum_{k'=k}^d w_{k'}^2},\end{aligned}$$

which implies (5.51). Using the same formula, we have

$$\begin{aligned}\frac{\partial\theta_k}{\partial x_j} &= \sqrt{\frac{\sum_{k'=k}^d x_{k'}^2}{\sum_{k'=k+1}^d x_{k'}^2}}\frac{x_k x_j}{(\sum_{k'=k}^d x_{k'}^2)^{3/2}} = \frac{x_k x_j}{(\sum_{k'=k}^d x_{k'}^2)\sqrt{\sum_{k'=k+1}^d x_{k'}^2}}\\ &= \frac{w_k w_j}{r(\sum_{k'=k}^d w_{k'}^2)\sqrt{\sum_{k'=k+1}^d w_{k'}^2}},\end{aligned}$$

which implies (5.52).

Using (5.53), for $j > l$, we have

$$\begin{aligned}F_{j,l} =& w_j\left(\sum_{k=1}^{l-1}\frac{w_k w_l}{\left(\sum_{k'=k}^d w_{k'}^2\right)\sqrt{\sum_{k'=k+1}^d w_{k'}^2}}\frac{\partial}{\partial\theta_k}\right.\\ &\left.+\frac{\sqrt{\sum_{k'=l+1}^d w_{k'}^2}}{\sum_{k'=l}^d w_{k'}^2}\frac{\partial}{\partial\theta_l}+w_l\frac{\partial}{\partial r}\right)\\ &-w_l\left(\sum_{k=1}^{j-1}\frac{w_k w_j}{\left(\sum_{k'=k}^d w_{k'}^2\right)\sqrt{\sum_{k'=k+1}^d w_{k'}^2}}\frac{\partial}{\partial\theta_k}\right.\\ &\left.+\frac{\sqrt{\sum_{k'=j+1}^d w_{k'}^2}}{\sum_{k'=j}^d w_{k'}^2}\frac{\partial}{\partial\theta_j}+w_j\frac{\partial}{\partial r}\right)\\ =&\sum_{k=j}^{l-1}\frac{w_j w_l w_k}{\left(\sum_{k'=k}^d w_{k'}^2\right)\sqrt{\sum_{k'=k+1}^d w_{k'}^2}}\frac{\partial}{\partial\theta_k}\\ &+w_j\frac{\sqrt{\sum_{k'=l+1}^d w_{k'}^2}}{\sum_{k'=l}^d w_{k'}^2}\frac{\partial}{\partial\theta_l}-w_l\frac{\sqrt{\sum_{k'=j+1}^d w_{k'}^2}}{\sum_{k'=j}^d w_{k'}^2}\frac{\partial}{\partial\theta_j},\end{aligned}$$

which implies (5.55).

Using (5.53) and (5.54), we have

$$
\begin{aligned}
F_{j,d} =& w_j\left(\sum_{k=1}^{d-2} \frac{w_k w_d}{\left(\sum_{k'=k}^{d} w_{k'}^2\right)\sqrt{\sum_{k'=k+1}^{d} w_{k'}^2}} \frac{\partial}{\partial \theta_k}\right.\\
&\left. + \frac{w_{d-1}}{w_{d-1}^2 + w_d^2}\frac{\partial}{\partial \theta_{d-1}} + w_d \frac{\partial}{\partial r}\right)\\
&- w_d\left(\sum_{k=1}^{j-1} \frac{w_k w_j}{\left(\sum_{k'=k}^{d} w_{k'}^2\right)\sqrt{\sum_{k'=k+1}^{d} w_{k'}^2}} \frac{\partial}{\partial \theta_k}\right.\\
&\left. + \frac{\sqrt{\sum_{k'=j+1}^{d} w_{k'}^2}}{\sum_{k'=j}^{d} w_{k'}^2}\frac{\partial}{\partial \theta_j} + w_j \frac{\partial}{\partial r}\right)\\
=& \sum_{k=j}^{d-2} \frac{w_k w_j w_d}{\left(\sum_{k'=k}^{d} w_{k'}^2\right)\sqrt{\sum_{k'=k+1}^{d} w_{k'}^2}} \frac{\partial}{\partial \theta_k}\\
&+ w_j \frac{w_{d-1}}{w_{d-1}^2 + w_d^2}\frac{\partial}{\partial \theta_{d-1}} - w_d \frac{\sqrt{\sum_{k'=j+1}^{d} w_{k'}^2}}{\sum_{k'=j}^{d} w_{k'}^2}\frac{\partial}{\partial \theta_j},
\end{aligned}
$$

which implies (5.56).

Exercise 5.4
We have

$$
\begin{aligned}
&\binom{n+d-1}{n} - \binom{n+d-3}{n-2}\\
=& \frac{(n+d-1)\ldots(n+1)}{(d-1)!} - \frac{(n+d-3)\ldots(n-1)}{(d-1)!}\\
=& \frac{[(n+d-1)(n+d-2) - n(n-1)](n+d-3)\ldots(n+1)}{(d-1)!}\\
=& \frac{(d-1)(2n+d-2)(n+d-3)\ldots(n+1)}{(d-1)!}\\
=& \frac{(2n+d-2)((n+d-3)\ldots(n+1))}{(d-2)!},
\end{aligned}
$$

which implies (5.60).

Exercise 5.5
Since

$$
\begin{aligned}
&(x_k\partial_l - x_l\partial_k)(x_{k'}\partial_{l'} - x_{l'}\partial_{k'})\\
=&x_k\partial_{l'}\delta_{l,k'} - x_l\partial_{l'}\delta_{k,k'} - x_k\partial_{k'}\delta_{l,l'} + x_l\partial_{k'}\delta_{k,l'}\\
&+ (x_k x_{k'}\partial_l\partial_{l'} - x_k x_{l'}\partial_l\partial_{k'} - x_l x_{k'}\partial_k\partial_{l'} + x_l x_{l'}\partial_k\partial_{k'}),\\
&(x_{k'}\partial_{l'} - x_{l'}\partial_{k'})(x_k\partial_l - x_l\partial_k)\\
=&x_{k'}\partial_l\delta_{l',k} - x_{l'}\partial_l\delta_{k',k} - x_{k'}\partial_k\delta_{l',l} + x_{l'}\partial_k\delta_{k',l}\\
&+ (x_k x_{k'}\partial_l\partial_{l'} - x_k x_{l'}\partial_l\partial_{k'} - x_l x_{k'}\partial_k\partial_{l'} + x_l x_{l'}\partial_k\partial_{k'}),
\end{aligned}
$$

we have

$$
[L_{k,l}, L_{k',l'}] = L_{k,l'}\delta_{l,k'} - L_{l,l'}\delta_{k,k'} - L_{k,k'}\delta_{l,l'} + L_{l,k'}\delta_{k,l'},
$$

which implies (5.62).

Exercise 5.6
Since

$$
\begin{aligned}
\frac{\partial^2(x_2+ix_1)^n}{\partial x_1^2} &= -n(n-1)(x_2+ix_1)^{n-2}\\
\frac{\partial^2(x_2+ix_1)^n}{\partial x_2^2} &= n(n-1)(x_2+ix_1)^{n-2}\\
\frac{\partial^2(x_2+ix_1)^n}{\partial x_j^2} &= 0,
\end{aligned}
$$

for $j \ge 3$, we have $\Delta_{\mathbb{R}^d}(x_2+ix_1)^n = 0$.

Exercise 5.7

$$
\begin{aligned}
&\Delta_{\mathbb{R}^d}\sum_{k=0}^{\lfloor\frac{n}{2}\rfloor}(-1)^k\binom{n}{2k}x_1^{n-2k}x_2^{2k}\\
=&\sum_{k=0}^{\lfloor\frac{n}{2}\rfloor-1}(-1)^k(n-2k)(n-2k-1)\binom{n}{2k}x_1^{n-2k-2}x_2^{2k}\\
&+\sum_{k=1}^{\lfloor\frac{n}{2}\rfloor}(-1)^k 2k(2k-1)\binom{n}{2k}x_1^{n-2k}x_2^{2k-2}\\
=&\sum_{k=1}^{\lfloor\frac{n}{2}\rfloor}x_1^{n-2k}x_2^{2k-2}\\
&\cdot\left((-1)^{k-1}(n-2k+2)(n-2k+1)\binom{n}{2k-2} + (-1)^k 2k(2k-1)\binom{n}{2k}\right)\\
=&0.
\end{aligned}
$$

Exercise 5.8
Let $x = \sin\theta_1 \cos\theta_2$. We have

$$\int_0^{2\pi} d\theta_1 \int_0^{\pi} d\theta_2 \int_0^{\pi} d\theta_3 \frac{(\sin\theta_1\cos\theta_2 + i\cos\theta_1)^n}{2(1-\cos\theta_1)} \sin^2\theta_1 \sin\theta_2$$
$$=\pi \int_0^{2\pi} d\theta_1 \int_0^{\pi} d\theta_2 \frac{(\sin\theta_1\cos\theta_2 + i\cos\theta_1)^n}{2(1-\cos\theta_1)} \sin^2\theta_1 \sin\theta_2$$
$$=-\pi \int_0^{2\pi} d\theta_1 \int_{\sin\theta_1}^{-\sin\theta_1} dx \frac{(x + i\cos\theta_1)^n}{2(1-\cos\theta_1)} \sin\theta_1$$
$$=\pi \int_0^{2\pi} d\theta_1 \int_{-\sin\theta_1}^{\sin\theta_1} dx \frac{(x + i\cos\theta_1)^n}{2(1-\cos\theta_1)} \sin\theta_1$$
$$=\pi \int_0^{2\pi} d\theta_1 \Big(\frac{1}{n+1} \frac{(\sin\theta_1 + i\cos\theta_1)^{n+1}}{2(1-\cos\theta_1)} \sin\theta_1$$
$$-\frac{1}{n+1} \frac{(-\sin\theta_1 + i\cos\theta_1)^{n+1}}{2(1-\cos\theta_1)} \sin\theta_1\Big)$$
$$=\pi \int_0^{2\pi} d\theta_1 \Big(\frac{1}{n+1} \frac{(\sin(n+1)\theta_1 + i\cos(n+1)\theta_1)}{2(1-\cos\theta_1)} \sin\theta_1$$
$$-\frac{1}{n+1} \frac{(-\sin(n+1)\theta_1 + i\cos(n+1)\theta_1)}{2(1-\cos\theta_1)} \sin\theta_1\Big)$$
$$=\frac{2\pi i^n}{n+1} \int_0^{2\pi} d\theta_1 \frac{\sin\theta_1}{2(1-\cos\theta_1)} \sin(n\theta_1).$$

Exercise 5.9
c We have

$$\int_0^{2\pi} \frac{\sin\theta\sin 2\theta}{1-\cos\theta} d\theta = \int_0^{2\pi} \frac{\sin\theta(1+\cos\theta)\sin 2\theta}{1-\cos^2\theta} d\theta$$
$$= \int_0^{2\pi} \frac{(1+\cos\theta)\sin 2\theta}{\sin\theta} d\theta = \int_0^{2\pi} 2(1+\cos\theta)\cos\theta d\theta$$
$$= \int_0^{2\pi} 2\cos^2\theta d\theta = \int_0^{2\pi} 1 + \cos 2\theta d\theta = 2\pi.$$

d Since $n \geq 2$, we have

$$\int_0^{2\pi} \frac{\sin\theta\sin(n+1)\theta}{1-\cos\theta} d\theta = \int_0^{2\pi} \frac{\sin\theta(1+\cos\theta)\sin(n+1)\theta}{1-\cos^2\theta} d\theta$$
$$= \int_0^{2\pi} \frac{(1+\cos\theta)\sin(n+1)\theta}{\sin\theta} d\theta$$
$$= \int_0^{2\pi} \frac{(1+\cos\theta)(\sin n\theta\cos\theta + \cos n\theta\sin\theta)}{\sin\theta} d\theta$$

$$= \int_0^{2\pi} \frac{(\cos\theta + \cos^2\theta)\sin n\theta}{\sin\theta} d\theta + \int_0^{2\pi} (1+\cos\theta)\cos n\theta d\theta$$
$$= \int_0^{2\pi} \frac{(\cos\theta + 1 - \sin^2\theta)\sin n\theta}{\sin\theta} d\theta + 0$$
$$= \int_0^{2\pi} \frac{(\cos\theta + 1)\sin n\theta}{\sin\theta} d\theta + \int_0^{2\pi} -\sin\theta \sin n\theta d\theta$$
$$= \int_0^{2\pi} \frac{(-\cos^2\theta + 1)\sin n\theta}{\sin\theta(1-\cos\theta)} d\theta + \int_0^{2\pi} -\sin\theta \sin n\theta d\theta$$
$$= \int_0^{2\pi} \frac{\sin^2\theta \sin n\theta}{\sin\theta(1-\cos\theta)} d\theta + 0 = \int_0^{2\pi} \frac{\sin\theta \sin n\theta}{1-\cos\theta} d\theta.$$

Exercise 5.10

$$T(f)(w) = \int_0^{2\pi} d\theta_1 \int_0^{\pi} d\theta_2 \int_0^{\pi} d\theta_3 \frac{\sin^2\theta_1 \sin\theta_2}{2(1-\cos\theta_1)}$$
$$= \int_0^{2\pi} d\theta_1 \int_0^{\pi} d\theta_2 \int_0^{\pi} d\theta_3 \frac{(1+\cos\theta_1)\sin\theta_2}{2}$$
$$= 2\pi \int_0^{\pi} d\theta_2 \int_0^{\pi} d\theta_3 \frac{\sin\theta_2}{2} = 2\pi^2 \int_0^{\pi} d\theta_2 \frac{\sin\theta_2}{2} = 2\pi^2.$$

Exercise 5.11
Using the commutation relation (5.2), we have

$$\Big[\sum_{j=1}^{3} P_j^2, Q_k P_k\Big] = -2i\hbar P_k^2, \quad \Big[\sum_{j=1}^{3} P_j^2, P_k Q_k\Big] = -2i\hbar P_k^2.$$

So, we have

$$\Big[\sum_{j=1}^{3} P_j^2, A\Big] = \frac{-i}{\mu} \sum_{k=1}^{3} P_k^2.$$

Also,

$$\Big[\frac{1}{r}, Q_k P_k\Big] = -i\hbar \frac{x_k^2}{r^3}, \quad \Big[\frac{1}{r}, P_k Q_k\Big] = -i\hbar \frac{x_k^2}{r^3}.$$

Since $\sum_{k=1}^{3} \frac{x_k^2}{r^3} = \frac{r^2}{r^3} = \frac{1}{r}$, we have

$$\Big[\sum_{j=1}^{3} P_j^2, A\Big] = i \frac{Ze^2}{4\pi\epsilon_0 r}.$$

Thus, we have the desired commutation relation

$$i[H, A] = \frac{1}{2\mu}\sum_{j=1}^{3} P_j^2 + H.$$

Exercise 5.12
We show (5.100) only for $k = 3$.

$$\sum_{j=1}^{4}\left(\frac{\partial w_j}{\partial p_k}\right)^2$$
$$=\frac{\left(\sum_{j=1}^{2}(4p_0 p_j p_3)^2\right) + (2p_0)^2 \cdot (p^2 - 2p_3^2 + p_0^2)^2 + (4p_0^2 p_3)^2}{(p^2 + p_0^2)^4}$$
$$=\frac{4p_0^2(4p_0^2(p^2 - p_3^2) + (p^2 - 2p_3^2 + p_0^2)^2 + 4p_0^2 p_3^2)}{(p^2 + p_0^2)^4}$$
$$=\frac{4p_0^2(p^2 + p_0^2)^2}{(p^2 + p_0^2)^4} = (\frac{2p_0}{p^2 + p_0^2})^2.$$

Next, we show (5.88) only for $k = 2, l = 3$.

$$\sum_{j=1}^{4}\frac{\partial w_j}{\partial p_2}\frac{\partial w_j}{\partial p_3}$$
$$=\frac{16p_0^2 p_1^2 p_2 p_3 - 8p_0^2(p^2 - 2p_2^2 + p_0^2)p_2 p_3 - 8p_0^2(p^2 - 2p_3^2 + p_0^2)p_2 p_3 16p_0^4 p_2 p_3}{(p^2 + p_0^2)^4}$$
$$=\frac{8p_0^2(2(p_0^2 + p_1^2)p_2 p_3 - p_2 p_3(2p^2 - 2p_2^2 - 2p_3^2 + 2p_0^2)}{(p^2 + p_0^2)^4}$$
$$=\frac{8p_0^2 p_2 p_3(2(p_0^2 + p_1^2) - (2p_1^2 + 2p_0^2)}{(p^2 + p_0^2)^4} = 0.$$

Exercise 5.13
(5.107) can be shown as

$$
\begin{aligned}
\langle \phi_1 | \phi_4 * \mathcal{F}_3[\phi_2] \rangle = & \int_{\mathbb{R}^3} \phi_1(x)^* \phi_4 * \mathcal{F}_3[\phi_2](x) d^3x \\
= & \int_{\mathbb{R}^3} \phi_1(x)^* \frac{1}{(2\pi)^{\frac{3}{2}}} \int_{\mathbb{R}^3} \phi_4(y) \mathcal{F}_3[\phi_2](x-y) d^3y d^3x \\
= & \frac{1}{(2\pi)^{\frac{3}{2}}} \int_{\mathbb{R}^3} \int_{\mathbb{R}^3} \phi_1(x)^* \mathcal{F}_3[\phi_2](x-y) \phi_4(y) d^3y d^3x \\
= & \frac{1}{(2\pi)^{\frac{3}{2}}} \int_{\mathbb{R}^3} \int_{\mathbb{R}^3} \phi_1(x)^* \mathcal{F}_3^{-1}[\phi_2](y-x) \phi_4(y) d^3y d^3x \\
\stackrel{(a)}{=} & \int_{\mathbb{R}^3} \frac{1}{(2\pi)^{\frac{3}{2}}} \left(\int_{\mathbb{R}^3} \phi_1(x) \mathcal{F}_3^{-1}[\phi_2](y-x)^* d^3x \right)^* \phi_4(y) d^3y \\
= & \int_{\mathbb{R}^3} (\phi_1 * \mathcal{F}_3^{-1}[\phi_2]^*)(y)^* \phi_4(y) d^3y = \langle \phi_1 * \mathcal{F}_3^{-1}[\phi_2]^* | \phi_4 \rangle,
\end{aligned}
$$

where (a) follows from Exercise 3.13.

Exercise 5.14

a We choose the spherical coordinate (r, θ_1, θ_2) given in (5.7). We assume that $p = (0, 0, |p|)$. Then, choosing $t := \cos\theta_1$, we have

$$
\begin{aligned}
& \int_{\mathbb{R}^3} e^{-\delta|x|} \frac{e^{-i(xp_1+yp_2+zp_3)}}{\sqrt{x^2+y^2+z^2}} dxdydz \\
= & \int_0^\infty dr \int_0^{2\pi} d\theta_1 \int_0^\pi d\theta_2 e^{-\delta r} \frac{e^{-i|p|r\cos\theta_1}}{r} r^2 \sin\theta_1 \\
= & 2\pi \int_0^\infty dr \int_0^\pi d\theta_1 e^{-\delta r} \frac{e^{-i|p|r\cos\theta_1}}{r} r^2 \sin\theta_1 \\
= & 2\pi \int_0^\infty dr \int_{-1}^1 dt r e^{-\delta r} e^{-i|p|rt} = e^{-\delta r} \frac{4\pi}{|p|} \int_0^\infty \sin(|p|r).
\end{aligned}
$$

b

$$
\begin{aligned}
\lim_{\delta\to 0} \frac{4\pi}{|p|} \int_0^\infty e^{-\delta r} \sin(|p|r) dr = & \lim_{\delta\to 0} \frac{4\pi}{i|p|} \int_0^\infty e^{-\delta r} (e^{i|p|r} - e^{-i|p|r}) dr \\
= & \lim_{\delta\to 0} \frac{2\pi}{i|p|} \left(\frac{0-1}{i|p|-\delta} + \frac{0-1}{i|p|+\delta} \right) = \frac{4\pi}{|p|^2}.
\end{aligned}
$$

c The combination of **a** and **b** implies (5.109).

Exercise 5.15

Now, we consider the case when the spin is $\uparrow$. Since the flavor system has the permutation symmetry, it is enough to show the orthogonality the vectors in the spin system when the vector in the flavor system is $|udd\rangle$. We have $\langle udd|n\rangle = |\uparrow'\rangle$, $\langle udd|\Delta^0\rangle = |\uparrow\uparrow\downarrow\rangle + |\uparrow\downarrow\uparrow\rangle + |\downarrow\uparrow\uparrow\rangle$, and $\langle \uparrow' |(|\uparrow\uparrow\downarrow\rangle + |\uparrow\downarrow\uparrow\rangle + |\downarrow\uparrow\uparrow\rangle) = 0$.

Exercise 5.16

Now, we consider the case when the spin is $\uparrow$. It is enough to show the orthogonality the vectors in the spin system when the vector in the flavor system is $|uds\rangle$. Since $\langle uds|\Lambda^0\rangle = |\uparrow\downarrow\uparrow\rangle - |\downarrow\uparrow\uparrow\rangle$, $\langle uds|\Sigma^0\rangle = |\uparrow\downarrow\uparrow\rangle + |\downarrow\uparrow\uparrow\rangle - 2|\uparrow\uparrow\downarrow\rangle)$, and $\langle uds|\Sigma^{*0}\rangle = |\uparrow\uparrow\downarrow\rangle + |\uparrow\downarrow\uparrow\rangle + |\downarrow\uparrow\uparrow\rangle$, they are orthogonal to each other.

Exercise 5.17

It is enough to show that $(a^*, b^*)J_\rho \begin{pmatrix} a \\ b \end{pmatrix} \geq 0$ for two complex numbers a and b.

$$(a, b)J_\rho \begin{pmatrix} a \\ b \end{pmatrix} = \mathrm{Tr}(a\mathsf{Q}_0 + a\mathsf{P}_0)^\dagger (a\mathsf{Q}_0 + a\mathsf{P}_0)\rho \geq 0.$$

Exercise 6.1

Let θ is the angle between α and β. Then, $\|\alpha\|\|\beta\| \cos\theta = (\alpha, \beta)$. So, $\langle \beta, \alpha\rangle = \frac{2(\beta,\alpha)}{(\alpha,\alpha)}$. Thus, $\langle \alpha, \beta\rangle\langle \beta, \alpha\rangle = 4\cos^2\theta$, which should be an integer. Hence, $\cos^2\theta$ is limited to $0, \frac{1}{4}, \frac{1}{2}, \frac{3}{4}, 1$, i.e., $\langle \alpha, \beta\rangle\langle \beta, \alpha\rangle$ is limited to $0, 1, 2, 3, 4$. When $\langle \alpha, \beta\rangle\langle \beta, \alpha\rangle \neq 4$, since $\|\alpha\| \leq \|\beta\|$, the pair $(\langle \alpha, \beta\rangle, \langle \beta, \alpha\rangle)$ is limited to $(0, 0)$, $(\pm 1, \pm 1)$, $(\pm 1, \pm 2)$, and $(\pm 1, \pm 3)$. When $\langle \alpha, \beta\rangle\langle \beta, \alpha\rangle = 4$, $\cos^2\theta = 1$, i.e., $\alpha = \pm\beta$. So, the pair $(\langle \alpha, \beta\rangle, \langle \beta, \alpha\rangle)$ is $(\pm 2, \pm 2)$.

Exercise 6.2

a Since $(\alpha, \beta) > 0$, $\langle \alpha, \beta\rangle$ and $\langle \beta, \alpha\rangle$ are positive. So, Exercise 6.1 shows that one of two is one, at least. When $\langle \alpha, \beta\rangle = 1$, $\alpha - \beta = \alpha - \langle \alpha, \beta\rangle\beta = \mathcal{W}_\beta(\alpha) \in \Phi$. When $\langle \beta\alpha = 1$, $\beta - \alpha \in \Phi$. So, $\alpha - \beta = -(\beta - \alpha) \in \Phi$.

b Substituting β into $-\beta$ in **a**, we obtain the desired statement.

Exercise 6.3

Consider liner transform that maps γ_j to e_j, where e_j is the vector that has non-zero component 1 only j-th entry. Since $\cap_{j=1}^r V^+(e_j)$ is not empty, $\cap_{j=1}^r V^+(\gamma_j)$ is also not empty.

Exercise 6.4

a Assume that there exists an element $\alpha \in \Phi^+(\gamma)$ such that it is not written as non-negative $\mathbb{Z}$-linear combination of $\Delta(\gamma)$. We choose $\alpha' := \mathrm{argmin}_\alpha(\alpha, \gamma)$, where the minimum is taken among elements $\alpha \in \Phi^+(\gamma)$ satisfying the above condition. Since α' is decomposable, there exists $\beta_1, \beta_2 \in \Phi^+(\gamma)$ such that $\alpha = \beta_1 + \beta_2$. We find $(\alpha, \gamma) = (\beta_1, \gamma) + (\beta_2, \gamma)$ and $(\beta_1, \gamma), (\beta_2, \gamma) > 0$. Due to the above minimum choice, β_1 and β_2 are written as non-negative $\mathbb{Z}$-linear combinations of $\Delta(\gamma)$. So, α is also written as a non-negative $\mathbb{Z}$-linear combination of $\Delta(\gamma)$, which is the contradiction of the assumption.

b Assume that $(\alpha, \beta) > 0$ and $\alpha \neq \beta$. So, $\beta \neq -\alpha$. Exercise 6.2 guarantees that $\alpha - \beta \in \Phi$. Thus, $\alpha - \beta$ or $\beta - \alpha$ belongs to $\Phi^+(\gamma)$. When $\alpha - \beta \in \Phi^+(\gamma)$,

$\alpha = (\alpha - \beta) + \beta$, which contradicts the assumption $\alpha \in \Delta(\gamma)$. When $\beta - \alpha \in \Phi^+(\gamma)$, $\beta = (\beta - \alpha) + \alpha$, which contradicts the assumption $\beta \in \Delta(\gamma)$.
c Assume that there exist real numbers r_α such that $\sum_{\alpha \in \Delta(\gamma)} r_\alpha \alpha = 0$. Define subsets $\Delta(\gamma)_1 := \{\alpha \in \Delta(\gamma) | r_\alpha > 0\}$ and $\Delta(\gamma)_2 := \{\alpha \in \Delta(\gamma) | r_\alpha < 0\}$. So, we have $a := \sum_{\alpha \in \Delta(\gamma)_1} r_\alpha \alpha = \sum_{\alpha' \in \Delta(\gamma)_2} -r_{\alpha'} \alpha'$. Thus, **b** guarantees that $(a, a) = \sum_{\alpha \in \Delta(\gamma)_1, \alpha' \in \Delta(\gamma)_2} r_\alpha (-r_{\alpha'})(\alpha, \alpha') \le 0$. So, $a = 0$. Thus, since $\alpha \in \Phi^+(\gamma)$, $0 = (a, \gamma) = \sum_{\alpha \in \Delta(\gamma)_1} r_\alpha (\alpha, \gamma) > 0$, which contradicts the assumption.
d **a** and **c** guarantee the conditions **(B2)** and **(B1)**, respectively.
e Due to Exercise 6.3, we can choose $\gamma \in \cup_{\alpha \in \Delta} V^+(\alpha)$. Due to condition **(B2)** and **a**, we have $\Phi^+ \subset \Phi^+(\gamma')$ and $\Phi^- \subset \Phi^+(-\gamma') = -\Phi^+(\gamma')$. So, $\Phi^+ = \Phi^+(\gamma')$. Thus, $\Delta \subset \Delta(\gamma')$. Since the number of elements of both sets are dimension, we have $\Delta = \Delta(\gamma')$.

Exercise 6.5
Firstly, we assume that $\alpha - \beta \in \Phi$. However, it contradicts **(B2)**. So, $\alpha - \beta$ is not a root. Next, we assume that $(\alpha, \beta) > 0$, which implies$\alpha \neq -\alpha$. Since $\alpha \neq \beta$, Exercise 6.2 guarantees that $\alpha - \beta \in \Phi$, which contradicts **(B2)**. So, we have $(\alpha, \beta) \le 0$.

Exercise 6.6
a Assume that $\beta \in \Phi^-$. $\beta = -\sum_{\alpha \in \Delta} r_\alpha \alpha$ with non-negative integer r_α. Choose $\alpha' \in \Delta$ such that $r_{\alpha'} \neq 0$. So, $\beta + \alpha' \in \Phi^-$, which contradicts the assumption.
b Assume that there exists an element $\alpha \in \Delta$ such that $(\beta, \alpha) < 0$, which implies $\beta \neq \alpha$. Since $\beta \in \Phi^+$, $\beta \neq -\alpha$. Exercise 6.2 guarantees that $\alpha + \beta \in \Phi$, which contradicts the assumption. So, we have $(\beta, \alpha) \ge 0$ for $\alpha \in \Delta$. Since β is not 0 and Δ spans the vector space V, there exists an element $\alpha \in \Delta$ such that $(\alpha, \beta) > 0$.
c Due to **a** and **(B2)**, we β is written as $\sum_{\alpha \in \Delta} r_\alpha \alpha$ with non-negative integer r_α. Define two subsets $\Delta_1 := \{\alpha \in \Delta | r_\alpha > 0\}$ and $\Delta_2 := \{\alpha \in \Delta | r_\alpha = 0\}$. Assume that Δ_2 is not empty. Exercise 6.5 guarantees that $(\beta, \alpha') = \sum_{\alpha \in \Delta_1} r_\alpha (\alpha, \alpha') \le 0$ for an element $\alpha' \in \Delta_2$. Since Φ is connected, there exist $\alpha \in \Delta_1$ and $\alpha \in \Delta_2$ such that $(\alpha, \alpha') \neq 0$. Exercise 6.5 guarantees that $(\alpha, \alpha') < 0$. Thus, $(\beta, \alpha') < 0$, which **b**.
d Due to **c**, we have $\beta = \sum_{\alpha \in \Delta} r_\alpha \alpha$ and $\beta = \sum_{\alpha' \in \Delta} r'_{\alpha'} \alpha'$ with $r_\alpha, r'_{\alpha'} > 0$. $(\beta, \beta') = \sum_{\alpha, \alpha' \in \Delta} r_\alpha r'_{\alpha'} (\alpha, \alpha')$, which is strictly positive due to **b**. When $\beta \neq \beta'$, Exercise 6.2 guarantees that $\beta - \beta' \in \Phi$, which implies $\beta \ge \beta'$ or $\beta' \ge \beta$, which contradicts the assumption.

Exercise 6.7
$X \in \mathfrak{h}$ satisfies

$$(X, [F^x_\alpha, F^y_\alpha])_\mathfrak{g} = ([X, F^x_\alpha], F^y_\alpha)_\mathfrak{g} = (-\alpha(X) F^y_\alpha, F^y_\alpha)_\mathfrak{g} = -(X, Z_\alpha)(F^y_\alpha, F^y_\alpha)_\mathfrak{g}.$$

So, as Killing form is non-degenerate, we have $[F^x_\alpha, F^y_\alpha] = -(F^y_\alpha, F^y_\alpha)_\mathfrak{g} Z_\alpha$. Similarly, we have $[F^x_\alpha, F^y_\alpha] = -(F^x_\alpha, F^x_\alpha)_\mathfrak{g} Z_\alpha$.

Exercise 6.8

The assumption yields that $\langle \boldsymbol{\alpha}_i, \boldsymbol{\alpha}_j \rangle = 0$. Then, we have $\langle \boldsymbol{\alpha}_i, \boldsymbol{\alpha}_i + \boldsymbol{\alpha}_j \rangle = \frac{\langle \alpha_i, \alpha_i \rangle}{\langle \alpha_i + \alpha_j, \alpha_i + \alpha_j \rangle} = \frac{\langle \alpha_i, \alpha_i \rangle}{\langle \alpha_i, \alpha_i \rangle + \langle \alpha_j, \alpha_j \rangle} < 1$. Since the above value is not a integer, $\boldsymbol{\alpha}_i + \boldsymbol{\alpha}_j$ does not belong to Φ.

Exercise 6.9

The set Φ^+ of positive roots is given as $\{\boldsymbol{\alpha}_{j,l}\}_{1 \le j < l \le r}$, where $\boldsymbol{\alpha}_{j,l} = \boldsymbol{\alpha}_j + \cdots + \boldsymbol{\alpha}_{l-1}$. Hence, $\boldsymbol{\delta} = \frac{1}{2} \sum_{1 \le j < l \le r} (\boldsymbol{\alpha}_j + \cdots + \boldsymbol{\alpha}_{l-1}) = \frac{1}{2} \sum_{j=1}^{r-1} j(r-j) \boldsymbol{\alpha}_j$.

Exercise 6.10

It is sufficient to show that $\frac{l-j+\sum_{t=j}^{l-1} \lambda_t}{l-j} = \frac{\langle \boldsymbol{\lambda}+\boldsymbol{\delta}, \boldsymbol{\alpha}_{j,l} \rangle}{\langle \boldsymbol{\delta}, \boldsymbol{\alpha}_{j,l} \rangle}$, for each root $\boldsymbol{\alpha}_{j,l} \in \Phi^+$. Exercise 6.9 guarantees that $\langle \delta, \boldsymbol{\alpha}_j \rangle = \frac{1}{2}(2j(r-j) - (j+1)(r-j-1) - (j-1)(r-j+1)) = 1$ for $j = 2, \ldots, r-2$, and $\langle \delta, \boldsymbol{\alpha}_1 \rangle = \langle \delta, \boldsymbol{\alpha}_r \rangle = \frac{1}{2}(2(r-1) - 2(r-2)) = 1$ for $j = 1, r-1$. So, $\langle \delta, \boldsymbol{\alpha}_{j,l} \rangle = j - l$. Also, $\langle \boldsymbol{\lambda}, \boldsymbol{\alpha}_{j,l} \rangle = \sum_{t=j}^{l-1} \lambda_t$.

Exercise 7.1

The desired statement follows from (7.9) and (3.13).

Exercise 7.2

When $\zeta' = \frac{q'+ip'}{\sqrt{2}}$, the distribution of the outcome is $|\langle \zeta | \zeta' \rangle|^2 \frac{dqdp}{2\pi} = e^{-|\zeta-\zeta'|^2} \frac{dqdp}{2\pi} = e^{-\frac{(p-p')^2+(q-q')^2}{2}} \frac{dqdp}{2\pi}$.

Exercise 7.3

$$
\begin{aligned}
& W_{|\alpha\rangle\langle\alpha|}(\zeta) = \langle \alpha | \mathsf{W}(\zeta) | \alpha \rangle = \langle 0 | \mathsf{W}(\alpha)^\dagger \mathsf{W}(\zeta) \mathsf{W}(\alpha) | 0 \rangle \\
=& \langle 0 | \mathsf{W}(-\alpha) \mathsf{W}(\zeta) \mathsf{W}(\alpha) | 0 \rangle = e^{\frac{1}{2}(\zeta\overline{\alpha} - \alpha\overline{\zeta})} \langle 0 | \mathsf{W}(-\alpha) \mathsf{W}(\zeta + \alpha) | 0 \rangle \\
=& e^{\frac{1}{2}(\zeta\overline{\alpha} - \alpha\overline{\zeta})} e^{\frac{1}{2}(-\alpha\overline{\zeta+\alpha} + (\zeta+\alpha)\overline{\alpha})} \langle 0 | \mathsf{W}(\zeta) | 0 \rangle = e^{\zeta\overline{\alpha} - \alpha\overline{\zeta}} e^{-\frac{|\zeta|^2}{2}}.
\end{aligned}
$$

Exercise 7.4

The distribution of the outcome is $|\langle \zeta | n \rangle|^2 \frac{dqdp}{2\pi} = e^{-|\zeta|^2} \frac{|\zeta|^{2n}}{n!} \frac{dqdp}{2\pi} = e^{-\frac{p^2+q^2}{2}} \frac{(\frac{p^2+q^2}{2})^{2n}}{n!} \frac{dqdp}{2\pi}$.

Exercise 7.5

The distribution of the outcome n is $|\langle \zeta | n \rangle|^2 = e^{-|\zeta'|^2} \frac{|\zeta'|^{2n}}{n!}$.

Exercise 7.6

$$
[Q \otimes I + I \otimes Q, P \otimes I - I \otimes P] = [Q \otimes I, P \otimes I] + [I \otimes Q, -I \otimes P] = iI - iI = 0.
$$

Exercise 7.7
Assume that $\alpha = \frac{x+iy}{\sqrt{2}}$ and $\beta = \frac{x'+iy'}{\sqrt{2}} = \zeta\sqrt{1-\eta} + \alpha\sqrt{\eta}$. Then, we have

$$\begin{aligned}
&\mathrm{Tr}_2\, \mathsf{U}(b_\eta)|\zeta\rangle\langle\zeta| \otimes \rho_{N/\eta}\mathsf{U}(b_\eta)^\dagger \\
=&\int_{\mathbb{C}} |\zeta\sqrt{1-\eta}+\alpha\sqrt{\eta}\rangle\langle\zeta\sqrt{1-\eta}+\alpha\sqrt{\eta}| e^{-\frac{|\alpha|^2\eta}{N}}\eta\frac{dxdy}{2\pi N} \\
=&\int_{\mathbb{C}} |\beta\rangle\langle\beta| e^{-\frac{|\beta-\zeta\sqrt{1-\eta}|^2}{N}}\frac{dx'dy'}{2\pi N} \to \Lambda_N(|\zeta\rangle\langle\zeta|).
\end{aligned}$$

Exercise 7.8

It is enough to show that $(a_1^*, \ldots, a_4^*)J_\rho \begin{pmatrix} a_1 \\ \vdots \\ a_4 \end{pmatrix} \geq 0$ for four complex numbers $a_1, \ldots, a_4$.

$$(a_1^*, \ldots, a_4^*)J_\rho \begin{pmatrix} a_1 \\ \vdots \\ a_4 \end{pmatrix} = \mathrm{Tr}(\sum_{i=1}^4 a_i\mathsf{A}_i)^\dagger(\sum_{i=1}^4 a_i\mathsf{A}_i)\rho \geq 0.$$

Exercise 7.9
The relation (7.66) follows as

$$\begin{aligned}
&\left(-i\frac{\partial}{\partial p} - \frac{q}{2}\right)e^{-\frac{|\zeta|^2}{2}+i(ap-qb)} = \left(-i\frac{\partial}{\partial p} - \frac{q}{2}\right)e^{-\frac{p^2+q^2}{4}+i(ap-qb)} \\
=&\left(a + \frac{-q+ip}{2}\right)e^{-\frac{p^2+q^2}{4}+i(ap-qb)} = \frac{1}{\sqrt{2}}(\alpha+\overline{\alpha}-\overline{\zeta})e^{-\frac{p^2+q^2}{4}+i(ap-qb)} \\
=&\mathcal{F}_{\mathsf{W}}^{-1}[\frac{\mathsf{a}+\mathsf{a}^\dagger}{\sqrt{2}}|\alpha\rangle\langle\alpha|](\zeta).
\end{aligned}$$

Similarly, the relation (7.67) follows as

$$\begin{aligned}
&\left(i\frac{\partial}{\partial q} - \frac{p}{2}\right)e^{-\frac{|\zeta|^2}{2}+i(ap-qb)} = \left(i\frac{\partial}{\partial q} - \frac{p}{2}\right)e^{-\frac{p^2+q^2}{4}+i(ap-qb)} \\
=&\left(b - i\frac{q+ip}{2}\right)e^{-\frac{p^2+q^2}{4}+i(ap-qb)} = -\frac{i}{\sqrt{2}}(\alpha-\overline{\alpha}+\overline{\zeta})e^{-\frac{p^2+q^2}{4}+i(ap-qb)} \\
=&\mathcal{F}_{\mathsf{W}}^{-1}[-i\frac{\mathsf{a}-\mathsf{a}^\dagger}{\sqrt{2}}|\alpha\rangle\langle\alpha|](\zeta).
\end{aligned}$$

Exercise 7.10
Choose θ such that $\begin{pmatrix} \cos\theta & -\sin\theta \\ \sin\theta & \cos\theta \end{pmatrix} \begin{pmatrix} \beta & \gamma \\ \gamma & \alpha \end{pmatrix} \begin{pmatrix} \cos\theta & \sin\theta \\ -\sin\theta & \cos\theta \end{pmatrix} = \begin{pmatrix} \beta' & 0 \\ 0 & \alpha' \end{pmatrix}$ with suitable two real numbers α' and β'. So, we have $\begin{pmatrix} \cos\theta & \sin\theta \\ -\sin\theta & \cos\theta \end{pmatrix} \begin{pmatrix} \beta & -\gamma \\ -\gamma & \alpha \end{pmatrix} \begin{pmatrix} \cos\theta & -\sin\theta \\ \sin\theta & \cos\theta \end{pmatrix} = \begin{pmatrix} \beta' & 0 \\ 0 & \alpha' \end{pmatrix}$.

Since the matrix $g = \begin{pmatrix} \cos\theta & 0 & 0 & -\sin\theta \\ 0 & \cos\theta & \sin\theta & 0 \\ 0 & -\sin\theta & \cos\theta & 0 \\ \sin\theta & 0 & 0 & \cos\theta \end{pmatrix}$ satisfies

$$g^T \begin{pmatrix} & 0 & \frac{i}{2} & 0 \\ 0 & 0 & 0 & \frac{i}{2} \\ -\frac{i}{2} & 0 & 0 & 0 \\ 0 & -\frac{i}{2} & 0 & 0 \end{pmatrix} g = \begin{pmatrix} & 0 & \frac{i}{2} & 0 \\ 0 & 0 & 0 & \frac{i}{2} \\ -\frac{i}{2} & 0 & 0 & 0 \\ 0 & -\frac{i}{2} & 0 & 0 \end{pmatrix},$$

the matrix g belongs to $\mathrm{Sp}(4,\mathbb{R})$. Thus, we have $g^T J_{\alpha,\beta,\gamma} g = J_{\alpha',\beta',0}$.

Exercise 7.11
Using (7.65), we have

$$Q_X(\zeta) = \langle\zeta|\Big(\int_{\mathbb{C}} W_X(\alpha)\mathsf{W}(-\alpha)\frac{dadb}{2\pi}\Big)|\zeta\rangle$$
$$= \int_{\mathbb{C}} W_X(\alpha)\langle\zeta|\mathsf{W}(-\alpha)|\zeta\rangle\frac{dadb}{2\pi} = \int_{\mathbb{C}} W_X(\alpha)e^{-\zeta\bar{\alpha}+\alpha\bar{\zeta}}e^{-\frac{|\zeta|^2}{2}}\frac{dadb}{2\pi}.$$

Exercise 8.1

$$\mathsf{W}_{\mathbb{Z}}(0,0) = I, \quad \mathsf{W}_{\mathbb{Z}}(1,0) = \begin{pmatrix} 0 & 0 & 1 \\ 1 & 0 & 0 \\ 0 & 1 & 0 \end{pmatrix},$$

$$\mathsf{W}_{\mathbb{Z}}(2,0) = \begin{pmatrix} 0 & 1 & 0 \\ 0 & 0 & 1 \\ 1 & 0 & 0 \end{pmatrix}, \quad \mathsf{W}_{\mathbb{Z}}(0,1) = \begin{pmatrix} 1 & 0 & 0 \\ 0 & e^{2\pi i/3} & 0 \\ 0 & 0 & e^{4\pi i/3} \end{pmatrix},$$

$$\mathsf{W}_{\mathbb{Z}}(1,1) = e^{\pi i/3}\begin{pmatrix} 0 & 0 & e^{4\pi i/3} \\ 1 & 0 & 0 \\ 0 & e^{2\pi i/3} & 0 \end{pmatrix}, \quad \mathsf{W}_{\mathbb{Z}}(2,1) = e^{2\pi i/3}\begin{pmatrix} 0 & e^{2\pi i/3} & 0 \\ 0 & 0 & e^{4\pi i/3} \\ 1 & 0 & 0 \end{pmatrix},$$

$$\mathsf{W}_{\mathbb{Z}}(0,2) = \begin{pmatrix} 1 & 0 & 0 \\ 0 & e^{4\pi i/3} & 0 \\ 0 & 0 & e^{2\pi i/3} \end{pmatrix}, \quad \mathsf{W}_{\mathbb{Z}}(1,2) = e^{2\pi i/3}\begin{pmatrix} 0 & 0 & e^{2\pi i/3} \\ 1 & 0 & 0 \\ 0 & e^{4\pi i/3} & 0 \end{pmatrix},$$

$$\mathsf{W}_{\mathbb{Z}}(2,2) = e^{4\pi i/3}\begin{pmatrix} 0 & e^{4\pi i/3} & 0 \\ 0 & 0 & e^{2\pi i/3} \\ 1 & 0 & 0 \end{pmatrix}.$$

Exercise 8.2

We have $\langle\langle \frac{1}{\sqrt{d}}\mathsf{W}_{\mathbb{Z}}(s',t')|\frac{1}{\sqrt{d}}\mathsf{W}_{\mathbb{Z}}(s,t)\rangle\rangle = \frac{1}{d}\operatorname{Tr}\omega_{\mathbb{Z}}^{s't-st'}\mathsf{W}_{\mathbb{Z}}(s-s',t-t') = \delta_{s,s'}\delta_{t,t'}$. So, $\{|\frac{1}{\sqrt{d}}\mathsf{W}_{\mathbb{Z}}(s,t)\rangle\rangle\}_{(s,t)\in\mathbb{Z}_d^2}$ forms a CONS on $\mathbb{C}^d\otimes\mathbb{C}^d$.

Exercise 8.3

It is sufficient to show that $(x,y) = (F(x),F(y))_{\mathbb{F}_4}$ for $x,y\in\mathbb{F}_2^2$. This relation can be shown as follows.

$$
\begin{aligned}
&(1,1)_{\mathbb{F}_4}=0,\ (1,\alpha)_{\mathbb{F}_4}=1,\ (\alpha,\alpha)_{\mathbb{F}_4}=1,\ (\alpha,\alpha+1)_{\mathbb{F}_4}=0,\\
&(\alpha+1,\alpha+1)_{\mathbb{F}_4}=1,\ (\alpha+1,1)_{\mathbb{F}_4}=1.
\end{aligned}
$$

Exercise 8.4

$\{(0,0,0,0),(1,1,0,0),(0,0,1,1),(1,1,1,1)\}$. It is easy to show that this subgroup is a self-orthogonal subgroup. The dimension of this subgroup is 2. So, its orthogonal group also has dimension 2. Thus, it is a strictly self-orthogonal group.

Exercise 8.5

Consider a matrix $\begin{pmatrix} a & b\\ c & d\end{pmatrix}$. Since

$$
\left\langle \begin{pmatrix} a & b\\ c & d\end{pmatrix}\begin{pmatrix} x\\ y\end{pmatrix}, \begin{pmatrix} a & b\\ c & d\end{pmatrix}\begin{pmatrix} x'\\ y'\end{pmatrix}\right\rangle = (ad-bc)(x'y-xy'), \tag{A.3}
$$

it belongs to $\mathrm{Sp}(2,\mathbb{F}_q)$ if and only if $ad-bc=1$.

Exercise 8.6

$$
\begin{aligned}
&\begin{pmatrix}1&0\\0&1\end{pmatrix}, \begin{pmatrix}1&0\\1&1\end{pmatrix}, \begin{pmatrix}1&0\\2&1\end{pmatrix}, \begin{pmatrix}1&1\\0&1\end{pmatrix}, \begin{pmatrix}1&2\\0&1\end{pmatrix},\\
&\begin{pmatrix}0&1\\2&0\end{pmatrix}, \begin{pmatrix}1&1\\2&0\end{pmatrix}, \begin{pmatrix}2&1\\2&0\end{pmatrix}, \begin{pmatrix}0&1\\2&1\end{pmatrix}, \begin{pmatrix}0&1\\2&2\end{pmatrix},\\
&\begin{pmatrix}0&2\\1&0\end{pmatrix}, \begin{pmatrix}1&2\\1&0\end{pmatrix}, \begin{pmatrix}2&2\\1&0\end{pmatrix}, \begin{pmatrix}0&2\\1&1\end{pmatrix}, \begin{pmatrix}0&2\\1&2\end{pmatrix},\\
&\begin{pmatrix}2&1\\1&1\end{pmatrix}, \begin{pmatrix}1&1\\1&2\end{pmatrix}, \begin{pmatrix}2&1\\1&2\end{pmatrix}.
\end{aligned}
$$

Exercise 8.7

Firstly, notice that the elements of $\mathrm{GL}(2,\mathbb{F}_2)$ corresponding to $\mathbb{F}_4$ via the $\mathbb{F}_2$-morphism F are limited to the following matrices.

$$
1\mapsto\begin{pmatrix}1&0\\0&1\end{pmatrix},\ \alpha\mapsto\begin{pmatrix}0&1\\1&1\end{pmatrix},\ \alpha+1\mapsto\begin{pmatrix}1&1\\1&0\end{pmatrix}. \tag{A.4}
$$

So, the matrix $g_0 := \begin{pmatrix} 1 & 0 \\ 1 & 1 \end{pmatrix} \in \mathrm{GL}(2, \mathbb{F}_2)$ is not contained in $\mathbb{F}_4$ in this sense.

Exercise 8.8
Use the element g_0 given in Exercise 8.7. The matrix $\begin{pmatrix} g_0 & 0 \\ 0 & g_0 \end{pmatrix}$ belongs to $\mathrm{Sp}(4, \mathbb{F}_2)$ but does not belong to $\mathrm{Sp}(2, \mathbb{F}_4)$ via the $\mathbb{F}_2$-morphism F.

Exercise 8.9
The following is an example of a generating matrix of a strictly self-orthogonal subgroup given in Exercise 8.4.

$$A(g) = \begin{pmatrix} 1 & 0 \\ 1 & 0 \\ 0 & 1 \\ 0 & 1 \end{pmatrix}.$$

Exercise 8.10
The space $\mathcal{H}^{\otimes 1:\mathrm{even}}$ is spanned by $\{|0\rangle, |1\rangle + |2\rangle\}$. The space $\mathcal{H}^{\otimes 1:\mathrm{odd}}$ is spanned by $\{|1\rangle - |2\rangle\}$.

Exercise 8.11

$$\left(\sqrt{\frac{2}{3}}\langle 0| + \sqrt{\frac{1}{3}}\langle 1|\right)\left(\sqrt{\frac{2}{3}}|0\rangle + e^{2\pi i/3}\sqrt{\frac{1}{3}}|1\rangle\right) = \frac{2}{3} + \frac{1}{3}e^{2\pi i/3} = \frac{1}{2} + \frac{1}{2\sqrt{3}}i.$$

So,

$$\left|\left(\sqrt{\frac{2}{3}}\langle 0| + \sqrt{\frac{1}{3}}\langle 1|\right)\left(\sqrt{\frac{2}{3}}|0\rangle + e^{2\pi i/3}\sqrt{\frac{1}{3}}|1\rangle\right)\right| = \sqrt{\frac{1}{3}}.$$

Similarly, we have

$$\begin{aligned}
&\left|\left(\sqrt{\frac{2}{3}}\langle 0| + \sqrt{\frac{1}{3}}\langle 1|\right)\left(\sqrt{\frac{2}{3}}|0\rangle + e^{4\pi i/3}\sqrt{\frac{1}{3}}|1\rangle\right)\right| \\
=&\left|\left(\sqrt{\frac{2}{3}}\langle 0| + e^{-2\pi i/3}\sqrt{\frac{1}{3}}\langle 1|\right)\left(\sqrt{\frac{2}{3}}|0\rangle + e^{4\pi i/3}\sqrt{\frac{1}{3}}|1\rangle\right)\right| = \sqrt{\frac{1}{3}}.
\end{aligned}$$

Also, we have

$$\left|\left(\sqrt{\frac{2}{3}}\langle 0| + \sqrt{\frac{1}{3}}\langle 1|\right)|1\rangle\right| = \left|\left(\sqrt{\frac{2}{3}}\langle 0| + e^{-2\pi i/3}\sqrt{\frac{1}{3}}\langle 1|\right)|1\rangle\right|$$

$$= \left|\left(\sqrt{\frac{2}{3}}\langle 0| + e^{-4\pi i/3}\sqrt{\frac{1}{3}}\langle 1|\right)|1\rangle\right| = \sqrt{\frac{1}{3}}.$$

References

1. W. Alltop, Complex sequences with low periodic correlations (corresp.). IEEE Trans. Inf. Theory **26**(3), 350–354 (1980)
2. A. Ambainis, J. Emerson, Quantum t-designs: t-wise independence in the quantum world, in *Twenty-Second Annual IEEE Conference on Computational Complexity (CCCf07)* (2007), pp. 129–140
3. D.M. Appleby, SIC-POVMs and the extended Clifford Group. J. Math. Phys. **46**, 052107 (2005)
4. D.M. Appleby, Symmetric informationally complete-positive operator valued measures and the extended Clifford group. J. Math. Phys. **46**, 052107 (2005)
5. D.M. Appleby, S.T. Flammia, C.A. Fuchs, The Lie algebraic significance of symmetric informationally complete measurements. J. Math. Phys. **52**, 022202 (2011)
6. D.M. Appleby, H. Yadsan-Appleby, G. Zauner, Galois automorphisms of a symmetric measurement. Quantum Inf. Comput. **13**(7–8) 0672–0720 (2013). http://arxiv.org/abs/1209.1813
7. D.M. Appleby, C.A. Fuchs, H. Zhu, Group theoretic, Lie algebraic and Jordan algebraic formulations of the SIC existence problem. Quantum Inf. Comput. **15**(1–2), 61–94 (2015)
8. E. Artin, *Geometric Algebra* (Interscience Publishers, New York, 1957)
9. K. Asano, H. Nagao, *Group Theory* (Iwanami shoten, Tokyo, 1965). (In Japanese)
10. J.J. Aubert et al., Experimental Observation of a Heavy Particle. J. Phys. Rev. Lett. **33**(23), 1404 (1974)
11. J.-E. Augustin et al., Discovery of a Narrow Resonance in e^+e^- Annihilation. Phys. Rev. Lett. **33**(23), 1406 (1974)
12. J. Baik, P. Deift, K. Johansson, On the distribution of the length of the longest increasing subsequence of random permutations. J. Amer. Math. Soc. 12(4), 1119–1178 (1999); J. Baik, P. Deift, K. Johansson, On the distribution of the length of the second row of a Young diagram under Plancherel measure. Geom. Funct. Anal. **10**(4), 702 (2000)
13. M. Bander, C. Itzykson, Group Theory and the Hydrogen Atom (I). Rev. Mod. Phys. **38**, 330 (1966)
14. S. Bandyopadhyay, P.O. Boykin, V. Roychowdhury, F. Vatan, A new proof for the existence of mutually unbiased bases. Algorithmica **34**(4), 512–528 (2002)
15. Eiichi Bannai, Etsuko Bannai, A survey on spherical designs and algebraic combinatorics on spheres. Eur. J. Combinator. **30**(6), 1392–1425 (2009)
16. E. Bannai, S.G. Hoggar, On tight t-designs in compact symmetric spaces of rank one. Proc. Japan Acad. **61A**, 78 (1985)
17. E. Bannai, S.G. Hoggar, Tight t-designs and squarefree integers. Eur. J. Comb. **10**, 113 (1989)

M. Hayashi, *Group Representation for Quantum Theory*,
DOI 10.1007/978-3-319-44906-7

18. B.J. Bjorken, S.L. Glashow, Elementary particles and SU(4). Physics Letters **11**(3), 255–257 (1964)
19. K. Chandrasekharan, *Arithmetical Functions* (Springer, Heidelberg, 1970)
20. H.-F. Chau, Unconditionally secure key distribution in higher dimensions by depolarization. IEEE Trans. Inf. Theory **51**(4), 1451–1468 (2005)
21. R. Cleve, D. Leung, L. Liu, C. Wang, Near-linear constructions of exact unitary 2-designs, arXiv:1501.04592 (2015)
22. M. Combescure, D. Robert, *Coherent States, and Applications in Mathematical Physics, Theoretical and Mathematical Physics* (Springer, Heidelberg, 2012)
23. C. Dankert, *Efficient Simulation of Random Quantum States and Operators*, Master thesis, University of Waterloo (2005). http://arxiv.org/abs/quant-ph/0512217
24. C. Dankert, R. Cleve, J. Emerson, E. Livine, Exact and approximate unitary 2-designs and their application to fidelity estimation. Phys. Rev. A **80**, 012304 (2009)
25. P. Delsarte, J.M. Goethals, J.J. Seidel, Bounds for systems of lines, and Jacobi poynomials. Philips Res. Rep. **30**, 91 (1975)
26. P. Delsarte, J.M. Goethals, J.J. Seidel, Spherical codes and designs. Geom. Dedicata **6**, 363–388 (1977)
27. D.P. DiVincenzo, D.W. Leung, B.M. Terhal, Quantum data hiding. IEEE Trans. Inf. Theory **48**(3), 580–598 (2002)
28. T. Durt, B.-G. Englert, I. Bengtsson, K. Życzkowski, On mutually unbiased bases. Int. J. Quant. Inf. **8**, 535 (2010)
29. J. Eisert, M.M. Wolf, Gaussian quantum channels, in *Quantum Information with Continous Variables of Atoms and Light N*, ed. by J. Cerf, G. Leuchs, E.S. Polzik (Imperial College Press, London, 2007), pp. 23–42
30. V. Fock, Zur Theorie des Wasserstoffatoms. Z. Physik **98**, 145 (1935)
31. G.B. Folland, *Harmonic Analysis in Phase Space, Annals of Mathematics Studies 122* (Princeton University Press, Princeton, 1989)
32. G.B. Folland, *A Course in Abstract Harmonic Analysis, Studies in Advanced Mathematics* (CRC Press, Boca Raton, 1995)
33. C.A. Fuchs, R. Schack, Quantum-Bayesian coherence: the no-nonsense version. Rev. Mod. Phys. **85**, 1693–1715 (2013)
34. H. Führ, *Abstract harmonic analysis of continuous wavelet transforms*. Lecture Notes in Mathematics, vol. 1863 (Springer, Berlin, 2005)
35. S.L. Glashow, J. Iliopoulos, L. Maiani, Weak Interactions with Lepton-Hadron Symmetry. Physical Review D **2**(7), 1285–1292 (1970)
36. C. Godsil, A. Roy, Equiangular lines, mutually unbiased bases, and spin models. Eur. J. Combinator. **30**, 246–262 (2009)
37. R. Goodman, N.R. Wallach, *Representations and Invariants of the Classical Groups, Encyclopedia of Mathematics and Its Applications*, vol. 68 (Cambridge University Press, Cambridge, 1999)
38. M. Grassl, M. Rotteler, T. Beth, Efficient quantum circuits for non-qubit quantum error-correcting codes. International Journal of Foundations of Computer Science (IJFCS) **14**(5), 757–775 (2003)
39. D. Gross, K. Audenaert, J. Eisert, Evenly distributed unitaries: On the structure of unitary designs. J. Math. Phys. **48**(5), 052104 (2007)
40. D. Gross, F. Krahmer, R. Kueng, A partial derandomization of phaselift using spherical designs. J. Fourier Anal. Appl. **21**(2), 229–266 (2015)
41. J. Haantjes, Equilateral point-sets in elliptic two- and three dimensional spaces. Nieuw Arch. Wisk. **22**, 355–362 (1948)
42. R.H. Hardin, N.J.A. Sloane, McLarens improved snub cube and other new spherical designs in three dimensions Discrete. Comput. Geom. **15**, 429 (1996)
43. G.H. Hardy, S. Ramanujan, Asymptotic formulae in combinatory analysis. Proc. Lond. Math. Soc. **17**, 75–115 (1918)

44. Harish-Chandra, Representation of Semisimple Lie group VI. Amer. Jour. Math., **78**(3) 564–628 (1956)
45. A.W. Harrow, R.A. Low, Random quantum circuits are approximate 2-designs. Commun. Math. Phys. **291**(1), 257–302 (2009)
46. M. Hayashi, *Quantum information: an introduction* (Springer, Heidelberg, 2006) (Originally published in Japanese in 2004)
47. M. Hayashi, Discrete realization of group symmetric LOCC-detection of maximally entangled state (2008). arXiv:0810.3381
48. M. Hayashi, Exponential decreasing rate of leaked information in universal random privacy amplification. IEEE Trans. Inf. Theory **57**, 3989–4001 (2011)
49. M. Hayashi, Fourier analytic approach to quantum estimation of group action. Commun. Math. Phys. **347**(1), 3–82 (2009)
50. M. Hayashi, *A Group Theoretic Approach to Quantum Information* (Springer, In preparation) (Originally published in Japanese in, 2014)
51. M. Hayashi, F. Sakaguchi, Subnormal operators regarded as generalized observables and compound-system-type normal extension related to su(1,1). J. Phys. A:Math. Gen. **33**, 7793–7820 (2000)
52. A. Hayashi, T. Hashimoto, M. Horibe, Reexamination of optimal quantum state estimation of pure states. Phys. Rev. A **72**, 032325 (2005)
53. M. Hayashi, S. Ishizaka, A. Kawachi, G. Kimura, T. Ogawa, *Introduction to Quantum Information Science*, Graduate Texts in Physics. (Springer, Heidelberg, 2015) (Originally published in Japanese in 2012)
54. M. Hillery, R.F. OfConnell, M.O. Scully, E.P. Wigner, Distribution functions in physics: fundamentals. Phys. Rep. **106**, 121–167 (1984)
55. S.G. Hoggar, t-designs in projective spaces. Eur. J. Comb. **3**, 233–254 (1982)
56. S.G. Hoggar, Parameters of t-designs in $\mathbb{F}P^{d-1}$. Eur. J. Comb. **5**, 29 (1984)
57. A.S. Holevo, *Probabilistic and Statistical Aspects of Quantum Theory* (North-Holland, Amsterdam, 1982) (Originally published in Russian 1980, 2nd Edition, Springer 2012.)
58. R. Howe, E.C. Tan, *Non-Abelian Harmonic Analysis: Applications of SL (2, R)* (Springer, Heidelberg, 1992)
59. J.E. Humphreys, *Introduction to Lie Algebras and Representation Theory, Graduate Texts in Mathematics 9* (Springer, Heidelberg, 1972)
60. K. Husimi, Some formal properties of the density matrix. Proc. Phys. Math. Soc. Jpn. **22**, 264–314 (1940)
61. W.M. Kantor, MUBs inequivalence and affine planes. J. Math. Phys. **53**(3), 032204 (2012)
62. E.H. Kennard, Zur Quantenmechanik einfacher Bewegungstypen. Zeitschrift fur Physik (in German) **44**(4–5), 326 (1927)
63. M. Khatirinejad, *Regular Structures of Lines in Complex Spaces*, Ph.D. thesis, Simon Fraser University (2008)
64. A.A. Kirillov, *Elements of the theory of representations* (Springer, New York, 1976) (Grundlehren der Mathematischen Wissenschaften 220, Berlin)
65. A.W. Knapp, *Lie Groups Beyond an Introduction*, vol. 140, 2nd edn., Progress in mathematics (Birkhauser, 2002)
66. A.W. Knapp, *Representation Theory of Semisimple Groups: An Overview Based on Examples*, vol. 36 (Princeton, Princeton Mathematics Series (Princeton University Press, 1986)
67. T. Komorowski, A criterion for essential self-adjointness of a symmetric operator defined by some infinite hermitian matrix with unbounded entries. Integral Equ. Oper. Theory (2015). doi:10.1007/s00020-015-2237-2
68. R. Kueng, D. Gross, Qubit stabilizer states are complex projective 3-designs (2015). arXiv:1510.02767
69. J. Lawrence, Č. Brukner, A. Zeilinger, Mutually unbiased binary observable sets on N qubits. Phys. Rev. A **65**, 032320 (2002)
70. P.W.H. Lemmens, J.J. Seidel, Equiangular lines. J. Algebra **24**, 494 (1973)

71. V. Levenshtein, Designs as maximum codes in polynomial metric spaces. Acta Appl. Math. **29**, 1 (1992)
72. V. Levenshtein, On designs in compact metric spaces and a universal bound on their size. Discrete Math. **192**(1–3), 251–271 (1998)
73. W. Lisiecki, Coherent state representations. *A* survey. Rep. Math. Phys. **35**, 327–358 (1995)
74. I.G. MacDonald, *Symmetric Functions and Hall Polynomials* (Clarendon Press, New York, 1979)
75. W. Matthews, S. Wehner, A. Winter, Distinguishability of quantum states under restricted families of measurements with an application to quantum data hiding. Commun. Math. Phys. **291**(3), 813–843 (2009)
76. A. Messiah, *Quantum Mechanics*, vol. I (North Holland Publishing Company, Amsterdam, 1967). (Translation from the French by G.M. Temmer)
77. A. Messiah, *Quantum Mechanics (Volume II)* (North Holland Publishing, New York, 1981)
78. K.H. Neeb, *Holomorphy and Convexity in Lie Theory*, De Gruyter Expositions in Mathematics 28, Walter de Gruyter (1999)
79. M. Neuhauser, An explicit construction of the metaplectic representation over a finite field. J. Lie Theory **12**, 15 (2002)
80. M. Neuhauser, Explicit construction of the metaplectic representation for finite cyclic groups, preprint: http://www.matha.rwth-aachen.de/~markus/
81. A. Perelomov, *Generalized Coherent States and Their Applications, Texts and Monographs in Physics* (Springer, Heidelberg, 1986)
82. A.M. Perelomov, V.S. Popov, The Lorentz group as a dynamic symmetry group of the hydrogen atom. J. Exp. Theor. Phys. **23**(1), 118 (1966). (Russian original - ZhETF, Vol. 50, No. 1, p. 179, 1966)
83. D. Petz, An invitation to the algebra of canonical commutation relations. Leuven Notes Math. Theor. Phys. **2**, (1990)
84. A.V.M. Prasad, Padmavathamma, The Hardy-Ramanujan-Rademacher expansion for $C\phi_{m,m'}(n)$ using ford circles. Indian J. Pure Appl. Math. **22**(11), 913–925 (1991)
85. H. Rademacher, On the partition function p(n). Proc. Lond. Math. Soc. **43**, 241–254 (1938)
86. A.V. Razumov, M.V. Saveliev, *Lie Algebras, Geometry, and Toda-type Systems* (Cambridge Lecture Notes in Physics, Cambridge, 1997)
87. M. Reed, B. Simon, *Methods of Modern Mathematical Physics I: Functional Analysis* (Academic Press, New York, 1980)
88. J.M. Renes, *Frames, Designs, and Spherical Codes in Quantum Information Theory,* Ph.D. thesis, University of New Mexico (2004)
89. J.M. Renes, R. Blume-Kohout, A.J. Scott, C.M. Caves, Symmetric informationally complete quantum measurements. J. Math. Phys. **45**, 2171 (2004)
90. H.P. Robertson, The uncertainty principle. Phys. Rev. **34**, 163–164 (1929)
91. R.T. Rockafellar, *Convex analysis* (Princeton University Press, Princeton, 1997)
92. A. Roy, *Complex lines with restricted angles*, Ph.D. thesis, University of Waterloo (2006). http://arxiv.org/abs/1306.0978
93. A. Roy, A.J. Scott, Unitary designs and codes. Des. Codes Cryptogr. **53**(1), 13–31 (2009)
94. A.J. Scott, Tight informationally complete quantum measurements. J. Phys. A: Math. Gen. **39**, 13507 (2006)
95. A.J. Scott, Optimizing quantum process tomography with unitary 2-designs. J. Phys. A: Math. Theor. **41**(5), 055308 (2008)
96. A.J. Scott, M. Grassl, SIC-POVMs: a new computer study. J. Math. Phys. **51**, 042203 (2010)
97. J.-P. Serre, *Linear Representation of Finite Groups*, 5th edn. (Springer, Heidelberg, 1996)
98. M. Suzuki, *Group Theory I* (Springer, Heidelberg, 1982) (Grundlehren der mathematischen Wissenschaften)
99. G. Teschl, *Mathematical Methods in Quantum Mechanics With Applications to Schrödinger Operators Graduate Studies in Mathematics*, vol. 99 (American Mathematical Society, Providence, 2014)
100. L.H. Thomas, The Motion of the Spinning Electron. Nature **117**, 514 (1926)

101. C.A. Tracy, H. Widom, Comm. Math. Phys. **159**, 151 (1994)
102. J.H. van Lint, J.J. Seidel, Equilateral point sets in elliptic geometry. Proc. Kon. Nederl. Akad. Wet. Ser. A: Indag. Math. **69**, 335–348 (1966)
103. V.S. Varadarajan, *Lie groups, Lie Algebras and Their Representations, Graduate Texts in Mathematics 102* (Springer, Heidelberg, 1984)
104. D.F. Walls, G.J. Milburn, *Quantum optics* (Springer, Heidelberg, 2008)
105. Z. Webb, The Clifford group forms a unitary 3-design (2015). http://arxiv.org/abs/1510.02769
106. H. Weyl, *Gruppentheorie und Quantenmechanik*, Leipzig: Verlag von S. Hirzel in Leipzig (1928) (English translation, *The Theory of Groups and Quantum Mechanics*, of the second (1931) ed. was reprinted by Dover, 1950)
107. E.P. Wigner, On the quantum correction for thermodynamic equilibrium. Phys. Rev. **40**, 749–759 (1932)
108. G. Wolf, *Mathieu Functions and Hillfs Equation* (2013). http://dlmf.nist.gov/28
109. W.K. Wootters, B.D. Fields, Optimal state determination by mutually unbiased measurements, Ann. Phys. (N.Y.) 191, 363–381 (1989)
110. G. Zauner, *Quantendesigns - Grundzuge einer nicht kommutativen Designtheorie*, Ph.D. thesis, University of Vienna (1999)
111. G. Zauner, Quantum designs: Foundations of a noncommutative design theory. Int. J. Quant. Inf. **9**, 445–507 (2011)
112. H. Zhu, SIC POVMs and Clifford groups in prime dimensions. J. Phys. A: Math. Theor. **43**, 305305 (2010)
113. H. Zhu, Quantum State Estimation and Symmetric Informationally Complete POMs. Ph.D. thesis, National University of Singapore (2012). http://scholarbank.nus.edu.sg/bitstream/handle/10635/35247/ZhuHJthesis.pdf
114. H. Zhu, Tomographic and Lie algebraic significance of generalized symmetric informationally complete measurements. Phys. Rev. A **90**, 032309 (2014)
115. H. Zhu, Quantum state estimation with informationally overcomplete measurements. Phys. Rev. A **90**, 012115 (2014)
116. H. Zhu, Multiqubit Clifford groups are unitary 3-designs (2015). http://arxiv.org/abs/1510.02619
117. H. Zhu, Permutation symmetry determines the discrete Wigner function (2015). http://arxiv.org/abs/1504.03773
118. H. Zhu, Mutually unbiased bases as minimal Clifford covariant 2-designs. Phys. Rev. A **91**, 060301 (2015)
119. H. Zhu, Super-symmetric informationally complete measurements. Ann. Phys. **362**, 311–326 (2015)
120. H. Zhu, B.-G. Englert, Quantum state tomography with fully symmetric measurements and product measurements. Phys. Rev. A **84**, 022327 (2011)

Curriculum Vitae

Masahito Hayashi

Graduate School of Mathematics
Nagoya University
Furocho, Chikusaku
Nagoya, 464-8602, Japan, and
Centre for Quantum Technologies
National University of Singapore
3 Science Drive 2, Singapore 117542.

Phone: + 81(52)789-2549
Fax: + 81(52)789-5397
Email: masahito@math.nagoya-u.ac.jp
Homepage:
`http://www.math.nagoya-u.ac.jp/~masahito/index_e.html`

Personal

Born on December 28, 1971.
Born in Tokushima, Japan.
Japanese citizen.

Education

Ph.D. Depart of Mathematics, Graduate School of Science, Kyoto University, 1999.
M.A. Depart of Mathematics, Graduate School of Science, Kyoto University, 1996.
B.S. Physics & Mathematics, Faculty of Science, Kyoto University, 1994

Employment (Full time)

Research Fellow, the Japan Society for the Promotion of Science, 1998–2000
Researcher, Laboratory for Mathematical Neuroscience (Amari's Laboratory), Brain Science Institute, RIKEN, 2000–2003
Research Manager, ERATO, Quantum Computation and Information Project, Japan Science and Technology Agency (JST), 2003–2006
Group Leader, Quantum Information Theory Group, ERATO-SORST Quantum Computation and Information Project, Japan Science and Technology Agency (JST), 2006–2007
Associate Professor, Graduate School of Information Sciences, Tohoku University, 2007–2012

M. Hayashi, *Group Representation for Quantum Theory*,
DOI 10.1007/978-3-319-44906-7

Full Professor, Graduate School of Mathematics, Nagoya University, 2012–current.

Employment (Part time)

Part-time Lecturer (Quantum Information), SOKENDAI (The Graduate University for Advanced Studies), 2002
Adjunct Associate Professor, Superrobust Computation Project, Information Science and Technology
Strategic Core (21st Century COE by MEXT), Graduate School of Information Science and Technology, The University of Tokyo, 2004–2007
Visiting Associate Professor, Graduate School of Information Sciences, Tohoku University, 2007
Visiting Research Associate Professor, Centre for Quantum Technologies, National University of Singapore, 2009–2012
Visiting Research Professor, Centre for Quantum Technologies, National University of Singapore, 2012–current.
Visiting Professor, Faculty of Engineering and Information Technology, University of Technology, Sydney, 2013

Honors, Awards, and Fellowships

1. ***12th Japan Academy Medal***: "Information Theory and Quantum Information Theory for Finite-Coding-Length" This prize is awarded by The Japan Academy to up to 6 prestigious Japanese researchers under 45 years old among all scienti_c areas. Awardees are selected from among the annual recipients of the JSPS Prize.
2. ***FY2015 JSPS PRIZE***: "Information Theory and Quantum Information Theory for Finite-Coding-Length" The JSPS PRIZE is meant to recognize at an early stage in their careers young researchers with fresh ideas who have the potential to become world leaders in their _elds, while helping to enhance their opportunities to advance their research and make breakthroughs. Twenty-_ve researchers under 45 years old were selected for this year's PRIZE. Their _elds of research run the spectrum from the humanities and social sciences to the natural sciences.
3. ***2011 IEEE Information Theory Society Paper Award***: "Information spectrum approach to second-order coding rate in channel coding" *IEEE Transactions on Information Theory*, Vol. 55, No. 11, 4947–4966 (2009). This prize is the most distinguished paper award in the information theory community.
4. *Senior Member of IEEE*, 2013.
5. *Japan IBM Prize in the Computer Science Section 2010*: "Universal protocol in quantum information and its application to quantum key distribution" This prize is one of the most distinguished prizes in Japan in information science for researchers under 45 across Japan.
6. *Funai Foundation for Information Technology Award in the Computer Science Category 2010*: "Universal quantum information protocol and its application to quantum cryptography".
7. *Research Fellowship for Young Scientists*, the Japan Society for the Promotion of Science, 1998.

8. *2001 SITA Encouragement Award* (by The Society of Information Theory and its Applications (SITA)): "Variable length universal entanglement concentration by local operations".
9. *16th TEPIA Video Award, 2006* (by Association for Technological Excellence Promoting Innovative Advances (TEPIA)): This prize is awarded to Public Video for ERATO Quantum computation and information project. I managed the production process as Research Manager.

Professional Activities

Editorial Board Member

1. Editorial Board Member, International Journal of Quantum Information (IJQI) World Scientific, 2002–Current.
2. Advisory Board Member, International Journal On Advances in Security, IARIA, 2009–Current. (Editorial Board Member: 2008-2009)

Organization of International Meetings

1. Organizing Committee Chair: *ERATO Conference on Quantum Information Science 2003 (EQIS 03)*, Doshiha Univ., Kyoto, Japan, September 4–6, 2003.
2. Organizing Committee and Program Committee: *COE Symposium on Quantum Information Theory*, Doshiha Univ., Kyoto, Japan, September 2–3, 2003.
3. Program Committee Vice-Chair: *ERATO Conference on Quantum Information Science 2004 (EQIS 04)*, Hitotsubashi-memorial hall, Tokyo, Japan, September 1–5, 2004.
4. Program Committee Vice-Chair: *ERATO Conference on Quantum Information Science 2005 (EQIS 05)*, JST-Museum Hall, Tokyo, Japan, August 26–30, 2005.
5. Organizing Committee Chair: *COE-Kakenhi Workshop on Quantum Information Theory and Quantum Statistical Inference*, University of Tokyo, Hongo, Tokyo, November 17–18, (2005)
6. Local Committee: *The 8th International Conference on Quantum Communication, Measurement, and Computing (QCMC2006)*, Tsukuba Epocal, Tsukuba, Japan, November 28–December 3, (2006).
7. Program Committee: *Asian Conference on Quantum Information Science 2006 (AQIS06)*, Beijing, China, September 1–4, 2006.
8. Technical Program Committee, *The First International Workshop on Quantum Security (QSEC2007)*, Guadeloupe, French Caribbean, France, January 2–6, 2007. Organized by IEEE France and IARIA.
9. Organizing Committee Chair: *Special seminar series on quantum information*, National Institute of Informatics (NII), Tokyo, February 1–March 21, 2007.
10. Program Committee: *The 3rd Workshop on Theory of Quantum Computation, Communication, and Cryptography (TQC2008)*, The University of Tokyo, Tokyo Japan, January 30–February 1, 2008.
11. Technical Program Committee Co-Chair: *ICQNM 2008, The Second International Conference on Quantum, Nano, and Micro Technologies*, Sainte Luce,

Martinique, France, February 10–15, 2008. Organized by IEEE France and IARIA.

12. Main Organizer: *GSIS & DEX-SMI Workshop on Quantum statistical inference and entanglement Graduate School of Information Sciences*, Tohoku University, February, 11–12, 2008
13. International Advisory Committee: *International School and Conference on Quantum Information (ISCQI - 2008)*, Institute of Physics (IOP), Bhubaneswar, Orissa, India, March 4–12, 2008
14. Program Committee: *Asian Conference on Quantum Information Science 2008 (AQIS08)*, Seoul, Korea, August 25–26, 2008.
15. Organizing Committee: *International Workshop on Statistical-Mechanical Informatics 2008 (IW-SMI2008)*, Sendai, Japan, September 14–17, 2008.
16. Organizing Committee Chair: *GSIS Workshop on Quantum Information Theory*, Sendai, Japan, November 5–7, 2008
17. Organizing Committee: *Multicritical Behaviour of Spin Glasses and Quantum Error Correcting Codes (MBQEC)*, Centennial Hall, Ookayama campus, Tokyo Institute of Technology, Tokyo, Japan, November 17–19 (2008)
18. Organizing committee chair: *DEX-SMI workshop on quantum statistical inference*, National Institute of Informatics (NII), Tokyo, Japan March, 2–4 (2009).
19. Program Committee: *ICQNM 2009, The Third International Conference on Quantum*, Nano and Micro Technologies, Cancun, Mexico, February 1–6 (2009)
20. Program Committee: *Workshop on Theory of Quantum Computation, Communication, and Cryptography (TQC2009)*, Waterloo, Canada, May 11–13 (2009)
21. Program Committee: *ICQNM 2010, The Fourth International Conference on Quantum, Nano and Micro Technologies*, St. Maarten, Netherlands Antilles, February 10–16 (2010).@
22. Program Committee (Special Area Chair): *ICQNM 2011, The Fifth International Conference on Quantum, Nano and Micro Technologies*, Nice, France, August 21–27 (2011).
23. Technical Program Committee: *ICQNM 2012, The Sixth International Conference on Quantum, Nano and Micro Technologies*, Rome, Italy, August 19–24 (2012)
24. Program Committee: *12th Asian Quantum Information Science conference (AQIS12)*, School of Physical Science and Technology, Soochow University, Suzhou, China, August 23–26 (2012).
25. Workshop Co-chair: *Japan-Singapore Workshop on Multi-user Quantum Networks*, Centre for Quantum Technologies, National University of Singapore, Singapore, September 17–20 (2012)
26. Special Area Chair: *ICQNM 2013, The Seventh International Conference on Quantum, Nano and Micro Technologies*, Barcelona, Spain, August 25–31 (2013)
27. Program Committee: *13th Asian Quantum Information Science conference (AQIS 13)*, IMSC Chennai, India, August 25–30 (2013)
28. Program Committee: The *7th International Conference on Information Theoretic Security (ICITS)*, NTU, Singapore, November, 28–30 (2013)

29. Program Committee: *2014 International Symposium on Information Theory (ISIT)*, Honolulu, Hawaii, June 29–July 4 (2014)
30. Program Committee: *14th Asian Quantum Information Science conference (AQIS 14)*, Kyoto, Japan, August 21–25 (2014)
31. Special Area Chair: *ICQNM 2014, The Eighth International Conference on Quantum, Nano and Micro Technologies*, Lisbon, Portugal, November 16–20 (2014)
32. Program Committee: *The 9th Conference on the Theory of Quantum Computation, Communication and Cryptography (TQC2014)*, National University of Singapore, Singapore, May, 21-23 (2014)
33. Organizer: *Australia-Japan Workshop on Multi-user Quantum Networks*, University of Technology, Sydney, October, 22–24 (2014)
34. Program Committee: *Quantum Information Processing (QIP 2015)*, Sydney January (2015)
35. Program Committee: *2015 International Symposium on Information Theory (ISIT)*, Hong Kong, June 14–19 (2015)
36. Special Area Chair: *ICQNM 2015, The Seventh International Conference on Quantum, Nano and Micro Technologies*, Venice, Italy, August 23–28, 2015
37. Program Committee: *2015 IEEE Information Theory Workshop (ITW)*, Jeju Island, Koreag, October 11–15, 2015
38. Special Area Chair: *ICQNM 2016, The Tenth International Conference on Quantum, Nano and Micro Technologies*, Nice, France, July 24–28, 2016
39. Program Committee: *16th Asian Quantum Information Science conference (AQIS'16)*, Academia Sinica, Taipei, Taiwan, August 29-September 1, 2016
40. Conference Co-Chair: *Beyond I.I.D. in Information Theory 2017*, Institute for Mathematical Sciences, National University of Singapore, Singapore, July 24–28, 2017

Review Committee for Ph.D. Thesis

1. 2005, Review Committee Member for Two Ph.D. Candidates, Graduate School of Information Science and Technology, The University of Tokyo, Japan.
2. 2006, Review Committee Member for a Ph.D. Candidate, Graduate School of Information Science and Technology, The University of Tokyo, Japan.
3. 2008, External examiner (referent), Faculty of Mathematics and Natural Sciences, University of Leiden, Netherlands
4. 2008, External examiner, National Institute of Education, Nanyang Technological University, Singapore.
5. 2008, External examiner, Department of Communication and Integrated Systems, Graduate School of Science and Engineering, Tokyo Institute of Technology.
6. 2012, External examiner, School of Computer Science, McGill University, Canada.
7. 2013, External examiner, Graduate School of Information Sciences, Tohoku University, Japan.
8. 2014, External examiner, Department of Electrical and Computer Engineering, National University of Singapore.

Other Kinds of Review Committees

1. 2015, External Reviewer of Research Committee, the Czech Science Foundation.
2. 2014, External Reviewer of Research Committee, University of Macau (UM) for the 2014 Multi-Year Research Grant (MYRG).
3. 2012, External Reviewer of TATA INSTITUTE OF FUNDAMENTAL RESEARCH.

Other Professional Activities

1. I am a founder of the *Asian Conference on Quantum Information Science (AQIS)* conference series, which is a major international conference series on quantum information and computation.
2. As Research Manager, I oversaw the ERATO Quantum Computation and Information Project, which at the time was the largest research group for quantum information and computation in Japan.
3. I have served as a referee for renowned international journals including *Nature Photonics, IEEE Transactions on Information Theory, Communications in Mathematical Physics, Physical Review Letters, Physical Review A, Journal of Physics A, New Journal of Physics, Quantum Information and Compu- tation, Journal of Mathematical Physics*, and many prestigious journals.

Membership of Academic Societies

IEEE (The Institute of Electrical and Electronics Engineers): Senior Member.
The Mathematical Society of Japan.
JPS (The Physical Society of Japan).
IEICE (The Institute of Electronics, Information and Communication Engineers).

Index

M. Hayashi, *Group Representation for Quantum Theory*,
DOI 10.1007/978-3-319-44906-7

MIX
Papier aus verantwortungsvollen Quellen
Paper from responsible sources
FSC® C105338
FSC
www.fsc.org

If you have any concerns about our products,
you can contact us on
ProductSafety@springernature.com

In case Publisher is established outside the EU,
the EU authorized representative is:
Springer Nature Customer Service Center GmbH
Europaplatz 3, 69115 Heidelberg, Germany

Printed by Libri Plureos GmbH
in Hamburg, Germany